Epitaxy and Applications of Si-Based Heterostructures

MATERIALS RESEARCH SOCIETY
SYMPOSIUM PROCEEDINGS VOLUME 533

Epitaxy and Applications of Si-Based Heterostructures

Symposium held April 13–17, 1998, San Francisco, California, U.S.A.

EDITORS:

Eugene A. Fitzgerald
Massachusetts Institute of Technology
Cambridge, Massachusetts, U.S.A.

Derek C. Houghton
SiGe Microsystems
Ottawa, Ontario, Canada

Patricia M. Mooney
IBM T.J. Watson Research Center
Yorktown Heights, New York, U.S.A.

Materials Research Society
Warrendale, Pennsylvania

Single article reprints from this publication are available through
University Microfilms Inc., 300 North Zeeb Road, Ann Arbor, Michigan 48106

CODEN: MRSPDH

Published by:

Materials Research Society
506 Keystone Drive
Warrendale, PA 15086
Telephone (724) 779-3003
Fax (724) 779-8313
Website: http://www.mrs.org/

Library of Congress Cataloging in Publication Data

Epitaxy and applications of si-based heterostructures : symposium held
April 13-17, 1998, San Francisco, California, U.S.A. / editors,
Eugene A. Fitzgerald, Derek C. Houghton, and Patricia M. Mooney.
        p.   cm. -- (Materials Research Society symposium proceedings ; v. 533)
        Includes bibliographical references and index.
        ISSN 0272-9172
        ISBN 1-55899-439-4
1. Semiconductors---Congresses. 2. Silicon alloys--Congresses.
3. Heterostructres--Congresses. 4. Epitaxy--Congresses. 5. Germanium alloys--
Congresses. I. Fitzgerald, Eugene A., II. Houghton, Derek C.,
III. Mooney, Patricia M. IV. Series : Materials Research Society symposium
proceedings ; v. 533.

TA7871.85.E655 1998   98--38883
621.3815'2--dc21   CIP

Manufactured in the United States of America

CONTENTS

PART I: TECHNOLOGIES AND DEVICES

PART II: DEVICES, PROCESSING, AND CHARACTERIZATION

*Invited Paper

*Invited Paper

vii

*Invited Paper

viii

PREFACE

This volume contains proceedings from the symposium entitled "Epitaxy and Applications of Si-Based Heterostructures," held April 13-17, at the 1998 MRS Spring Meeting in San Francisco, CA. The intention of the organizers was to gather researchers in the area of Si-based heterostructures to discuss advances in science and technology.

The symposium offered an interesting mix of SiGe device and circuit technology, the latest SiGe materials development, and new developments in SiGeC alloys. It opened with papers on radio-frequency applications of SiGe devices. The dominant application involved SiGe graded-base and heterojunction bipolar transistors integrated with Si CMOS. To address future heterojunction field-effect transistor applications, calculations showed that velocity overshoot in strained Si can potentially increase FET performance.

An early report on $Si_{1-y}C_y$ MOSFETs was presented, and SiGeC alloy results continue to define the field. SiGe materials reports tended to concentrate on defect introduction and control of threading dislocations. High-quality Ge on Si was discussed, indicating that Ge technology on Si substrates is feasible.

Integration of optoelectronics on Si using Si-based waveguide technologies and III-V integration were reported, showing that the desire to integrate photonics for optical interconnects remains, and progress is continuing. In addition, much work was presented and reviewed on the evolution of Ge and SiGe on Si. Through the use of *in situ* techniques and imaging, knowledge of crystal surface evolution in mismatched systems and SiGe/Si continues to develop.

The symposium consisted of 49 oral presentations and 37 in-room posters.

We would like to acknowledge the aid of Anabela Afonso in the preparation of this proceedings volume, which required much attention to detail and a continuous effort at collecting the revised manuscripts and materials.

Eugene A. Fitzgerald
Derek C. Houghton
Patricia M. Mooney

June 1998

MATERIALS RESEARCH SOCIETY SYMPOSIUM PROCEEDINGS

MATERIALS RESEARCH SOCIETY SYMPOSIUM PROCEEDINGS

Prior Materials Research Society Symposium Proceedings available by contacting Materials Research Society

Part I

Technologies and Devices

SiGe RF – ELECTRONIC:
DEVICES, CIRCUITS, COMPETITORS, MARKETS

U. König
Daimler-Benz AG, Research Center Ulm, D-89081 Ulm, Germany

ABSTRACT

The outstanding performance of the SiGe/Si heterosystem and its compatibility to the dominating Si-technology, opens perspectives for a new generation of high volume microelectronic components, just for communication markets. This paper reviews device and circuit results from experiments and simulations for SiGe HBTs and for SiGe HFETs, and compares those to device results made from competitive materials. Figures of merit considered are gains, transconductances, frequencies, noise, power, delays and bandwidths.

1. INTRODUCTION

Since SiGe/Si layer structures can be grown reproducibly and even commercially with sufficient quality even those in the "nm" or atomic scale by means of MBE (e.g. /1/), LPCVD (e.g. /2/) or UHVCVD (e.g. /3,4/) the SiGe community becomes larger from year. It includes now Daimler-Benz, GEC, Hitachi, IBM, Intel, Maxim, Mitsubishi, National Semiconductors, NEC, Northrop Grummon, Philips, Siemens, Temic, SGS Thomson and probably some others which do not tell.

The development strategy in the past was to show first the potential of the novel SiGe/Si heterostructures for discrete devices. Essential results are presented in Section 2. Then chip-manufacturers have been included. The focus of the activities shifted to circuits. Section 3 reports on outstanding circuit functions. Section 4 discusses in which respect the data achieved with SiGe differ from competitors (Si, III/V). Finally system partners have been interested. Topics like application and turn-over became dominant. Section 5 looks at those aspects.

2. SiGe DEVICES

Surprisingly first the most complex SiGe hetero field effect transistor or modulation doped field effect transistor (HFET, MODFET), has been realized, already in 1985 /5,6/. Then the development stagnated 7 years. Mid of 1992 an improvement occured, proved by transconductances around 300 mS/mm for n-HFETs. After the first report of RF-data in 1994 the frequency grows step by step up to f_T = 46 GHz and f_{max} = 92 GHz /7/ for n- HFETs and up to f_T = 70 GHz for p-HFETs /8/. The demonstration of first SiGe hetero bipolar transistor (HBT) only has been in 1988 /9/. Since 1989 there was a rapid increase in frequencies, from around 30 GHz to f_T = 130 GHz and f_{max} = 160 GHz /10,11/. Today the SiGe HBT is the best developed SiGe device and on the step to production.

In the following subsections the status of the SiGe devices will be reviewed. Prospects of the device performance are presented, too. Main problems but also approaches for the solution are briefly described.

2.1 HETEROBIPOLAR TRANSISTORS (HBT)

A silicon-based HBT can be built by using a SiGe layer as the base. High Ge contents up to 50% can be incorporated /12/ which yield extremely high current gains. Compared with a conventional Si-BJT (bipolar junction transistor) the base may be as thin as a few nm which helps to decrease base transit time and therefore to rise the device cut-off frequency. In addition, the doping in the base may be extremely high, above 10^{20} cm^{-3} which reduces the base sheet resistance. Values below 1 kΩ/sq have been reported /13/ favoring very high maximum oscillation frequencies.

Mat. Res. Soc. Symp. Proc. Vol. 533 © 1998 Materials Research Society

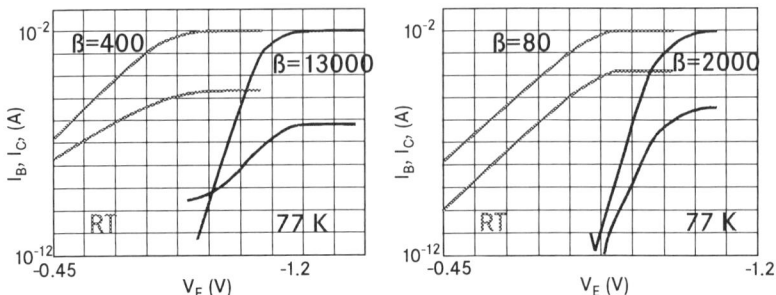

Fig. 1: Gummel plots of SiGe HBTs at room temperature and 77 K a) non passivated research HBTs, b) passivated HBTs from a Si-line (Temic)or a foundry line

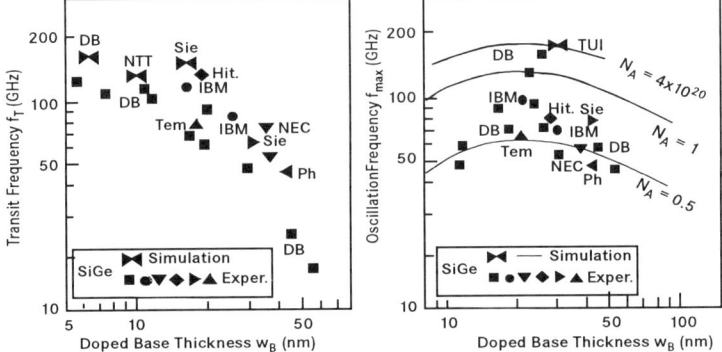

Fig. 2: Transit frequencies (a) and maximum oscillation frequencies (b) of SiGe HBTs, measured and simulated (N_A=0,5 or $1 \cdot 10^{20} cm^{-3}$) by various labs

The DC performance of a SiGe HBT is best represented by Gummel plots. Fig. 1a shows typical research results with high RT-current gains (here 400, record 5000 /14/) and extremely high 77K values up to 13000 /15/. Nonparallel collector and base currents versus base-emitter voltage are due to nonpassivated Mesa-HBTs. Passivated HBTs instead, coming out of a Si-production or a foundry line exhibit (Fig. 1b) nearly constant current gains over 9 decades /16/. The application potential of high current gains is presently not really exploited, but perhaps those will become important for low noise amplifiers in the lower GHz-range and can further rise the dynamic input resistance. Fig. 2a and b show transit frequencies f_T and maximum oscillation frequencies f_{max} from various companies (Daimler-Benz, Hitachi, NEC, Philips, Siemens, Temic, see e.g. /10,11,17/). Base thicknesses below 50 nm are required to rise f_T above 20 GHz. High values f_T of 116 GHz /18/ or 130 GHz /10/ were achieved with W_B=7 nm or 20 nm respectively, the record f_{max} of 160 GHz with W_B=37 nm /11/. A too thin base yields lower f_{max} values, due to a distinctly higher base resistance inspite of increasing f_T. According to simulations (the intrinsic RF potential is up to 1 THz, deduced from the base transit τ_B) the SiGe HBTs might reach around 250 GHz, when the external elements R_B, R_E, C_{BE} are further scaled down (Fig. 3). The f_T/f_{max} ratio strongly depends on the collector design (Fig. 4). A thin, heavily doped collector favours higher f_T values. HBTs with equal f_T and f_{max} of about 70-80 GHz can be fabricated. The frequencies increase with the collector current (Fig. 5a, b) up to a rapid fall off, which is attributed to the Kirk effect. Non- and passivated samples showed this. /16,18/. Furthermore f_{max} grows with increasing collector-emitter voltage V_{CE}, due to a decreasing collector-base capacitance. E.g. V_{CE}=1-3V yield typically 70-90 GHz, but V_{CE}=6V even 160 GHz /11/.

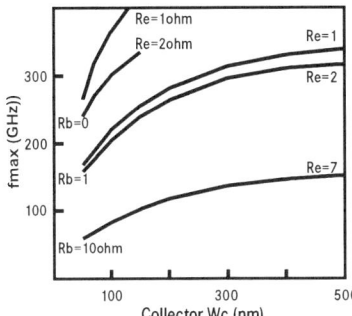

Fig. 3: Intrinsic RF Potential of SiGe HBTs can't be exploited due to the dominance of external elements (R_B, C_{CB})

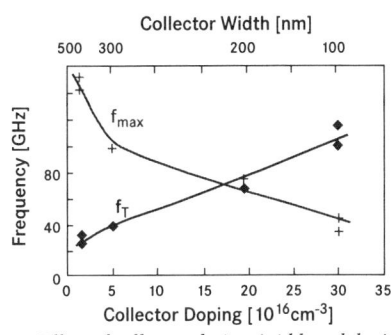

Fig. 4: Effect of collector design (width and doping) on transit and maximum oscillation frequency or on their ratio respectively /1/

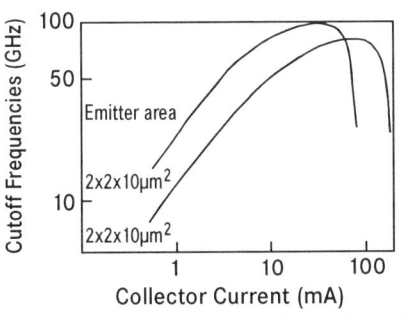

Fig. 5: Collector current or current density dependence a) for mesa-like HBTs of different emitter size and b) for planar, passivated HBTs /16,18/

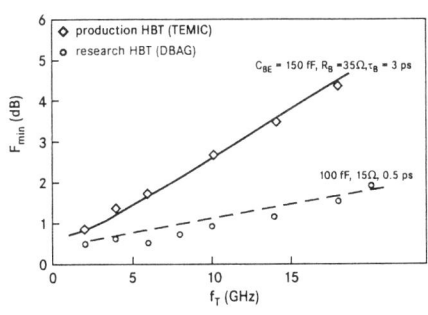

Fig. 6: High frequency noise of mesa chip HBTs and packaged HBTs. Simulated influence of base - emitter capacitance, base resistance and base transit time

Fig. 7: Noise simulations point to a strong effect of current gain (a) in the lower GHz- and of base-collector capacitance (b) in the higher GHz-range (Univ. Ulm)

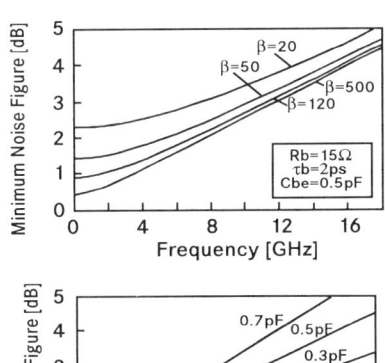

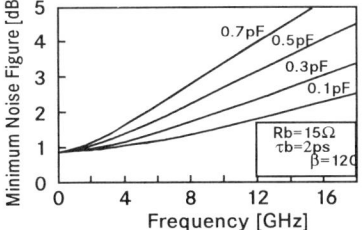

Essential figures of merit are the high and low frequency noise. Fig. 5 shows F_{min} for frequencies above 2 GHz, for research HBTs and for marketable packaged (in SOT 343 or 143) HBTs /21,22/. The noise figure at 2 GHz is below 0,5 dB, at 10 GHz still below 0,9 dB. Respective associated gains are 24 or 13 dB. The layer design of the HBTs used in the packages is not optimized for the 5-10 GHz range so far. But according to simulations shown in Figs. 6a, b /23/, there is a chance to reduce the noise, further, e.g. at lower frequencies by means of $\beta > 100$ and at higher frequencies by means of $C_{BE} < 0.5$ pF. Emitter resistances $R_E < 15\Omega$ and base transit times $\tau_B < 2$ ps would yield additional positive effects on the noise. The best low frequency noise behaviors are shown in Fig. 7. The corner frequency f_C, i.e. the transition point from the 1/f-noise to the shot-noise is below 300 Hz, even for the packaged HBT /21,24/. f_C decreases with the collector current. The Gummel plot gives an idea of the size of f_C (Fig. 8). Parallel collector and base currents, i.e. the ratio $\beta_{max}/\beta = 1$ guarantees corner frequencies below 1 kHz /25,26/. Fig. 9 relates f_c for various devices from IBM and our house to the operation frequency f_T /27/.

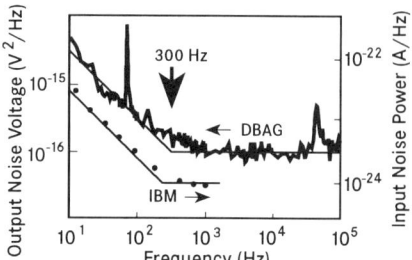

Fig. 8: Low-frequency noise show low corner frequencies of SiGe HBTs still at collector currents of 2 mA /19,21/

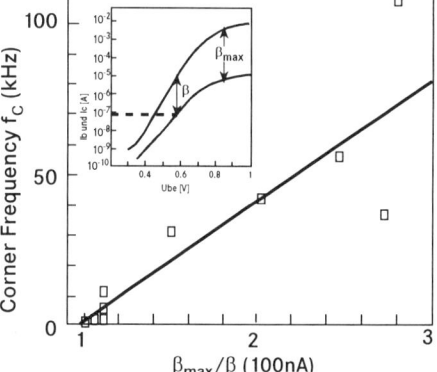

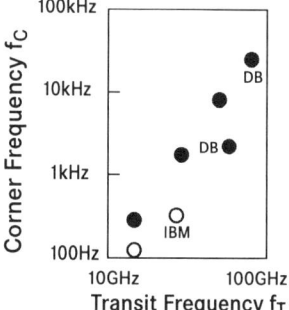

Fig. 9: Corner frequency related to the non-ideality of the Gummel plot (ratio of gain at high and at low currents)

Fig. 10: Corner frequencies related to the transit frequency of the HBTs

SiGe power HBTs intended for 1.9 GHz applications either using 10 or 60 emitter fingers exhibited a collector-emitter breakdown voltage of 4.5 V along with f_T and f_{max} values of 16 and 11 GHz, respectively, at a collector current of 400 mA and a V_{CE} of 3V. Class A load pull measurements at 1.9 GHz (Fig. 10a) demonstrated a power added efficiency PAE=44% at 1 W output power. A driver HBT exhibited a PAE of 72% at 900 MHz (Fig. 10b) for class A/B operation /22/. These excellent data are achieved without any thermal shunt precautions in the contacts and without substrate thinning, which reflects the benefits of a Si-substrate.

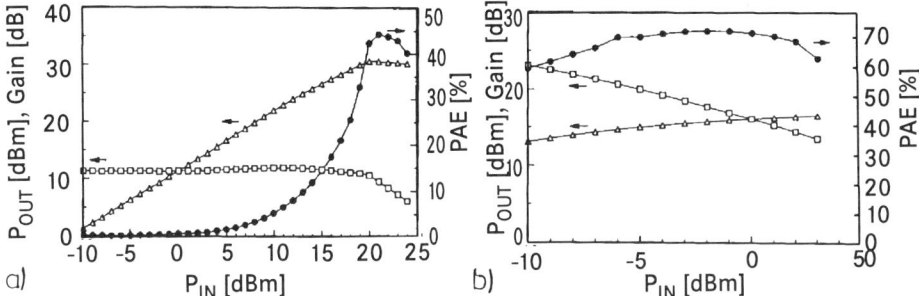

Fig.11: Gains and efficiencies of power SiGe HBTs a) class A, b) class A/B /22/

Inspite of the success of the SiGe HBTs some drawbacks still remain. In the HBT version with a lightly doped Si emitter, a higher doped SiGe base and with an abrupt box shaped Ge- and doping profile, the structure reacts sensitively to heat treatments, implantation's and thermal oxidation procedures. Certainly the mesa structure is advantageous in that case, but it creates another problem, i.e. defects in the base, due to high implantation energies necessary for reducing the base emitter spacing by means of a double spacer. Around the emitter area a parasitic bipolar junction transistor can occur due to outdiffusion, which reduces frequency and rises noise. A drawback of the HBT version with gradual Ge and dopant in the base is the high base resistance which lowers f_{max}.

Several solutions are offered: The selective low pressure CVD growth yields a selfaligned, down scaled formation of the active area. At least three groups are presently going this way /23-25/. On the other hand a layout with an exside-outside spacer process and with an amorphous Si-emitter allows to omit the emitter mesa, precisely to set the base emitter spacing, and to avoid critical implantation energies. A technique to suppress the lateral outdiffusion around the emitter is to incorporate a small amount of carbon into the layer structure during the growth /26/. Ideal Gummelplots and no indication of a parasitic BJT have been found.

2.2 HETERO FIELDEFFECT TRANSISTORS (HFET)

In a SiGe/Si Hetero FET the carriers are confined in a quantum well channel composed of Si or SiGe or even Ge. The carriers move nearly collision free owing to reduced coulomb scattering (separated from the dopant atoms, modulation doping MODFET) like in a two dimensional electron or hole gas (2DEG, 2DHG). High mobilities up to 2900 cm²/Vs for n-channels and up to 1800 cm²/Vs for p-channels have been found /27,28/ recently, even for high sheet carrier concentrations of $2-5 \cdot 10^{12}$cm^{-2}. In addition, there is evidence for an increased velocity overshoot due to strain in the channels. High frequencies can be expected /29/. Finally in the case of hetero MOSFETs, where the channel is buried under the gate, one profits from the elimination of interface scattering due to a more perfect Si/SiGe interface.

Fig. 12 summarizes best transconductances of n-HFETs, all of them with Schottky gates. The adjustment of the operation mode - depletion or enhancement - is made by the doping in supply layers above, below or even within the channel and by the layer thicknesses, e.g. the cap layers above the channel. Furthermore a gate recess can yield an enhancement mode FET from a layer structure which was originally designed for depletion operation. IBM and our house have reached extrinsic transconductances g_{me}=290 to 310 mS/mm for depletion HFETs at RT and IBM even 600 mS/mm at 77 K /7,30/. For enhancement FETs we have recently reached 480 to 510 mS/mm at RT and up to 780 mS/mm at 77K. High currents of about 320 mA/mm for depletion and 200 mA/mm for enhancement FETs have been found around V_{DS}=2V. Transconductances of p-HFETs are usually smaller (e.g. around 250 mS/mm for 0.2-0.25 μm gates). Theory predicts values like for n-HFETs when using a Ge-channel in p-HFETs. But high currents above 600 mA/mm have been obtained for MOS-gated (3,7 nm SiO₂) p-HFETs (see Fig. 13 /31/). Fig. 14 makes a comparison of the transconductance status with the expectations. Experiments have not established the significant gate length dependence (a,b,c) expected from theory /29/.

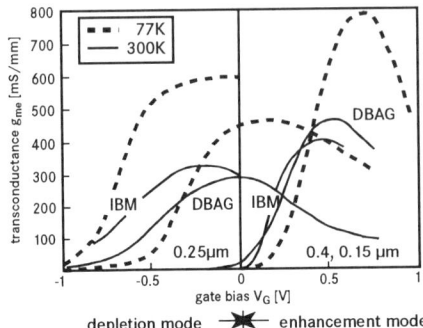

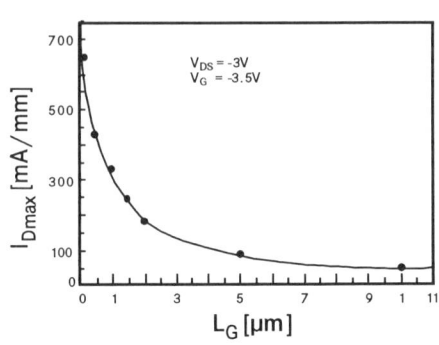

Fig. 12: Transconductances at room temperature and 77 K of depletion and enhancement mode SiGe HFETs

Fig. 13: p-HFETs with 3.7 nm MOS gate. Gate length dependence of the maximum drain current

Fig. 14: Gate length dependence of the transconductance. Best data reported compared to simulations

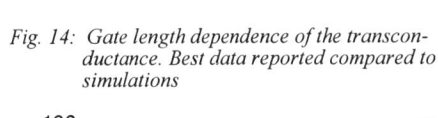

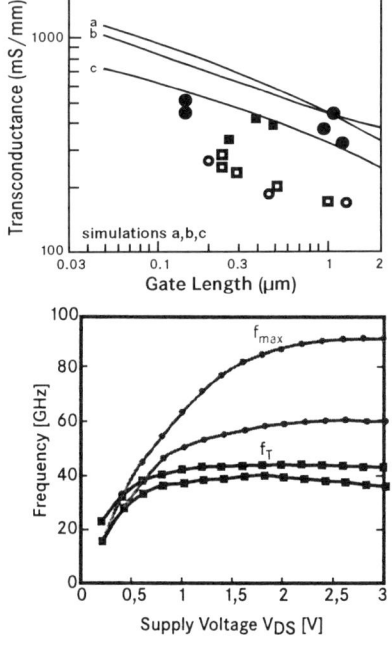

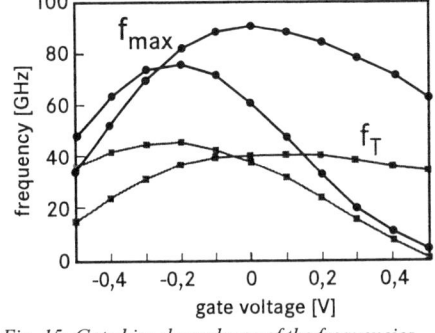

Fig. 15: Gate bias dependence of the frequencies for two SiGe HFETs /7,32/

Fig. 16: Drain-source supply voltage dependence of frequencies

Cut off frequencies f_{max} of 78 to 92 GHz and f_T of 43 to 46 GHz had been extrapolated from the gains. Samples are depletion n-HFETs, one with a frequency maximum at -0.2 V gate bias, the other at -1 V (Fig. 15). We have also realized devices operating around V_G=0 V with frequencies of the same order. The low-power potential of SiGe-HFETs elucidates Fig. 16. Drain voltages

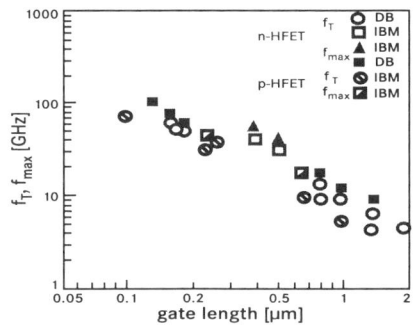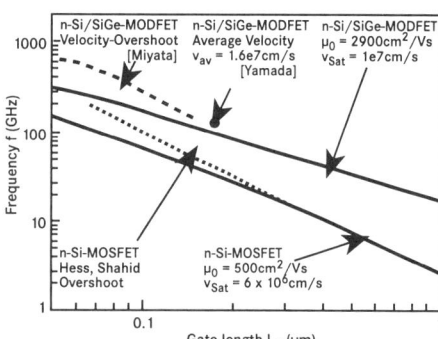

Fig. 17: Gate length dependence, a) measurements of n- and p-type HFETs and b) simulations for n-SiGe HFETs with and without velocity overshoot /31/

need to be only just above 1V to get a voltage independent operation. One benefits from the high mobilities which yield a rapid current - and thus frequency - slope with the supply voltage.

In Fig. 17a most of the frequencies f_T and f_{max} achieved so far for n- and p-HFETs (IBM and Daimler-Benz) are plotted /7,8,32,33/. There is a gate length dependence visible, but accurate control with simulation like those shown in Fig. 17b hint to a hardly exploited potential especially at gate length below 0.3 μm. Obviously layer structure and layout are not optimized in respect to doping and low parasitics. The range $L_G < 0.10$ μm, has not yet been investigated for SiGe HFETs. That promises frequencies up to 200 GHz, or even above (>400 GHz), if velocity overshoot really exists /29,31/. On the other hand HFET devices with a relaxed gate layout of 0.35 to 0.8 μm have a better performance than Si-MOSFETs and benefit from the manufacturability in a standard Si-MOS production line.

Though just in the last 3 to 5 years the performance of SiGe HFETs has made a large step, we are not at the end of the ladder. One obtains high mobilities, but the individual layers are with 10 to 50 nm too thick for devices which demand for high transconductance and high currents in the channel. With p-type layer structures the forecasted high mobilities close to those of n-type structures have not been obtained. Furthermore a trade off between a large bandgap difference (i.e. high Ge content, desired for an ideal carrier confinement) and the crystallographic quality of the structure has not been established. The dopant diffusion in the nm-thin modulation doped structures at typical device process temperatures is not under control. Perhaps it is possible to waive on additional doping like in standard MOSFETs which would allow to rise the thermal budget of the process again. Can we realize a proper gate oxide on SiGe/Si heterostructures with low fixed and interface charges?

More or less all that can be solved by an improved growth process just for the buffer layer and by low thermal budget runs. Numerous proposals for a strain relaxed buffer between the Si-substrate and the SiGe heterostructure are known (see e.g. /34/). A buffer has to be thin from an economic growth point of view and has to suppress dislocations, a severe technological challenge for the future. The reduction of the thermal budget e.g. by RTA (rapid thermal activation) or by LTO (low thermal oxide) is a trend in standard CMOS, too. But for SiGe HFETs the conditions are stronger: undoped structures tolerate about 850°C, modulation doped 650°C only. Several LTOs are under investigation, but also here no solution can be offered so far.

3. SiGe-CIRCUITS

Since five years only some companies like IBM, NEC, Siemens and Temic together with our company are making the step from discrete SiGe HBTs to circuits. Though the HBT is presently the best developed SiGe device it is an ambitious and time consuming goal to implement such a new technology into the well established Si-production line. According to optimistic statements of some companies we already should have SiGe circuits on the market. More realistic is to have those available at the end of 1998, but only for special system customers, with whom the chip manufacturers presently cooperate. Taking into consideration that the first III/V-IC was published more than 20 years ago /35/ and that real high-volume production of III/V is still absent, the 6-years transfer time for SiGe sounds promising.

The next device on the way to IC-production should be the SiGe HFETs. Technologically compatible up to 4 or even 8 inch Si-wafer production those can perhaps become a low cost alternative to III/V HEMTs, at least for some applications in the lower GHz range. Apart from analog ICs in a long-term view the digital potential of SiGe HFETs can be more attractive. If we succeed in the realization of a novel SiGe Hetero CMOS generation we will get the chance to win attractive market segments of the huge CMOS-mainstream. However, up to this point we have to overcome some technological problems. In the following the status and the potential of SiGe HBT- and SiGe HFET-circuits is reviewed.

3.1 HBT circuits

Various HBT-ICs have been reported so far, some with outstanding, some at least with promising performances. Fig. 18 shows a chip with circuits for 5 to 40 GHz operation realized on semiinsulating Si-substrate. More production like are circuits on 20 Ωcm substrates (one example in Fig. 19), which Temic has under development /36/. ECL ring oscillators realized by Siemens, Philips, IBM, NEC, or Temic together with our house (DBAG) exhibit, 11 to 20 ps /17,22,24/, with the very recent record of 9.3 ps by Hitachi /37/. A 12 bit DAC operating at 1 GHz has been demonstrated by IBM together with ADI /37/. NEC has reported on D-type flip flop for 20 Gbit/s, a selector for 30 Gbit/s and 33 ps, a 2:1 multiplexer for 20 Gbit/s and a preamplifier with 19 GHz and 36 dBΩ /25,38/. Multiplexer and demultiplexer with 28 Gbit/s have been realized by the Ruhr University Bochum using samples from DBAG /39/. A wideband amplifier capable of 9.5 dB gain with 18 GHz bandwidth while drawing only 50 mW from a 3 V supply and

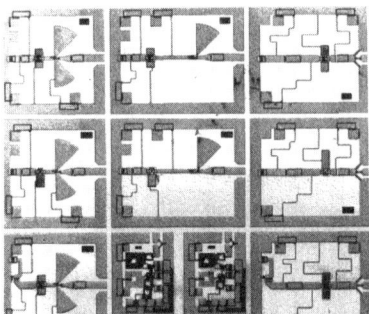

Fig. 18: SiGe HBT chip with circuits for 5-40 GHz

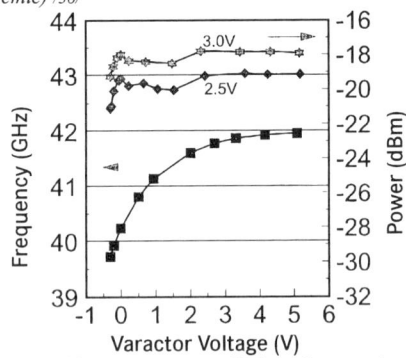

Fig. 19: SiGe HBT switches for 2 GHz with spiral inductors (Temic) /36/

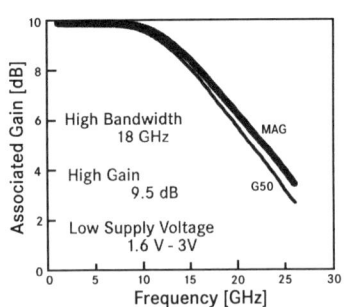

Fig. 20: a) Low power consumption broad-band amplifier, and b) varactor controlled oscillator with SiGe HBTs /40,42/

10

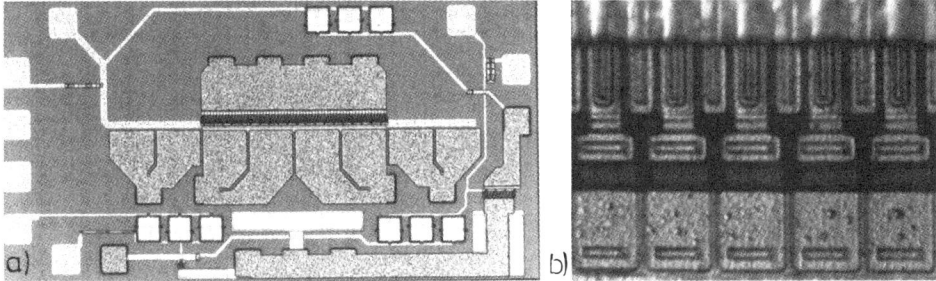

Fig. 21: a) SiGe HBT power amplifier, and b) enlarged emitter area

operating even at 1.6 V was realized (Fig. 20a) /40/. Varactor controlled oscillators for different frequency ranges of 1.8, 11, 26, 28, 40 GHz (see Fig. 20b) were presented by Nortel together with IBM, by Temic, by IBM and by DBAG /41,42/. Power amplifiers for 0.9 to 2 GHz have been realized by Temic together with DBAG (Fig. 21a, b) and by Philips /22,41/. LNAs with an f_{min} of 1.7 to 1.9 dB came from Temic together with DBAG and the University of Ulm (Fig. 22) /43/ and from IBM. A frequency divider of 42 Gbit/s and a 60 Gbit/s demultiplexer were recently realized by Siemens /44,45/. The MSI and LSI potential was demonstrated by means of arrays of small SiGe HBTs with more than 1000 up to 30000 devices by DBAG, Temic and IBM.

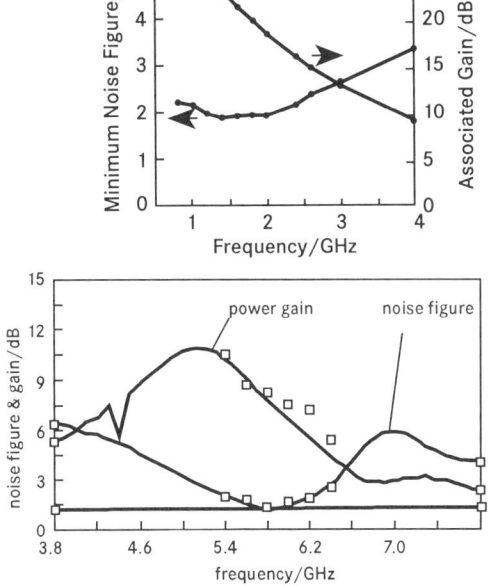

Fig. 22: Low noise amplifier with HBTs having 6 and 4 emitter fingers (Temic, Univ. Ulm) /43/

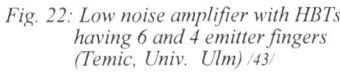

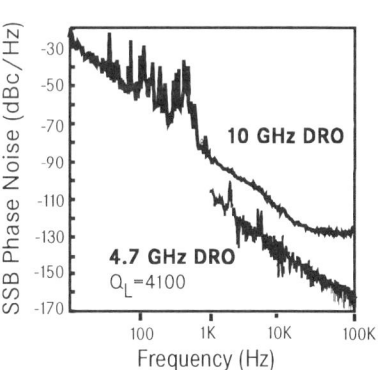

Fig. 23: Hybrid active antenna with a SiGe HBT yielding low noise at the receiving frequency of 5.8 GHz (Univ. Ulm) /46/

Fig. 24: Phase noise of a 4.7 GHz and a 10 GHz DRO /48/

We have used SiGe HBTs for hybrid integration, too. Very recently an active antenna for a 5.8 GHz receiver with the excellent noise figure of 1.4 dB was reported by the University of Ulm using a device of DBAG (Fig. 23) /46/. Extremely low phase noise has been demonstrated for 4.7 GHz oscillators. Only –135 dBc at 10 kHz. A 10 GHz oscillator showed –115 dBc at 10 kHz off. Dielectric resonator oscillators (DRO) for 4.7 and 10 GHz (Fig. 24) and a 8-12 GHz VCO were reported by DBAG together with Dornier and CNET /47,48/.

3.2 HFET CIRCUITS

These type of circuits are still in a less developed status. Digital and analog demonstrators are under investigation. A digital chip contains ring oscillators (23 and 43 steps), inverters (Fig. 25), level shifters (Fig. 26) and some other test circuits. The chip technology is based on e-beam gates from 0.8 to 0.15 μm and on two interconnect levels /49/.

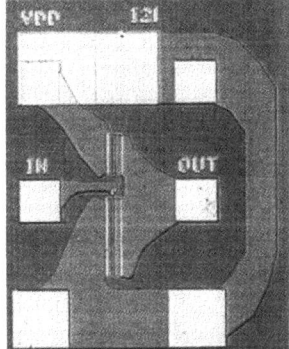

Fig. 25: Inverter circuit with a resistor as load and an n-
type depletion MODFET as driver   /49/

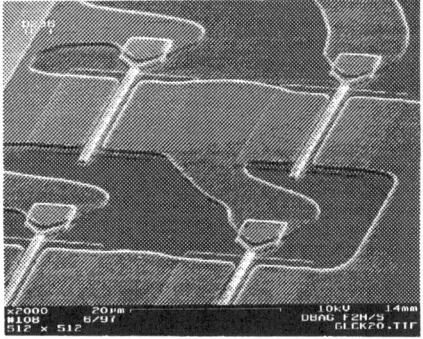

Fig. 26: Level shifter with n-HFETs

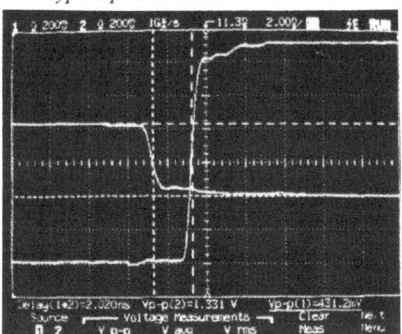

Fig. 27: Gate delay and maximum output voltage
swing of the inverter /49/

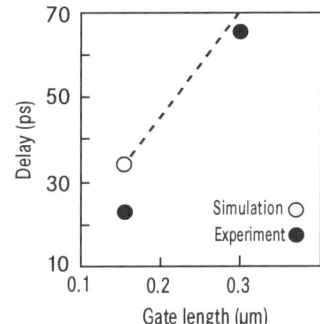

Fig. 28: Gate delays for n-HFET inverters

First results are promising. Large signal measurements of the inverter with an input signal of 600 ps rise time at a supply voltage of $V_{dd} = 2$ V show a gate delay of 70 ps for the 300 nm gate length (Fig. 27). The maximum output voltage swing is 430 mV. Measurements with an input signal of 150 ps rise time show a gate delay of 22 to 25 ps for the 150 nm gate length inverter circuit (Fig. 28). The gate delay was determined from the delay between the 50% input and output signal value, taking into account the different RC-delays of the measurement equipment. For digital logic design the inverter circuit of Fig. 25 needs a second stage to shift output to input levels, as shown in Fig. 26.

A transimpedance amplifier (Fig. 29) consists of two stages /50/. In this very preliminary phase we chose a flexible design in order to accept both depletion and enhancement HFETs as well as technology-related scatter in parasitic elements. Amplifiers passivated with a CVD oxide showed a high transimpedance of 56 dBΩ with a –3 dBΩ frequency of 1.8 GHz (Fig. 30). Circuits with sputter oxide passivation yield a higher transimpedance of up to 72 dBΩ with 1 GHz bandwidth.

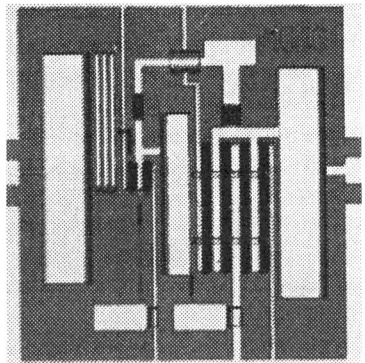

Fig. 29: Microphotograph of a two-stage transimpedance amplifier

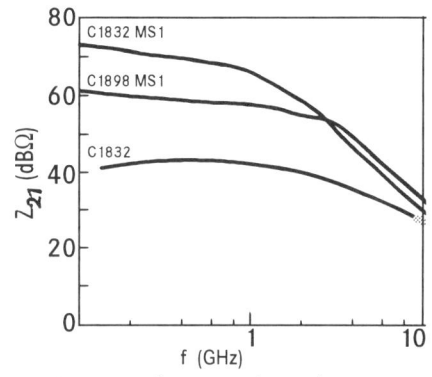

Fig. 30: Measured transimpedance characteristics of amplifiers with 2.34 kΩ and 540 Ω feedback resistors sputter-passivation, 2.34 kΩ CVD-passivation, 540Ω /50/

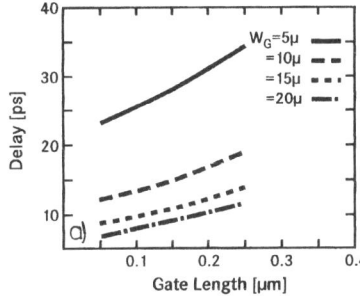

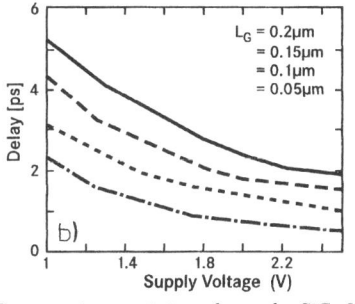

Fig. 31: Speed potential (simulations) of a novel CMOS generation consisting of n- and p-SiGe HFETs

The integration of n-HFETs with p-HFETs offers a new generation of CMOS circuits. SiGe Hetero CMOS enables the same mobility and velocity for both types of HFETs with equal geometry. Simulations (Fig. 31a,b) predict gate delays down to 1 ps in unloaded and around 3 ps in loaded operation /29/. Modular circuit concepts allow the integration of HFET circuits with standard VLSI-CMOS. The HFETs can be used in fast digital or fast analog processing circuits. The technology starts with the conventional CMOS process including the poly gate, leaving blank the areas for the heterodevices. Then follows the HFET processing (i.e. low temperature epitaxy, patterning). Reintroduced into the CMOS-line the circuit receives contacts, interconnections and passivation. We have already experimentally checked the feasibility of such a heterointegration /51/.

The chances for a SiGe HFET circuit fabrication are high. When the SiGe technology is implemented in Si-IC factories, this infrastructure may be used for HFET circuits, too. Even the modular combination with CMOS has the best chance to be realized in medium term, presuming the CMOS line will accept wafers which have seen different labs. The strongest technological challenge for the future is the realization of the ideal SiGe Hetero CMOS concept. Problems with the p-HFET technology have to be overcome, with low thermal budget oxides and with defect densities related to the buffer layers. However, the outstanding performance perspectives justify the effort.

4. COMPETITORS TO SiGe

SiGe represents a young generation of microelectronics and it has to compete with standard Si and with III/V. In general the SiGe performance fits into the gap between Si and GaAs (e.g. in frequency) and is even better in some aspects (e.g. phase noise). Cost advantages over III/V is another benefit (see section 5). A brief summary of the performance comparison is given in this section, a more detailed one in /17,34/.

Let us look at HBTs first. The transit frequency depends on the various transit and charging times. The additional barrier ΔE_v of the HBT in the valence band, minimizes τ_E in comparison to standard Si bipolar transistors (BJTs). The difference between III/V and SiGe is mainly due to the base transit. The minority carrier mobility in heavily doped ($N_A > 10^{19}$ cm^{-3}) GaAs is about 1000 cm²/Vs, about three times higher than in SiGe. Consequently one needs a thinner base in SiGe to achieve high f_T values (compare Fig. 2a). The present record in III/V is around 200 GHz and in SiGe around 120 GHz. The maximum oscillation frequency of III/Vs has reached 200-250 GHz, whereas SiGe is at 160 GHz. The theory explains the difference by the lower base resistance R_B of III/Vs, because the product $\mu_p N_A$ is larger in III/V. Simulations for different base doping N_A show an optimum base thickness around 20 nm. Above, f_{max} decreases due to the increasing base resistance. Below, f_{max} decreases due to the falling transit frequency. If we choose a base doping of 5.10^{20} cm^{-3} in future SiGe HBTs – which is technologically possible – 200 GHz and more are possible. Concerning the RF noise, SiGe HBTs are better than all III/V HBTs (at least atup to 10 GHz). Hawkins law explains this by the internal transit frequency through the base $f\tau_B$: the transit in a thin SiGe base below 20 nm is faster than in a thick III/V base of more than 50 nm. Furthermore β is usually higher in SiGe HBTs, which makes them superior to standard Si BJTs(see the effect of β on F_{min} in Fig. 7). The low-frequency noise is best represented by the corner frequency f_C, the transition point from the 1/f to the shot noise. The burst noise, some times observed in III/V HBTs, is related to deep level recombination and is absent in well-processed silicon devices due to the high quality oxide passivation. Si BJTs and SiGe HBTs are therefore best in corner frequency f_C, followed by III/V HEMTs and III/V HBTs with typical values in the upper kHz range. In digital applications the switching speed of HBTs demands both a low base resistance and base-collector capacitance. The power delay product depends on the logic swing, and this demands a high transconductance for a small swing. The supply voltage V_S is limited by the threshold voltage V_{th}. Together with V_S also V_{th} has to decline in future low-power-dissipation circuits. The threshold voltage is fixed by the material system. III/Vs are worse than Si and SiGe in this respect. That is confirmed by outstanding delays around 10 ps of SiGe HBT ring oscillators, recently reported /10/. III/Vs reach low values only with sophisticated structures (e.g. grading), which reduce the bandoffsets at the EB-junction.

The second type of devices are HFETs: InP based MODFETs have reached extremely high transconductances of 1700 mS/mm at 50 nm gate length. The transconductance increases inversely with the gate length L_G and with the gate to channel distance d_{GC}, i.e. with the capacitance C_{GS}. SiGe HFET data up to 500 mS/mm with 150 nm gates are located in the range of GaAs-HFETs. However, they correspond to thinner d_{GC} values of around 15 nm. Si MOS has made an exciting progress in the last two years represented by transconductances around 1100 mS/mm with 60 nm gates /52/. However, simulations for SiGe HFETs point to higher transconductances, especially when taking the velocity overshoot into account, which will be more pronounced in strained SiGe than in pure Si /29/. The transit frequency of HFETs can be expressed either by the g_m/C_{GS} ratio or by the drift time V/L_G and by the sheet carrier concentration n_S. The maximum oscillation frequency depends on numerous design parameters, i.e. parasitic resistances and capacitances. Record frequencies for GaAs HEMTs are 350 GHz, for InGaAs/InP HEMTs even 600 GHz. Si MOSFETs have recently demonstrated $f_T \approx 200$ GHz for 60 nm gates /53/. SiGe HFETs with $f_{max} > 90$ GHz are located above Si MOSFETs at equal gatelengths. According to simulations SiGe HFETs might reach distinctly higher frequencies (see Fig. 17b). An exclusive property of SiGe is that p-type HFETs with a pure Ge or a SiGe channel will nearly reach the frequency of n-SiGe HFETs, today p-type SiGe HFETs are about a factor of 2 higher in frequency than p-type Si MOSFETs. This reflects the potential of a complimentary arrangement of n- and p-SiGe HFETs.

14

5. MARKETS FOR SiGe RF

The turnover of microelectronic and optoelectronic devices and circuits is rapidly growing, from $ 45 billion in 1990 to $ 77 billion in 1993 over $ 154 billion in 1995. One estimates about $ 350 billion in 2000. Looking at the market shares of the semiconductor materials, technologies and products (Fig. 32) one finds a dominance of Si with 97%. 72% to about 80% is and will be covered by CMOS mainly for microcontrolers and memories. Though the percentage share of bipolars will decrease from today to the year 2000, it nevertheless means an increase in turnover. This market trend is one of our motivations to develop Si-based heterodevices. SiGe HBTs and SiGe HFETs or HCMOS fit best into the respective Si-markets. SiGe HBTs will be introduced in 2000, while we hope to have SiGe HFETs ready to the market then. Which size of the market segment they can win, depends on their performance.

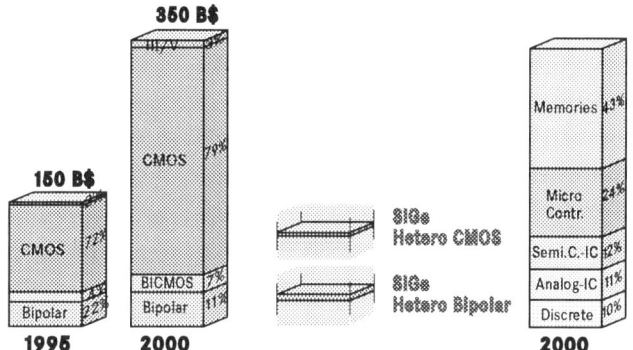

Fig. 32: Turn-over for microelectronic technologies and products. SiGe HBT and SiGe HFETs or HCMOS fit into bipolar and CMOS markets

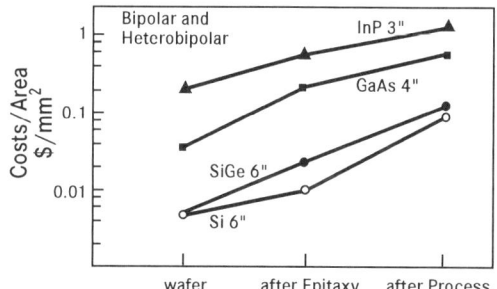

Fig. 33: Cost estimations and comparisons for bipolar and heterobipolar circuits

A new technology wins market segments if it is economic, too. SiGe meets this condition. SiGe needs no new fab, as Si-lines can be used, a fact that saves investment and processing costs. SiGe can use large Si-wafers which lowers material and area-related processing costs. Cost estimations in Fig. 33 confirm this for SiGe HBTs. Ranking the total costs of bipolar transistors there is SiGe with about $ 0.12/mm². Not essentially higher is Si with $ 0.09 mm² while GaAs and InP costs go up to $ 0.5/mm² or $ 1.2/mm². SiGe heterodevices and circuits will be produced by large chip manufacturers. Indeed, companies like IBM, Siemens, NEC, Philips, Hitachi and Temic together with Daimler-Benz are active in SiGe. Relevant foundries can open new markets with attractive prototypes and act as supplier for small and medium size system customers.

Fig. 34: Scenario for applications of SiGe/Si heterodevices in the high volume communication market

Low cost, high performance SiGe ICs are best suited for high volume markets. Those are various communication services (Fig. 34): The mobile communication (MOBICOM) transmits audio and voice via handy phones at 0.9 to 2 GHz. The wireless local area networks (WLAN) at 2.4 to 5.8 GHz connects PCs. The satellite communication (SATCOM) at different bands ku (10-14 GHz) or ka (and 25 GHz) or even higher supply low-infrastructure areas and mobile users. The wideband communication via cables presently by coax and in future increasingly by optical fibres (FIBRECOM) transmit from 3 to 40 Gbit/s mainly in hubs of conurbation and in intercontinental networks. Each of these services will have 50 to 100 million users or terminals in the year 2005. With module/system costs between $ 100 and $ 700 one expects market shares of 20 to 60 billion Dollar each /54/. Further markets are seen in navigation of mobile objects, e.g. global positioning (GPS, ~1.5 GHz), satellite navigation (>10 GHz), defense and landing radar (20-40 GHz) and collision avoidance of cars (~70 GHz), robotic and industry sensors (20-50 GHz) etc. Market analysis also expects 100 Mio modules for both fields. Even computer and consumer electronics, which increasingly demand faster signal processing might be a market for SiGe chips. The share of microelectronic components in the systems is 20 to 40%, which makes these markets so attractive for chip manufacturers. Collecting studies of different companies and market research groups one can estimate for SiGe chips a turnover above $10 bill.

6. ACKNOWLEDGEMENT

The author would like to thank several co-workers in house and colleagues elsewhere, who have supported me with information's and simulations, i.e. D. Behammer, M. Glück, A. Gruhle, T. Hackbarth, H. Presting at Daimler-Benz Ulm, H. Schumacher, U. Erben, W. Dürr, G. Höck, M. Birk at University Ulm, T. Ostermann, R. Hagelauer at University Linz and J. Arndt, H. Dietrich, C. Mähner, A. Schüppen at Temic Heilbronn. Most of the reported work was conducted in the frame of the German BMBF initiative "Nanoelektronik", whose support is gratefully acknowledged.

7. REFERENCES

/1/   E. Kasper, H. Kibbel, F. Schäffler, J. Electrochem. Soc., 136, 4, 1154 (1989)
/2/   LPCVD Centura, Applied Material Technology
/3/   B.S. Meyerson, Solid State Technology 53, Feb. (1994)
/4/   SiGe Microsystems Inc., http://www.sige.com

/5/   H. Dämbkes, H. J. Herzog, H. Jorke, H. Kibbel and E. Kasper, IEDM 85, 768 (1985)
/6/   T.P. Pearsall, J.C. Bean, R. People et.al. Proc. 1 Int. Symp. Si-MBE, 400 (1985)
/7/   M. Glück, T. Hackbarth, U. König, M. Birk, A. Haas, E. Kohn, Proc. MSS 8. July 1997, St. Barbara
/8/   M. Arafa, K. Ismail, J.O. Chu, B.S. Meyerson and I. Adesida, IEEE-EDL 17, 586 (1996)
/9/   G.L. Patton, S.S. Iyer, S.L. Delage, S. Tiwari et.al., IEEE-EDL 9, 165 (1988)
/10/   K. Oda, E. Ohne, M. Tanabe, H. Shimamoto, T. Onai, K. Washio, IEDM Techn. Digest (1997)
/11/   A. Schüppen, U. Erben, A. Gruhle et.al. IEDM 95, 743 (1995)
/12/   A. Gruhle, A. Schüppen, H. Kibbel and U. König, Proc. Int. Symp. Comp. Semic. (1995)
/13/   E. Kasper, H. Kibbel and U. König, MRS Symp. Proc. 220, 451 (1991)
/14/   H.U. Schreiber and B.G. Bosch, IEDM 89, 643 (1989)
/15/   A. Gruhle, H. Kibbel, U. König et.al., IEEE-EDL 13, 206 (1992)
/16/   A. Schüppen, H. Dietrich, S. Gerlach et.al., IEEE-BCTM 8.2, 130 (1996)
/17/   U. König, A. Gruhle and A. Schüppen, IEEE GaAs-IC Symp. 14 (1995)
/18/   A. Schüppen, A. Gruhle, H. Kibbel, U. Erben and U. König, Electron. Lett 30, 1187 (1994)
/19/   C. Mähner, A. Gruhle, Electronic Letters, 33, 24, 2050 (1997)
/20/   H. Schumacher, U. Erben and A. Gruhle SOTOPOCS XVIII (1993)
/21/   C. Kermarrec, T. Tewksbury, G. Dawe et.al., IEEE-BCTM, 155 (1994)
/22/   A. Schüppen, S. Gerlach, H. Dietrich et.al., IEEE Microw. Lett. 6, 341 (1996)
/23/   A. Pruijmboom, D. Terpstra, C.E. Timmering et.al. IEDM 95, 747 (1995)
/24/   T.F. Meister, H. Schäfer, M. Franosch et.al., IEDM 95, 739 (1995)
/25/   F. Sato et.al., IEEE-BCTM, 82 (1995)
/26/   B. Heinemann, D. Knoll, G. Fischer et.al., ESSDERC (1997)
/27/   S.F. Nelson, K. Ismail et.al., Appl. Phys. Lett. 63, 367 (1993)
/28/   U. König and F. Schäffler, SSDM, 201 (1993)
/29/   R. Hagelauer, T. Ostermann et.al., Electron. Letters 33, 208 S(1997)
/30/   K. Ismail, B.S. Meyerson, S. Rishton, J. Chu et.al. IEEE-EDL 13, 229 (1992)
/31/   U. König, M. Glück, G. Höck, Proc. Silicon Heterostructures from Phys. to Devices, Engineering
        Foundation, Barga Italy (Sept. 1997)
/32/   M. Glück, T. Hackbarth, U. König et.al., Electron Lett. 33, 335 (1997)
/33/   K. Ismail, IEDM 95, 509 (1995)
/34/   U. König, Physica Scripta T 68, 90 (1996)
/35/   see G. Bechtel, chairman at IEEE GaAs IC Symp. (1995)
/36/   A. Schüppen, H. Dietrich, J. Arndt, unpublished
/37/   D.L. Harame, J.M.C. Stork et.al., IEDM 93, 71 (1993)
/38/   T. Suzaki, Proc. Ultrafast Elec. and Optoel. Conf., 91 (1995)
/39/   W. Geppert and H.U. Schreiber, Electron. Lett. 33, 447 (1996)
/40/   H. Schumacher, A. Gruhle, U. Erben et.al., IEEE-BCTM (1995)
/41/   IEEE-BCTM 1996 various papers
/42/   A. Gruhle, A. Schüppen et.al., IEDM 95, 725 (1995)
/43/   H. Schumacher et.al. unpublished
/44/   M. Wurzer, T.F. Meister et.al., IEEE Int. Sol. State Circ. Conf., 122 (1997)
/45/   A. Felder, M. Möller, M. Wurzer, M. Rest, T.F. Meister, H.M. Rein, Electr. Letters 33, 1984(1997)
/46/   W. Dürr, W. Menzel and H. Schumacher, IEEE Microw. Lett. 7, 63 (1997)
/47/   A. Gruhle, H. Kibbel and R. Speck, Proc. EU MC-Conf. 648 (1994)
/48/   B. van Haaren, M. Regis, O. Llopis, L. Escotte, A. Gruhle et.al., 28th Conf. European Microwave
        (1998)
/49/   T. Ostermann, M. Glück, R. Hagelauer et.al., Proc. Int. Semicond. Device Research Symposium
        (Dec. 1997)
/50/   M. Saxarra, M. Glück, J.N. Albers et.al., Electron. Letters 34, 4, 499 (1998)
/51/   U. König and H. Dämbkes, Sol. State Electron. 38, 1595 (1995)
/52/   Y. Mii, S. Rishton, Y. Taur, D. Kern et.al., IEEE-EDL 15, 28 (1994)
/53/   H. S. Momose, S. Nakamura, Y. Katsumata et.al. ESSDERC, 133 (1997)
/54/   deduced from various market studies, e.g. Alcatel, Dasa, Daimler-Benz, ESA, Temic

17

TOWARD RF SYSTEM-LEVEL INTEGRATION: PROCESS INTEGRATION ISSUES IN SIGE BICMOS

G. Freeman, K. Schonenberg, D. Ahlgren, S-J. Jeng, D. Nguyen-Ngoc, K. Stein, D. Colavito, S. Subbanna, D. Harame*, B. Meyerson**
IBM Microelectronics Division, Hopewell Junction, New York
* IBM Microelectronics Division, Essex Junction, Vermont
** IBM Research Division, Yorktown Heights, New York

ABSTRACT

The SiGe HBT, integrated with CMOS devices on the same chip, will be the first integrated device combination to realize the long-standing technology goal of fabricating RF systems on a chip. It has been demonstrated that the HBT device can replace the standard GaAs front-end of RF systems and take advantage of the reduced cost available from Si technologies in 200-300mm wafers. However, the SiGe HBT can only be part of a large-scale RF system on a chip when, in the same technology, NFETs and PFETs are provided for low power, low frequency digital logic, and a suite of resistors, capacitors, diodes, and inductor passive elements are provided for the high frequency analog circuitry. Furthermore, all these elements must be manufacturable defect-free at medium and high levels of integration.

This paper covers key process integration issues confronting technologists when integrating a SiGe HBT device with the requisite CMOS and passive elements and at the same time maintaining very high GaAs-like performance. Topics to be discussed are 1) a review of high-performance HBT device integration schemes employed to date and integration issues with each scheme, 2) integration issues in epitaxial cleaning and growth techniques, 3) integration issues influencing crystal defects, and 4) integration issues with passive elements. Status of the IBM SiGe BiCMOS technology will be presented to illustrate the first successful integration of this set of devices into a manufacturable process.

INTRODUCTION

For much the same reasons that digital integrated circuits are ever placing more functionality on a smaller chip area, decreasing size and increasing functionality are also the principle trends in today's wireless marketplace. Consumers of cell phones, personal digital assistants, laptop computers, etc. are expecting next year's product not only to be smaller and cheaper, but also to have a greater functionality than this year's product. Riding the coattails of the exponential growth in digital functionality, wireless is viewed as key to providing the critical link of convenient connectivity to a world of information. Making these devices ever smaller and cheaper is the challenge of RF semiconductor technologies.

Dominating the greater than 1 GHz downconverter and upconverter/power amplifier functions (where the RF signal is converted to and from the information-containing baseband signals) has been the GaAs MESFET, and, more recently, the GaAs PHEMT. In frequency capability, the MESFET has been able to stay several steps ahead of silicon MOSFET devices because of the higher electron mobility in GaAs compared to silicon. Silicon BJT devices, on the other hand, have typically dominated the marketplace at frequencies up to 1GHz. Thus, in today's 900MHz RF products, one finds a mix of GaAs and BJT technologies, but in the rapidly growing market for 2 GHz and above RF products, one finds the downconverting and upconverting functions

Mat. Res. Soc. Symp. Proc. Vol. 533 © 1998 Materials Research Society

managed by small integration-level GaAs devices, and the baseband functionality managed by conventional low cost and large-scale integrated silicon technologies.

The question for the future is which technology will dominate in the next generation of higher-frequency, higher-integration RF appliances. While today's GaAs technologies are well suited for today's wireless applications, the following trends are challenging GaAs's abilities compared to silicon: 1) increasing RF frequency device counts, 2) increasing system functionality on a single chip, and 3) decreasing costs. Small wafer sizes, high wafer cost, lower yields, poor repeatability, and low thermal conductivity have long been the drawbacks in GaAs integrated circuits.

It is generally agreed today that silicon-based devices will be capable of providing some functionality in the low GHz range. However, it is not so clear whether the HBT or CMOS device will predominate. Some researchers consider it a matter of destiny that increasingly scaled silicon CMOS technologies will eventually fill the low GHz RF market, principally because of the low cost associated with volume CMOS technologies [1][2]. Other researchers have pointed out that factors important to RF circuits, such as low noise, high transconductance, low knee voltage (lower potential power supply voltages), higher power densities, and lower cost lithography point to the BJT or HBT as the preferred device [3][4].

In practice, it is the SiGe HBT that has recently shown the potential to meet stringent RF requirements and may dominate the RF 1 - 30 GHz market for a number of years. Numerous circuits utilizing the SiGe HBT have been shown to comfortably operate in the 1 - 10 GHz range, some of which incorporate a large (>1000) HBT device count [5][6]. Furthermore, it has been demonstrated that the HBT may be integrated on the same chip with standard silicon CMOS without sacrificing performance or yield of the CMOS or HBT [7]. Low cost is achievable since the device may be manufactured in a standard 200mm CMOS fabrication facility with few unique tools, with excellent repeatability and high yields.

Integration involves managing a set of constraints that one set of fabrication steps or films imposes of another set of fabrication steps or films, with successful integration resulting in simultaneously functioning devices on a chip. In a larger sense, it is also the set of constraints imposed by one device on another. Before embarking on the technology integration issues, it should be pointed out that certain business and cost constraints are often defined as a direct result of the fact that the same manufacturing line is to be used for multiple technologies, including prior generation BiCMOS and as well as CMOS technologies. Besides the manufacturing process, the critical and complex chip design infrastructure is also in place for a given CMOS process. Development of unique process steps and design infrastructure should be minimized so as to minimize costs and maximize prior CMOS design availability. Thus the principle constraint on the CMOS device is that it be electrically identical to, and process steps including isolation, lithography, planarization, etch, and hot process be the same as the baseline CMOS process. In contrast, constraints on the HBT device are largely performance related. The HBT must contain a narrow base width, and demonstrate low parasitic capacitances, low series resistances, high output impedance, and low noise (among other attributes). In addition, the electrical characteristics of this device must be repeatable and low in defects. These HBT constraints are often at odds with the thermal cycles defined in the CMOS process, and with the isolation stresses common in CMOS isolation. Thus from the practical constraints imposed by the CMOS process commonality decision, the CMOS device fabrication, including tooling and process steps, is largely predetermined. The difficulty lies in the integration of the epitaxial-base HBT device with that CMOS process.

The focus of this paper is the interaction of HBT fabrication process steps with the CMOS device structure and interaction of CMOS process steps with the HBT device structure. Central to this set of interactions is the integration scheme, since this determines the deposition and removal sequence of device films, the types of films, and the required thermal cycles. This topic will be covered in the next section, limiting the discussion to HBT integration schemes employed for high-performance self aligned HBTs. Although few integrated CMOS and self aligned HBT processes have been reported in the literature, the discussion will be instructive in revealing the key tradeoffs in temperature, film compatibility, and topography between the various schemes. Specifically, critical constraints relating to the epitaxial SiGe film are determined by the integration scheme. Following the integration discussion, we will discuss these materials-related SiGe film integration issues, including pre-clean, growth, and stress-related issue in integration. Lastly, because inclusion of passive elements into the integrated chip is necessary for RF single-chip solutions, and these elements many times also influence integration choices, tradeoffs in passive element integration will be briefly covered.

EPITAXIAL BASE INTEGRATION

In the HBT, collector-base capacitance C_{CB} and base resistance R_B are typically the most critical and difficult parasitic components to minimize. Self-aligned technologies, where the low resistance, heavily doped extrinsic base is aligned to both the emitter and the moderately doped collector, minimize these parasitic components and are consequently utilized in state-of-the-art bipolar transistors. By ensuring that the distance between the heavily-doped extrinsic base edge and the emitter edge is fixed and not subject to lithographic variations, emitter-base self-alignment allows a reduction of this distance (and therefore the base resistance) and ensures that current flow to the emitter is uniform from all sides. And, by aligning the moderately doped collector to the extrinsic base, overlap between these regions (and therefore collector-base capacitance) is minimized. Several broadly categorized integration schemes have been used for achieving self alignment in epitaxial-base HBTs.

Double-poly self-aligned

The most common silicon bipolar self-aligned integration scheme, known as double-poly self aligned, has been retrofitted for SiGe HBTs using a selective epitaxial base by NEC, Hitachi, Siemens, and Phillips [8][9][10][11]. The essential elements of the structure are shown in Fig. 1. Identifying features of this integration scheme are 1) the pre-epitaxial patterned boron-doped polysilicon extrinsic base on insulator, with inside spacer and 2) selective epitaxy in the patterned opening. The principle advantage of the double-poly self aligned epitaxial base scheme is its compatibility with prior generation Bipolar and BiCMOS technologies in a manufacturing line. It also has potential to require fewer thermal cycles, since the extrinsic base is deposited prior to the intrinsic base. An additional potential advantage is the reduction of the collector-base capacitance since the extrinsic base forms a junction with single-crystal silicon only in the link-up region of the device, and otherwise is isolated by insulator over the single-crystal collector.

Difficulties arise principally from the non-planarity of the emitter polysilicon (which is aggravated with lateral shrink of the device) and the linkup of the extrinsic base with the selective epitaxial intrinsic base. Due to the topography, implanted emitter polysilicon requires a long thermal cycle to diffuse the dopant and thus reduce emitter resistance [12]. These thermal cycles have the negative effect of widening the base and reducing performance. For this reason, in-situ phosphorus-doped polysilicon is often used as an emitter electrode in high performance double poly self aligned HBTs [8][9]. In-situ doped polysilicon is not common in CMOS fabrication, and therefore this film cannot be reused as a CMOS film, and specialized tools are required. In

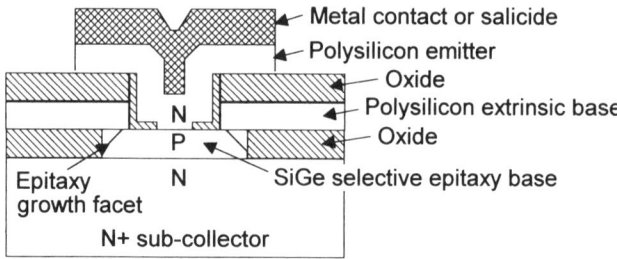

Fig. 1: Essential structure of double-poly self-aligned selective epitaxial-base HBT.

addition, when using selective epitaxy to form the base, careful attention must be paid to the growth of the material in the extrinsic base link-up region. Faceting and growth of the epitaxy in the vicinity of the polysilicon extrinsic base requires that a substantial undercut of the spacer be defined in order to sufficiently distance the faceting edges from the active junction and avoid anomalous base leakage current. As a result of the distance required, base resistance may be relatively high, and because growth must occur under the spacer, specialized structures may be required to enhance the linkup and reduce the base resistance [9]. Furthermore, selective epitaxy is known to exhibit local loading effects, which can cause variable electrical characteristics between different devices due to growth area and proximity-related film thickness.

Whereas it is useful to describe the double poly self-aligned integration scheme because of its prominence and likely marketability, such a device with an epitaxial base has not been integrated with MOS devices. This is likely due to the relative complexity of the HBT fabrication linked with choices of non-common films and isolation schemes compared with CMOS. In particular, it is difficult to see how the films in the double poly self aligned scheme can be shared in the CMOS devices, as the requirements for those films are very different. This forces the HBT device to be fabricated following the CMOS devices, resulting in substantial process complexity. Toshiba has recently described a *non*-self aligned *non*-selective epitaxial base double poly technology [13] which is integrated with CMOS. This is achieved through base and emitter film deposition following the CMOS device fabrication, but using the same isolation and metallization schemes in both sets of devices. Although not described in [13], significant integration issues certainly must arise from the deposition and removal of the full set of base and emitter films over the CMOS device topography. Similar solutions appear to be possible in the self aligned version, but as of this writing have not been published.

High base-concentration HBT

A departure from the historic integration schemes is taken by Temic [14]. This approach takes advantage of the exponential effect that high Ge concentration at the emitter-base junction has on increasing the electron injection into the base of the HBT without affecting the hole injection into the emitter. In this approach, germanium concentration is increased to 26% across the width of the base, which permits the dramatic increase of base concentration and reduction of emitter concentration. The result is an intrinsic base sheet resistance which is roughly ten times

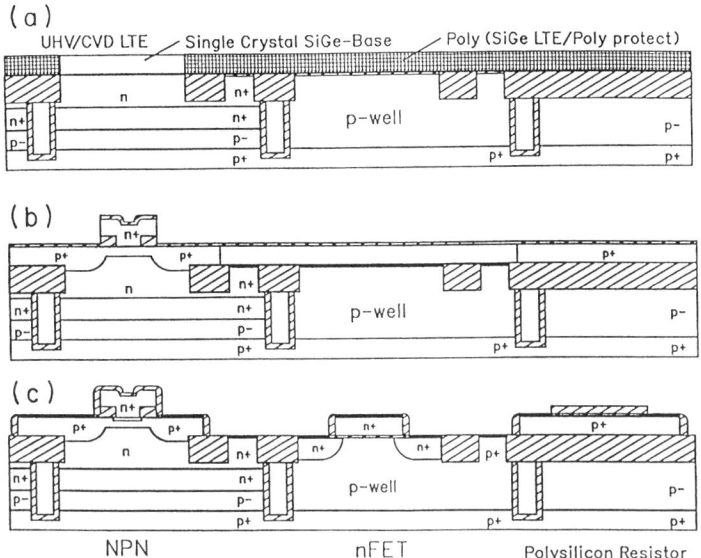

(a)

UHV/CVD LTE — Single Crystal SiGe-Base — Poly (SiGe LTE/Poly protect)

(b)

(c)

NPN   nFET   Polysilicon Resistor

Fig. 2: Gate=Base with emitter mandrel structure process sequence. Cross-section marked (a) shows structure after UHV/CVD epi-base deposition. Structure (b) is following emitter poly patterning, and structure (c) shows the completed devices with self-aligned titanium silicide prior to contact stud and wiring levels.

lower than other techniques. This, together with the epitaxial in-situ lightly-doped emitter, permits more flexibility in the self-alignment scheme. The net result is a *total* base resistance and single-transistor performance comparable to the self-aligned schemes but with less complexity in the alignment scheme. Tradeoffs are yield issues stemming from instability in the epitaxial film, which is likely to severely limit integration levels and thermal cycles following the base deposition, as well as strong temperature dependence of current gain on operating temperature and self heating. These issues will be discussed in later sections.

Gate=Base with Emitter Mandrel

To date, the only reported self-aligned HBT BiCMOS process has been implemented by IBM and utilized an approach described here as gate=base with emitter mandrel [7][15]. Like the previously described double-poly self-aligned processes, this process also employs two polysilicon films, but utilizes a much different self-alignment scheme. In this scheme, the single-crystal intrinsic base is epitaxially deposited over field-oxide-bounded single crystal active area. Simultaneously, the polysilicon extrinsic base is deposited over planarized field oxide and the polysilicon MOS gate structure is deposited over thin gate oxide. The self-aligned extrinsic base implant is defined by a series of layers and a spacer process, through which a sacrificial mandrel film defines the position of the emitter opening. The resulting structure, shown in Fig. 2(c), has several key advantages relative to the issues raised in the double-poly self aligned scheme. First, emitter topography is minimized, as the extrinsic base is not defined in the high topography structure of an etched polysilicon layer. Instead, it is defined through implantation into the same single-crystal epitaxial film as the intrinsic base. Thus the emitter polysilicon is deposited over minimized topography. This permits the use of a CMOS tooling-compatible intrinsic (as deposited) polysilicon and arsenic implanted emitter and acceptable drive-in thermal cycles, together with a non-silicided emitter polysilicon (as was required in [9] for low emitter resistance in a narrow emitter opening). Furthermore, through the use of blanket epitaxy and implanted extrinsic base, structural issues in the link-up of intrinsic and extrinsic bases are avoided. Finally, integration with CMOS is straightforward, as the epitaxial base film is the same as the gate polysilicon, and thus topography issues related to deposition and removal of emitter films is simplified.

Difficulties arise from thermal cycles necessary to produce the emitter mandrel and the MOS devices. These include formation of the sacrificial emitter films, oxidation of the MOS device sidewall, and anneals required for the MOS device extensions and source/drains. It has been shown that these required CMOS thermal cycles are consistent with those that are needed for arsenic emitter drive-in and activation of extrinsic base and emitter dopants [7] and therefore result in little performance loss.

EPITAXIAL FILM ISSUES

At the heart of any SiGe HBT integration scheme are issues related to the epitaxial SiGe film. Pre-epitaxial deposition surface treatment, complex growth conditions, film structure, and film stability, all either influence or are influenced by structures already existing on the wafers at the time of growth, or are constrained by post processing required for proper functionality of the HBT, CMOS, or passive devices. Selection of the working set of operating conditions for all devices is clearly one of the largest challenges. Furthermore, processes which promote the fewest constraints will likely have the largest process window (i.e.: be the least susceptible to normal process variability) and be the most successful in a manufacturing environment.

Surface treatment

It is generally accepted, because of the stacking of layers and thermal cycles involved, that integration of the epitaxial base with conventional isolation techniques (LOCOS, shallow trench, or deep trench), requires the isolation to be in place prior to the epitaxial growth. Furthermore, with BiCMOS integration, well implants and gate oxidation must also occur prior to the epitaxial growth so that the thermal cycle effects on boron diffusion in the base are avoided. Traditional high temperature pre-baking of silicon surfaces prior to epitaxy clearly is inconsistent with the presence of these structures.

Relatively recent medium-temperature surface treatment developments have made possible HBT integration with standard CMOS-like processes. An in-situ medium temperature (850°C - 900°C, 1 - 10 minute), medium vacuum (10 torr) H2 cleaning has been discussed in [10][16]. The principle integration issue with this approach is temperature effect on prior processed devices. Thus the integration scheme becomes important, in that the scheme determines if the critical CMOS diffusions are in place prior to this surface treatment. Comparing the schemes discussed, the double-poly self aligned process requires the CMOS, including the LDD (lightly-doped drain) and extension implants, to be in place prior to the epitaxy deposition, whereas the gate=base process requires the source/drain and extensions to be implanted after the epitaxy. It is well known that long times at even intermediate temperatures of 700 - 800 degrees can readily move dopants, substantially affecting FET characteristics. Further complicating the situation, boron penetration of gate oxides is known to be sensitive to anneals in the presence hydrogen. With adequately low thermal cycles, CMOS dopant diffusion during the HBT process steps must be accounted for in the design of the MOS devices. Larger thermal cycles (e.g.: > 900°C), particularly in the presence of hydrogen, clearly limit the fabrication of high performance MOS devices together with the HBT.

Epitaxy deposition

Numerous factors in the growth of the SiGe film must simultaneously meet the stringent set of integration constraints in order to result in a manufacturable, controllable process. The integration issues that the various growth techniques, including MBE, UHV/CVD, APCVD, and RPCVD, must address are briefly mentioned here without detail. Tradeoffs clearly exist among process conditions of temperature, pressure, and gas flows, and the resulting throughput, control of dopant and Germanium profile, thermal diffusion of base and well dopant profiles, defect level, faceting, and, for selective epitaxy, the selectivity, loading effects, and extrinsic base linkup.

Low temperature clearly is most beneficial for the base profile and for minimal influence on CMOS dopant profiles. At temperatures below 600°C, substantial control in the boron, and most importantly, the germanium profiles, results from the lower film growth rate. In addition, dopant diffusion is extremely slow at these temperatures.

Low pressure conditions also favor integration and control. Complex flow and reaction characteristics which result in non-uniformities across wafers, area-dependent loading effects, and faceting are minimized at such process conditions [10].

Longer deposition time is the price paid for the benefits listed above in a low temperature, low pressure process. Single wafer, low temperature, low pressure environments such as found in MBE require long deposition times (on the order of many hours per wafer) and therefore are extremely cost-prohibitive. CVD techniques may also benefit from low temperature and low pressure, and therefore require long processing times, but such process techniques support batch processing, and so the processing time per wafer decreases with the quantity of wafers in a batch. Because of the multiple benefits of control, low thermal cycle impact on the base profile and CMOS dopants, low defect, and minimal film non-uniformities, CVD techniques are preferred for high performance integrated processes.

Whereas gas flows are of critical importance for all epitaxy processes, they are substantially more critical for the double-poly self aligned HBT device because of the selectivity requirement. Typically, HCl flow is varied to control the growth selectivity of the Si and SiGe films between the silicon and oxide or nitride surfaces. However, because concentration of HCl also influences the growth rate and loading effects, these flow conditions must also be chosen to optimize each of these three effects together. It has been noted that to find one set of conditions that satisfies all constraints may be extremely difficult [10]. In the case that loading effects are to be accepted as part of the normal process, the process constraints are in effect transferred to the electrical modeling and device layout, causing the design to be more difficult, unpredictable, and possibly unmanufacturable.

Film Stability

Because of the flexibility and control available through the epitaxial process, germanuim profiles may be defined in one of many possible shapes. This control together with the well established electrical properties of the SiGe alloy gives the device designer a great deal of flexibility in the optimization of the HBT characteristics. The design aspects are covered in detail elsewhere [17], and thus will not be discussed at length here. Referencing Fig. 3, it is simply noted that the following parameters are principally affected:

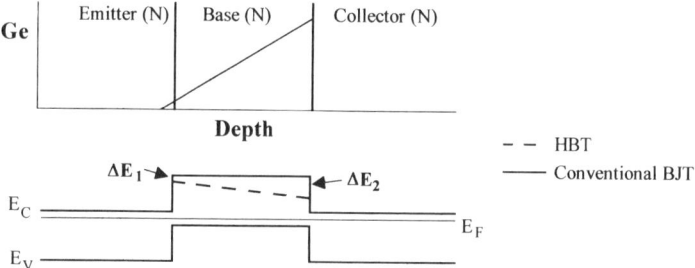

Fig. 3: Germanium concentration and band structure schematic for a graded band-gap SiGe double heterojunction transistor. Device properties are dependent on bandgap offsets at emitter-base and base-collector junctions.

1. Electron injection into the base is exponentially enhanced by the presence of germanium at the emitter-base junction. In other words, the enhancement in current gain $\beta = I_C/I_B$ over a BJT roughly increases exponentially with increasing ΔE_1.
2. The Early Voltage V_A, representing the DC output impedance of the device, is enhanced relative to a BJT, and increases exponentially with increasing $\Delta E_2 - \Delta E_1$.
3. The transit time through the base τ_b, a critical performance factor in the operating speed of a device, is aided by the drift field in the base resulting from the graded profile. Theoretically, this time decreases exponentially with increasing $\Delta E_2 - \Delta E_1$.
4. Current gain β relative to a BJT is exponentially related to $\Delta E_1/T$ (T= Temperature). When properly designed, the temperature dependence compensates the normal BJT characteristic of increasing β with temperature for critical applications such as power amplifiers or applications with widely varying temperature extremes.

Clearly, in the absence of other constraints, one would interpret from the first three of these dependencies that a larger germanium concentration would produce a higher performance device, with a larger β, larger V_A, and lower τ_b. Limits are found in the electrical behavior of the device (as in the thermal response) as well as in the increased probability of structural failure with higher Germanium concentration. Understanding these limits is important in the task of germanium profile optimization for maximum device performance. This understanding is substantially complicated by several factors, including inadequacies in the current theories, reduction of defect likelihood in small areas, and additional stresses induced by adjacent isolation.

It must be understood that, because silicon's lattice constant is 4% smaller than that of germanium, a commensurate SiGe layer is always under some degree of compressive strain. If the strain becomes sufficiently large because of thickness or concentration, the film will no longer be stable, and misfit dislocations, formed between the film and the substrate, together with terminating threading dislocations will form. Formation and propagation of these defects will take place either during the growth or through subsequent heat treatment.

It has been argued that some amount of defects or instability can be acceptable. Whereas this may be true for small scale integration, transistor yields would be very unpredictable with this built-in defect probability, and large scale integration would be unlikely. Because the misfit

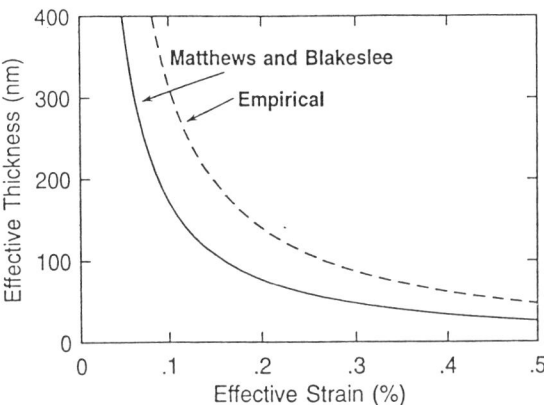

Fig. 4: Derived and empirical critical thickness for blanket SiGe alloy grown pseudomorphic to a silicon substrate. Empirical represents blanket UHV/CVD layers annealed at 900C [20].

dislocations travel at the SiGe/Si interface (i.e.: the onset of the germanium profile), which is contained in the collector of the device, it is possible that devices contain misfit dislocations and that these defects have a minor impact on the device behavior. The defects are undesirable because each one has a finite probability of ending in a threading dislocation which travels through the active device, resulting in high collector-emitter leakage ("pipes") and in general causing low HBT yields. Some have also argued that it may be possible to yield devices with very high concentration metastable Ge profiles by depositing the film at a sufficiently low temperature and limiting the subsequent thermal processing to below 600°C [14]. This however is not a practical solution in that virtually all integrated processes require film depositions, oxidations, anneals, silicide formations, and other thermal cycles in excess of 600°C toward the end of the process.

The limits of thickness and concentration were explored by multiple authors, most notably by Matthews and Blakeslee [18] and Tsao and Dodson [19]. Although this has been a field of considerable study, researchers agree that the determination of this "critical thickness" is highly dependent on a number of factors, including the presence of pre-existing dislocations, and the characterization techniques used to assess the level of defects (for a discussion of this topic, see [20]). In any case, there is sufficient evidence to support the general "critical thickness" dependence proposed by these authors (shown in Fig. 4). In practice, depending in part on the deposition technique, the surrounding structure and its accompanying stresses, the implant and anneal history of the device, and the temperatures the device experiences, the practical limit may be found to be somewhat different in each integrated process. This practical limit is extremely difficult to ascertain since the driving factors are numerous and highly variable. Complicating matters is the difficulty in measurement of these defects, since a large sample is required to adequately cover the statistical range of variables.

Some generalizations can be made about incorporation of SiGe films into real devices and have recently been reported in [21] and [22]. Both authors observe substantial reduction in linear density of misfit dislocations for small areas compared with larger areas. This "small area effect" points to a displacement of the critical thickness curve for realistic devices as compared to blanket films. Counteracting this effect is the fact that small areas are also more susceptible to isolation stresses resulting from oxidation of trench sidewalls. This stress can add to the stresses in the alloy film to potentially decrease the effective critical thickness. The counteracting effects and variability of the different sources of stress make it difficult to determine the germanium profile limitations.

INTEGRATION OF PASSIVE ELEMENTS

As previously stated, a critical requirement for integrated RF circuit technologies is to support passive elements as well as the active HBT and CMOS devices. For the most part, the integration of these devices is a matter of utilizing pre-existing films and creating layouts that are useable for resistors, capacitors and inductors. Many times, the use of these layers for passive components increases the requirements for control of those films that make up the passive components.

An example of this is described in [23] for the case of using an extrinsic base polysilicon for a resistor. The increased control requirements result from film resistance being a more dominant component of the resistor device electrical characteristics compared to the HBT device (where the resistance of the film is one of many components of the total base resistance). Optimization of the implant conditions is required for the resistor device for improved control, and impact to the HBT must be minimized.

Integration of spiral inductors (patterned in a thick metal wiring layer over a region of unpatterned substrate) into silicon technologies are receiving a great deal of attention in the literature (see for example [24]). The essential fact is that the coupling of the inductor fields with the substrate necessitate a lightly doped substrate in order to minimize losses at high frequencies [25]. This is often at odds with requirements in CMOS for more heavily doped substrates to minimize latchup. Therefore, CMOS wells may require special designs compared to a heavily doped substrate to compensate for the reduced latchup protection.

CONCLUSIONS

A successful integration is one where all devices are robust to process variations and the process steps used in creation of each device, and structures in place for prior-processed devices, minimally constrain each other. Many of these constraints are unavoidable and must be adjusted for in the standard process steps or taken into account in the integration scheme: CMOS isolation requires that the HBT process be robust to patterned, partially oxide covered and stressed wafers; topography from gate stacks may complicate removal of HBT films from CMOS areas of the chip; thermal cycles from CMOS post gate patterning (e.g.: gate passivation and anneals) may adversely affect HBT base width; and pre-clean and epitaxy deposition temperatures could adversely influence high performance CMOS drain and well profiles.

Attesting to the integration difficulties is the fact that only one self-aligned BiCMOS SiGe HBT integrated process has been reported in the literature [7]. This singularity may be due to the fact that the IBM integration scheme is likely the most CMOS compatible scheme because of numerous factors touched on in this paper. Foremost is that the integration scheme was conceived with CMOS integration in mind. Components that make it CMOS compatible are its planarity in emitter definition, dual use of the epitaxial film in the gate and base (which avoids the complex film removal problems and provides more flexibility in the epitaxy pre-clean techniques), lack of extrinsic base link-up issues due to the implanted self-aligned extrinsic base, and blanket low temperature, low pressure epitaxy. Results achieved are remarkable: high yielding HBTs, with expected acceptable yields in the 10s of thousands of devices [26]; high yielding CMOS, with expected acceptable yields into the millions of devices [7]. As important are the electrical characteristics. The HBT characteristics are well suited for the high frequency RF market: high f_T

and f_{MAX} (48 and 69GHz), low noise, and capable of high power, high power added efficiency and low noise [27]. Furthermore, HBT device DC and AC characteristics are well controlled [26]. Not to be underestimated, the CMOS electrical characteristics are identical to another IBM CMOS-only process, permitting the design tools and predefined and verified digital designs and design building blocks (e.g.: gates, multiplexers, adders, etc.) to be utilized with the high performance RF devices.

Certainly this will not be the last self-aligned HBT BiCMOS process. Other schemes, some of which are described in this paper, should be CMOS compatible and may provide benefits in parasitic and thermal cycle reduction over the IBM approach. Difficulties in the integration, and likely past business misgivings of a new technology are to be overcome with time and many SiGe BiCMOS technologies may soon be revolutionizing our wireless marketplace.

REFERENCES

1. S.P. Voinigescu, S.W. Tarasewicz, T. MacElwee and J. Ilowski, IEEE Proc. IEDM 1995, pp 721-724.

2. M. Saito, M. Ono, R. Fujimoto, H. Tanimoto, N. Ito, T. Yoshitomi, T. Ohguro, H. Momose, H. Iwai, IEEE Trans. on Electron Devices, Vol. 45, No. 3, March 1997.

3. F. Ali, A. Gupta and A. Higgins, IEEE Microwave and Millimeter Wave Circuits Symposium, 1996 pp. 61-66.

4. D. Harame, Proc. IEEE BCTM, Sept. 1997, pp 36-43.

5. D.L. Harame, J.M.C. Stork, B.S. Meyerson, K.Y.-J Hsu, J. Cotte, K.A. Jenkins, J.D. Cressler, P. Restle, E.F. Crabbe, S. Subbanna, T.E. Tice, B.W. Scharf, J.A. Yasaitis, IEEE IEDM, Dec 1993, pp 71-73.

6. P. Xiao, K. Jenkins, M. Soyer, H. Ainspan, J. Burghartz, H. Shin, M. Dolan, D. Harame, IEEE ISSCC, v 40 Feb. 1997, pp. 124-125.

7. D. Ahlgren, G. Freeman, S. Subbanna, R. Groves, D. Greenberg, J. Malinowski, D. Nguyen-Ngoc, S.J. Jeng, K. Stein, K. Schonenberg, D. Keisling, B. Martin, S. Wu, D.L. Harame, B. Meyerson, IEEE BCTM, Sept. 1997, pp. 195-197.

8. T. Tashihiro, T. Hashimoto, F. Sato, Y. Hayashi, T. Tatsumi, , IEICE Transactions on Electronics E80-C n5, May 1997, pp 707-713.

9. K. Washio et. al., IEEE IEDM, Dec 1997, pp 795-798.

10. D. Terpstra, W. De Boer and J. Slotboom, Solid State Electronics, Vol. 41, No. 10, pp. 1493-1502 1997.

11. T.F. Meister, H. Schafer, M. Franosch, M. Molzer, K. Aufinger, U. Scheler, C. Walz, M. Stolz, S. Boguth, J. Bock, IEEE IEDM Technical Digest 1995, p739.

12. J.N.Burghartz, J.Y.-C.Sun, C.L.Stanis, S.R.Mader, J.D.Warnock, IEEE Trans. Electron Devices, 39 (6), 1992, pp 1477-1489.

13. H. Nii, T. Yoshino, N. Itoh, H. Nakajima, H. Sugaya, H. Naruse, Y. Katsumata, H. Iwai, IEEE BCTM, Sept. 1997, pp 68-71.

14. A. Schuppen, A. Gruhle, U. Konig, Journal of Materials Science: Materials in Electronics 6 (1995) 298-305.

15. D. Harame, J. Comfort, J. Cressler, E. Crabbe, J. Sun, B. Meyerson and T. Tice., IEEE Trans. on Electron Devices, Vol. 42, no. 3, March 1995, pp 469-482.

16. K. Oda and Y. Kiyota, J. Electrochem. Soc. 143, (1996) 2361.

17. D. Harame, J. Comfort, J. Cressler, E. Crabbe, J. Sun, B. Meyerson and T. Tice., IEEE Trans. on Electron Devices, Vol. 42, no. 3, March 1995, pp 455-468.

18. J. W. Matthews and A.E. Blakeslee, Journal of Crystal Growth 27 (1974) pp. 118-125

19. J.Y. Tsao and B.W. Dodson, Applied Physics Letters. 53 (10) September 1988.

20. S.R. Stiffler, J.H. Comfort, C.L. Stanis, D.L. Harame, E. de Fresart, J. Applied Physics. 70 (3) August 1991.

21. L. Vescan, T. Stoica, C. Dieker and H. Luth, Mat. Res. Soc. Symp. Proc. Vol. 298. 1993 pp. 45 - 50.

22. K. Schonenberg, S-W Chan, D. Harame, M. Gilbert, C. Stanis and L. Gignac, J. Mater. Res., Vol 12, No. 2, Feb 1997 pp. 364-370.

23. S.J. Jeng, D.C. Ahlgren, G.D. Berg, B. Ebersman, G. Freeman, D.R. Greenberg, J. Malinowski, D. Nguyen-Ngoc, K.T. Schonenberg, K.J. Stein, D. Colavito, M. Longstreet, P. Ronsheim, S. Subbanna, D.L. Harame, Proc. IEEE BCTM, Sept. 1997, pp 187-190.

24. J.N.Burghartz, M.Soyuer, K.A.Jenkins, M.Kies, P.Dolan, K.Stein, J.Malinowski, D.L.Harame IEEE Journal on Solid State Circuits, vol. 32, no. 9, pp. 1440-1445, 1997.

25. D. Harame, L. Larson, M. Case, S. Kovacic, et al., IEEE IEDM 1995 Technical Digest, pp 731-734.

26. D.C. Ahlgren, M. Gilbert, D. Greenberg, S.J. Jeng, J. Malinowski, D. Nguyen-Ngoc, K. Schonenberg, K. Stein, G. Groves, K. Walter, G. Hueckel, D. Colavito, G. Freeman, D. Sunderland, D. Harame, B. Meyerson, IEEE IEDM 1996 Technical Digest, p859

27. D.R. Greenberg, M. Rivier, P. Girard, E. Bergeault, J. Moniz, D. Ahlgren. G. Freeman, S. Subbanna, S.J. Jeng, K. Stein, D. Nguyen-Ngoc, K. Schonenberg, J. Malinowski, D. Colavito, D.L. Harame, B. Meyerson, IEEE IEDM 1997 Technical Digest, pp 799-802.

CARRIER TRANSPORT AND VELOCITY OVERSHOOT IN STRAINED SI ON SIGE HETEROSTRUCTURES

DAVID K. FERRY*, GABRIELE FORMICONE**, DRAGICA VASILESKA*
*Department of Electrical Engineering, Arizona State University, Tempe, AZ 85287-5706
**Motorola SPS, 2100 E. Elliot Road, Tempe, AZ 85284

ABSTRACT

We examine the velocity overshoot effect in strained Si on Si_xGe_{1-x} heterostructures. We also investigate the performance of surface-channel strained-Si MOSFETs for devices with gate lengths representative of the state-of-the-art technology. The Ensemble Monte Carlo method, self-consistently coupled with the 2D Poisson equation solver, is used in the investigation of the device performance. Our simulations suggest that, in short-channel devices, velocity overshoot is very important. In fact, when velocity overshoot occurs, it greatly affects the carrier dynamics and the current enhancement factor of both surface-channel strained-Si and conventional Si MOSFETs.

INTRODUCTION

The continued growth of the silicon-based VLSI technology over time has led to orders of magnitude of improvement in performance, device density, and cost [1-3]. The Semiconductor Industry Association (SIA) projects a steady-state improvement in most key parameters up to year 2012, when the leading edge devices will employ 50 nm gate lengths and have operating voltages below 0.6 V. However, recent experimental efforts in fabricating MOSFETs with very small dimensions [4] have demonstrated that it will be very difficult to go to sub-50 nm gates and make MOSFETs that operate reliably. This is related to the fact that the gate oxide thickness has already reached the 3-4 nm range (the tunneling leakage current limit for the gate insulator), and the substrate impurity concentrations are already in the range from 5×10^{17} cm^{-3} to 1×10^{18} cm^{-3} (the leakage current limit for the source/drain to substrate *p-n* junctions).

The strive for better circuit performance and the difficulties associated with further MOSFET miniaturization have been the primary motivation to look for alternatives. The idea of band-gap engineering within the Si material system has proven to be a viable alternative. It has been anticipated that, if realized, this would bring together the superior properties of the III-V devices and the processing maturity of the Si technology [5]. This became reality in the late 1980's with the introduction of new epitaxial growth techniques, which involved grading of the Ge composition profile in the Si/SiGe heterosystem [6]. This provides relaxed $Si_{1-x}Ge_x$ layers with lower surface defect densities, suitable for device fabrication. Low-temperature mobilities of up to 500,000 cm^2/V-s have been reported [7]. A variety of devices utilizing $Si_{1-x}Ge_x$ have been studied, including the heterojunction bipolar transistor (HBT). In the latter, the introduction of Ge into the base layer of an otherwise all-silicon bipolar transistor created a significant improvement in the operating frequency, current noise, and power capabilities, while maintaining the key advantages of a state-of-the-art BiCMOS process technology, including high integration level and economy of scale [8-10]. Recent studies suggest that circuits fabricated in this technology are capable of fulfilling application requirements for RF analog in the 1-5 GHz range, and for high-speed digital circuits at or above the 10 Gb/s range, with potentially lower power, lower cost and higher reliability when compared to other high-speed RF technology counterparts [11]. Strained $Si_{1-x}Ge_x$ alloys were found to be very attractive for optoelectronic applications as well [12].

Mat. Res. Soc. Symp. Proc. Vol. 533 © 1998 Materials Research Society

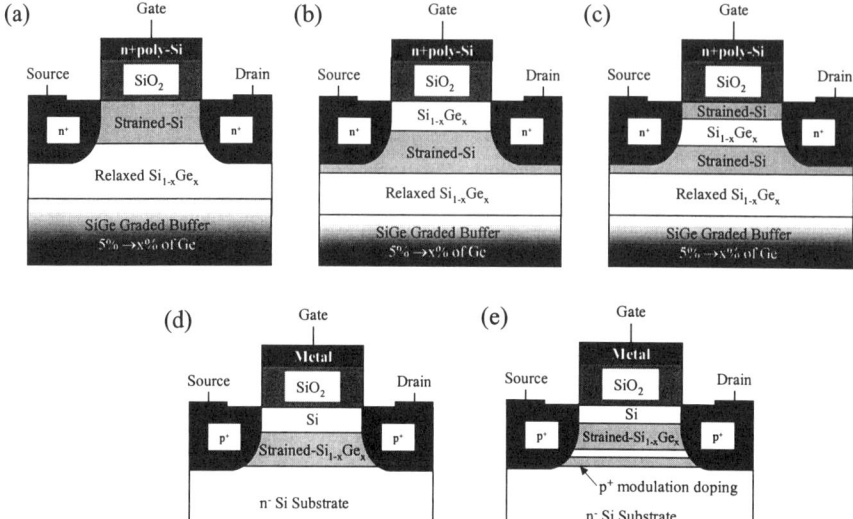

Figure 1. (a) *n*-type MOSFET employing strained-Si electron channel at the surface, (b) *n*-type MOSFET employing strained-Si channel below the $Si_{1-x}Ge_x$ cap, (c) *n*-type buried strained-Si MOSFET with a thin strained-Si layer at the surface which suppresses the avalanche breakdown at anomalously low drain voltages at low temperatures and, at the same time, extends the device performance enhancement at room temperature to higher gate voltages, (d) *p*-MOSFET with a buried $Si_{1-x}Ge_x$ hole channel, and (e) modulation-doped *p*-MOSFET where the thin *p*-type doping supplies carriers to the $Si_{1-x}Ge_x$ channel layer, thus extending the voltage range over which transport is dominated by carriers in this higher mobility layer.

Various *n*- and *p*-channel MOSFET structures that involved: (1) surface- or buried-channel strained-Si layer, and (2) buried-channel strained $Si_{1-x}Ge_x$ layers were also proposed [13-17]. Out of the variety of structures summarized in Fig. 1, surface-channel and buried-channel strained silicon-MOSFETs are probably the most attractive candidates for replacing the conventional Si devices in VLSI technology when compared to, for example, buried channel strained-$Si_{1-x}Ge_x$ devices. Both hole and electron mobilities can be increased with respect to standard Si devices. On the contrary, only hole mobility can be increased in strained-$Si_{1-x}Ge_x$ buried-channel devices. Second, the same circuit configuration as in standard Si technology can be used.

The main effect of the strain in strained-Si that leads to enhanced hole and electron low-field mobilities, occurs in the energy band structure: there is a splitting of the two-fold degenerate heavy and light hole bands, which leads to corresponding modifications of the hole effective masses in the valence band. In addition, the sixfold-degenerate conduction-band valleys split into two separate sets of valleys: a two-fold degenerate, perpendicular Δ_2-band and a four-fold degenerate, in-plane Δ_4-band. To first order, the ellipsoidal shape of each band in k-space is not deformed, so unlike the valence band case, the effective mass of the conduction band remains unchanged. However, the relative energies of each band do shift. The tensile strain leads to energy splitting of the Δ_2 and Δ_4 bands $\Delta E_{cond} \cong 0.67 \cdot x_{Ge}$ [18], which leads to smaller effective in-plane mass and reduced inter-valley scattering. This translates into higher low-field electron mobilities and consequently, to higher current drive capability of n-MOSFETs. The strain also causes a change in the bandgap E_G, which is empirically given by $E_G(x_{Ge})=1.11-0.4x_{Ge}$ eV [18].

Within the Nanostructures Research Group at Arizona State University, during the course of several years we have done extensive studies on the electron transport properties of strained-Si layers by an ensemble Monte Carlo technique and a real-time Green's functions formalism [19]. The performance of the modulation doped strained-Si/Si$_{1-x}$Ge$_x$ FET structures with 0.18 μm gate length was studied with a quantum hydrodynamic simulator using the Monte Carlo results [20]. Quite recently, a 2D Ensemble Monte Carlo Particle Simulator was also developed [21]. It is currently being used in the investigation of the performance enhancement in surface-channel strained Si with respect to conventional n-channel Si MOSFETs with gate-lengths representative of the state-of-the art technology.

QUANTUM HYDRODYNAMIC EQUATION MODEL FOR DEVICE SIMULATION

A set of quantum hydrodynamic equations is used in the simulation. The equations, which explicitly include quantum corrections and describe the particle conservation, momentum conservation and energy conservation, discussed in detail in [22] are the following:

$$\frac{\partial n}{\partial t} + \nabla \cdot (n\mathbf{v}) = 0, \tag{1}$$

$$\frac{\partial \mathbf{v}}{\partial t} + \mathbf{v} \cdot \nabla \mathbf{v} = -\frac{q\mathbf{E}}{m^*} - \frac{1}{nm^*}\nabla(nk_B T_q) - \frac{\mathbf{v}}{\tau_m}, \tag{2}$$

and

$$\frac{\partial T}{\partial t} + \frac{1}{3\gamma}\mathbf{v} \cdot \nabla T_q = -\frac{2}{3\gamma}\nabla \cdot (\mathbf{v}T_q) + \frac{m^*\mathbf{v}^2}{3\gamma k_B}\left(\frac{2}{\tau_m} - \frac{1}{\tau_w}\right) - \frac{T - T_0}{\tau_w}, \tag{3}$$

where n is the average electron density, $\mathbf{v}$ is the average electron velocity, T is the effective electron temperature, m^* is the effective electron mass, $\mathbf{E}$ is the electric field, τ_m is the momentum relaxation time, τ_w is the energy relaxation time , and T_q is given by

$$T_q = \gamma T + \frac{2}{3k_B}U_q \tag{4}$$

with

$$U_q = -\frac{\hbar^2}{8m^*}\nabla^2 \ln(n), \tag{5}$$

where U_q is the quantum correction. The explicit quantum correction involves the second order space derivative of the log of the density. Hence, it tends to smooth the electron distribution, especially where the electron density has sharp changes. The factor γ is the degeneracy factor [23] given by

$$\gamma = \frac{F_{3/2}(\mu_f / k_B T)}{F_{1/2}(\mu_f / k_B T)}, \tag{6}$$

where μ_f is the Fermi energy measured from the conduction band edge and is introduced as a correction to the total average electron kinetic energy

$$w = \frac{1}{2}m * \mathbf{v}^2 + \frac{3}{2}\gamma k_B T + U_q \; . \tag{7}$$

The relaxation times τ_m and τ_w are functions of energy, and are determined by fitting the homogeneous hydrodynamic equations to the velocity-field and energy-field relations from Monte Carlo simulations.

2D ENSEMBLE MONTE CARLO DEVICE SIMULATOR

The Ensemble Monte Carlo (EMC) method is used to solve the Boltzmann Transport Equation (BTE) directly [24,25]. Briefly, the random motion of the charged carriers in the phase-space is modeled in terms of *free-flights* and *collisions* that are assumed to occur *instantaneously* in time and are *local* in space. Particle motion during the free-flights is determined by an electric field acting on the particle at every point along its trajectory. The duration of the carrier free-flight and the scattering events are selected stochastically in accordance with some given probabilities describing the microscopic processes. After the collision event, the particle momentum is modified depending upon the type of the scattering process selected, and the particle is then continued on a new free-flight.

The Monte Carlo model for studying the transport properties in strained-Si/relaxed $Si_{1-x}Ge_x$ system is based on the usual unstrained Si band structure for three-dimensional electrons in a set of six nonparabolic Δ valleys with energy-dependent effective masses. To be able to model the smaller transverse-mass transport properly, the explicit inclusion of the longitudinal and transverse masses is important and this is done in the program using the Herring-Vogt transformation [26]. The six conduction band valleys are included through a three valley model: lowered pair 1 pointing in the (100) direction, raised pair 2 in the (010) direction, and raised pair 3 in the (001) direction. Electron intravalley scattering is limited to acoustic phonon, ionized impurity and surface-roughness scattering. Intervalley scattering is taken into account through zero- and first-order *f-* and *g*-phonon scattering. A standard set of coupling constants for the phonon modes is adopted so that measured velocity-field characteristics and the electron low-field mobility of bulk unstrained Si (≈ 1500 cm^2/Vs) are recovered. We use a soft-ionization model [25,27] for impact ionization, which leads to quadratic energy dependence of the scattering rate. For the ionization threshold of strained-Si we use $E_{th} = 1.18 E_{gap}$ in order to fit the experimental data for the ionization rate. When ionization occurs, a generated electron-hole pair is discarded in the simulation and the original electron loses kinetic energy equal to E_{gap}. Nonparabolicity is included using its dependence on E_{gap}, and therefore it depends on the Ge concentration in the substrate [20,28,29].

To be able to study real device structures, our 2D Poisson equation solver is coupled to the EMC solver and is called frequently to provide self-consistent potential for the charge distribution given directly by the EMC procedure. The basic steps involved in the EMC Device Simulator (EDS) are the following [30,31]: (1) *Initialize data*-Set up the geometry of the device, doping profile and discretization scheme, (2) *Charge assignment*-The charge of each particle is assigned to the neighboring mesh points, (3) *Potential solution*-The 2D Poisson's equation is solved in order to determine the electric field profile within the device, (4) *Flights*-Each particle, now treated as an individual electron, undergoes the standard MC sequence of scattering and free-flights, subject to the local field previously determined from the solution of the Poisson's equation. At self-consistency, the steady-state current is calculated through the net number of particles crossing one contact per unit time. By performing several computer runs, we construct the current-voltage characteristics of the device under investigation.

For the numerical solution of the linear system of equations arising from the finite-difference discretization of the Poisson equation on a tensor-product mesh, we use the Successive-Over-Relaxation (SOR) method. The typical grid size used in the simulations has 150×80 points along the x-axis (source-to-drain) and y-axis (depth), respectively. The final number of simulated particles within the device depends upon the doping density and the applied gate bias. Besides the appropriate number of particles used, the proper coupling between charged particles and Coulomb forces is also required to maintain zero self-force and good spatial accuracy of forces. These issues have been dealt with extensively by Hockney and Eastwood [30] and quite recently by Laux [32]. At present, we use the nearest-grid-point (NGP) scheme for the charge assignment and force interpolation.

Yet another issue that poses serious problems in EDS is the proper modeling of the ohmic contacts, partially due to the limited knowledge of the physics of the contacts. In the present version of the code, the ohmic contacts are modeled as ideal ohmic contacts, i.e. as a region of the device that is in thermal equilibrium even when current is flowing through it. To satisfy this requirement, at each time step we check for charge neutrality in the ohmic contact regions. When required, an appropriate number of thermal carriers with wavevectors k_y pointing into the device are introduced to ensure charge neutrality at both source and drain ohmic contacts. The geometry for the source and drain regions is 0.3μm×0.2μm, doped to $N_D=10^{17}$ cm^{-3}. The spatial extension of the ohmic contact regions is limited to the row of potential mesh cells adjacent to the electrodes.

In our EDS, we also model holes. The scattering processes included in our EMC code for holes account only for standard silicon, which means that our current simulator is currently not suitable for simulating strained-Si p-MOSFETs. Holes are needed in the simulations to properly account for the substrate depletion region underneath the gate. The depth of the substrate below the source and drain ohmic regions is 0.2 μm. The substrate doping is $N_A=10^{17}$ cm^{-3}. We start our simulation with 20,000 super-electrons and 40,000 super-holes. At steady-state, we end-up with approximately 12,000 super-electrons and about 28,000 super-holes. When solving the Poisson equation, the maximum error tolerance for the potential update is 10^{-5} V, which is achieved in about 90 iterations. The time-of-flight used in the EMC and the time step at which we update the electric field profile (through the solution of the 2D Poisson equation) equals 1 fs.

RESULTS AND DISCUSSION

Bulk Ensemble Monte Carlo Results

Typical bulk EMC results are the velocity-field characteristics shown in Fig. 2, where one can see the effect of the strain for different energy splittings ΔE (in eV). We account for phonon scattering only (Fig. 2a) or both phonon and surface roughness scattering (Fig. 2b). From the results shown, it is obvious that strained-Si exhibits larger low-field mobility (phonon limited mobility in strained-Si is about 2800 cm^2/V-s and the phonon limited bulk mobility in Si is about 1500 cm^2/V-s). In addition, the phonon-limited mobility in strained-Si does not significantly depend upon the energy splitting when the germanium content in the buffer is raised to more than 10 %. The addition of surface-roughness scattering leads to significantly lower values for the low-field mobility in both samples (The low field mobility in Si is now about 500 cm^2/V-s and in strained Si is about 900 cm^2/V-s). Again, a saturation effect in the mobility enhancement due to strain is observed. For both cases, the ratio of the drift velocity in strained-Si with respect to unstrained-Si decreases at higher electric fields. At first sight, it appears that high electric fields will degrade the performance gain of strained-Si. However, this is not really a problem because shrinking the device size is always accompanied with a reduction of the applied voltages, which means that the peak electric fields in the channel regions of deep-submicron devices are not expected to be significantly different.

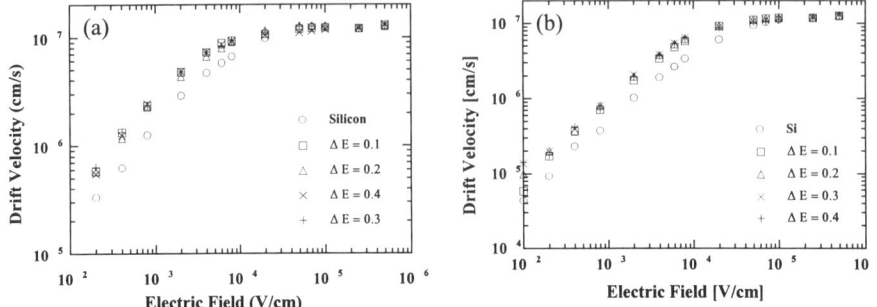

Figure 2. (a) Phonon-limited drift velocity in Si and strained Si at various electric fields. (b) Phonon and surface roughness-limited drift velocity in Si and strained Si at various electric fields. The effective transverse field used in these simulations is $E_{eff} = 4.8 \times 10^5$ V/cm. We use L_c = 3 nm for the roughness correlation length and $\Delta_c = 0.4$ nm for the root-mean square height of the roughness.

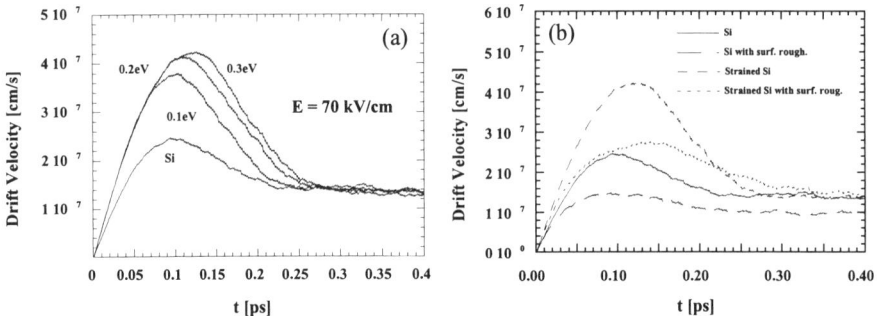

Figure 3. (a) Transient overshoot velocity with a sudden application of a 70 kV/cm electric field at room temperature for various energy splitting (surface-roughness scattering is not included). (b) Transient overshoot velocity with a sudden application of a 70 kV/cm electric field in Si and strained Si ($\Delta E = 0.3$ eV) with and without surface-roughness scattering.

The time evolution of the average carrier velocity with a sudden application of 70 kV/cm electric field in both regular and strained-Si samples with different strain levels is shown in Fig. 3. Significant velocity overshoot is observed in the first 0.25 ps. The peak velocity in unstrained Si is 2.4×10^7 cm/s, whereas strained Si with $\Delta E = 0.3$ eV exhibits a peak velocity of about 4.2×10^7 cm/s (75% higher). When surface-roughness is included in the model, it leads to an overall reduction of the peak velocity and an increase of the peak velocity enhancement ratio from ~1.7 to 2.

In Fig. 4 we show how the magnitude of the electric fields affect the velocity overshoot effect in conventional and strained Si samples (with $\Delta E = 0.3$ eV). In both samples, the velocity peak occurs at shorter times with increasing strength of the field. This is due to the fact that higher electric fields accelerate electrons more and, as they gain energy from the field, they start emitting optical phonons faster. When the emission of optical phonons balances the energy gained, the drift velocity reaches its peak value. However, the ratio of the peak velocity in strained Si with respect to regular Si decreases at higher electric fields. For example, it equals 1.81 at 50 kV/cm and drops to 1.43 at 500 kV/cm. A closer look at the results shown in Figs. 4a and 4b reveals that, for applied electric field of 20 kV/cm, the velocity overshoot is totally absent in Si. In strained Si, the peak velocity is approximately twice the saturation velocity and the overshoot regime extends for about 0.8 ps (not shown in the figure). In this regime, strained Si has a 100% gain with respect to Si.

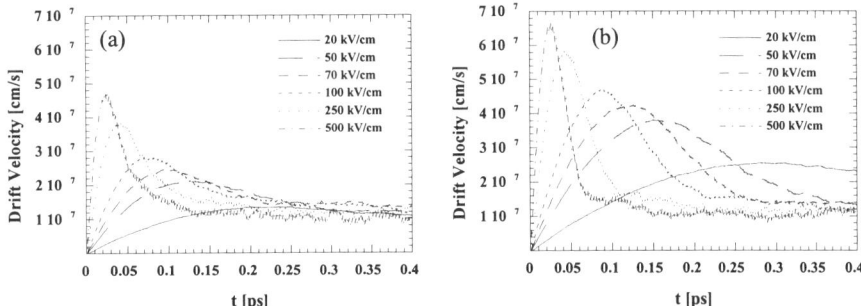

Figure 4. (a) Transient overshoot velocity in Si at various electric fields (surface-roughness scattering is not included). (b) Transient overshoot velocity in strained Si ($\Delta E = 0.3$ eV). Again, surface-roughness scattering is not included in the model.

2D Simulation of Quantum-Well Devices

The quantum hydrodynamic model is used in simulation of a 0.18 μm gate-length, quantum well device with a modulation-doped structure of $Si_{0.7}Ge_{0.3}/Si/Si_{0.7}Ge_{0.3}$. The device structure is shown in Fig. 5a. The doping of the top $Si_{0.7}Ge_{0.3}$ layer is 3.5×10^{18} cm^{-3}, and a doping of 1×10^{14} cm^{-3} is used in the $Si_{0.7}Ge_{0.3}$ substrate. The simulation domain is 1μm×0.09μm. The thickness of the top $Si_{0.7}Ge_{0.3}$ layer is 19 nm, and the strained-Si channel is 18 nm. The modulation structure in [33] is more complicated, but the active regions of both devices are very similar.

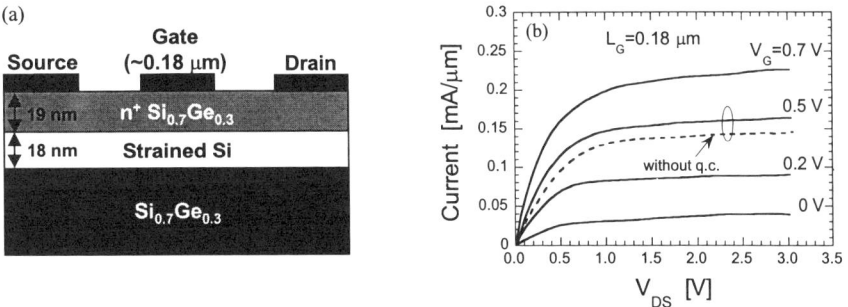

Figure 5. (a) Device structure. (b) I-V Characteristics of a 0.18 μm SiGe device.

In Fig. 5b we show the simulated *I-V* characteristics. The peak transconductance is about 300 mS/mm, and a good saturation with a drain conductance of 4.6 mS/mm is obtained for 0.5 V on the gate. These simulation results are comparable with the experimental measurements from Ref. [33]. It is interesting to note that the transconductance of this SiGe device approaches the same order of magnitude as that of a AlGaAs/GaAs device with same geometry, although the transconductance of the SiGe device is about three times smaller. The inclusion of the quantum corrections leads to about 15% current increase for gate voltage of 0.5 V. By inspecting the density distribution along the channel region of the device (Fig. 6a), one can see that this is due to the rapid change in the electron density at the gate end close to the drain contact within a region that is much shorter than the gate length. The inclusion of the quantum potential also leads to increase of the electron density in the channel. A velocity overshoot, with a peak velocity of 2.6×10^7 cm/s is also observed in the strained-Si channel, and is very important in achieving the observed transconductance (Fig. 6b). In other words, the velocity overshoot causes large average velocity for the electron to travel through the channel. The first velocity peak is due to the model structure we used for the change of the interface discontinuity. Although it is not practical, it suggests that the structure can increase the electron velocity through the device and enhance the device performance.

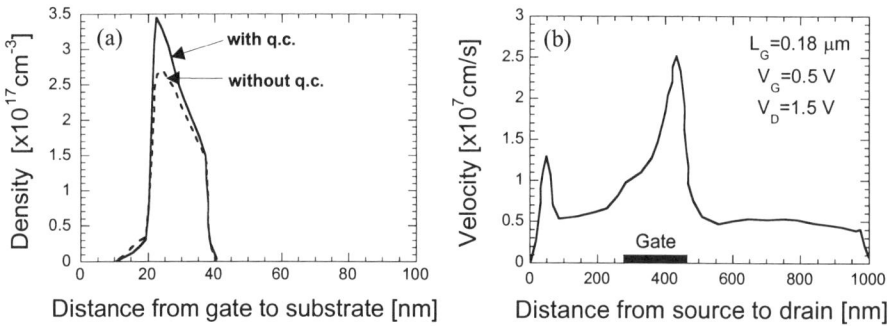

Figure 6. (a) Electron density along the channel. (b) Longitudinal velocity in the quantum well for gate voltage V_G=0.5 V and drain voltage V_D=1.5 V.

2D Ensemble Monte Carlo Simulations of *n*-MOSFETs

The main purpose of this study is to estimate the performance enhancement of surface-channel strained-Si MOSFETs (Fig. 1a) with respect to conventional Si devices with gate-lengths L_G representative of the state-of-the-art technology. We consider devices with L_G equal to 0.35, 0.25, 0.18 and 0.15 μm. To have better estimate, we use the same substrate doping and gate thickness, which equal to N_A=10^{17} cm^{-3} and t_{ox}=80 Å, respectively. This means that the devices that we are simulating are not going to show optimum performance. However, they still show the true difference in material performance regardless of the improvements that can be achieved with careful doping profile design. Keeping the doping profile the same, means that all the devices that we are simulating have identical threshold voltages. We only scale the effective gate-lengths and the drain voltages. In other words, the applied gate voltage is V_G=3.3 V. We use V_D=3.3, 2.5, 1.8 and 1.5 V for the device with 0.35, 0.25, 0.18 and 0.15 μm gate length, respectively. For the surface-channel strained Si MOSFETs, we assume that $\Delta E=0.2$ eV, which corresponds to 20% Ge in the $Si_{1-x}Ge_x$ buffer.

In Fig. 7 we show the average electron velocity and the average electron energy along the channel for all devices considered in this study. We want to point out that Coulomb and surface roughness scattering are not included in the present version of our 2D EDS. Taking a closer look at our bulk EMC results shown in Fig. 3b, it is quite clear that surface-roughness scattering leads to approximately 30% decrease in the peak velocity in Si. This explains the higher peak-velocity obtained for the conventional Si device with 0.25 μm gate length when compared to the simulation results reported in [34]. However, the overall variation of the carrier velocity and energy along the channel is quite similar to that reported in [34].

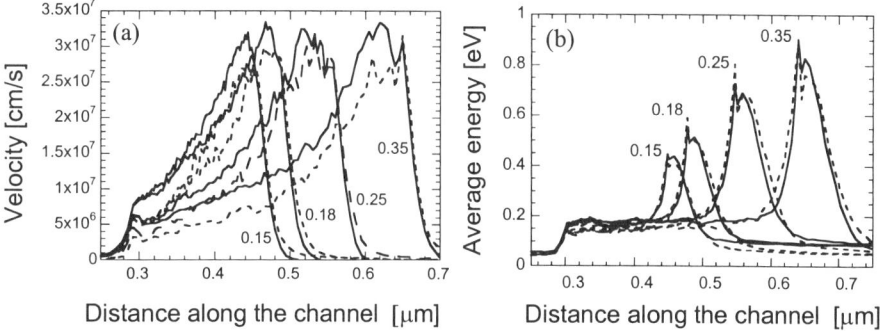

Figure 7. (a) Average electron velocity along the channel. (b) Average electron energy along the channel. Solid (dashed) lines correspond to surface-channel strained-Si devices (conventional Si devices) with different gate-lengths.

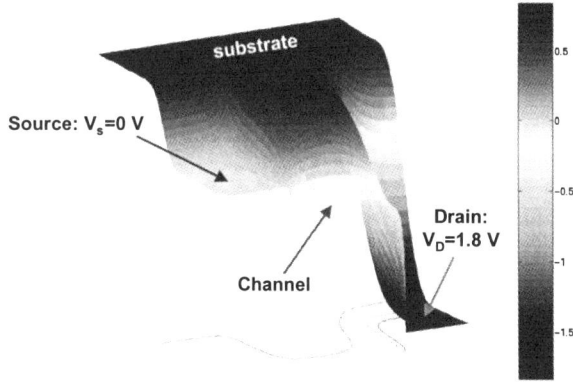

Figure 8. Conduction band profile for the surface-channel strained Si device with 0.18 μm gate length. We use $V_G=3.3$ V, $V_S=0$ V and $V_D=1.8$ V.

The conduction band profile for the device with 0.18 μm gate-length is shown in Fig. 8. It is quite clear that the substrate doping density is sufficient to prevent the punch-through effect.

Finally, in Fig. 9, we show the saturation currents for both conventional and surface-channel strained-Si MOSFETs with 0.35, 0.25, 0.18 and 0.15 μm gate-lengths. The current increase with decreasing gate-length is related to the mobility and increased field at the source end of the channel (see Fig. 7a).

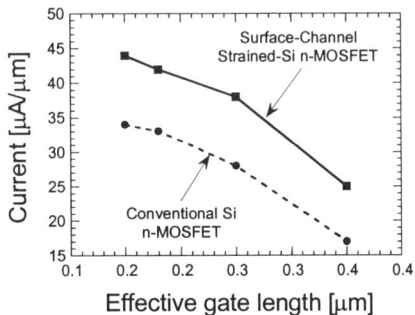

Figure 9. Performance enhancement versus gate-length.

CONCLUSIONS

We investigated the importance of the velocity overshoot effect in SiGe/strained-Si/SiGe FET structure and surface-channel strained-Si n-MOSFETs. The simulations for the FET structure predict that the SiGe devices will have similar performance as AlGaAs/GaAs HEMTs. The transconductance derived from the simulated I-V characteristics is comparable to the experimentally determined one [33]. The overshoot regime is found to be very important in achieving such high transconductance value. In the case of surface-channel strained Si n-MOSFETs, the velocity overshoot regime is also found to be responsible for the obtained performance improvement when compared to standard Si n-channel devices. A performance decrease is observed with decreasing the effective channel length down to 0.15 μm. For devices with gate lengths below 0.1 μm, we expect to observe the opposite trend, and this is left for future investigations.

ACKNOWLEDGEMENTS

The work at ASU is supported in part by the Office of Naval Research.

REFERENCES

1. R.H. Dennard, F.H. Gaensslen, H.-N. Yu, V.L. Rideout, E. Bassous, and A.R. LeBlanc, IEEE J. Solid-State Circuits 9, 256 (1974).

2. J.R. Brews, W. Fichtner, E.H. Nicollian, and S.M. Sze, IEEE Electron Device Lett. 1, 2 (1980).

3. D.K. Ferry, and L.A. Akers, IEEE Circuits and Devices Magazine 13, 41 (1997).

4. M. Ono, M. Saito, T. Yoshitomi, C. Fiegna, T. Ohguro and H. Iwai, IEDM Tech. Dig., 119 (1993); IEEE Trans. Electron Devices 42, 1822 (1995).

5. G. Abstreiter, H. Brugger, T. Wolf, H. Jorke, and H.J. Herog, Phys. Rev. Lett. **54**, pp. 2441-2444 (1985).

6. B.S. Meyerson, K.J. Uram and F.K. LeGoues, Appl. Phys. Lett. **53**, 2555 (1988).

7. K. Ismail, M. Arafa, K.L. Saenger, J.O. Chu, and B.S. Meyerson, Appl. Phys. Lett. **66**, 1077 (1995).

8. S.S. Iyer, G.L. Patton, J.M.C. Stork, B.S. Meyerson, and D.L. Harame, IEEE Trans. Electron Devices **36**, 2043 (1989).

9. D.L. Harame, J.H. Comfort, J.D. Cressler, E.F. Crabbé, J.Y.-C. Sun, B.S. Meyerson, and T. Tice, IEEE Trans. Electron Devices **42**, 455 (1995); IEEE Trans. Electron Devices **42**, 469 (1995).

10. J.D. Cressler, IEEE Spectrum **32**, 49 (1995).

11. J.R. Long, M.A. Copeland, S.J. Kovacic, D.S. Malhi, D.L. Harame, and J.H. Wuorinen, 1996 IEEE International Solid-State Circuits Conference (ISSCC), Digest of Technical Papers, (San Francisco, CA), 82, 423.

12. T.P. Pearsall, CRC Critical Rev. Solid State Material Sci. **15**, 551 (1989).

13. J. Welser, J.L. Hoyt, and J.F. Gibbons, IEDM Tech. Dig., 1000 (1992); IEEE Trans. Electron Devices **40**, 2101 (1993); IEDM Tech. Dig., 545 (1993).

14. J. Welser, J.L. Hoyt, S. Takagi, and J.F. Gibbons, IEDM Tech. Dig., 373 (1994).

15. K. Rim, J. Welser, J.L. Hoyt, and J.F. Gibbons, IEDM Tech. Dig., 517 (1995).

16. S. Takagi, J.L. Hoyt, J.J. Welser, and J.F. Gibbons, J. Appl. Phys. **80**, 1567 (1996).

17. K. Bhaumik, Y. Shacham-Diamand, J.-P. Noël, J. Bevk, and L.C. Feldman, IEEE Trans. Electron Devices **43**, 1965 (1996).

18. M. M. Rieger and P. Vogl, Phys. Rev. B **48**, 14 276 (1993).

19. G.F. Formicone, D. Vasileska, and D.K. Ferry, Solid-State Electronics **41**, 879 (1997).

20. T. Yamada, J.-R. Zhou, H. Miyata, and D.K. Ferry, IEEE Trans. Electron Devices **41**, 1513 (1994).

21. G.F. Formicone, D. Vasileska, and D.K. Ferry, Phys. Stat. Sol. (b) **204**, 531 (1997).

22. J.-R. Zhou, and D.K. Ferry, IEEE Trans. Electron Devices **39**, 473 (1992).

23. J.-R. Zhou, and D.K. Ferry, IEEE Trans. Electron Devices **40**, 421 (1993).

24. C. Jacoboni, and L. Reggiani, Rev. Mod. Phys. **55**, 645 (1983).

25. D. K. Ferry, Semiconductors (Macmillan Publishing Company, New York, 1991).

26. C. Herring and E. Vogt, Phys. Rev. **101**, 944 (1956).

27. B.K. Ridley, Quantum Processes in Semiconductors (Oxford University Press Inc., New York, 1993).

28. H. Miyata, T. Yamada and D. K. Ferry, Appl. Phys. Lett. **62**, 2661 (1993).

29. T. Yamada and D. K. Ferry, Solid-State Electronics **38**, 881 (1995).

30. R.W. Hockney and J.W. Eastwood, Computer Simulation Using Particles (McGraw-Hill, Maidenhead, 1981).

31. P. Lugli, IEEE Trans. Computer-Aided Design **9**, 1164 (1990).

32. S.E. Laux, IEEE Trans. CAD Integr. Circ. Syst. **15**, 1266 (1996).

33. K. Ismail, B.S. Meyerson, S. Rishton, J. Chu, S. Nelson, and J. Nocera, IEEE Electron Device Lett. **13**, 229 (1992).

34. M.V. Fischetti, and S.E. Laux, Phys. Rev. B **38**, 9721 (1988).

Characteristics of Surface-Channel Strained $Si_{1-y}C_y$ n-MOSFETs

K. Rim, T.O. Mitchell, J.L. Hoyt, G. Fountain*, and J.F. Gibbons
Solid State Electronics Laboratory, Stanford University, Stanford, CA 94305
*Research Triangle Institute, Research Triangle Park, NC 27709

ABSTRACT

The first demonstration of n-MOSFETs fabricated using strained $Si_{1-y}C_y$ surface channels is reported. Tensile-strained $Si_{1-y}C_y$ layers with substitutional carbon contents up to 0.8 atomic percent were epitaxially grown on <100> Si substrates by rapid thermal chemical vapor deposition, using silane and methylsilane as the silicon and carbon precursors. n-MOSFETs were fabricated using standard MOS processing with reduced thermal exposure to minimize the possibility of strain relaxation. A remote plasma CVD oxide was employed to form the gate oxide. The $Si_{1-y}C_y$ devices exhibit electrical characteristics that are typical for Si n-MOSFETs, with good turn-on and subthreshold characteristics. MOS capacitance-voltage analysis demonstrates comparable oxide interface qualities for the $Si_{1-y}C_y$ and Si control devices. No carbon-related leakage current is observed in source and drain diode junctions. Characterization of the MOSFET electron inversion layer mobility at room temperature shows comparable mobilities, within the sensitivity of the measurement, for the $Si_{1-y}C_y$ and Si control devices. This is in contrast to the mobility enhancement observed in n-MOSFETs fabricated using tensile-strained Si grown on relaxed $Si_{1-x}Ge_x$ layers. At low temperatures, the inversion layer mobility of $Si_{1-y}C_y$ devices is lower than that of the Si controls, and appears to be affected by Coulomb and possibly random alloy scattering.

INTRODUCTION

Substitutional incorporation of carbon in Si and $Si_{1-x}Ge_x$ has attracted much attention in electronic materials research in recent years for the possibility of new column IV heterostructure devices. Because of the shorter bond length between Si and C, substitutional incorporation of carbon in epitaxial Si, grown on Si substrates, introduces biaxial tensile strain in the resulting $Si_{1-y}C_y$ layer. Carbon incorporated in $Si_{1-x}Ge_x$ has been shown to compensate the compressive strain in the film [1]. Progress in substitutional incorporation of carbon in Si and in its material quality has been reported recently by various groups [2,3]. Basic device structures and band alignment of strained $Si_{1-y}C_y$ to Si have also been investigated. Experiments employing modulation-doped Hall effect structures [4] and MOS C-V methods [5] have shown that the conduction band energy of strained $Si_{1-y}C_y$ is lower than that of bulk Si, as predicted by the theoretical calculations for $Si_{1-y}C_y$ with small carbon contents [6]. The strained $Si_{1-y}C_y$/Si conduction band line-up is analogous to the band alignment of strained Si grown on relaxed $Si_{1-x}Ge_x$, where the tensile strain splits the six-fold degeneracy of the Si conduction band, and lowers the conduction band edge with respect to the underlying relaxed $Si_{1-x}Ge_x$. In strained $Si_{1-y}C_y$, the magnitude of band splitting due to tensile strain is expected to be 67 meV per atomic % carbon, similar to the conduction band splitting in strained Si for the equivalent amount of lattice mismatch. The energy lowering of the two-fold degenerate Δ_2 valleys (see inset, Figure 1) contributes to the lowering of the conduction band edge of strained $Si_{1-y}C_y$ relative to that of the

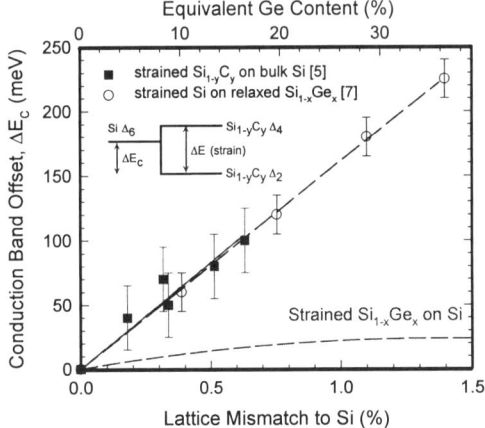

Figure 1. Conduction band offset ΔE_C vs. lattice mistmatch. For a given amount of mismatch, strained $Si_{1-y}C_y$ provides larger ΔE_C than strained $Si_{1-x}Ge_x$. (after [5] and [7]). Inset shows schematic conduction band alignment for strained $Si_{1-y}C_y$ on Si.

underlying bulk Si substrate. Figure 1 compares the magnitude of conduction band offset ΔE_C as a function of lattice mismatch to the respective substrates for three different column IV het1erostructure systems: strained $Si_{1-x}Ge_x$ on Si, strained Si on relaxed $Si_{1-x}Ge_x$, and strained $Si_{1-y}C_y$ on Si. Note that the band offsets for $Si_{1-y}C_y$ and strained Si are as determined by MOS C-V techniques [5, 7]. For a given amount of strain (or mismatch to the substrate), strained $Si_{1-y}C_y$ on Si results in a considerably larger conduction band offset than compressively strained $Si_{1-x}Ge_x$ on Si, and in a similar magnitude of band offset as strained Si on relaxed $Si_{1-x}Ge_x$.

The conduction band splitting in strained Si also results in enhanced carrier transport. As a result of the degeneracy splitting, and the consequent re-population of the conduction band valleys, the electron mobility of strained-Si MOSFETs increases as a function of the strain [8]. However, such structures require growth of a thick relaxed $Si_{1-x}Ge_x$ layer which tends to be of poor material quality with high defect densities. The relaxed $Si_{1-x}Ge_x$ layer also has low thermal conductivity, which results in performance degradation of short channel devices due to self-heating effects [8,9], similar to those observed in SOI devices.

Strained $Si_{1-y}C_y$ on Si provides a Si-compatible, tensile-strained layer without the need for growing such thick relaxed $Si_{1-x}Ge_x$ layers. An analogous mobility enhancement may be expected in strained $Si_{1-y}C_y$ if the presence of carbon in the MOSFET channel does not introduce additional scattering mechanisms that dominate the electron mobility. The lattice mismatch between Si and β-SiC (silicon carbide) is approximately 5 times larger than the mismatch between Si and Ge. Therefore, in strained $Si_{1-y}C_y$, only 1 % of atomic carbon content is required to achieve the same amount of strain as in strained Si grown on relaxed $Si_{1-x}Ge_x$ layers with 10 % germanium content. This report discusses the first demonstration of surface-channel n-MOSFETs employing strained $Si_{1-y}C_y$ channel, and its basic electrical characteristics.

FABRICATION AND DEVICE CHARACTERIZATION

All epitaxial layers were grown on <100> Si substrates in a RT-CVD reactor, using silane (SiH$_4$) and methylsilane (SiH$_3$CH$_3$) as the silicon and carbon precursors. First, roughly 1.2 μm-

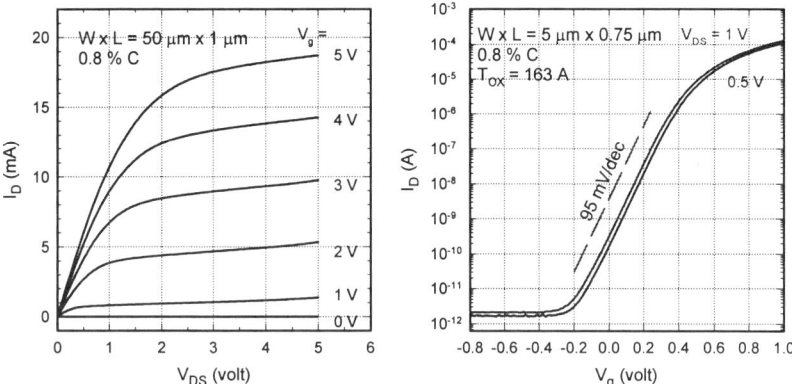

Figure 2. Turn-on and subthreshold characteristics of surfaced channel strained $Si_{1-y}C_y$ *n*-MOSFETs. Devices exhibit typical characteristics of Si *n*-MOSFETs.

thick Si layers were grown at 1000°C with *in-situ* diborane doping, resulting in a doping level of $4\sim5\times10^{16}$ cm^{-3}. The strained $Si_{1-y}C_y$ channel layer (~300 Å) was then grown at 550°C with a silane partial pressure of 300 mTorr, followed by a very thin sacrificial layer of Si (~30 Å) which was consumed during the various cleaning steps during the MOS processing. These growth conditions have previously been shown to result in essentially fully substitutional incorporation of carbon in the $Si_{1-y}C_y$ layer, within the accuracy of the measurements [2].

A standard NMOS process was used to fabricate the strained $Si_{1-y}C_y$ MOSFET structures. In order to preserve strain in the $Si_{1-y}C_y$ layer, thermal exposure during the processing was limited to 600°C. A remote plasma CVD oxide [10] was deposited at 400°C as the gate dielectric with thickness of 160~190 Å. The gate electrode was formed by *in-situ* doped n+ polycrystalline Si. As implants (45 keV and 85 keV), followed by a dopant activation rapid thermal anneal (RTA) for 2 min. at 600°C, were used to form the source and drain junctions. The strain and carbon content in the channel layer were confirmed by high-resolution x-ray diffraction (XRD) after the processing.

Figure 2 shows the turn-on and subthreshold characteristics of $Si_{0.992}C_{0.008}$ *n*-MOSFETs. The devices exhibit characteristics of typical Si *n*-MOSFETs, and the presence of carbon in the channel layer does not result in any degradation of the MOSFET I-V behavior. The threshold voltage for $Si_{1-y}C_y$ *n*-MOSFETs ranged from 0.23~0.24 V, and for all devices including the CZ Si control, the typical subthreshold slope was 90~95 mV/dec, which is consistent with the relatively thick gate oxide and the substrate doping level.

Figure 3 shows the characteristics of source/drain diodes. The ideality factor in forward bias current, *n*, was nearly unity for all samples, and the reverse leakage current was comparable among the samples. The reverse bias breakdown voltage ranged from 11~12 V. The presence of carbon in the surface layer did not cause an extra reverse bias leakage or non-ideality in forward characteristics, confirming the good electronic quality of the material. The high frequency and quasi-static *C-V* curves for MOS capacitors with a $Si_{0.992}C_{0.008}$ channel layer are shown in Figure 4. The excellent agreement between the HF and QS *C-V*'s indicates that the interface quality of the remote plasma CVD oxide is reasonable. A 1-D device simulation (SEDAN) [11] and fitting were used to verify that the doping under the gate was in the range of 5×10^{16} to 1×10^{17} cm^{-3}. The extracted oxide fixed charge, Q_f was 2 to 5×10^{-11} cm^{-3} for these plasma CVD gate oxides.

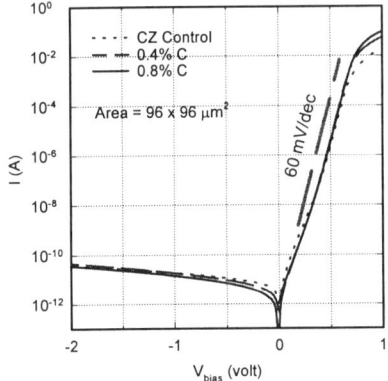

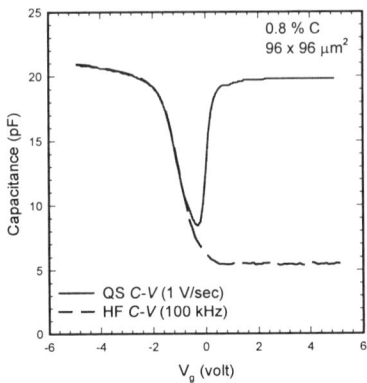

Figure 3. Source/drain diode characteristics. All diodes show near ideal forward characteristics. Reverse leakage currents are independent of carbon content.

Figure 4. MOS *C-V* for strained $Si_{0.992}C_{0.008}$ channel. Good agreement between QS and HF *C-V*'s indicate reasonable gate oxide interface quality.

EFFECTIVE ELECTRON MOBILITY

As mentioned in the introduction, strain-induced inversion layer mobility enhancement has been experimentally observed in strained Si *n*-MOSFETs [8]. Analogous to strained Si, the biaxial tensile strain in $Si_{1-y}C_y$ splits the 6-fold degeneracy in the conduction band, and lowers the Δ_2 valleys with respect to Δ_4 valleys. This band splitting suppresses phonon-assisted intervalley scattering between Δ_2 and Δ_4 valleys. In addition, the preferential carrier population of the lowered Δ_2 transversal valleys, which have smaller intrinsic transport mass, reduces the effective transport mass of electrons. As the result, the phonon-limited mobility of strained $Si_{1-y}C_y$ is expected to increase as a function of carbon content and strain in the layer. Figure 5 summarizes the measured effective mobilities of strained $Si_{1-y}C_y$ *n*-MOSFETs with 0.4 and 0.8 % carbon content, and that of the CZ Si control device. Effective mobility was calculated from drain conductance g_d at low drain bias (V_{DS} = 10 mV) by the following expression,

$$\mu_{eff} = g_d \cdot \frac{L}{W} \cdot \frac{1}{Q_{inv}} \tag{1}$$

where inversion charge, Q_{inv}, was found by integrating gate-to-channel capacitance from split *C-V* measurements. Effective field is calculated from Q_{inv} and bulk depletion charge Q_b:

$$E_{eff} = \frac{1}{\varepsilon_{Si}} \cdot \left(Q_b + \frac{1}{2} Q_{inv} \right) \tag{2}$$

Also shown are the *n*-MOSFET mobility data for strained Si on $Si_{0.9}Ge_{0.1}$ and the corresponding CZ Si control device, as reported by Welser, et. al. [8] Note that strained Si MOSFETs were fabricated with thermally grown gate oxide, while a CVD oxide was employed in strained $Si_{1-y}C_y$ MOSFETs. Within the experimental accuracy, the mobilities of strained $Si_{1-y}C_y$ MOSFETs are comparable to that of the CZ Si control device at intermediate to high vertical effective field, while at low field they appear to be degraded by additional scattering mechanisms. Unlike the strained Si MOSFETs, where mobility enhancement by almost 50 % is observed even at 10% Ge content in the substrate, no strain-induced mobility enhancement is observed for strained $Si_{1-y}C_y$ *n*-MOSFETs for the carbon content up to 0.8 %.

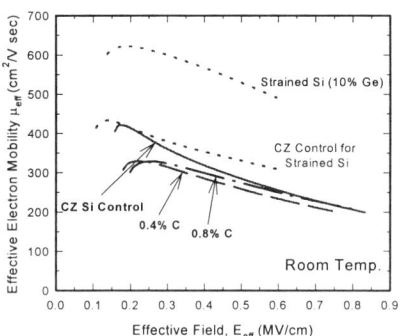

Figure 5. Measured room-temperature effective mobility of strained $Si_{1-y}C_y$ n-MOSFETs. Dotted lines are the mobility data for strained Si MOSFET and its CZ control device [8]. Unlike strained Si, no mobility enhancement is observed in strained $Si_{1-y}C_y$.

Figure 6 shows the effective mobility on a log-log scale, at room temperature and at 77 K, as function of effective field. The universal mobilities for MOSFET inversion layers for devices with state of the art gate oxides [12] are also shown for comparison. At room temperature, the mobilities of our devices, including the CZ Si control, are limited by the gate oxide quality, and are degraded in comparison to the mobility of the state of the art MOSFETs. At 77 K, the mobilities of strained $Si_{1-y}C_y$ MOSFETs are no longer comparable to the mobility of CZ Si control device, and appear to be additionally degraded.

For Si n-MOSFETs, phonon scattering limited mobility is empirically known to have $T^{-1.75}$ dependence [12]. At low temperature, therefore, phonon scattering is typically not the limiting factor for carrier transport in MOSFET inversion layers, and the mobility is dominated by other scattering mechanisms. First principle approximations indicate that alloy scattering has $T^{-0.5}$ dependence [13] while Coulomb scattering is proportional to T^{α}, where α varies from -0.5 to 1, depending on the strength of the screening by free carriers [14]. Thus, at low temperature, random alloy scattering and Coulomb scattering in $Si_{1-y}C_y$ may be degrading the mobility of the strained $Si_{1-y}C_y$ MOSFETs. Such Coulomb scattering may be associated with the small residual amount of non-substitutional carbon, which may sit at interstitial sites or form complexes with other impurity species, such as oxygen. These off-lattice-site carbon atoms are known to form both charged and neutral impurity scattering sites [15].

Ershov et. al. has shown, by Monte Carlo calculations, that the magnitude of alloy scattering potential, U_0, has a strong influence on the mobility of strained $Si_{1-y}C_y$ [16]. For low value of U_0, calculations show that the mobility of strained $Si_{1-y}C_y$ is enhanced with increasing carbon content (and thus the strain in the layer), much like the enhanced mobility of strained Si. However, for higher values of U_0, the mobility enhancement is suppressed by random alloy scattering. For $U_0 = 2.2$ eV, the calculated mobility of strained $Si_{1-y}C_y$ decreases monotonically with increasing carbon content. Based on the lack of mobility enhancement at room temperature and degradation at low temperature observed in the present work, it appears that in strained $Si_{1-y}C_y$, random alloy scattering and Coulomb scattering are offsetting the enhancement of phonon-scattering-limited mobility and dominate the electron mobility at low temperature. The observed degradation of mobility is also consistent with the Hall mobility measurements on MBE grown samples reported by Osten et. al. [17]

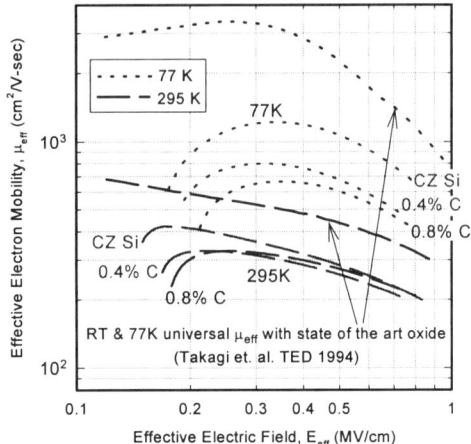

Figure 6. Comparison of strained $Si_{1-y}C_y$ n-MOSFET mobility to the state of the art inversion layer mobility at 77 K and 295 K [12]. At 77 K, mobility of strained $Si_{1-y}C_y$ is degraded by random alloy scattering and/or carbon-related Coulomb scattering.

CONCLUSION

For the first time, n-MOSFETs with strained $Si_{1-y}C_y$ surface channel have been demonstrated. The devices exhibit normal Si n-MOSFET characteristics with good turn-on and subthreshold behavior, and the MOS C-V and source/drain diodes show no carbon-related degradation. At room temperature, the electron mobility extracted from devices with strained $Si_{1-y}C_y$ channels is comparable to that measured in CZ Si control devices. At 77 K, however, the mobilities of $Si_{1-y}C_y$ MOSFETs are degraded by random alloy scattering and/or Coulomb scattering. It is postulated that the strain-induced enhancement of phonon scattering limited mobility is offset by additional scattering mechanisms in the $Si_{1-y}C_y$ channel, and that the total mobility is limited by such scattering.

REFERENCES

1. K. Eberl, S.S. Iyer, S. Zollner, J.C. Tsang, and F.K. LeGoues, Appl. Phys. Lett. **60**, 3033 (1992).
2. T.O. Mitchell, J.L. Hoyt, and J.F. Gibbons, Appl. Phys. Lett. **71**, 1688 (1997).
3. S.S. Iyer, K. Eberl, M.S. Goorsky, F.K. LeGoues, J.C. Tsang, and F. Cardone, Appl. Phys. Lett. **60**, 356 (1992).
4. W. Faschinger, S. Zerlauth, G. Bauer, and L. Palmetshofer, Appl. Phys. Lett. **67**, 3933 (1995).
5. K. Rim, T.O. Mitchell, D.V. Singh, J.L. Hoyt, and J.F. Gibbons, Appl. Phys. Lett., to be published.
6. A.A. Demkov and O.F. Sankey, Phys. Rev. B **48**, 2207 (1993).
7. J.J. Welser, PhD Thesis, Stanford University, (1995).
8. J.J. Welser, J.L. Hoyt, S. Takagi, and J.F. Gibbons, Proc. 1994 IEDM, p.373-6 (1994).
9. K. Rim, J.J. Welser, J.L. Hoyt, and J.F. Gibbons, Proc. 1995 IEDM, p.517-20 (1995).
10. G.G. Fountain, R.A. Rudder, S.V. Httangady, R.J. Markunas, P.S. Lindorme, J. Appl. Phys. **63**, 4744 (1988).
11. Z. Yu and R.W. Dutton, <u>SEDAN III</u>, Stanford University, Stanford, CA, (1985).
12. S. Takagi, A. Toriumi, M. Iwase, and H. Tango, IEEE Trans. Elect. Dev. **41**, 2357 (1994).
13. J. Singh, <u>Physics of Semiconductors and Their Heterostructures</u>, p. 372, (McGraw-Hill, New York, 1993).
14. M. Lundstrom, <u>Fundamentals of Carrier Transport</u>, p. 49, (Addison-Wesley, 1990).
15. G. Davis and R.C. Newman, in <u>Handbook of Semiconductors Vol.3</u>, edited by T.S. Moss (Elsevier Science, New York, 1994).
16. M. Ershov and V. Ryzhíi, J. Appl. Phys. **76**, 1924 (1994).
17. H.J. Osten and P. Gaworzewski, J. Appl. Phys. **82**, 4977 (1997).

CHARACTERIZATIONS OF Zr/ $Si_{1-x-y}Ge_xC_y$ AFTER RAPID THERMAL ANNEALING

V. Aubry-Fortuna, M. Barthula, F. Meyer, A. Eyal[*], C. Cytermann[*], M. Eizenberg[*], O. Chaix-Pluchery[**]

Institut d'Electronique Fondamentale, CNRS URA 22, Bât. 220, Université Paris Sud, 91405 Orsay Cedex, France, fortuna@ief.u-psud.fr
[*] Solid-State Institute, Technion, Haifa 32000, Israel
[**] LMGP, CNRS UMR 5628, ENSPG, BP75, 38402 St Martin d'Hères, France

ABSTRACT

In this work, we have investigated the reaction between Zr and SiGeC alloys after Rapid Thermal anneals performed at 800°C for 5 min. The interactions of the metal with the alloy have been investigated by X-Ray diffraction. Four crystal X-Ray diffraction was also performed to measure the residual strain in the epilayer. The final compound of the reaction is the C49-$Zr(Si_{1-x}Ge_x)_2$ phase. The C49 film contains the same Ge concentration as in the as-deposited $Si_{1-x-y}Ge_xC_y$ layer. This suggests that no Ge-segregation occurs during annealing. Only a small strain relaxation is detected in the unreacted SiGe epilayer during the reaction. The addition of C in the epilayer prevents any strain relaxation. These results are in contrast with those observed in systems with Ti and Co, and show that the system Zr-Si-Ge is much more stable. Schottky barrier heights have been also measured: annealing leads to a slight decrease of the barrier without any degradation of the contact. The resistivity of the C49 film is about 80 $\mu\Omega$cm. These results indicate that Zr may be a good candidate for contacts on IV-IV alloys in term of thermal stability.

INTRODUCTION

In recent years, SiGe alloys have generated a great interest for applications in microelectronic and optoelectronic devices, such as heterojunction bipolar transistors and infra-red photodetectors. However, the lattice mismatch between Si and Ge lcads to SiGe layers under compressive strain, which can relax by dislocation nucleation above a so-called critical thickness. This strain can be far reduced or totally compensated by adding C ($\approx$ 1% for 10% of Ge) onto substitutional sites [1]. The ternary alloy SiGeC presents a better thermal stability, compared to the SiGe alloy [2]. The band-gap only shows a small increase of 25 meV per %C [3]. Therefore, SiGeC films can be used in heterostructure devices.

However, the implementation of these IV-IV alloys in devices requires reliable ohmic and rectifying contacts. For contacts in Si devices, Ti, Co and W are the most often used metals. It has been shown that $TiSi_2$ and $CoSi_2$ can be used in high performance SiGe HBT's or MODFET's [4]. However, many studies of these metals on SiGe alloys have shown that the silicide formation is strongly affected by the presence of Ge [4-8]. The germanosilicide formed contains less Ge than in the as-deposited alloy, and this leads to Ge-segregation, either at the interface, or at the surface or at the grain boundaries in the germanosilicide film. For the Ti-Si-Ge system, the reported Ge-segregation may be related to material transport. Indeed, the reaction path varies with the alloy composition: Ti/Si leads to C49-$TiSi_2$ followed by the final phase C54-$TiSi_2$, while Ti/Ge forms Ti_6Ge_5 and finally C54-$TiGe_2$ [8]. In addition, the enthalpies of formation of $TiSi_2$ and $TiGe_2$ differ by about 8 kJ per mole atoms [9]. Ge-segregation in the case of the reaction Co/ SiGe is explained by the non miscible character of $CoGe_2$ in $CoSi_2$, according to the Co-Si-Ge ternary phase diagram [10]. For a W contact, W reacts to only form WSi_2 [5]. In addition, few studies have focussed on the stability of the unreacted SiGeC layer during the reaction. It has been shown that a strained SiGe layer can relax up to 90% depending on the initial Ge-content in the as-deposited alloy [6,11,12]. Adding C improves the stability and limits the strain relaxation after reaction with Ti: $\approx$20% for a nearly compensated layer [11].

Ge-content in the as-deposited alloy [6,11,12]. Adding C improves the stability and limits the strain relaxation after reaction with Ti: ≈20% for a nearly compensated layer [11].

In spite of a higher resistivity, recent studies have pointed out an attractive system, with Zr as the metal [9,14]. These previous experimental results suggest that the Zr-Si-Ge system is more stable to Ge-segregation than the Ti-Si-Ge one [9]. In this study, Zr/ SiGeC contacts were annealed in a RTA furnace at 800°C for 5 min, which we can be considered to be severe conditions for thermal stability. The phase formation and the Ge stoichiometry in both the unreacted alloy and the metallic film were followed for different Ge- and C- contents. We have also investigated the strain retained in the SiGeC unreacted alloy. Finally, electrical measurements have been performed to determine the resistivity of the metallic films, and Schottky barrier height of the contacts.

EXPERIMENT

The p-type (B-doped) SiGeC layers were grown by Rapid Thermal Chemical Vapor Deposition on a (100) oriented p-type (B-doped) Si substrate. A first set of samples have been used to investigate in detail the phase formation. A 0.6-μm-thick $Si_{1-x}Ge_x$ layer (x = 17 and 33%) was grown on top of a relaxed graded composition $Si_{1-y}Ge_y$ layer (y from 0 to x). A detailed description of the deposition procedure can be found in Ref. 15. The second set consists of 100-nm strained $Si_{1-x-y}Ge_xC_y$ alloys deposited at 550°C (x = 10% and y = 0 or 1.3%) and has been used to follow the relaxation of the unreacted layer after annealing.

Prior to metal deposition, each SiGeC sample was cleaned by using a standard chemical degreasing procedure, followed by a dip in diluted HF for 30 s and a final rinsing in de-ionised water. 16-80 nm-thick-Zr films were deposited in a dc-magnetron sputtering chamber. For Schottky barrier height measurements, contacts of different areas were defined by photolithography and subsequent selective chemical etch.

Heat treatments were performed in a RTA system under Ar/H₂ atmosphere at 800°C for 5 min. During annealing, the sample was maintained between two other samples of Zr/ Si. In that case, the sample surface is protected and is not directly in contact with the furnace atmosphere. This procedure limits the oxidation. The different phases formed during annealing were determined by X-Ray diffraction (XRD) using a Siemens D500 θ-2θ diffractometer ($\lambda = Cu_{K\alpha} = 1.5608$ Å). Four crystal X-Ray diffraction (FCD) was performed to measure the residual strain and the lattice parameters of the unreacted SiGeC layer, with the symmetric (004) and the assymetric (115) and (113) reflections. The barrier height values were determined by I-V measurements at room temperature, using a procedure described previously [16]. The resistivity of the films was determined from sheet resistance measurements using a four point probe.

RESULTS

Figure 1 exhibits the XRD spectra for annealed 80 nm-thick-Zr films on Si and $Si_{1-x}Ge_x$ samples (x = 17 and 33%). The peaks are mainly related to the C49-$ZrSi_2$ phase, indicating that the final germanosilicide is C49-$Zr(Si_{1-z}Ge_z)_2$. ZrSi-like peaks have also been detected, mainly in samples with a 16-nm-thick-Zr film for any Ge- and C-contents. Contrary to what we thought, the reaction is not completely achieved after a 800°C treatment of 5 min [14]. It can be seen from Fig.1 that the diffraction peaks of C49-$Zr(Si_{1-z}Ge_z)_2$ for the (131) planes are located at lower angles when the Ge-content increases. This behavior also exists for the (060) and (002) planes, and is explained by the variation of lattice planes with Ge-content. A plot of the corresponding d-spacings as a function of Ge-content shows a good agreement between these values and the behavior described by the Vegard's law. This agreement suggests that the Ge-content (z) in the C49 phase is equal to that in the alloy (x), and that no Ge-segregation has occured. Further XRD investigations in another range of the spectra indicated that only peaks related to the Si substrate and to the unreacted $Si_{1-x}Ge_x$ layer were detected. These results differ from those obtained after reaction with Ti, Co and W. These systems always lead to a final Ge-segregation, even if at the early stages the stoechiometry is conserved [7]. The enthalpies of formation of $ZrSi_2$ and $ZrGe_2$ only differ by 5 kJ per mole atoms, which corresponds to a lower driving force for Ge-

50

segregation than for the Ti-Si-Ge system (8 kJ per mole atoms) [9]. The reaction paths are similar: Zr/ Si and Zr/ Ge leads to $ZrSi_2$ and $ZrGe_2$ respectively, with identical intermediate phases. According to Wang *et al.*[9], a combination of lower atomic mobility (higher average enthalpy of formation), a lower driving force and similar material transport may make C49-$Zr(Si_{1-z}Ge_z)_2$ more stable to Ge-segregation than C54-$Ti(Si_{1-z}Ge_z)_2$. These thermodynamical considerations are in agreement with the experimental results of this study. In addition, we did not observe any modification of the Ge-content in the germanosilicide due to the presence of C in the as-deposited alloy. The corresponding d-spacing is again in agreement with Vegard's law. We did not see any delay of the reaction due to C-incorporation. A delay of the formation is usually observed with Ti [11] and Co [13], probably due to the accumulation of C at the interface between the (germano) silicide layer and the epilayer. Up to now, preliminary SIMS analyses have shown a C contamination ($\approx$ 2%) in the as-deposited Zr films. This contamination prevents us from following the transfer of C from the epilayer to the metallic film during annealing.

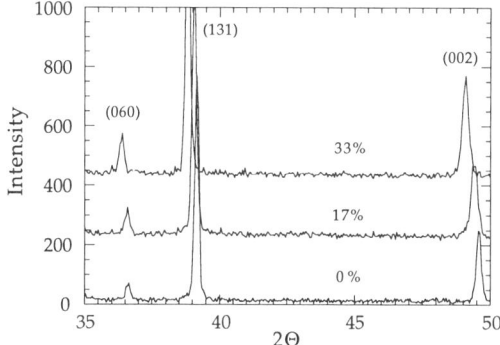

Fig. 1: X-Ray diffraction scans of Zr(80 nm)/ SiGe samples annealed at 800°C during 5 min, for x = 0, 17 and 33%.

We have also focused on the relaxation of strained $Si_{1-x-y}Ge_xC_y$ layers associated with the thermal processing. The X-Ray rocking curves of 16-nm Zr on $Si_{90\%}Ge_{10\%}$ sample before and after annealing are plotted on Fig. 2. Computer simulations of these spectra showed that the as-deposited SiGe layer is completely strained. The epilayer with 1.3% C is compensated and its peak (not shown here) is overlapped by the Si peak. As can be seen in Fig. 2 by the slight shift of the SiGe peak after annealing, only a small relaxation of about 6% has occured in the SiGe layer. A comparison of strain retained in a $Si_{90\%}Ge_{10\%}$ layer after reaction with Ti [12], Co [6] and Zr is shown on Fig. 3. After a 700°C heat treatment for 40 s, the $CoSi_2$ formation is accompanied by the formation of stacking faults and misfit dislocations [6]. According to FCD measurements, the $Si_{90\%}Ge_{10\%}$ layer has relaxed by a factor of about 60%. In the case of the sample Ti/$Si_{90\%}Ge_{10\%}$ annealed at 800°C for 10 min, the formation of the germanosilicide leads to a relaxation of 40% [12]. It seems that the relaxation is always observed in addition to the Ge-segregation. The peak corresponding to the epilayer with 1.3% C does not move, which suggests that the strain (see Fig.3), as well as the amount of C on substitutional sites, are preserved. Results with Ti [11] have shown that the relaxation is limited (compared to layers without C), but still exists, since the presence of C does not prevent the Ge-segregation.

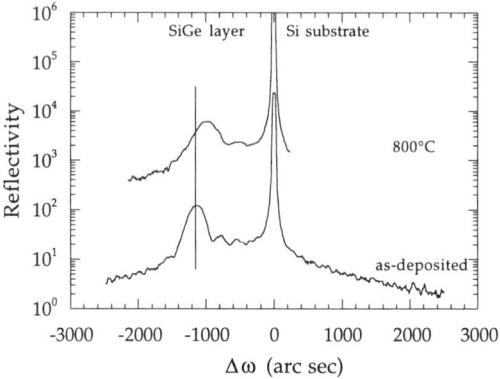

Fig. 2: X-Ray rocking curves of Zr(16nm)/ $Si_{90\%}Ge_{10\%}$ before and after annealing.

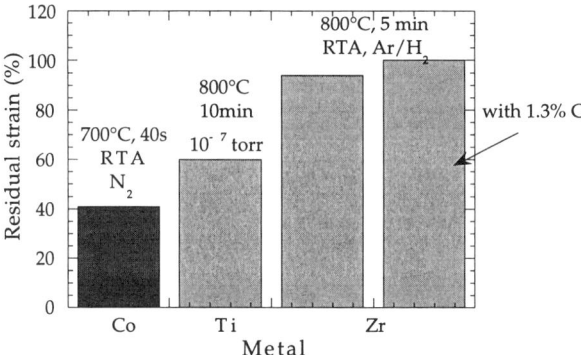

Fig. 3: Residual strain in a $Si_{90\%}Ge_{10\%}$ epilayer after reaction with Co, Ti and Zr.

Some electrical measurements have been performed. Schottky barrier heights have been determined from reverse *I-V* measurements at room temperature, using a procedure described previously [16]. Figure 4 shows the effective barrier heights before and after annealing for the relaxed p-type $Si_{83\%}Ge_{17\%}$ and the two strained p-type $Si_{90\%-y}Ge_{10\%}C_y$ alloys. The as-deposited contacts have different barrier heights. In previous papers [16], we have shown that the Schottky barrier heights for W on p-type $Si_{1-x}Ge_x$ follow with x the same trends as the band-gap of the corresponding strained or relaxed alloys. Adding C leads to a large increase of the barrier [16]. RTA at high temperature results only in a slight decrease of the barrier. In the case of 10% of Ge, the barrier height after annealing is about 0.3 eV and the current is limited by the series resistance (due to the Si substrate). Indeed the Schottky barrier can not be determined accurately. The current is still governed by thermionic emission. This means that the annealing does not degrade the Schottky contact. Usually, a barrier to SiGe with a silicide can show either a lower [17] or a higher [18] barrier than that with the pure metal. In the case of Ge-segregation, the barrier is modified dramatically [18].

Sheet resistance measurements have also been performed. The sheet resistance decreases with annealing, which is correlated to the phase formation and to the increase in the film thickness due to the reaction. A 80-nm Zr film exhibits a resistivity of about 80 μΩcm, as well as the corresponding 207-nm C49 film obtained after annealing. We did not notice any influence of the Ge- and C-contents on these values. These values are higher than those reported for optimized heating procedures in vacuum (35-40 μΩcm for 160-nm C49 films [19]) and than those obtained on C54-TiSi$_2$ films (13-25 μΩcm).

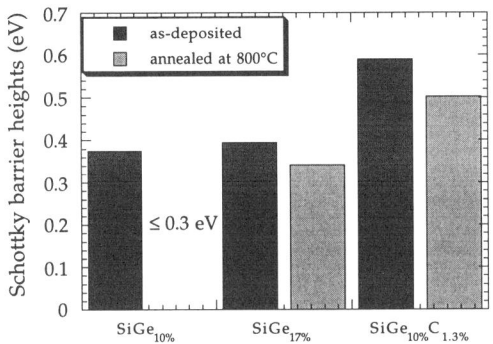

Fig. 4: Effective barrier heights of Zr/ p-type SiGeC before and after annealing.

CONCLUSIONS

Severe annealing conditions (800°C, 5min) of Zr/ Si$_{1-x-y}$Ge$_x$C$_y$ contact lead to a final C49-Zr(S$_{1-z}$Ge$_z$)$_2$ compound and do not result in Ge-segregation. The Zr-Si-Ge system is much more stable to Ge-segregation than the Ti-Si-Ge, Co-Si-Ge and W-Si-Ge ones. The reaction leads to a slight strain relaxation in the SiGe epilayer. The presence of C further improves the stability of the system. It seems that a dramatic strain relaxation occurs only in systems when Ge-segregation takes place. The Schottky contact is not degraded by the annealing: we observed a slight decrease of the barrier compared to the as-deposited values. The resistivity of thick pure Zr or C49 films is equal to 80 μΩcm. According to these results, Zr offers some advantages to Ti, Co and W in terms of thermal stability.

ACKNOWLEDGEMENTS

The authors are indebded to G. Tremblay and V. Mathet for the metal deposition and the technical assistance.

REFERENCES

[1] H.J. Osten, E. Bugiel, P. Zaumseil, Appl. Phys. Lett. **64**, 3440 (1994).
[2] P. Warren, J. Mi, F. Overney, M. Dutoit, J. Cryst. Growth **157**, 414 (1995).
[3] P. Boucaud, C. Francis, F.H. Julien, J.-M. Lourtioz, D. Bouchier, S. Bodnar, B. Lambert, J.L. Regolini, Appl. Phys. Lett. **64**, 875 (1994).
[4] M. Glück, A. Schüppen, M. Rösler, W. Heinrich, J. Hersener, U. König, O. Yam, C. Cytermann, M. Eizenberg, Thin Solid Films **270**, 549 (1995).

[5] V. Aubry, F. Meyer, R. Laval, C. Clerc, P. Warren, D. Dutartre, Mat. Res. Soc. Symp. Proc. Vol. **320**, 299 (1994).
[6] O. Nur, M. Willander, L. Hultman, H.H. Radamson, G.V. Hansson, M.R. Sardela, J.E. Greene, J. Appl. Phys. **78**, 7063 (1995).
[7] D.B. Aldrich, Y.L. Chen, D.E. Sayers, R.J. Nemanich, S.P. Ashburn, M. Öztürk, J. Appl. Phys. **77**, 5107 (1995).
[8] D.B. Aldrich, Y.L. Chen, D.E. Sayers, R.J. Nemanich, S.P. Ashburn, M. Öztürk, J. Mat. Res. **10**, 2849 (1995).
[9] Z. Wang, D.B. Aldrich, R.J. Nemanich, D.E. Sayers, J. Appl. Phys. 82 (1997) 2342.
[10] N. Boutarek, R. Madar, Appl. Surf. Sci. **73**, 209 (1993).
[11] A. Eyal, R. Brener, R. Beserman, M. Eizenberg, Z. Atzmon, D.J. Smith, J.W. Mayer, Appl. Phys. Lett. **69**, 64 (1996).
[12] M. Lyakas, M. Beregovsky, I. Moskowitz, M. Eizenberg, Mat. Res. Soc. Symp. Proc. Vol. **402**, 475 (1996).
[13] R.A. Donaton, K. Maex, A. Vantomme, G. Langouche, Y. Morciaux, A. St Amour, J.C. Sturm, Appl. Phys. Lett. **70**, 1266 (1997).
[14] V. Aubry-Fortuna, M. Barthula, J.-L. Perrossier, F. Meyer, V. Demuth, H.P. Strunk, O. Chaix-Pluchery, accepted for publication in J. Vac. Sci. Technol. B, May/June 98.
[15] D. Dutartre, P. Warren, F. Provenier, F. Chollet, A. Pério, J. Vac. Sci. Technol. A **12**, 1009 (1994).
[16] F. Meyer, M. Mamor, V. Aubry-Fortuna, P. Warren, S. Bodnar, D.Dutartre, J.L. Regolini, J. Elect. Mat. **25**, 1748 (1996).
[17] O. Nur, M. Willander, R. Turan, M.R. Sardela, H.H. Radamson, G.V. Hansson, J. Vac. Sci. Technol. B **15**, 241(1997).
[18] X. Xiao, J.C. Sturm, S.R. Parihar, D. Meyerhofer, S. Palfrey, F.V. Shallcross, IEEE Elec.Dev. Lett. **14**, 199 (1993).
[19] M. Setton, J. van der Spiegel, J. Appl. Phys. **70**, 193 (1991).

ANALYSIS OF SiGe FET DEVICE STRUCTURES ON SILICON-ON-SAPPHIRE SUBSTRATES BY X-RAY DIFFRACTION

P.M. Mooney*, J.O. Chu*, J.A. Ott*, J.L. Jordan-Sweet*, W.B. Dubbelday**+, K.L. Kavanagh+
I. Lagnado,** and B.S. Meyerson*
*IBM Research Division, T.J. Watson Research Center, Yorktown Heights, NY 10598
**Naval Space and Warfare Systems Center, San Diego, CA 92152
+ECE Department, UC San Diego, La Jolla, CA 92093

ABSTRACT

Si/Si$_{1-x}$Ge$_x$ heterostructures on improved silicon-on-sapphire substrates were grown epitaxially by ultra-high vacuum chemical vapor deposition for application as p-channel field effect transistors. High-resolution triple-axis x-ray diffraction was used to analyze these structures quantitatively and to evaluate the effects of device fabrication processes on them. Out-diffusion of Ge from the Si$_{1-x}$Ge$_x$ quantum well was observed after fabrication as was the change in thickness of the Si cap layer due to wafer cleaning and gate oxidation at 875 °C.

INTRODUCTION

Silicon-on-sapphire (SOS) is one of the most mature silicon-on-insulator (SOI) technologies. The low dielectric loss as well as better thermal properties and radiation hardness of SOS compared to other SOI technologies make SOS an excellent candidate for high-reliability microwave circuit applications [1]. Several years ago it was shown that the performance of p-channel field effect transistors (FETs) is enhanced by using a strained Si$_{1-x}$Ge$_x$ channel to confine the holes [2]. Comparable performance enhancements have now been achieved with similar heterostructures grown epitaxially on SOS substrates [3][4]. Unlike pseudomorphic Si/Si$_{1-x}$Ge$_x$ heterostructures grown epitaxially on bulk Si substrates, structures grown on SOS substrates contain high densities of extended defects. These defects apparently do not degrade the performance of FETs. However, their presence makes it impossible to use double-crystal x-ray rocking curves, a standard method for the characterization of pseudomorphic heterostructures grown on bulk Si or other semiconductor substrates [5], to obtain quantitative information on the thickness and alloy composition of the various layers in these device structures.

Reciprocal space mapping [5] using a triple-axis configuration has been successfully used to study strain-relaxed structures having various concentrations of misfit dislocations and layer thicknesses typically 100 nm or greater [6]. Here we present a quantitative analysis of Si/Si$_{1-x}$Ge$_x$ heterostructures on SOS substrates using triple-axis x-ray diffraction measurements. In these structures layer thicknesses of <10 nm were measured and changes in the thickness of the SiGe quantum well on the order of 2 nm were observed after device fabrication, due to out-diffusion of Ge from the strained SiGe quantum well. The consumption of the Si cap layer by wafer cleaning and gate oxidation was also determined.

EXPERIMENT

The heterostructures analyzed in these experiments were grown epitaxially by ultra-high vacuum chemical vapor deposition (UHV/CVD) [7] on both Si(001) and improved silicon-on-sapphire (SOS) substrates at 540 °C [8]. Transmission electron micrographs (TEM) show that properly improved substrates have a high density of threading defects, typically 10^8 cm^{-2}, but that the density of stacking faults is below the detection limit ($<10^4$ cm^{-2}). The nominal thickness of the SOS layer is 100 nm; however, in some cases the SOS layer was thinned by oxidation prior to epitaxial growth. The epitaxial structures grown for Si$_{1-x}$Ge$_x$ FETs consist of three layers: first a 15 nm-thick Si layer, then a strained 10 nm-thick Si$_{1-x}$Ge$_x$ layer, and finally a 15 nm-thick Si cap layer, as shown in Fig. 1. Structures consisting of a 40 nm-thick Si layer were also grown in order to determine the performance enhancement due to the SiGe quantum well.

The devices were fabricated using a CMOS process, which was altered slightly to accommodate the sapphire substrate. No specific steps were taken to reduce the thermal budget. The gate oxidation was performed in a dry oxygen ambient at 875 °C. The polysilicon gate electrode was deposited at low temperature and then doped with phosphorous using POCl$_3$ at 875 °C. Rapid thermal annealing was used to form the self-aligned titanium disilicide; however, furnace anneals at 850 °C were used to activate the boron and arsenic source drain implants. During device fabrication the wafers were at T≥850 °C for a total of 2.5 hours.

Two different high resolution x-ray diffraction systems were used for these experiments. X-ray rocking curves were taken using a Philips MRD diffractometer mounted on a rotating anode x-ray source. The incident beam was conditioned by a four-crystal asymmetrically cut Ge(220) monochromator which selects the Cu$_{K\alpha}$ line with a divergence of 25". The detector aperture was 4°. Triple-axis x-ray diffraction spectra were taken at beamline X20A at the National Synchrotron Light Source. A double-crystal Ge(111) monochromator set to the energy of the Cu$_{K\alpha}$ line, the sample, and a Si(111) analyzer were positioned in an approximately non-dispersive geometry [9].

RESULTS

Fig. 2 shows 004 x-ray rocking curves of a bare SOS wafer and a UHV/CVD-grown heterostructure on an SOS substrate (G15). The large amount of mosaic broadening of the Si peak, which has a full width at half maximum of 0.24°, arises from the high density of extended defects in SOS substrate. Note also that the Bragg angle of the Si peak is smaller than that of bulk Si(001), indicating that the SOS layer is compressively strained. This strain occurs when the wafer is cooled after the improvement process [8], due to the difference in the thermal expansion coefficients of Si and sapphire. The thickness of the SOS layer varies across the wafer, typically in a bull's eye pattern with the thickest region near the center of the wafer. The nominal thickness of the SOS layer is 100 nm; however, the variation in the intensity of the x-ray peak indicates that the thickness variation is

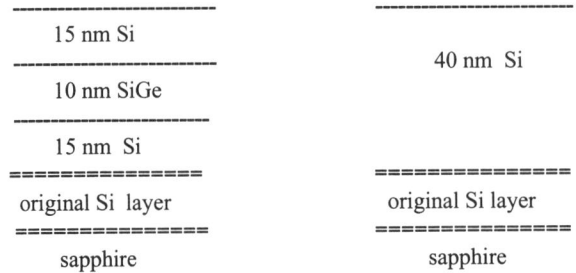

Fig. 1. P-channel FET structures grown by UHV/CVD with nominal layer thicknesses indicated.

The Fig. 2. 004 x-ray rocking curve for an SOS substrate and a Si/Si$_{1-x}$Ge$_x$/Si on SOS structure (G15 as-grown).

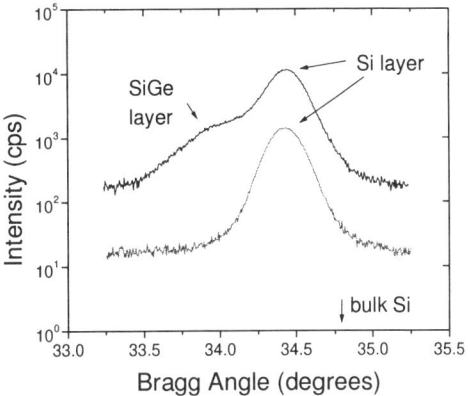

about 30 nm. The threading defects in the SOS layer are, of course, replicated in the epitaxial layers. Despite the mosaic broadening of the x-ray peaks, the ~10 nm-thick Si$_{1-x}$Ge$_x$ layer can be observed as a shoulder on the low-angle side of the Si peak.

To obtain quantitative information about heterostructures grown on SOS substrates, it was necessary to use a triple-axis x-ray configuration to eliminate the line broadening due to the defects [5]. While there was sufficient x-ray intensity from the rotating anode generator to observe the relatively intense Si peak, the intensity of rest of the spectrum was too weak to be observed above the detector noise, making is necessary to use a synchrotron x-ray source for the triple-axis measurements. Fig. 3 shows experimental and simulated 004 spectra for sample G15 (as-grown), the same structure shown in Fig. 2. Here we see the thickness fringes from the underlying Si layer superimposed on the diffraction peaks arising from the strained Si$_{1-x}$Ge$_x$ layer and the Si cap layer. The RADS [10] simulation for this structure requires five parameters: the strain and thickness of the underlying Si layer, the composition and thickness of the Si$_{1-x}$Ge$_x$ layer and the thickness of the Si cap layer. It was assumed that the Si$_{1-x}$Ge$_x$ and Si cap layers remain fully pseudomorphic to the underlying Si layer, i.e. that any strain relaxation occurring during the UHV/CVD growth results

Fig. 3. 004 triple-axis x-ray diffraction spectra (heavy line) shown with simulated spectra (fine line) for a Si/Si$_{1-x}$Ge$_x$/Si on SOS structure (G15 as-grown).

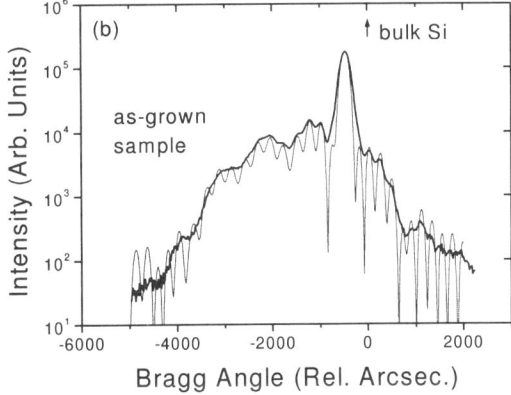

in misfit dislocations lying at the interface between the original Si layer and the sapphire substrate. The validity of this assumption was verified by plan-view TEM measurements. Changes in the layer thickness of 0.1 nm or in Ge mole fraction of 0.002 yielded observable changes in the simulated curve. We estimate the uncertainty in the layer thicknesses to be 0.2 nm and that of the Ge mole fraction to be 0.01.

The thermal stability of these structures was investigated by annealing pieces of several wafers at 850 °C in an inert atmosphere (dry He gas) for 1 hr. Significant out diffusion of Ge from the SiGe quantum well was observed as previously reported [11]. The same amount of out-diffusion of Ge was found to occur in heterostructures grown simultaneously on SOS and bulk substrates indicating that the high densities of threading defects present in device structures on SOS substrates play a negligible role in the diffusion process, consistent with a previous report that interdiffusion of Si and Ge in superlattices is not changed in samples partially relaxed by misfit dislocations [12].

Since the wafers spend 2.5 hours at T=850 or 875 °C during device fabrication, out-diffusion of Ge from the SiGe quantum well is expected to occur. In addition, the thickness of the Si cap layer is reduced by wafer cleaning and gate oxidation. To quantify these effects, pieces of as-grown wafers were measured and were compared with pieces of wafers from the same epitaxial growth run after device fabrication. Fig. 4(a) shows the measured triple-axis spectra for an as-grown structure and for a processes wafer. The Bragg angle of the main Si peak, and thus the strain in the SOS layer, is unchanged; however, there are measurable differences in the other features of the spectrum. The measured spectra are shown individually with simulated spectra in Figs. 4(b) and 4(c). The

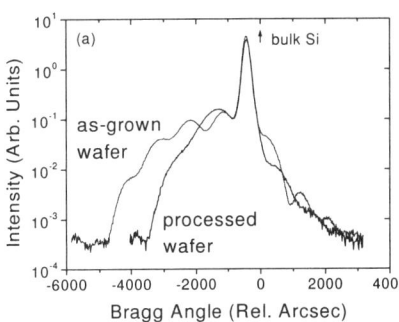

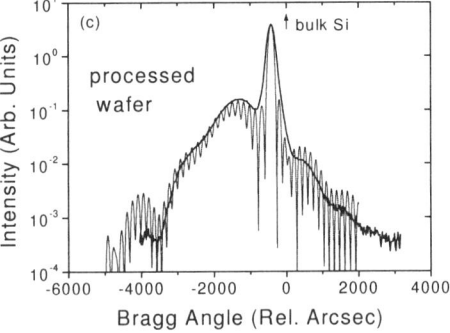

Fig.4. (a) Measured 004 triple-axis x-ray diffraction spectra for as-grown and processed samples from epitaxial growth run G19. (b) measured (heavy line) and simulated (fine line) spectra for the as-grown sample, and (c) measured and simulated spectra for the processed sample.

Table I: Simulation parameters for 004 x-ray diffraction spectra of as-grown and processed samples.

Sample	Si layer thickness (nm)	Si layer strain (%)	SiGe layer thickness (nm)	SiGe layer composition, x	Si cap layer thickness (nm)
G20 as-grown	71.0	0.31	9.0	0.22	15.0
G20 processed	72.0	0.31	11.6	0.15	6.5
G19 as-grown	100	0.31	8.7	0.22	14.5
G19 processed	115	0.31	11.5	0.15	6.2
G15 as-grown	66.0	0.34	9.8	0.19	14.7
G15 processed	60.0	0.34	12.5	0.14	8.5
G10 as-grown	80	0.34	10.0	0.17	14.5
G10 processed	100	0.33	13.0	0.13	6.0

parameters used for the simulations are given in Table I. Comparing the as-grown and processed samples, we see that the Ge fraction of the $Si_{1-x}Ge_x$ layer decreased from 0.22 to 0.15. The SiGe layer thickness increased by 2.8 nm and the thickness of the Si cap layer decreased from 14.5 to 6.2 nm. Changes in the thickness of the underlying Si layer cannot be determined from these measurements, since the variation in thickness across each wafer is about 30 nm.

Data for four sets of wafers processed in three different fabrication runs are given in Table I. Wafers G19 and G20 were processed together. Comparing data for these two sets of wafers we see that the changes are essentially identical. Wafers G15 and G10 were different in that thay have quantum wells with lower initial Ge concentrations. The decrease in the average Ge concentration due to out-diffusion of Ge was smaller in these wafers, however, thus making the Ge concentration of the quantum wells of all the fabricated devices comparable. We note that the Si cap layer was about 2 nm thicker in the G15 processed wafer than the other processed wafers. This was due to a modification in the wafer cleaning sequence, specifically one HF dip was omitted. The run-to-run reproducibility of the Si cap layer thickness was good for the other two fabrication runs.

CONCLUSION

We have used triple-axis x-ray diffraction measurements along with RADS simulations to obtain quantitative information about the effects of device processing on $Si/Si_{1-x}Ge_x/Si$ structures grown epitaxially on SOS substrates. The triple-axis configuration eliminates the mosaic broadening of the x-ray peaks caused by the high density of threading defects in these heterostructures. The high sensitivity of the triple axis measurements to small changes in the heterostructure enabled us to measure the out-diffusion of Ge occurring during processing at $T \geq 850$ °C as well as the final Si cap layer thickness. Our measurements show that these quantum well structures are degraded by the processes used here and that lower process temperatures are required to maintain the integrity of the as-grown structures.

ACKNOWLEDGMENTS

This work was partially supported by SPAWAR Systems Center, SD/Office of Naval Research contract No. N66001-95-C-6011, and by Department of Energy contract No. DE-AC02-76CH00016.

REFERENCES

1. P.R. de la Houssaye, C.E. Chang, B. Offord, R. Johnson, P.M. Asbeck, G.A. Garcia and I.Lagnado, IEEE Electron Device Lett. **16**, 289 (1995).

2. S. Verdonct-Vandebroek, E.F. Crabbe, B.S. Meyerson, D.L. Harame, P.J. Restle, J.M.C. Stork and J.B. Johnson, IEEE Trans. Elect. Devices **41**, 90 (1994).

3. S.J. Mathew, G. Niu, W.D. Dubbelday, J.D. Cressler J.A. Ott, J.O.Chu, P.M.Mooney, K.L.Kavanagh, B.S. Meyerson, and I. Lagnado, submitted to Electron Device Lett.

4. S.J. Mathew, G. Niu, W.D. Dubbelday, J.D. Cressler, J.A. Ott, J.O. Chu, P.M. Mooney, K.L.Kavanagh, B.S. Meyerson and I. Lagnado, submitted to Electron Device Letters.

5. P.F. Fewster, Semicond. Sci. Technol. **8**, 1915 (1993).

6. P.M. Mooney, J.L. Jordan-Sweet, J.O. Chu and F.K. LeGoues, Appl. Phys. Lett. **66**, 3642 (1995); P.M. Mooney, J.L. Jordan-Sweet, K. Ismail, J.O. Chu, R.M. Feenstra and F.K. Legoues, Appl. Phys. Lett. **67**, 2373 (1995).

7. B.S. Meyerson, Appl. Phys. Lett. **48**, 797 (1986).

8. S.S. Lau, S. Matteson, J.W. Mayer, P. Revesz, J. Gyulai, J. Roth, T.W. Sigmon and T. Cass,Appl. Phys. Lett. **34**, 76 (1979).

9. P.M. Mooney, J.L. Jordan-Sweet, G.B. Stephenson, F.K. LeGoues and J.O. Chu, Advances in X-Ray Analysis **38**, 181 (1995).

10. Rocking Curve Analysis by Dynamical Simulation, Bede Scientific, Inc.

11. P.M. Mooney, J.A. Ott, J.O. Chu and J.L. Jordan-Sweet, submitted to Appl. Phys. Lett.

12. P. Boucaud, L. Wu, C. Guedj, F.H. Julien, I. Sajnes, Y. Campidelli and L. Garchery, J. Appl. Phys. **80**, 1414 (1996).

Part II

Devices, Processing, and Characterization

RAMAN SPECTROSCOPY OF EPITAXIAL Si/Si$_{1-x}$Ge$_x$ HETEROSTRUCTURES

RAN LIU, STEFAN ZOLLNER, MING LIAW*, DAVID O'MEARA**, NIGEL CAVE**
Arizona Technology Laboratories, Motorola, Inc., 2200 W. Broadway Rd., Mesa, AZ
* Materials Research and Strategic Technologies, Motorola, Inc., 2100 E. Elliot Rd., Tempe, AZ
** Advanced Products Research and Development Laboratory, Motorola Inc., Austin, TX 78721

ABSTRACT

Raman scattering studies were carried out on epi Si/Si$_{1-x}$Ge$_x$ (x = 0.1 to 0.3) heterostructures consisting of a thin Si cap layer (100 - 400 Å), a grade-down Si$_{1-x}$Ge$_x$ layer, a constant Si$_{1-x}$Ge$_x$ buffer layer and a grade-up graded Si$_{1-x}$Ge$_x$ layer on (100) oriented Si substrates. Different Ge composition, Si$_{1-x}$Ge$_x$ layer thicknesses and thermal treatment were used to achieve different relaxation in the Si$_{1-x}$Ge$_x$ layers. It has been revealed that, to a very good approximation, the absolute strains in the cap Si and constant Si$_{1-x}$Ge$_x$ layers follow a simple sum-rule that is imposed by the lattice mismatch between unstrained Si and completely relaxed Si$_{1-x}$Ge$_x$. This sum rule can be used to determine the Ge composition and stresses in both cap Si and constant Si$_{1-x}$Ge$_x$ layers. Excellent agreement was found between the theoretical curve obtained with LO phonon strain coefficient b=-930cm^{-1} and the experimental total strain for all samples, regardless of the degree of the relaxation of the grade-up Si$_{1-x}$Ge$_x$ layer.

INTRODUCTION

The Ge composition and stresses are two most important parameters affecting the electric properties in the Si/Si$_{1-x}$Ge$_x$ heterojunction bipolar transistor (HBT) and strain-enhanced-mobility field effect transistor (SEMFET) devices. The biaxial stresses created by the lattice mismatch between Si and Si$_{1-x}$Ge$_x$ split the degeneracy of the light-hole and heavy-hole bands at the top of the valence band and the degeneracy of the equivalent conduction band minima and thus enhance the hole and electron mobilities, respectively. As revealed in this study, the maximum stress of the epi Si layer achieved when the Si$_{1-x}$Ge$_x$ layer is fully relaxed is set by the Ge composition. The conventional methods to measure the Ge composition are RBS and SIMS. However, RBS is not a very accurate technique to measure very thin Si$_{1-x}$Ge$_x$ layer and SIMS is a destructive method. Both techniques can not be used as in-line monitoring tools. Recently, we succeeded in developing Raman technique to determine composition as well as stresses in the epi Si and flat Si$_{1-x}$Ge$_x$ layers.

EXPERIMENT

The samples used in this work were gown at temperature between 700 and 800°C by reduced pressure chemical vapor deposition and radiantly heated chemical vapor deposition. The epi Si/Si$_{1-x}$Ge$_x$ structures consisted of a grade-up Si$_{1-x}$Ge$_x$ layer followed by a flat Si$_{1-x}$Ge$_x$ buffer layer and then a grade-down layer and finally a thin Si cap layer. Different Ge composition in the range from 0% to 30% was obtained by controlling the flow rates of germane (GeH$_4$) and silane (SiH$_4$) or dichlosilane (SiH$_2$Cl$_2$). The grade-up layer thickness varied from 0 to 1.5 μm, the buffer layer thickness from 0.05 to 2 μm, the grade-down layer thickness from 0 to 600 Å and the Si epi layer thickness from 100 to 1000 Å. Some of the samples underwent different rapid thermal annealing processes at temperatures between 900 and 1100°C for 20 to 100 seconds. For simplicity, Table I only lists the structural parameters of 5 samples out of the total 25 samples studied in this work.

Table I. Structural parameters and strain-induced Si-Si phonon frequency shifts in the cap Si and buffer $Si_{1-x}Ge_x$ of five samples grown by reduced pressure CVD. The x values obtained from RBS and Raman are also listed for comparison.

Wafer	Structure	x (RBS)	x (Raman)	Frequency shift (cm^{-1})		
				Si	SiGe	Total
#4	200 Å Si cap			|-7.23|		9.71
	300 Å grade-down $Si_{1-x}Ge_x$					
	5000 Å buffer $Si_{1-x}Ge_x$	0.277	0.275		2.48	
	1500 grade-up $Si_{1-x}Ge_x$					
#5	200 Å Si cap			|-5.99|		8.43
	300 Å grade-down $Si_{1-x}Ge_x$					
	5000 Å buffer $Si_{1-x}Ge_x$	0.243	0.240		2.44	
	1500 grade-up $Si_{1-x}Ge_x$					
#6	200 Å Si cap			|-4.94|		7.56
	300 Å grade-down $Si_{1-x}Ge_x$					
	5000 Å buffer $Si_{1-x}Ge_x$	0.229	0.216		2.62	
	1500 grade-up $Si_{1-x}Ge_x$					
#7	200 Å Si cap			|-3.17|		6.12
	300 Å grade-down $Si_{1-x}Ge_x$					
	5000 Å buffer $Si_{1-x}Ge_x$	0.172	0.176		2.95	
	1500 grade-up $Si_{1-x}Ge_x$					
#8	200 Å Si cap			|-6.52|		7.70
	300 Å grade-down $Si_{1-x}Ge_x$					
	10000 Å buffer $Si_{1-x}Ge_x$	N/A	0.220		1.18	
	15000 grade-up $Si_{1-x}Ge_x$					

The Raman spectra were acquired using a Dilor XY 800 triple-grating spectrometer. An Ar laser beam at 4579 Å was focused by a 100x microscope objective down to about 1 μm in diameter onto the sample surface. The incident laser power was kept below 5 mW to avoid effects due to laser heating. The scattered light was collected by the same objective, analyzed by the spectrometer with high-dispersion additive configuration and detected by a liquid nitrogen cooled Spectra-One CCD camera from Instrument ISA.

RESULTS AND DISCUSSION

The optical penetration depth at the laser wavelength (4579 Å) is about 4200 Å and decreases in $Si_{1-x}Ge_x$ with increasing x. For few samples with a total thickness of the epi Si and $Si_{1-x}Ge_x$ layers less than 4000 Å, the substrate Si LO peak is also seen and makes the analysis of the x and stresses at low Ge composition ($x<0.05$) difficult. For most samples studied in this work, the buffer $Si_{1-x}Ge_x$ layer is thicker than 5000 Å and thus no Si substrate phonon peak will be seen. Furthermore, the graded $Si_{1-x}Ge_x$ layers are relatively thin and with rapid change of x and contribute only some broad background to the Raman spectra. Therefore, the Raman spectra from our samples contain dominantly the Si-Si vibrational peaks from the Si epi layer and the $Si_{1-x}Ge_x$ buffer layer, the Si-Ge and Ge-Ge peaks in the $Si_{1-x}Ge_x$ layer. All these peaks can be used to characterize the Ge composition and stresses. However, the intensities of the Ge-Ge and Si-Ge peaks are much weaker than those of the Si-Si peaks for our samples with Si rich $Si_{1-x}Ge_x$ layers and it is difficult to obtain accurate values for x and stresses using these peaks. In this work, we report a method to derive x and the stresses in both the cap Si and the $Si_{1-x}Ge_x$ buffer layer by only using the frequencies of the Si-Si LO peaks from these two layers.

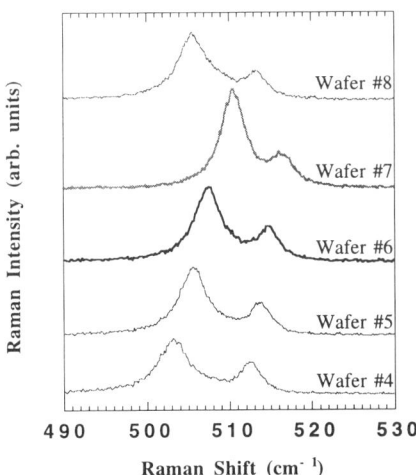

Figure 1. Si-Si phonon Raman spectra of the samples listed in Table I.

Figure 1 shows the Raman spectra of the Si-Si LO phonons from the samples listed in Table I. Since the epi Si layer is psudomorphically grown on the $Si_{1-x}Ge_x$ layer, the Si-Si LO phonon frequencies in these layers can be written as

$$\omega_{Si} = \omega_0 + \Delta\omega_{Si} = \omega_0 + b_{Si}\varepsilon_{Si} = \omega_0 + b_{Si}\frac{a - a_0}{a_0} \qquad (1)$$

$$\omega_{SiGe} = \omega(x) + \Delta\omega_{SiGe} = \omega(x) + b_{SiGe}\varepsilon_{SiGe} = \omega(x) + b_{SiGe}\frac{a - a(x)}{a(x)}, \qquad (2)$$

where, $\omega_0 = 520.7$ cm^{-1} and $a_0 = 5.431$ Å are phonon frequency and lattice constant for unstrained Si, ε is the strain, b is the strain-phonon coefficient, a is the real lattice constant of both Si and flat $Si_{1-x}Ge_x$ epi layers, and

$$\omega(x) = 520.7 - 69 \bullet x \ (cm^{-1}) \qquad (3)$$

and

$$a(x) = 5.431 + 0.2x + 0.026x^2 \qquad (4)$$

are the phonon frequency and lattice constant[1] of unstrained $Si_{1-x}Ge_x$. The linear function of the unstrained frequency was obtained by fitting the experimental LO frequencies in totally relaxed $Si_{1-x}Ge_x$ films deposited on Si substrate. The strain-induced frequency shift of the LO phonon in the epi Si layer can be obtained from the measured LO phonon peak frequency. The strain-induced frequency shift in the $Si_{1-x}Ge_x$ layer can be derived using the unstrained Si-Si LO phonon frequency which was calculated using Equation (3) and x values obtained from RBS measurements. Figure 2 displays the strain-induced frequency shifts in both Si and $Si_{1-x}Ge_x$ layers. It was found that, although the relative frequency shifts and thus the strains of the two layers changes from sample to sample due to different x and thickness, the sum of the values of the two layers seems to conserve as a single-value function of Ge composition. This useful sum-rule can be theoretically proved by following equation

$$|\varepsilon| = |\varepsilon_{Si}| + |\varepsilon_{SiGe}| = \frac{a - a_0}{a_0} + \left| \frac{a - a_0(x)}{a_0(x)} \right| = \left| \frac{a_0(x) - a_0}{a_0} \frac{a}{a_0(x)} \right| \approx \frac{a_0(x) - a_0}{a_0}. \qquad (5)$$

Therefore, the total strain of the two layers is controlled by the lattice mismatch between unstrained Si and $Si_{1-x}Ge_x$ with very good accuracy. If one assumes a x-independent b, then a sum-rule also holds for the total strain-induced frequency shift. Indeed, a very good agreement between experiment data and the theoretical curve can be obtained using

$$|\Delta\omega| = |b| (|\varepsilon_{Si}| + |\varepsilon_{SiGe}|) \approx |b| \frac{a_0(x) - a_0}{a_0} \qquad (6)$$

and $b_{Si} = b_{SiGe} = b = -930 cm^{-1}$ (see Figure 2a). These two sum-rules are very robust and the only condition for it to hold is the coherent interface between the Si and $Si_{1-x}Ge_x$ layers. The Si-Si phonon strain constant $b_{Si} = b_{SiGe} = b = -930 cm^{-1}$ agrees with Halliwell et al.,[2] but is considerably larger than other reported values.[3,4,5] Moreover, Lockwood and Baribeau[5] also proposed a linear dependence of b on Ge composition. However, their experimental values of b were rather sample dependent and scattered as function of x. These different b values seem to result from the sample difference (bulk samples versus thin epi films versus partially relaxed films). In case of the work on bulk materials, there might be some relaxation of applied stress near the surface and thus the b value depends on the probing depth at the used laser wavelength.[3] The stresses, σ, in the Si and $Si_{1-x}Ge_x$ layers can be derived from

$$\Delta\omega = b\varepsilon = b(S_{11} + S_{12})\sigma , \qquad (7)$$

where the compliance constants S_{11} and S_{12} are function of Ge composition. The stress in the Si layer was obtained by, again, using $b = -930$ cm^{-1} and $(S_{11} + S_{12})^{-1} = 180$ GPa, and thus $\sigma = 0.194\Delta\omega$ GPa/cm^{-1}. Using a linear extrapolation, $(S_{11} + S_{12})^{-1} = 180 - 42x$ GPa, from the pure Si value to the pure Ge value (138 GPa), the stress in the SiGe layer was evaluated (see Figure 2b). Although there is no sum rule for the total stress due to the composition dependence of $(S_{11} + S_{12})^{-1}$, one can still calculate the maximum and minimum total stress for a given Ge composition by either assuming completely strained Si layer or completely strained $Si_{1-x}Ge_x$ layer. The calculated curves of the maximum and minimum total stresses are plotted in Figure 2b and match the experimental data well.

Although good agreement has been achieved between theory and experiment, some deviation of the experimental data from theory is clearly evident. These discrepancies are probably caused by rather big error bar in using RBS for Ge composition measurements in such complicated structures (thin sandwiched $Si_{1-x}Ge_x$ layer and existence of the graded layers). We found that better agreement can be achieved by using the x values determined from the two Si-Si LO phonon frequencies. Using equations (1) and (2), the difference of these two frequencies can be written as

$$\omega_{Si} - \omega_{SiGe} \approx b \frac{a(x) - a_0}{a_0} - 69x = -4.45x^2 + 35x. \tag{8}$$

The solutions to this equation are

$$x = 3.89 \pm \sqrt{15.12 - 0.222(\omega_{Si} - \omega_{SiGe})}, \tag{9}$$

where only the solution with minus sign is physical and can be used to determine the Ge composition. Figures 3a and 3b show the total strain-induced frequency shift and stress as functions of the Ge composition obtained using the Raman method. Excellent agreement can be clearly seen and proves that the Raman technique has better accuracy than the RBS method in case of thin epi Si/ $Si_{1-x}Ge_x$ structures.

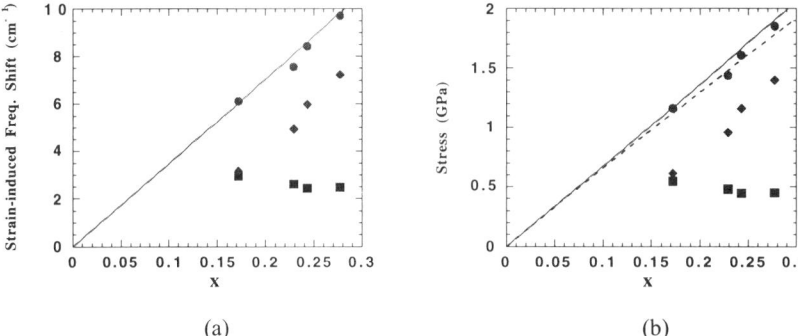

(a)   (b)

Figure 2. Strain-induced frequency shifts (a) and stresses (b) in epi Si and SiGe layers. The x values were obtained from RBS measurements. The squares, diamonds and circles denote the values for the Si and $Si_{1-x}Ge_x$ layers and the total value. The solid line in (a) is the calculated total frequency shift. The solid (dash) line in (b) is the calculated maximum (minimum) stress.

As can be expected from the sum-rules of the strains and strain-induced LO phonon shifts, this Raman method can be applied to all epi Si/ $Si_{1-x}Ge_x$ structures without significant dislocations near the interface between Si and $Si_{1-x}Ge_x$. Figure 4a and 4b summarize the results obtained from all samples studied. In can be seen that, despite differences in deposition process, thickness and thermal treatment, the self-consistency between the experiment data and the theoretical curves is always observed.

CONCLUSIONS

Raman spectroscopy can be used for characterization and in-line monitoring of both stresses and Ge composition in epi $Si/Si_{1-x}Ge_x$ structures. In such structures, the absolute coherent strains in the Si and $Si_{1-x}Ge_x$ layers follow a sum rule imposed by the lattice mismatch between the unstrained Si and $Si_{1-x}Ge_x$ and, thus, is a conserved function of x. This sum rule can be extended to the total absolute strain-induced phonon frequency shifts in the epi Si and $Si_{1-x}Ge_x$ layers if a composition-independent strain-phonon coefficient, b. This allows to calculate the Ge content and stress in the $Si_{1-x}Ge_x$ buffer layer in addition to the stress in the epi Si layer using the measured Si-Si LO phonon frequencies in the Si epi layer and the $Si_{1-x}Ge_x$ buffer layer. This assumption has been confirmed self-consistently by the excellent agreement between the experimental total strain-induced frequency shift as function of x in the range $0.1 < x < 0.3$ with the theoretical curve using a constant $b = -930$ cm^{-1}.

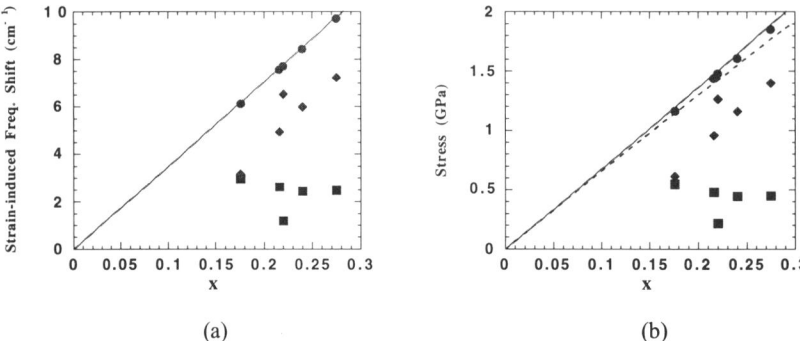

(a)   (b)

Figure 3. Strain-induced frequency shifts (a) and stresses (b) in epi Si and SiGe layers. The x values were obtained using the Raman technique. Symbols and lines: see Figure 2.

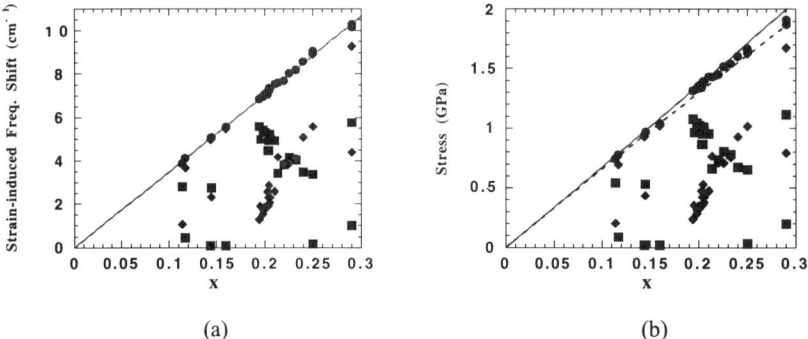

(a)   (b)

Figure 4. Strain-induced frequency shifts (a) and stresses (b) in epi Si and SiGe layers for all samples. The x values were obtained using the Raman technique. Symbols and lines: see Figure 2.

REFERENCES

1. J. P. Dismukes, L. Ekstrom, and R. J. Paff, J. Phys. Chem. **10**, 3021 (1964).
2. M. A. G. Halliwell, M. H. Lyons, S. T. Davey, M. Hockly, C. G. Tuppen, and C. J. Gibbings, Semicon. Sci. Technol. **4**, 10 (1989).
3. E. Anastassakis, A. Cantarero, and M. Cardona, Phys. Rev. B **41**, 7529 (1990).
4. E. Anastassakis, A. Pinczuk, E. Burstein, F. H. Pollack, and M. Cardona, Solid State Commun. **8**, 133 (1970).
5. D. J. Lockwood and J. -M. Baribeau, Phys. Rev. B **45**, 8565 (1992).

INVESTIGATION OF PLASTIC RELAXATION IN SI$_{1-x}$GE$_x$/SI DEPOSITED BY SELECTIVE EPITAXY

S. WICKENHÄUSER AND L. VESCAN
Institut für Schicht- und Ionentechnik (ISI), Forschungszentrum Jülich GmbH (KFA),
D-52425 Jülich, Germany

ABSTRACT

Si$_{1-x}$Ge$_x$/Si heterostructures with different layer thicknesses grown by selective LPCVD epitaxy at different growth temperatures were investigated with regard to plastic relaxation. AFM and optical micrograph were performed to determine the dislocation density. From the analysis of the misfit dislocations at the initial stage of relaxation of the samples grown at 700°C it was possible to determine the nucleation site density and an activation energy of 2.8 eV for the heterogeneous nucleation of misfit dislocations. While the critical thickness h$_c$ for a given Ge content increases with decreasing growth temperature between 800°C-680°C one observes a dramatic decay of h$_c$ at a growth temperature of 625°C. For growth at 625°C it was found that this activation barrier is drastically decreased.

INTRODUCTION

Relaxed Si$_{1-x}$Ge$_x$ buffer layers become more and more interesting for applications such as virtual substrates for strained Si layers [1]. Below a certain stress Si$_{1-x}$Ge$_x$ layers relax plastically, that is, by formation of misfit dislocations [2-9]. During this process threading dislocations usually are created, which would degrade any device. Therefore, in order to avoid the disturbing threading dislocations it is necessary to study this relaxation mechanism in detail. One possibility to reduce the threading dislocation density is the growth of a low temperature Si buffer layer, buried beneath the Si$_{1-x}$Ge$_x$ layer. This Si layer acts as a reservoir for nucleation sites [10,11].

In this work the various stages of relaxation are explored by using selective epitaxy. By variation of the layer thickness and the sample lateral dimension it was possible to explore the nucleation of the dislocations in detail [6]. From the analysis of misfit dislocations within squares of different dimensions, it was possible to separate the three relaxation mechanisms. While in small structures the relaxation process has just begun (only nucleation and propagation of misfit dislocation are of importance) the interaction process has already started in larger structures. Each layer thickness and each square dimension yields a different degree of relaxation and therefore we were able to examine the different relaxation mechanisms independently. Low and high temperature grown Si$_{1-x}$Ge$_x$ layers have been investigated with respect to the critical thickness, which reflects the degree of metastability due to activation barriers of the three relaxation mechanisms, namely nucleation, propagation and multiplication. The relaxation process of these two different growth temperatures have been compared.

EXPERIMENT

p- and n-type (100) Si (1000 Ωcm) wafers with 600 nm thick thermal oxide were patterned using standard optical lithography and reactive ion etching. The resulting pattern consisted of squares of various dimension (10-300 μm) aligned parallel to the <110>direction. The patterned substrates were cleaned by standard RCA-cleaning and a subsequent HF-dip. Selective epitaxy was carried out in a cold wall load-locked, high vacuum Low Pressure Chemical Vapour Deposition (LPCVD) system. The deposition was performed at 0.12 Torr at temperatures of 700°C and 625°C, with $SiCl_2H_2$ and GeH_4 as source gases and H_2 as carrier gas. The samples grown at 700°C consisted of a 350 nm Si buffer, a $Si_{0.84}Ge_{0.16}$ layer of varying thickness (105 - 430 nm) and a 5 nm Si cap layer. The first sample had a thickness in the range of the critical thickness (h_c) so that the relaxation had just started. Increasing the $Si_{1-x}Ge_x$ layer leads to stronger relaxation due to the advance of the epitaxial growth, and due to simultaneously annealing at the epitaxial temperature. The samples grown at 625°C consisted of a single $Si_{0.92}Ge_{0.08}$ layer of different layer thicknesses (50 – 140 nm). The layer thickness and the composition were determined by Rutherford backscattering. Diluted Schimmel etch (4 parts 50% HF : 5 parts 0.3 M CrO_3) and Yang etch (1 part 50% HF : 1 part 1.5M CrO_3) was used to reveal the misfit dislocations at the interface due to the anisotropic etch rate. A Nomarski microscope and a Digital Instruments Nanoscope IIIa AFM in Tapping Mode were used to determine the linear misfit dislocation density (ρ_{MD}).

RESULTS

We determined ρ_{MD} for the $Si_{0.84}Ge_{0.16}$ layers grown at 700 °C for each square (10-300 µm) and for the different layer thicknesses (215-430 nm). In Fig. 1, ρ_{MD} is plotted versus square dimension (length of square side) for the different layer thicknesses. For the three thinnest samples (215-300 nm) one observes a linear increase of with increasing square dimension up to 300 µm, whereas for the thicker samples ρ_{MD} increases linearly only for small structures and deviates from this linear behaviour increasing supralinearly for larger structures. The onset of this deviation decreases with increasing layer thickness. In addition one has to remark that all squares of dimension 10 µm were completely strained.

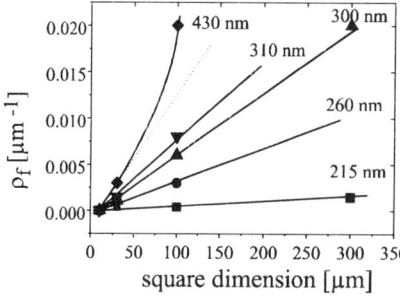

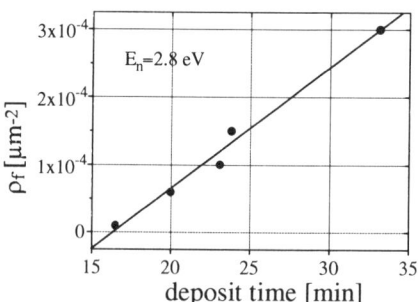

Fig. 1: Dislocation density ρ_{MD} versus square dimension for different layer thicknesses. From the slope of the lines it is possible to determine the nucleation site density ρ_f.

Fig. 2: Nucleation site density ρ_f versus deposition time, which is equivalent to the layer thickness. From the slope it is possible to determine a nucleation activation energy of 2.8 eV

To explain these results we make following assumptions: (1) a constant nucleation site density in all squares and (2) for small structures and small layer thicknesses the multiplication process has not yet started so that only the nucleation and propagation of misfit dislocations are of importance. With increasing square dimension the relaxation advances, because more nucleation sites are found within a square and the nucleated misfit dislocations become longer. In this stage we observe a linear increase of ρ_{MD} with square dimension (Fig.1). Following Fitzgerald *et al.* [12] we assume for the initial stage of relaxation:

$$\rho_{MD} = \frac{1}{2}\rho_f L \qquad (1)$$

where ρ_{MD} [μm^{-1}] is the linear misfit dislocation density, ρ_f [μm^{-2}] the nucleation site density and L[μm] the dimension of the squares. In the model of Fitzgerald *et al.* ρ_{MD} decreases linearly with decreasing structure dimension and extrapolates to zero for L = 0 μm. In our case ρ_{MD} becomes zero already for squares of 10 μm. This deviation indicates that elastic strain relaxation at the edges of the 10 μm structures must have become important.

For larger square dimensions the misfit dislocation density deviates from the linear behavior and increases supralinearly (Fig.1), due to the onset of misfit dislocation multiplication leading to a more relaxed state. With the aid of Eq. (1) we were able to determine the nucleation site density ρ_f for different layer thicknesses from the linear slope in Fig. 1. In Fig. 2 ρ_f is plotted versus the deposition time t, which is related over the growth rate R with the layer thickness (h = R·t). It is obvious that ρ_f increases with increasing layer thickness or deposition time. A possible explanation for the increase of the nucleation site density might be the fact that there were not only nucleation sites at the interface, but that nucleation sites are produced during the epitaxial growth in the SiGe layer. For example, localized stress concentration due to fluctuation in alloy composition were proposed to act as heterogeneous sources [3]. However, this fact can not explain the great increase of dislocation nucleation sites (factor 30 for a thickness increase of factor 2) observed in our experiments. In order to explain these results we assume that the density of active nucleation sites increases with the time interval the layer was exposed to the growth temperature. This idea was forwarded by Houghton [5], who gave a semiempirical expression for the first stage of relaxation, when only the nucleation process is active:

$$\frac{d\rho_f(t)}{dt} = B\rho_{f_0} \left(\frac{\sigma_{eff}}{\mu}\right)^{2,5} \exp\left(\frac{E_n}{kT}\right) \qquad (2)$$

where ρ_f (t)[μm^{-2}] is the density of active nucleation sites at time t, ρ_{f_0} [μm^{-2}] the nucleation site density at time t=0, due to defects in the interface or in the SiGe layer or residual substrate precipitates, B a material constant ($10^{18} s^{-1}$)[6], μ the shear modulus (64 GPa) [6], σ_{eff} [GPa] the effective strain and E_n the activation energy of heterogenous nucleation. It is obvious from Eq (2) that the density of active nucleation sites ρ_f increases linearly with respect to the annealing time compared to ρ_{f_0}, indicating a thermal activation of new sources. This was verified by Houghton in a series of iso-thermal anneals performed in the initial stage of relaxation. Equation (2) was proposed for *post growth anneal*. It gives the number of nucleation sites after the sample was annealed at a certain temperature for a certain time scale. During the anneal there is no change in both layer thickness and Ge concentration, and therefore the effective strain is constant. In the present experiments the sample is exposed during epitaxy to the growth temperature for a certain time, but the layer thickness continues to increase. The effective stress is not constant in this case. However, evaluating the effective stress (0.6 GPa) for the investigated thickness range (215-430 nm) we found that it is approximately constant with a deviation in the order of 4 % (0.02 GPa). Integrating Eq. (2), using $\rho_{f_0}=10^3$ cm^{-2} which is the data evaluated on the thinnest sample (similar values are suggested by Perovic *et al.*[13] for CVD experiments), we determine an activation barrier for heterogeneous nucleation of misfit dislocations of 2.8 eV from the slope in Fig. 2. This value agrees quite well with the value of 2.5 eV from Houghton [4,5]. The differences might be explained by the fact that in our case we cannot neglect a possible increase of the nucleation sites in the SiGe layer during epitaxy, whereas Houghton examined the samples after epitaxy. In addition in our experiment other experimental conditions (CVD, substrate cleaning) are given.

Figure 3 shows the experimental critical thickness for $Si_{1-x}Ge_x$ layers grown at different temperatures [7,9] and the equilibrium curve of Matthews & Blakeslee [8]. For growth temperatures between 680°C and 800°C all measured critical thicknesses lie well above the equilibrium curve due to the metastabile state of the layers. One observes a decrease of the critical thickness with increasing growth temperature and therefore an approach to the equilibrium curve, indicating that the relaxation process is not an equilibrium process, but kinetically limited due to activation barriers of nucleation and propagation. Figure 3 also shows the experimental results of the $Si_{0.92}Ge_{0.08}$ layers grown at 625°C. The unfilled points correspond to layer thicknesses of relaxed samples while the filled one corresponds to a sample completely strained. Therefore the onset of the relaxation process starts above 45 nm and the critical thickness for these $Si_{0.92}Ge_{0.08}$ layers lies between 45 and 60 nm. This value does not agree with the expected critical thickness of 300 – 400 nm for a $Si_{0.92}Ge_{0.08}$ layer grown at 625°C following the trend observed for 680°C-800°C. This dramatical decrease of the critical thickness and therefore the earlier start of the relaxation mechanism indicates that the relaxation process proceeds nearer

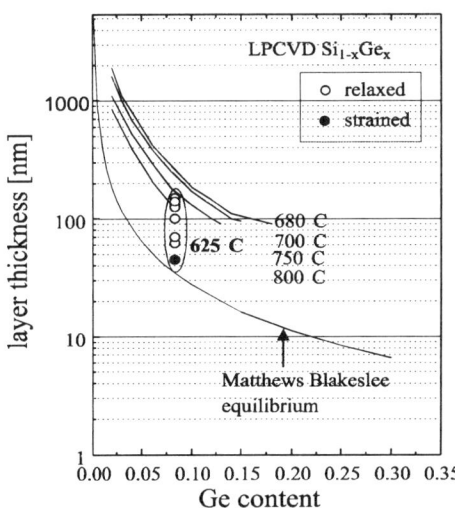

Fig.3: Critical thickness for strain relaxation by introduction of misfit dislocations [7,9]. The equilibrium curve is from [8]. Unfilled points correspond to relaxed samples, the filled one corresponds to the strained sample

to the equilibrium state than at a growth temperature of 680°C. The relaxation starts shortly above the equilibrium critical thickness (for x=0.08, h_{cequil}=35 nm), indicating lower activation barriers for the relaxation process than at higher growth temperatures.

Figure 4 shows a Normarski micrograph and an AFM image of samples grown at 700°C and 625°C respectively. The sample grown at 700° C shows long misfit dislocations (> 800 µm) indicating that the relaxation limiting process is the nucleation of the misfit dislocations rather than the gliding process (E_{nuc}>E_{gl}). A nucleated dislocation can glide very easily and therefore it becomes very long. In contrast, the sample grown at 625 °C reveals many short dislocations (2-3 µm), indicating that the relaxation limiting process is the propagation of the dislocations (E_{nuc}<E_{gl}).

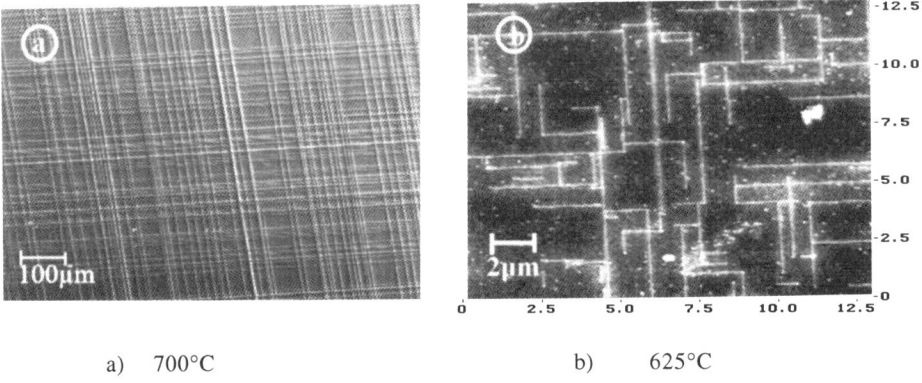

a) 700°C   b)   625°C

Fig.4: Misfit dislocations revealed by Yang etch. a) Normaski micrograph of a 250 nm $Si_{0.925}Ge_{0.075}$ layer, b) AFM image of a 80 nm $Si_{0.92}Ge_{0.08}$ layer

Just above the equilibrium critical thickness (Fig. 4b) many dislocations are already nucleated, but they are still short.
We determined the misfit dislocation density ρ_{MD} for the different layer thicknesses of the samples grown at 700°C (215 – 430 nm) and 625°C (60 – 140 nm), respectively. In Fig. 5 ρ_{MD} is plotted versus the layer thickness. For the $Si_{0.84}Ge_{0.16}$ layers grown at 700°C one observes the expected relaxation behaviour. Once the critical layer thickness (90 nm for x=0.16) is reached, the misfit dislocation density first increases linearly (from 100 to 150 nm) with increasing layer thickness, only nucleation and propagation of the dislocations is important. Further on, when many dislocations were created, the multiplication process starts and the dislocation density increases supralinearly. In the initial state of the relaxation the rate limiting process is the nucleation of the dislocations. In the case of the $Si_{0.92}Ge_{0.08}$ grown at 625°C (Fig. 5b) one observes a completely different relaxation behaviour. Above the critical thickness (for x=0.08, h_c=50 nm) the dislocation density rises very rapidly in contrast to the slow increase of ρ_{MD} at 700°C. It has to be remarked that the Ge content in both cases is not the same and one cannot compare the critical thicknesses for both cases.

The experimental critical thickness of 50 nm of the $Si_{0.92}Ge_{0.08}$ layer grown at 625°C has to be compared with 230 nm of $Si_{0.92}Ge_{0.08}$ layer grown at 700°C. The different relaxation behaviour remains still obvious.

One can explain these results for layers grown at 625°C by assuming that first the nucleation activation energy is lowered and second the density of nucleation sites is increased, in comparison to layers grown at higher temperature. The first assumption is supported by the fact that on the one hand h_c is drastically decreased, the relaxation proceeds near equilibrium and on the other hand the relaxation limiting process is the propagation of the dislocations. The second assumption will be clear by regarding Fig. 5b. Only a large density of nucleation sites can explain this drastic increase of ρ_{MD}.

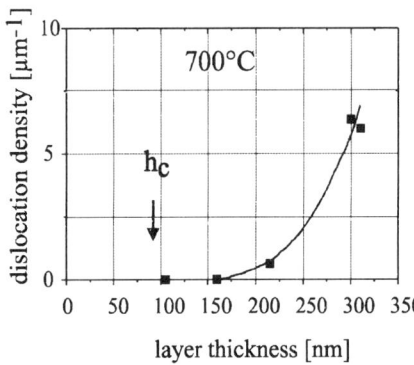 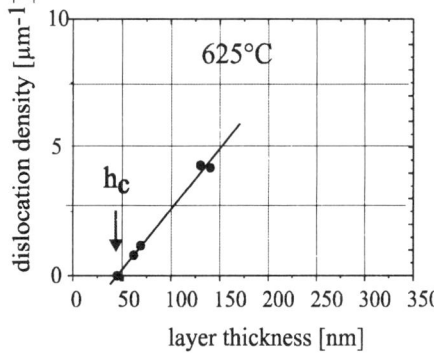

Fig. 5: Dislocation density versus layer thickness a) for $Si_{0.84}Ge_{0.16}$ grown at 700°C and b) for $Si_{0.92}Ge_{0.08}$ grown at 625 °C. The plots reveal the different relaxation behaviour

CONCLUSION

In conclusion, it was possible to determine the activation barrier for heterogeneous nucleation for a growth temperature of 700°C with the aid of the misfit dislocation density in selectively grown pads. The obtained value of 2.8 eV is in good agreement with results of other authors [4,5]. We used different layer thicknesses and different square dimensions to realize the various relaxation stages and this simplicity is the advantage of our method. Besides we demonstrated that the critical thickness is drastically decreased below a certain temperature (< 680°C) compared with layers grown at higher temperatures. At a growth temperature of 625°C the resulting epitaxial layer contains many nucleation sites which are able to nucleate dislocations with a lower activation energy than those in a layer grown at higher temperatures.

Acknowledgement: We are thankful to Erich Kasper for helpful discussions.

REFERENCES

1. K. Ismail, B.S. Meyerson, P.J. Wang, Appl. Phys. Lett., **58**, 2117 (1991)
2. R. Hull, J.C. Bean, D.C. Werder, R.E. Leibenguth, Phys. Rev. B, **40**, 1681 (1989)
3. R. Hull, J.C. Bean, J. Vac. Sci. Technol. **A7**, 2580 (1989)
4. D.C. Houghton, Appl. Phys. Lett., **57**, 2124 (1990)
5. D.C. Houghton, J. Appl. Phys., **70**, 2136 (1991)
6. S. Wickenhäuser, L.Vescan, K. Schmidt, H.Lüth, Appl. Phys. Lett., **70**, 324 (1997)
7. L. Vescan, (unpublished)
8. J.W. Matthews, A.E. Blakeslee, J. Cryst. Growth, **27**, 118 (1974)
9. H.P. Tang, L. Vescan, J. Cryst. Growth, **114**, 1 (1992)
10. H. Chen, L.W. Guo, Q. Cui, Q Hu, Q Huang, J.M. Zhou, J. Appl. Phys., **79**, 1167 (1996)
11. K.K. Linder, F.C. Zhang, J.-S. Rieh, P. Bhattacharya, D. Houghton, Appl. Phys. Lett., **70**, 3224 (1997)
12. E.A. Fitzgerald, G.P. Watson, R.E. Proano, D.G. Ast, P.D. Kirchner, G.D. Pettit, J.-M. Woodall, J. Appl. Phys., **65**, 2220 (1989)
13. D.D. Perovic, D.C. Houghton, phys. stat. sol. (a), **138**, 425 (1993)

MOSAIC CRYSTAL TILTS AND THEIR RELATIONSHIP TO DISLOCATION STRUCTURE, SURFACE ROUGHNESS AND GROWTH CONDITIONS IN RELAXED SiGe LAYERS

D.J. WALLIS, D.J. ROBBINS, A.J. PIDDUCK, G.M. WILLIAMS, A. CHURCHILL AND J. NEWEY.
EOMC, DERA, St Andrews Road, Malvern, Worcestershire, WR14 3PS, U.K.

ABSTRACT.

In recent years the growth of virtual substrates using graded SiGe buffer layers has shown great promise for the development of high performance devices. Whilst significant progress has been made in the control of growth conditions to produce low threading dislocation densities of the order suitable for commercial exploitation, several technological problems still have to be overcome. An example of such problems are cosmetic surface defects such as pits and the cross hatched surface roughness associated with mosaic crystal tilts. The work described here utilises a variety of techniques, including X-ray diffraction reciprocal space maps, TEM, AFM and SIMS to provide a comparison between several SiGe virtual substrates grown using low pressure-CVD at high ($\approx 800^{\circ}C$) and low ($\approx 600^{\circ}C$) temperatures, and at different grade rates (5-50% Ge μm^{-1}). The growth conditions are seen to have a strong effect on the crystal tilts present in the layers with the low temperature layers showing a much larger spread of mosaic tilts. The origin of these tilts is seen to occur during the early stages of the relaxation process irrespective of growth temperature and at similar Ge fractions for all samples. TEM imaging close to the initial growth interface shows that dislocation pileups occur in this region and also suggest that the pileups have a characteristic spacing of 1-2μm. A similar characteristic length scale is also observed in the surface roughness by AFM, the form of which is seen to depend upon the growth conditions.

INTRODUCTION.

The continuous demand for increased Si electronic device performance has meant that novel techniques have to be developed to satisfy this need. One approach which has already been proven is band gap engineering using tensile strain[1]. This allows confinement of electrons in a 2-dimension electron gas providing access to the high electron mobilities possible by remotely doping these low dimensional structures[2]. In order to facilitate this technology and to provide the necessary band-offsets, substrates with variable lattice parameters are required. For the SiGe/Si system, the usual approach to obtaining these *virtual substrates* is to grow a graded composition SiGe film, epitaxially, onto a defect free Si substrate and above this, a constant composition buffer layer of the required lattice parameter (Ge fraction). By grading the composition, misfit strain is dispersed through the layer thickness and thus the dislocations required to relieve the misfit strain are vertically separated. As a result of this vertical separation, the interaction between dislocations is reduced and they are able to move more easily without becoming pinned which could leave large numbers of threading segments. The purpose of the constant composition layer is to ensure complete relaxation of the structure and also to spatially separate the active device layers from the dislocation network. An important aim of this approach is to promote nucleation limited relaxation of the film[3,4] and therefore provide low threading dislocation densities. For example, threading dislocation densities in the range of 10^5-10^6 cm^{-2} have been achieved for virtual substrates. This is compared to threading dislocation densities of 10^{11} cm^{-2} for constant composition layers deposited directly onto Si[5]. Utilising this technology very high electron mobilities of the order of 300000cm^{2}V^{-1}s^{-1} have been achieved in tensile strained Si quantum wells[6-10].

Despite the progress described above, several technological problems still have to be overcome before this technology can be fully exploited. One such problem is the surface roughness of the virtual substrates which is likely to be detrimental during lithographic processing and will pose an ultimate limit on the minimum scale of devices which can be produced. This surface roughness is invariably comprised of a pronounced <110> crosshatch pattern which can be identified on many different length scales. Surface pits can also be identified. Lutz et. al.[11] have shown that the surface crosshatch may be associated with the dislocation network in the graded layers. Here the effects of different grading rates and growth temperatures on the dislocation network and hence the surface cross-hatch are investigated using a variety of experimental techniques.

EXPERIMENTAL.

A range of layers have been studied spanning different growth conditions from slow grade rate (approx. 5% Ge μm^{-1}) and low growth temperature (560-610°C) to high grade rate (approx. 50% Ge μm^{-1}) and high growth temperature (800°C). All the layers examined in this study were grown by low pressure chemical vapour deposition[12] (LP-CVD) onto chemically cleaned 100mm diameter Si wafers after in-situ oxide desorption at 850-900°C in 130Pa H$_2$. Layers also had a multilayer structure grown on them consisting of a modulation-doped Si quantum well to allow electron mobility measurements to be made. Table 1 lists the various growth conditions employed for each sample.

Table 1. Growth conditions for virtual substrates.

| Sample | Graded Composition Layer | | | Constant Composition Buffer Layer | | Surface Quality |
	Tg (°C)	Grade Rate (%Geμm^{-1})	Form of Grade	Ge Fraction	t (μm)	RMS Roughness (nm)
6A8	800-650	45	6 steps	0.28	0.33 (650°C)	20-25
6A9	800	42	6 steps	0.27	0.33 (650°C)	23-25
6A47	560	36	10 steps	0.29	0.09	2-3
6A58	560-605	52	9 steps	0.26	0.87	3-4
6B34	800	5	Linear	0.24	0.88	2-6
6B35	800	5	Linear	0.24	0.88	3-6
6B121	800	7	Linear	0.23	0.68	2-6

Characterisation of the virtual substrates surface roughness was carried out by Atomic Force Microscopy (AFM), in air, using a Digital Instruments Dimension 3000 with microfabricated Si cantilevers. Details of the strain state of the layers and the dislocation network were obtained by high resolution X-ray diffraction space mapping on a Bede D3 diffractometer with a rotating copper anode source. The high intensity of X-ray photons available from this source enables reasonably large areas of diffraction space to be mapped and fairly weak features to be observed with a typical counting time of 15 hours. Transmission Electron Microscopy (TEM) was also performed on a JEOL 4000EX electron microscope operated at 400keV to allow the form and position of dislocations to be studied. For many of the layers examined in TEM a continuous grade rate from 0% Ge was used making it very difficult to observe the epi-layer/substrate interface. It was therefore necessary to use Secondary Ion Mass Spectrometry on a Cameca 4f to provide composition as a function of depth and for location of the exact depth of the epilayer/substrate interface.

RESULTS AND DISCUSSION.

Previous studies[13] have shown that layers grown at high temperature and grade rate undergo roughening due to elastic relaxation of the surface. This mode of relaxation is driven by the surface strain and high Tg. RMS roughness values of 20-25nm are seen for such layers and thus layers grown in this regime are clearly unsuitable for use as virtual substrates.

Surface morphology of the remaining layers may be roughly divided into two categories according to growth temperature: Figure 1 shows AFM images typical of these two regimes taken from layers 6A58 and 6B34 which were grown at low and high temperatures respectively. The surface morphology for the low temperature (high grade rate) layer (figure 1a) has a distinctly tiled appearance. In contrast, high temperature layers (figure 1b) have a much more undulating surface morphology. In all cases this surface roughness has a distinct minimum wavelength in the range of 0.5-2μm. Cross-sectional TEM of the virtual substrates show pile-ups of dislocations along {111} slip planes with a very similar spacing to that seen in the surface morphology (figure 2), strongly implying that these two features are linked as has been previously suggested[11,14]. For the high grade rate layers, such as that shown in figure 2a, the dislocation pile ups are seen to extend into the Si substrate, however, for those with slower grade rates the dislocation pileups are mainly confined to the graded composition region.

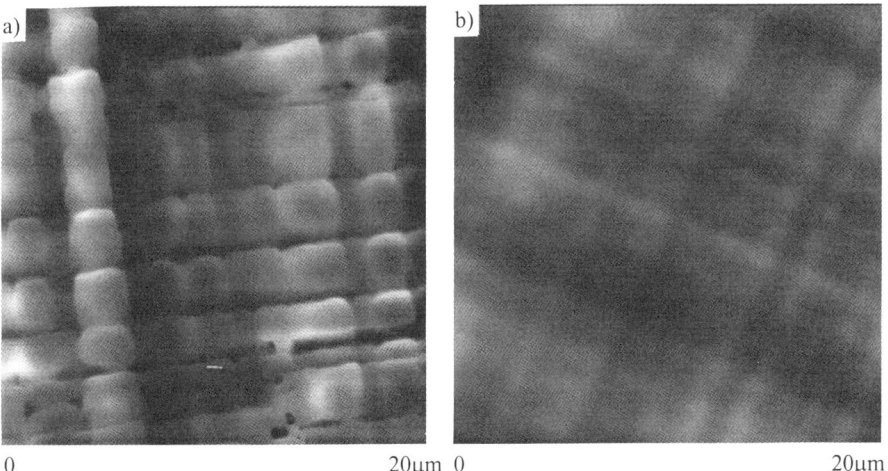

a)	b)
0 20μm	0 20μm

Figure 1. 20μm square AFM images typical of virtual substrate surfaces grown at a) low temperatures (Sample 6A58) and b) high temperatures (Sample 6B34). Gray scale 0-20nm.

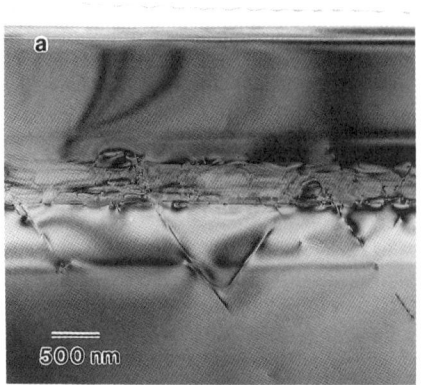 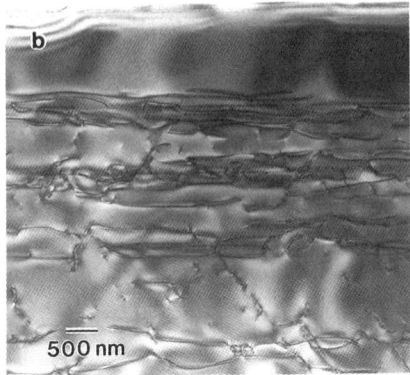

Figure 2. TEM images typical of virtual substrates grown at a) low temperatures (Sample 6A58) and b) high temperatures (Sample 6B121).

X-ray diffraction space maps provide information about the size and nature of the dislocation pile-ups. Figure 3 shows (004) diffraction space maps taken with the X-ray beam parallel to one of the [110] directions for virtual substrates 6A58 and 6B35. Along the ordinate such maps show changes in Bragg angle, and thus the lattice parameter of these fully (>95%) relaxed epilayers. By combining this information with a knowledge of the composition profile obtained via SIMS, the diffraction space maps provide depth sensitive information. The abscissas of the space maps are sensitive to the tilt of the epilayer relative to the substrate and can thus be interpreted in terms of a preference for dislocations of a particular Burgers vector since a tilt of the layer necessarily implies the presence of more dislocations with a particular Burgers vector.

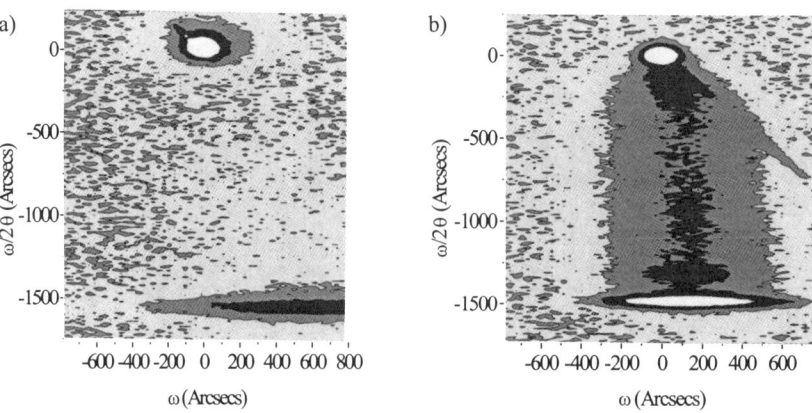

Figure 3. (004) X-ray diffraction space maps typical of virtual substrates grown at a) low temperatures (Sample 6A58) and b) high temperatures (Sample 6B35).

In both the high and low temperature cases the constant composition buffer layer is seen to have a net average tilt relative to the substrate. It has been previously suggested that the tilting of the epilayers is due to the preference of a specific slip system which favours dislocations of a particular Burgers vector[4]; the preferred slip system resulting from the initial miscut of the substrate and acting to return the surface of the epilayer towards (001). Accurate measurements of the tilt and miscut of several wafers shows that this is indeed the case for the layers grown at high temperature. However, layers grown at lower temperature have a tilt which operates at approximately 90° to the wafer miscut. The reason for this apparent anomaly is unknown, although one slip system is still clearly preferred since tilting of the virtual substrates occurs. From figure 3, tilting of the layers occurs during growth of the initial third of the graded Ge region. Cross-sectional TEM images of the layers show that this is the same region where the majority of the large dislocation pileups with the 0.5-2μm spacing occur. Above this region the dislocation network appears to be more random (figure 2).

The width of the tilt distribution seen in the diffraction space maps is related to the mosaic spread of tilts in the crystal and contains information about the number of dislocations with a particular Burgers vector in each pileup. This can be simply understood since each dislocation can only give rise to a specific value of tilt and so the large tilts seen must be due to the accumulation of many dislocations with identical Burgers vectors. If a pileup were to occur with a random distribution of Burgers vectors, this would lead to a very narrow distribution of mosaic tilts as their Burgers vectors would cancel. Presumably, the pileups of dislocations with the same Burgers vectors that are seen in these samples, result from the operation of heterogeneous dislocation sources. The width of the tilt distributions for the low temperature layers is much larger (1300 arcsecs FWHM) than for the high temperature layers (400 arcsecs FWHM). This indicates that at low temperatures a relatively small number of these sources is activated leading to larger pileups of dislocations with identical Burgers vectors. At higher temperatures, the number of sources is increased and the pileup size is reduced. This increase in the number of dislocation sources may be expected as the thermal energy available to activate them will increase with increasing growth temperature.

The size of the dislocation pileups present in the samples may be correlated to the form of the surface roughness observed in the AFM images. Lutz et. al.[11] have shown that each 60° dislocation is associated with a 2.5A height surface step and that the pileup of many dislocations with the same Burgers vectors results in a tiled surface as seen for the low temperatures layers. For layers grown at higher temperatures the size of the pileups is significantly reduced and thus the tiling at the surface is expected to be much weaker. However, for high temperature growth the mobility of surface atoms will be higher which is likely to encourage the onset of strain driven roughening in the presence of any residual strain fields. Such strain fields are likely to exist in these samples since despite the fact that all the layers are measured to be greater than 95% relaxed, the presence of large dislocation pileups containing a majority of one Burgers vector means that locally the strain associated with the misfit dislocations will not cancel[14]. It is believed likely that such strain driven roughening is the origin of the undulating surface morphology seen for the high temperature layers. The 0.5-2μm scale of this roughness arises from the spacing of the dislocation pileups which result in the residual strain fields at the epilayer surface.

It is interesting to note that in both high and low temperature cases the origin of surface roughening is the presence of dislocation pileups, even through the roughness occurs through different mechanisms. The diffraction space maps and TEM images demonstrate that the nature of these pileups is determined during the very early stages of layer relaxation. Thus in order to control the surface roughness it is the earliest stages of the growth which must be controlled.

At much larger scales (100-1000μm) gross disruptions in the surface morphology of these layers are also seen under optical microscopy. These domain boundaries are characterised as the points where sets of cross-hatch lines are discontinuous. It is suggested that such features result at the intersection of two domains in which a different slip system is preferred. Such domains are often observed in cross-sectional TEM where adjacent dislocation pileups occurring on the same slip system dominate (see figure 2). This is also seen in AFM of the tiled surfaces (figure 1a) where the tiles are seen to run in the same direction over large areas.

CONCLUSIONS.

The virtual substrates examined may be divided into two categories according to their growth conditions. At low growth temperatures (irrespective of grade rate) a relatively small number of heterogeneous dislocation sources are activated and this results in the presence of large dislocation pileups containing a particular Burgers vector. These pileups give rise to a large spread of mosaic crystal tilts. At high growth temperatures, more dislocation sources are activated and the pileup size and hence spread in mosaic crystal tilt is reduced. In both cases the presence of these dislocation pileups controls the surface morphology, although through different mechanisms. Large pileups of 60° dislocations result in a tiling of the surface at low temperatures because of the surface disruption caused by the tilt component of the dislocation's Burgers vector. This effect is reduced at higher growth temperatures, as pileup size is reduced, and the surface morphology is dominated by strain driven roughening due to the residual strain field around dislocation pileups and the increased surface mobility afforded by higher temperature growth. The origins of these roughening mechanisms are found to occur during the very earliest stages of relaxation suggesting that it is this part of the growth which must be controlled to remove these effects.

REFERENCES.

1. K Ismail, et. al. Appl. Phys. Lett. **58** 2117 (1991).
2. A Sadek and K Ismail , Solid State electron. **38** 1731 (1995).
3. FK LeGoues, BS Meyerson, J.F Morar and PD Kirchner, Appl. Phys. **71** 4230 (1992).
4. FK LeGoues, PM Mooney and JO Chu, Appl. Phys. Lett. **62** (2) 140 (1993).
5. PM Mooney, FK LeGoues, JO Chu and SF Nelson, Appl. Phys. Lett. **62** (26) 3464 (1993).
6. AC Churchill, DJ Robbins, DJ Wallis, N Griffin, DJ Paul and AJ Pidduck, Semicond. Sci. Technol. **12** 943 (1997).
7. K Ismail, FK LeGoues, KL Saenger, JO Chu and BS Meyerson, Phys. Rev. Lett. **73** 3447 (1994).
8. K Ismail, M Arafa, KL Saenger, JO Chu and BS Meyerson, Appl. Phys. Lett. **66** 1077 (1995).
9. K Ismail, J. Vac Sci. Technol. **B14** 2776 (1996).
10. P Weitz, RJ Maug, K vonKlitzing and F Schaffier, Surf. Sci. **361/362** 542 (1996).
11. MA Lutz, RM Feenstra, FK LeGoues, PM Mooney and JO Chu, Appl. Phys. Lett. **66** (6) 724 (1995).
12. DJ.Robbins and IM Young, Appl. Phys. Lett. **50** 1575 (1987).
13. AJ Pidduck, DJ Robbins, DJ Wallis, GM Williams, AC Churchill, JP Newey, C Crumpton and PW Smith, Inst. Phys. Conf. Ser. No 157, 135-144 (1997).
14. E. Fitzgerald,YH Xie, D Monroe, PJ Silverman, JM Kuo, AR Kortan, FA Thiel and BE Weir, J. Vac. Sci. Technol. B **10** (4) 1807-1819 (1992).

DEVICE AND FABRICATION ISSUES OF HIGH PERFORMANCE SI/SIGE FETS

M. ARAFA*, I. ADESIDA*, AND K. ISMAIL**
*Coordinated Science Laboratory and the Center of Compound Semiconductor Microelectronics, Department of Electrical and Computer Engineering, University of Illinois at Urbana-Champaign, Urbana, IL 61801, arafa@capone.ccsm.uiuc.edu.
**IBM T. J. Watson Research Center, Yorktown Heights, NY 10598.

ABSTRACT

A review of the latest results on high performance Si/SiGe FETs grown on relaxed buffer is presented. A discussion of the fabrication issues facing the achievement of these devices is also included.

INTRODUCTION

In the last decade, advances in Si/SiGe heterostructure growth have brought to fore a silicon-compatible technology having the potential to compete favorably with III-V devices in application-specific designs requiring low power consumption such as cellular telephone and other portable and wireless electronics. The outstanding performance demonstrated by bandgap-engineered devices such as HBTs and MODFETs on GaAs and InP has provided the impetus to study them using the Si/SiGe materials. The excellent transport properties achieved for both electrons and holes led to record high performance devices. High performance MODFETs have been demonstrated by several groups. A record unity current gain cutoff frequency of 70 GHz was achieved for p-type MODFETs at a relatively low drain bias[1]. The performance of the fabricated p-type SiGe MODFETs is higher than any p-type FETs achieved on Si and III-V compound semiconductors. A record maximum frequency of oscillation of 81 GHz was achieved for n-type MODFETs [2]. There are still many ways to improve the performance by using a better quality material and employing a more sophisticated fabrication techniques.

HIGH PERFORMANCE SI/SIGE MODFETS

SiGe based heterostructures add an extra dimension to device design. It offers several attractive features such as the possibility of forming atomically abrupt discontinuities in either the conduction- or valence-band leading to a well-controlled buried channels for electrons and holes. Also, a reduction in the supply bias can be adopted due to the enhancement of the carrier mobility. Moreover, a larger improvement in the holes transport properties, compared to electrons, leads to a more symmetric complementary structure.

MODFETs are much easier to fabricate than MOSFETs. They also tend to have relatively higher gate leakage current and drain off-current. A lower background doping in the epitaxial Schottky layer should result in a lower gate current. This can be achieved by optimizing the growth condition. Another alternative is to cap the modulation-doped structure with a thin oxide. This approach results in a lower carrier control efficiency due to the low dielectric constant of the oxide. Extra care need to be taken in the heterostructure design to prevent the formation of a parasitic channel at the oxide interface. In this case, a high quality oxide interface does not need to be achieved, allowing the use of low-temperature deposited oxides as a gate insulator. Tables 1-3 summarizes the recent published results for Si/SiGe FETs.

	L_g μm	Dop.	Ohmic	Gate	g_{mext} mS/mm RT	g_{mext} mS/mm 77	g_{mint} mS/mm RT	g_{mint} mS/mm 77	Gate Leakage RT	Gate Leakage 77	f_T GHz	f_{max} GHz	Notes
Daembkes et. al.	1.6	Sb	AuSb (330 C)	Pt/Ti/Au	40		70				2.2[1]		
König et. al.	1.4		P+ impl.(800 C) AuSb (340 C)	Pt/Ti/Au	80		88						
König et. al.	1.4		P+ impl.(750 C) AuSb (340 C)	Pt/Au	340	670	380	800	high[2]				pinchoff only at 77K
Ismail et. al	0.25	P	AuSb (120 C)	Pt	330	600			0.5 mA[3]	5μA			
König et. al.	1.0	Sb	P+ impl. AuSb (340 C)	Pt/Au	135	590							
Jackson et. al.	1.5	none	P+ impl. (600C) Al	WSi$_2$		100							Self-aligned
	2.5					150[4]							
Ismail et. al.	0.5	P	Au/Sb (120C)	Pt	390	520			1 μA/mm				0.3 V knee voltage[5]
Kuznstov et al.	0.5	Sb	P+ Impl. Al	Pt	130						7.5		310mS/mm after recess
Glück et al.	0.18	Sb	P+ Impl. Al	Pt/Au	270						46	81	D-mode
Welser et al.	10	NA	As+ Impl. Al	12 nm ox. N+ Poly	58								Surface MOSFET
Ismail et. al.	0.5	P	P+ Impl.	Pt and Pt/Ti/Al	417 327				<1 μA/mm				Integrated E-/D-mode

Table 1: Summary of published results for Si/SiGe n-type MODFETs [3].

---

[1] Estimated value.
[2] Mentioned only that the gate leakage current at room temperature was high.
[3] High gate current prevented complete pinch off.
[4] This device has a lower insulator (SiGe upper barrier) thickness compared to the 1.5 μm gate-length.
[5] This device have a highly doped cap layer which reduces R_s.

	L_g μm	Dop.	Ohmic	Channel	g_{mext} mS/mm RT	g_{mext} mS/mm 77	Mode Enh.	Mode Dep.	Gate Structure Oxide	Gate Structure Gate mat.	f_T GHz	f_{max} GHz	Notes
Nayak et. al.	0.7		B+ (770C) Al (400C)	buried SiGe	64		✓		5 nm therm.	poly Si			
V.-Vandebroek et. al.	0.9	B	Sb preamorph. B (600C)	buried SiGe			✓		7 nm PECVD	n+ poly Si			reported mobility
Kesan et. al.	0.25	None	Sb preamorph. B (600C) Ti/Al	buried SiGe	167	201[1]	✓		7.1 nm Thermal & PECVD	TiSi$_2$			
Nayak et. al.	10 / 1		B (550C-850C) Al-1%Si (400C)	buried SiGe	18 / 80			✓	6.5 nm WRTO[2]	p+ poly			SIMOX substrate
Bhaumik et. al.	0.15	B	BF$_2$ (800C)		83				6 nm wet therm.	n+ poly Si	23	35	
Murakami et. al.	4	Ga	AuGa (330C)	buried Ge		50[3]	✓		50 nm CVD	Al			Ge substrate
Li et. al.	1		BF$_2$ (700C) Al	surface SiGe	48	60	✓		10 nm ECR[4]	Al			

Table 2: Summary of published results for Si/SiGe p-type MOSFETs [4].

[1] Measured at 82 K
[2] Wet Rapid Thermal Oxidation
[3] This was a g_{mint} that was calculated from unmentioned g_{mext}.
[4] Electron Cyclotron Resonance Plasma

	L_g μm	Dop.	Ohmic	Gate	g_{mext} mS/mm RT	g_{mext} mS/mm 77	g_{mint} mS/mm RT	g_{mint} mS/mm 77	Gate Leakage RT	Gate Leakage 77	f_T GHz	f_{max} GHz	Notes
Pearsall et. al.	3	B	BF$_2$ (700C) Al	Ti	2.5								
König et. al.	1.4	B	BF$_2$ (700C) Ti/Au	Ti/Au	34	67	45[1]	90[1]					
	0.55				37	103	60[1]	150[1]					
König et. al.	1.2	Ga	BF$_2$ (620C) Cr/Au	Ti/Au	125	290			high leakage[2]	reasonable[2]			pure Ge channel
Arafa et. al.	1.5	B	Pd/Al/Pd (450 C)	Ti/Au	95		138		few μA/mm		2.1		
Arafa et. al.	1.0 0.7	B		Ti/Au	105	205			4 nA/mm		5.3 9.5	10 17.8	
Arafa et. al.	0.25	B	Pt (350 C)	Ti/Pt/Au	230		278				24	37	
Adesida et. al.	0.25	B	Pt (350 C)	Ti/Mo/Au	250		285		<1 μA/mm		40	45	self-aligned
Arafa et. al.	0.1	B	Pt (350 C)	Ti/Mo/Pt/Au	257		295		<1 μA/mm		70	55	self-aligned

Table 3: Summary of published results for Si/SiGe p-type MODFETs [5].

---

[1] Values are estimated from a graph.
[2] From I-V characteristics.

A record high performance was achieved for p-type MODFETs. This was accomplished for a T-gate depletion-mode device with a shadow-evaporated self-aligned source and drain contacts. The enabling processes for such a device was the possibility of obtaining a low access resistance using a shallow low anneal temperature Pt contacts. The gate metal stack was engineered to provide a low resistance by having a T-shape with a gold top layer. A Mo barrier layer in the stack was found to prevent Au indiffusion at the temperature required to anneal the contacts. For this device a high performance was achieved at a relatively low drain bias. Figure 1 shows the dependence of the unity current gain cutoff frequency (f_T) for self-aligned p-type SiGe MODFETs with various gate lengths on the drain bias. It is clear that the high frequency performance of the fabricated Si/SiGe devices is maintained at low V_{ds}. For example, a unity current gain cutoff frequency of 60 GHz can be achieved for a 0.1 μm self-aligned SiGe MODFET at a drain bias of only -0.3 V. This is explained by the fact that the critical field at which carrier velocity saturates is much lower for strained SiGe. Below this field the hole velocity is 4-5 times higher than Si. For high longitudinal electric field this enhancement in the hole velocity starts to diminish [6]. As the length of the devices decreases the region having low electric field decreases. This puts more weight on the high-field part of the velocity field characteristics, hence, getting the performance of SiGe MODFETs closer to their MOSFET counterparts. Figure 2 compares the maximum extrinsic transconductance of a SiGe MODFET to state-of-the art Si PMOS. Although deep submicron SiGe MODFETs may not show enhanced transconductance when compared to Si MOSFETs, they are capable of

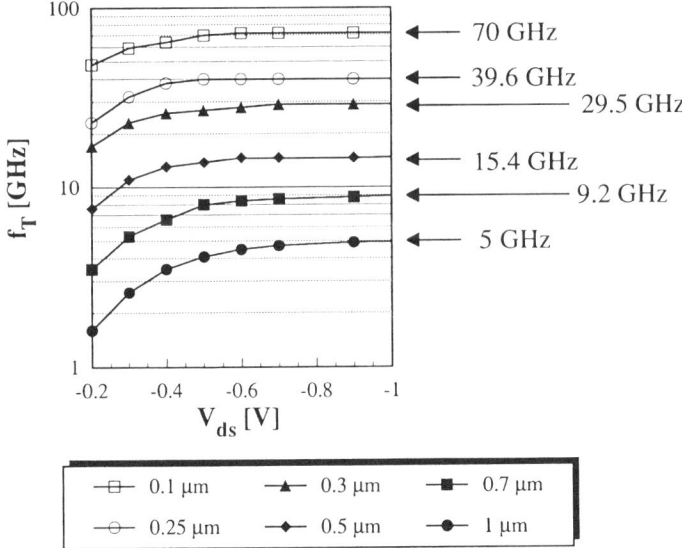

Figure 1:   The dependence of the unity current gain cutoff frequency (f_T) on the drain bias (V_{ds}) for various gate lengths of fabricated self-aligned SiGe p-type MODFETs.

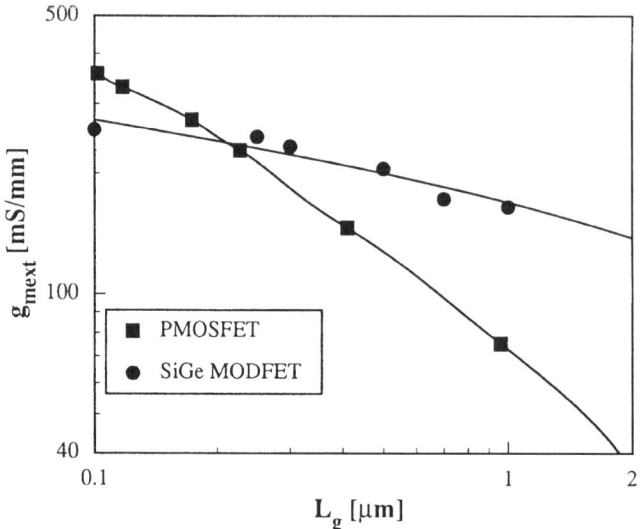

Figure 2: Comparison of the extrinsic transconductance of the self-aligned p-type SiGe MODFETs and state-of-the-art MOSFETs with 35 Å gate oxide [7]

achieving a high performance at a much lower drain bias. This allows the scaling of the power supply voltage which improves the reliability of the devices and leads to a power consumption savings.

MATERIAL ISSUES

Achieving the previously mentioned performance for large scale circuits, requires a significant amount of breakthroughs to solve material related problems. Most of the devices having extremely high performance are fabricated on relaxed SiGe buffers (virtual substrate). These buffers are necessary for various reasons:

- They allow for the growth of strained Si channels which are necessary for achieving high electron mobility.
- They allow for an increase of the Ge content of the channel without the need to rely on psuedomorphic layers (strained layers having a thickness higher than the critical thickness). This results in a better confinement as well as a higher hole mobility.
- The integration of high-performance p-type and n-type devices requires a substrate that can accommodate both strained n- and p-type devices. Figure 3 illustrates the variation in the conduction- and valence-band discontinuity for strained layers grown on relaxed $Si_{0.7}Ge_{0.3}$. By growing a thin strained Si channel a difference of the conduction-band edge can be achieved, while the growth of a higher Ge concentration thin layer results in a difference in the valence-band edge. The difference in the band offset is between 220 meV and 300 meV. These values are high enough to confine both electrons and holes to their respective channels at room temperature.

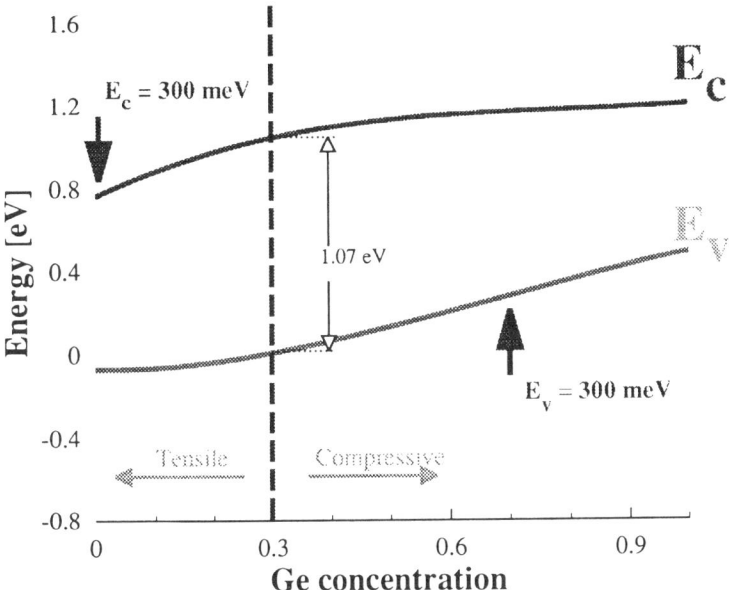

Figure 3: Band alignment for strained $Si_{1-x}Ge_x$ grown on top of relaxed $Si_{0.7}Ge_{0.3}$ [8]

The relaxed buffers impose a big manufacturability problem. First, The occurrence of dislocation is eminent because of the difference in lattice constant between the buffer and the substrate. If these dislocations were to propagate to the active layer of the device they may have significant impact on the carrier mobility. Strain engineering can be used to reduce the number of threading dislocations and misfit dislocation at the channel interface [9]. Although the effect of all kinds of dislocations on mobility will vanish as the average distance between dislocations becomes higher than the carriers mean free path, these dislocations will alter the characteristics of devices when they fall within their active areas. This may result in an increase in leakage currents or a shift in the threshold voltage for the affected devices hence hampering the realization of large scale integration.

Ideally misfit dislocations can be made to terminate on the outer edge of the wafer by growth conditions optimization [10] or by using selective epitaxy on small islands [11]. Even if these dislocations were to be prevented completely from threading to the active layer, the accumulation of these dislocations at the bottom interface of the relaxed buffer leads to a periodic stress concentration resulting in surface roughness [12]. Figure 4 demonstrate the extent of this roughness for a heterostructure grown on a relaxed $Si_{0.7}Ge_{0.3}$ buffer and having a hole mobility of 700 cm^2/V sec at room temperature. A clear cross hatching pattern is visible with optical microscopes. The periodicity of this pattern is 2.2 μm and the difference in height between peaks and valleys is approximately 300 A. A significant reduction in the surface roughness is required before high-performance strained Si/SiGe MODFETs can be used for circuit applications.

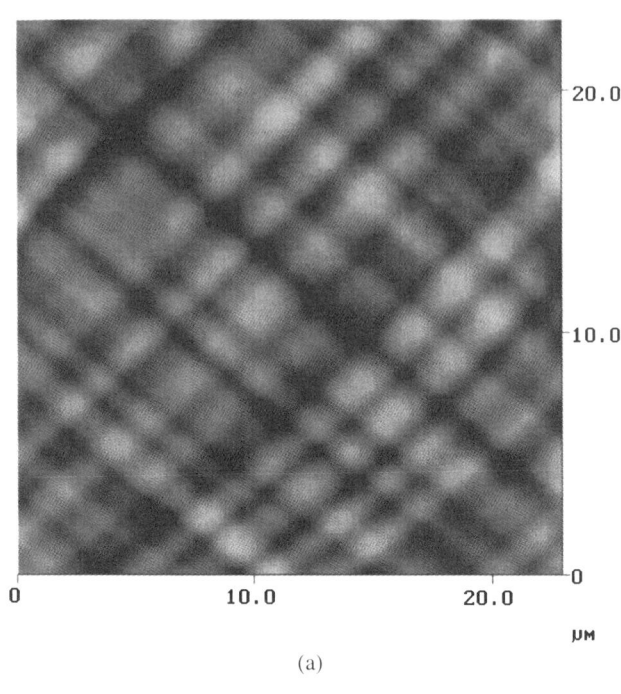

(a)

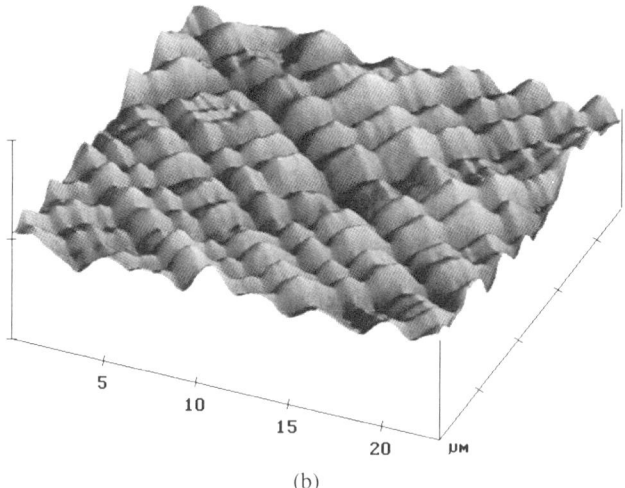

(b)

Figure 4: (a) Two-dimensional and (b) Three-dimensional plots of an atomic force microscopy scan of the surface of the heterostructure used for the fabrication of record-high speed SiGe p-type MODFETs.

CONCLUSION

A summary of reported results for p- and n-type Si/SiGe FETs was presented. The main advantages and limitations of current strained Si/SiGe heterostructure was discussed. As the FETS are scaled towards smaller dimensions, the enhancement in mobility allows the scaling of the power supply while maintaining comparable performance to Si CMOS biased at a relatively high drain voltage. This feature will improve both the reliability and the power consumption of the fabricated circuits.

ACKNOWLEDGEMENT

This work was funded by NSF Grant ECS 97-10418 and DARPA Grant N66001-97-1-8906 (Program Monitor -- Dr. Cynthia Hanson).

REFRENCES

1. M. Arafa, K. Ismail, J. O. Chu, B. S. Meyerson, and I. Adesida, IEEE Electron Dev. Lett., **17** (12), 586 (1996); M. Arafa, Ph.D. thesis, University of Illinois at Urbana-Champaign, 1997.

2. M. Glück, T. Hackbarth, U. König, A. Haas, G. Hock, E. Kohn, Electronics Lett., **33** (4), 335, (1997).

3. Heinrich Daembkes et al. , IEEE Trans. on Electron Devices, **ED-33**(5), 63 (1986); U. König et al., Electronics Lett., **27** (16), 1405 (1991); U. König et al., Electronics Lett., 160 (1992); K. Ismail et al., IEEE Trans. on Electron Devices, **13** (5), 229 (1992); U. König et al., IEEE Electron Device Lett., **14** (3), 97 (1993); T. N. Jackson et al., IEEE Trans. on Electron Devices, **40** (11), 2104 (1993); K. Ismail et al., IEEE Electron Device Lett., **14** (7), 348 (1993); V. I. Kuznestov et al., J. Vac. Sci. Technol. B, **13** (6), 1353 (1995); M. Glück et al., Electronics Lett., **33** (4), 335 (1997); J. Welser et al., IEEE Electron Device Lett., **15** (3), 100 (1994); K. Ismail et al., IEEE Electron Device Lett., **18** (9), 435 (1994).

4. T. P. Pearsall et al., IEEE Electron Device Lett., **EDL-7** (5), 308(1986); U. König et al., Electronics Lett., **29** (5), 486 (1993); U. König et al., IEEE Electron Device Lett., **14** (4), 205 (1993); M. Arafa et al., Electronics Lett., **31** (8), 680 (1995); M. Arafa at al., IEEE Electron Device Lett., **17** (3), 124 (1996); M. Arafa et al., IEEE Electron Device Lett., **17** (9), 449 (1996); I. Adesida et al., Microelectronic Engineering, **35**, 257 (1997); M. Arafa et al., IEEE Electron Dev. Lett., **17** (12), 586 (1996).

5. D. K. Nayak et al., IEEE Electron Device Lett., **12** (4), 154 (1994); Sophie Verdonckt-Vandebroek et al., IEEE Electron Device Lett., **12** (8), 447 (1991); Sophie Verdonckt-Vandebroek et al., IEEE Trans. on Electron Devices, **41** (1), 90 (1994); V. P. Kesan et al., Proceedings of IEDM-91, 25 (1991); D. K. Nayak et al., Proceedings of IEDM-92, 777 (1992); D. K. Nayak et al., IEEE Electron Device Lett., **14** (11), 520 (1993); Kaushik Bhaumik et al., Proc. of the 1993 Int. Semiconductor Device Research Symposium ISDRS, 349 (1993); E. Murakami et al., IEEE Trans. on Electron Devices, **41** (5), 857 (1994).

6. K. Ismail, IEDM'95, Washington DC (unpublished).

7. Y. Taur, S. Cohen, S. Wind, T. Lii, C. Hsu, D. Quinlan, C. Chang, D. Buchanan, P. Agnello, Y. Mii, C. reeves, A. Acovic, V. Kesan, Proceedings of IEDM'92, 901-904 (1992).

8. Martin M. Reiger, and P. Vogl, Physical Review B, **48** (19), 14276-14278 (1993).

9. K. Ismail, Solid State Phenomena, **47-48**, 409-418 (1996).

10. F. K. LeGoues, MRS Bulletin /April, 38-44 (1996).

11. H. Richter, A. Fisher, G. Kissinger, D. Kruger, Proceedings of the International Semiconductor Conference, CAS, **1**, 41-50 (1996).
12. R. M. Feenstra, M. A. Lutz, Frank Stern, K. Ismail, P. M. Mooney, F. K. LeGoues, C. Stanis, J. O. Chu, and B. S. Meyerson, J. of Vac. Science and Technology B, **13** (4), 1608-1612 (1995).

EVIDENCE OF INTERDIFFUSION EFFECT IN STACKED POLYCRYSTALLINE SiGe/ Si LAYERS FOR CMOS GATE APPLICATION

C. HERNANDEZ, Y. CAMPIDELLI, I. SAGNES, E. HENRISEY, F. MARTIN*,
D. BENSAHEL
France Telecom, Branche Développement, CNET, BP 98, 38243, Meylan-Cedex, France
* CEA-LETI, rue des Martyrs, 38054, Grenoble-Cedex 9, France

ABSTRACT :
Poly-SiGe stacked gates with Ge content ([Ge])varying between zero and 100% have been fabricated using an industriel single-wafer machine. These poly-SiGe layers were characterised and fully integrated in a 0.18 μm CMOS process. Interdiffusion of Si and Ge upon subsequent annealing of the structure has been observed and studied. This interdiffusion effect was found to be responsible for the discrepancy observed between theoretical and practical values of the Ge workfunction ϕ_{ms} evaluated from our electrical measurements and from those of different authors. A technique for the limitation of this interdiffusion effect has then been developped and is described.

INTRODUCTION :
Polycrystalline $Si_{1-x}Ge_x$ (hereafter poly-SiGe) has been proposed recently as a candidate in advanced CMOS technologies [1-2]. Its workfunction ϕ_{ms} in a MOS architecture, typically a $Si_{substrate}$/gate oxide/gate material technological structure, can be theoretically tuned between 5.2eV (p^+poly-Si, i.e. [Ge]=0%) and 4.7 eV (p^+poly-Ge, i.e. [Ge]=100%). In this last case, a mid-gap material is obtained with respect to Si, i.e. the Fermi level in p+poly-Ge being pinned to mid-gap of Si, resulting for instance, in fewer technological steps in CMOS process integration. However, for an industrial application, besides the general interest of this new mid-gap gate material, several issues must be addressed such as : (i) a sequence of material fabrication performed in an actual industrial machine, and (ii) a process flow as much as possible remaining compatible with conventional silicon CMOS technology. This paper presents relevant solutions of both (i) the key material issues, i.e. the layer fabrication in an acceptable time and that with a minimised surface roughness, and (ii) the influence of the thermal budget on the electrical properties of this new generation of gate material. It should be noted that other pertinent parameters such as etching of poly-SiGe and its passivation for spacers formation in a CMOS process have been discussed elsewhere [3-4] and will not be included in the present paper.

MATERIAL ISSUES
- Wetting properties :
Poly-SiGe layers are deposited as poly-Si ones by a CVD technique. However, since Ge is prone to agglomerate on an oxidised surface [5-8], a classical deposition scheme, i.e. a direct deposition of the SiGe from the gaseous precursors SiH_4 and GeH_4 with H_2 as the carrier gas, results in a too high surface roughness of the layer. A Si precursor layer must be deposited first as shown by several authors [5-8]. Figures 1a and 1b present the effect of the thickness of such a layer on the morphology of a pure poly-Ge, all the CVD parameters being the same for the two deposited poly-Ge layers. In our case, the reproducibility of this thin Si layer was assessed through a Contact Angle technique as shown in Figure 2 : the contact angle evolution of two different media (water and diiodomethane) reflects the progressive covering of the SiO_2 surface with Si deposit. This technique seems more appropriate than ellipsometry (dramatic effect for

the diiodomethane in Figure 2), since no assumption is made on the surface state, roughness and optical indexes.

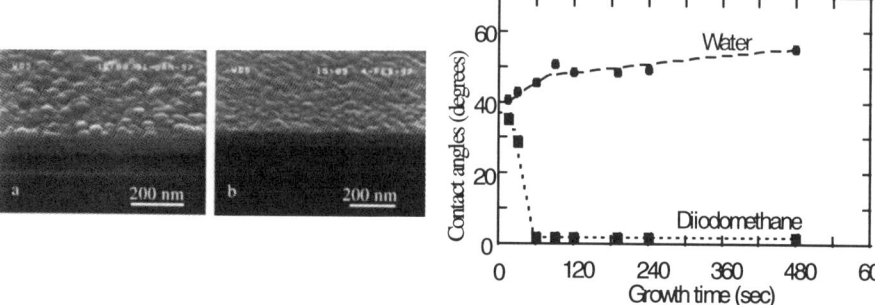

Figure 1 : Roughness of poly-Ge layer with two different times of a wetting Si layer (a) 15 s, (b) 30 s.

Figure 2 : Contact angle measurements as a function of wetting Si layer deposition time.

Surface roughness is not only due to the starting state of the layer. Even with a "wetting" layer, poly-SiGe deposition brings about a surface roughness. Figure 3 presents the "rms" values of several poly-SiGe layers of different [Ge] and thickness. The "rms" values increase both with [Ge] and layer thickness. As shown in Figure 3, it is then preferable to use the lowest SiGe thickness (lowest surface roughness), since the ϕ_{ms} is established at the interface SiGe/SiO$_2$ or very nearby. For practical application in a CMOS gate, poly-SiGe layers will be typically lower than 60 nm.

- Arrhenius plot :

Figure 4 presents the Arrhenius plot of several poly-SiGe layers with different Ge contents. Depositions are performed in a dilute gaseous phase at atmospheric pressure in order to increase the growth rate. In the low temperature surface activated regime, a decrease of the apparent activation energy with increasing [Ge] is observed, in accordance with classical theories : the regime is governed by hydrogen desorption catalysed by the Ge presence. As not discussed here but described in Ref [7], for higher temperature the GR of poly SiGe with Ge content up to 55% is greater than that of poly-Ge. Since Ge is very sensitive to oxygen presence (the GeO$_x$ sub-oxides being mostly volatile), it was found necessary to cap the poly-SiGe layer by a Si deposited layer for further technological processing. Poly-SiGe growth rate is higher than Si deposition at the same temperature and it was found necessary to control precisely both the different SiGe and Si growth rate and the crystallinity state (polycrystalline, amorphous, columnar...) of the two layers. To fulfil this point, a single-wafer deposition system, in our case, an ASM Paragon machine [9] in which temperature can be rapidly switched between two adjacent steps was used. The final resulting structure developed for the gate was then the following : a "wetting" Si layer, a poly-SiGe layer lower than 50 nm thick, and a capped Si layer, amorphous or polycrystalline for the completion of an entire 200 nm total thickness of the gate. This balance between the two layers also strongly reduces the problems occurring in gate etching definition and spacer formation [3-4].

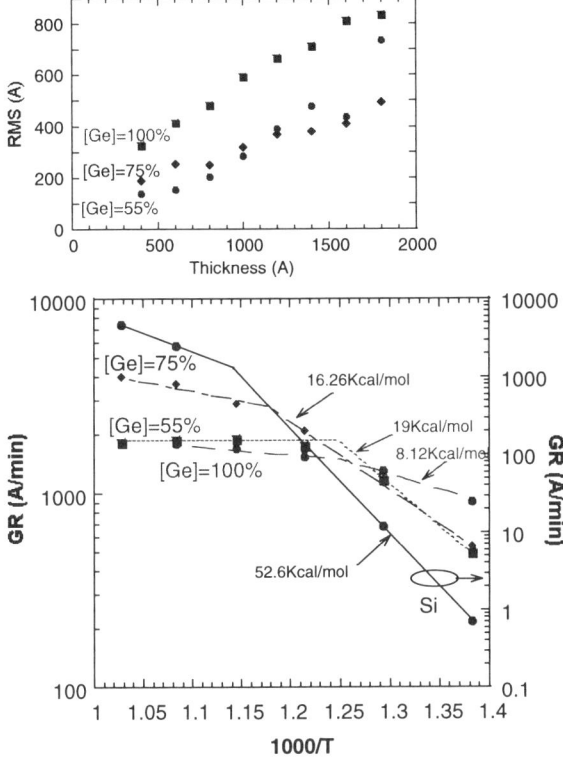

Figure 3 : Surface roughness measured by atomic force microscopy as a function of layer thickness.

Figure 4 : Arrhenius plot for poly-Si and poly-SiGe versus the reciprocal temperature.

THERMAL BUDGET ISSUES

In a CMOS process, front-end steps such as the gate stack, undergo several thermal treatments : deposition of spacers, dopant drive-in and activation, back-end processing for interconnections. In Figures 5a, 5b and 5c, SIMS profiles of as-deposited and annealed stacked structures are presented. The thermal budget was the following : 800°C/1 h (simulating the drive-in and different deposited layers) plus a Rapid Thermal Annealing >900°C/30 sec (simulating Source and Drain activation). Figure 5a presents the case of an as-deposited poly-Ge/Si cap stack implanted with boron to simulate a CMOS gate, and Figures 5b and 5c present an equivalent structure annealed, the difference between 5b and 5c being the state of the Si cap layer, polycrystalline and amorphous respectively. First of all, boron is well redistributed and no problem is then expected from this point of view. The main striking effect is the observation of a Ge tail extending deeply in the Si layer, the Ge extension being less effective in the case of amorphous-Si cap deposition than in the polycrystalline case. Meanwhile, a diffusion of Si from the cap into the polycrystalline Ge layer is also observed, resulting in the final stage at an apparent lowering of the [Ge] content in the layer. TEM analysis (not presented here) were carried out on these samples. The whole crystallisation of the stack is observed, and grain growth enhanced in the case of poly-crystalline Si for the cap layer. Moreover, the Ge profile was recorded by local XRays-TEM analysis showing a redistribution of Ge. This leads us to conclude that Ge was diffusing through the grain boundaries of the Si cap (and vice-versa), this diffusion being more impeded in the case of the amorphous Si cap.

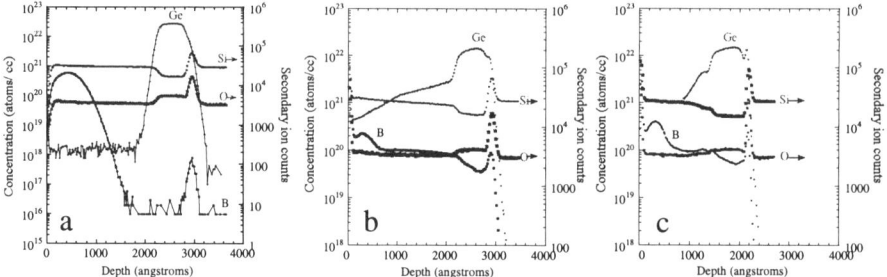

Figure 5: SIMS profiles of SiGe/ Si bilayer (a) before annealing and (b), (c) after annealing (800°C/ 1h + 950°C/ 30 sec) with a (b) poly-Si cap and (c) an amorphous Si cap. The Si cap thickness is not the same for all samples. The total [Ge] at the left Y-axis is only qualitative.

This interdiffusion has several drawbacks : as shown below, the lowering of the [Ge] will affect the electrical properties. But it also affects the integration of this stack in a CMOS process : we have proven that the presence of Ge at the surface of the cap layer had deleterious effect on the Ti silicidation by Physical Vapor Deposition, the worst being a break-up of the $TiSi_2$ formed film.

EFFECT OF Si-Ge INTERDIFFUSION ON THE ELECTRICAL PROPERTIES

Several capacitors and batches of 0.18 μm CMOS wafers were fabricated using the stacked gate described above. Different [Ge] in the SiGe layer were used : 0, 55 and 100%. Figure 6 presents the ΔV_{fb} variation deduced from capacitance measurements of a pure p^+ poly-Si gate and a p^+poly-SiGe/Si cap one. The values fit with theoretical values for 55% [Ge]. For the 100% [Ge] gate, we observe a decrease of ΔV_{fb} (compared to theoretical values) due to the Si-Ge interdiffusion effect : in order to amplify this effect, we choose a high thermal budget for the three 100% poly-Ge samples. SIMS profiles confirm that interdiffusion was present and even in the highest case, a slight penetration of Ge into the thin oxide gate (4nm thick). Same results were observed on CMOS 0.18 μm transistors as shown in Figure 7, which presents the threshold voltage values of pMOS transistors. To our opinion, these results enlighten the discrepancies observed in earlier published results on the use of the poly-SiGe gate [10] : for high [Ge] in the poly-SiGe layer (say, above 50%), the different ϕ_{ms} values reported vary and this is due to the Si-Ge interdiffusion effect as demonstrated above.

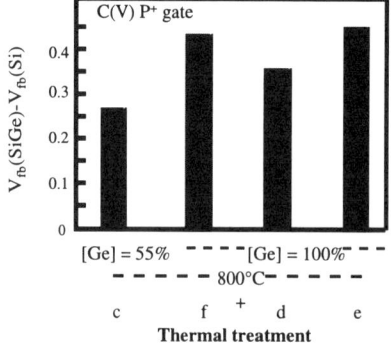

Figure 6 : Flatband voltage shift as function of [Ge] and thermal budget. All wafers received an 800°C/ 1h anneal. Additional anneal time and temperature are shown below each bar.

c=1000°C/20 sec
d=950°C/30 sec
e=900°C/30 sec
f= no extra anneal

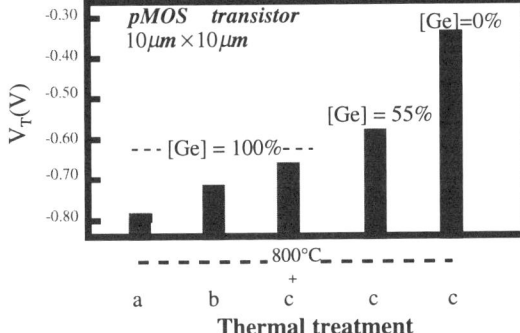

Figure 7 : Threshold voltage as a function of Ge content and thermal budget on pMOS transistors. All wafers received an 800°C/ 1h anneal. Additional annealing time and temperature are shown below each bar.

a=900°C/20 sec
b=950°C/20 sec
c=1000°C/20 sec

REDUCTION OF THE INTERDIFFUSION EFFECT

Under our annealing conditions [(800°C, 1 h) + (900°C, 60 sec)], a maximum of $\Delta\phi_{ms}$ of 440 meV has been obtained, instead of around 550 meV as predicted by theory. In order to reach this value, the easiest way is to decrease the thermal budget. However, even if the doping of the gate can be obtained with in-situ doping and low thermal budget, the source and drain activation need so far moderately elevated thermal budget. It is then of interest to develop a process to limit the interdiffusion effect. For this purpose, we deliberately introduce a reproducible, thin layer of oxide between the two layers (poly-SiGe and Si cap) <0.8nm. Such an approach has been used by H. Ito et al. [11] in the case of a pure poly-Si doped gate. In our case, the layer fabrication is described elsewhere [12] and Figure 8 presents the SIMS profiles obtained after annealing at (800°C/ 1h + 950°C/ 30 sec). We observe a serious limitation of the interdiffusion effect as depicted by the rapid decrease of the Ge profile compared to the case where the interfacial layer is not present. This layer has no influence on the boron diffusion. Work is in progress to electrically test the viability of this approach.

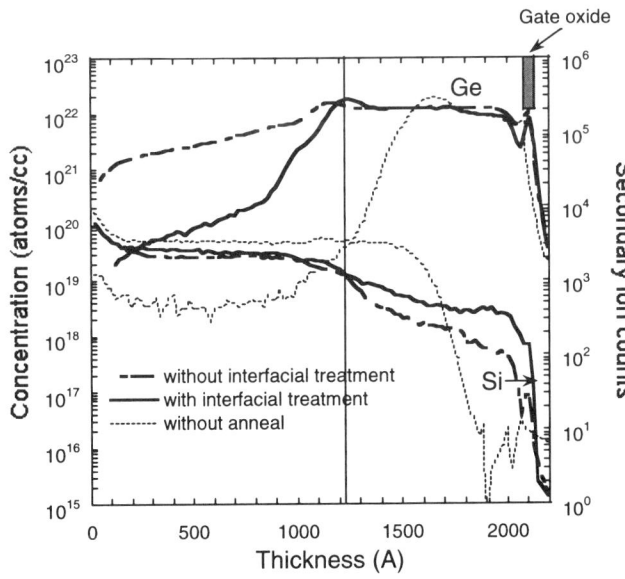

Figure 8 : SIMS profiles of Si/ SiGe bilayer before and after annealing (800°C/ 1h + 950°C/ 30 sec) with and without interfacial treatment. The vertical line indicates the interfacial layer position.

CONCLUSION :

We have shown that a stacked poly-SiGe/cap Si layer structure can be used advantageously as a new gate material in CMOS processing. This stack structure is compatible with classical Si technological steps. With increasing [Ge], interdiffusion of Si and Ge is found to occur upon further annealing through the grain boundaries of the different layers and this effect decreases the maximum values of $\Delta\phi_{ms}$ targeted. Thermal budget must then be reduced in order to obtain high values of $\Delta\phi_{ms}$. A technique devoted at the limitation of the interdiffusion effect has been proposed and has been successfully implemented and tested by physical characterisation.

ACKNOWLEDGEMENTS :

This work has been carried out within the GRESSI consortium between CEA-LETI and France Telecom-CNET. It has also been partly founded by the european SEA N° 20.331 SIDOSI and N° 23.806 ULTRA projects. C. Patel is also kindly acknowledged for critical reading of the script.

REFERENCES :

[1] T. J. King, J. R. Pfiester, J. D. Shott, J. P. McVittie, K. C. Saraswat, IEDM Techn. Digest, 253 (1990).

[2] T. Skotnicki, P. Bouillon, R. Gwoziecki, A. Halimaoui, C. Mourrain, I. Sagnes, J.L.Regolini, O. Joubert, M. Paoli, P. Schiavone, ESSDERC'97, 216 (1997).

[3] C. Monget, S.Vallon, F. H. Bell, L.Vallier, and O. Joubert, J. Electrochem. Soc. 144, 2455 (1997).

[4] S.Vallon, C. Monget, O. Joubert, L.Vallier, F. H. Bell, M. Pons, J.-L. Regolini, C. Morin, and I. Sagnes, J. Vac. Sci. Technol. A 15(4), 1874 (1997).

[5] M. C. Öztürk, D. T. Grider, J. J. Wortman, M. A. Littlejohn, Y. Zhong, D. Batchelor and P.Russell, Journal of Electronis Materials 19, 1129 (1990).

[6] V. Z-Q. Li, M. R. Mirabedini, R. T. Kuehn, D. Gladden, D. Batchelor, K. Christenson, J.J.Wortman, M. C. Ozturk, and D. M. Maher in Polycrystalline Thin Films : Structure, Texture, Properties, and Applications II, edited by H. J. Frost, M. A. Parker, C. A. Ross, E.A.Holm, (Mater. Res. Soc. Proc., Pittsburgh, PA, 1996) pp.333-338.

[7] S. Bodnar, C. Morin, J. L. Regolini, Thin Solid Films, 294, 11 (1997).

[8] H. C. Lin, C. Y. Chang, W. H. Chen, W. C. Tsai, T. C. Chang, T. G. Jung, and H. Y. Lin, J.Electrochem. Soc. 141, 2559 (1994).

[9] D. Bensahel, Y. Campidelli, C. Hernandez, F. Martin, I. Sagnes, D. J. Meyer, Solid State Technology, 41(3), S5 (1998).

[10] J. Alieu, R. Gwoziecki, M. Paoli, T. Skotnicki, C. Hernandez, F. Martin, C. Mourrain, D.Bensahel, M.-T. Basso, J. Galvier and M. Haond, VLSI Symp. Tech. Dig. (1997). (to be published).

[11] H. Ito, M. Sasaki, N. kimizuka, K. Uwasawa, N. Nakamura, T. Ito, Y. Goto, A. Tsuboi, S.Watanuki, T. Ueda, T. Horiuchi, IEDM Techn. Digest, 635 (1997).

[12] A. Halimaoui, E. Henrisey, C. Hernandez, J. Martins, M. Paoli, M. Regache, L. Vallier, D.Bensahel, B. Blanchard, D. Rouchon, F. Martin, presented at 1998 MRS Spring Meeting, San Francisco, CA (1998). (to be published).

IMPACT OF THE Ge CONTENT ON THE RADIATION HARDNESS OF HETERO-JUNCTION DIODES IN SiGe STRAINED LAYERS

H. Ohyama*, E. Simoen**, C. Claeys**, Y. Takami***, K. Hayama*, T. Hakata*,
J. Tokuyama*, K. Kobayashi*, H. Sunaga****, J. Poortmans** and M. Caymax**

*Kumamoto National College of Technology, 2659-2 Nishigoshi Kumamoto, 861-1102 Japan
**IMEC, Kapeldreef 75, B-3001 Leuven, Belgium
***Rikkyo University, 2-5-1 Nagasaka Yokosuka Kanagawa, 240-0101 Japan
****Takasaki JAERI, 1233 Watanuki Takasaki Gunma, 370-1207 Japan

ABSTRACT

The degradation of the electrical performance of strained $Si_{1-x}Ge_x$ epitaxial diodes by 220-MeV carbon particles is reported and compared with the effect of 20-MeV alpha rays and 20-MeV protons. The macroscopic damage is studied in a broad fluence range and for different Ge contents, ranging from 8 to 16 %. It is shown that the radiation damage of carbon irradiated diodes is about one order of magnitude larger than that for alpha ray irradiation, which can be explained by considering the difference of the nonionizing energy loss (NIEL). It is observed that the reverse current at a fixed bias increases with increasing fluence, while the rate of increase decreases with increasing fluence and/or Ge content. The fact that a close to square root dependence exists between the boron deactivation in the diode depletion region, derived from capacitance-voltage measurements and the reverse current increase suggests that the device degradation is dominated by radiation induced deep levels associated with interstitial boron complexes.

INTRODUCTION

$Si_{1-x}Ge_x$ based heteroepitaxial devices offer some clear performance advantages compared with their silicon counterparts. For example, the higher hole mobility enables to fabricate high-speed devices and circuits. Hereby is the study of the radiation damage in $Si_{1-x}Ge_x$ epitaxial layers important for the realization of high-performance devices which have to operate in a radiation-rich environment. It is also important to understand the basic mechanisms related to the formation of processing induced defects, resulting for example from ion implantation in such layers. Although the results on electron, neutron and proton radiation damage in $Si_{1-x}Ge_x$ diodes and HBTs have been reported in the past [1,2], little is known on the damage induced by high energy particle irradiation, like e. g. alpha and carbon particles.

In the present paper, the radiation damage of strained $Si_{1-x}Ge_x$ epitaxial diodes, which are irradiated by 220-MeV carbon, is investigated. The degradation of device performance of strained layer $Si_{1-x}Ge_x$ diodes, subjected to 20-MeV alpha rays and protons, is also studied as a function of fluence and germanium content. It is shown that there exists a close correlation between the macroscopic diode degradation - in case the degree of B deactivation and the reverse current increase - and the microscopic damage, corresponding to the introduction rate of the deep-levels. The observed experimental trends are further supported by the calculation of the energy transfer per particle.

EXPERIMENTAL

Strained layer n^+-Si/p^+-$Si_{1-x}Ge_x$ epitaxial diodes have been fabricated on CZ silicon substrates, with a B doping density of about 10^{15} cm^{-3}. The strained $Si_{1-x}Ge_x$ epitaxial layers were grown on the silicon substrates using an ultra high vacuum chemical vapour deposition system (UHV CVD) at a growth temperature of 630 °C and a deposition pressure of 0.26 Pa. The B concentration was 7 x 10^{17} cm^{-3}. The germanium content of the epitaxial layer with nominal thickness 100 nm was x = 0.08, 0.12 and 0.16, respectively. A layer of n^{+} polysilicon was used to form the n^{+} region for diodes. The diode area was 10^{2} μm^{2}.

After dicing in chips of 5 x 5 mm^{2} size the devices were irradiated at room temperature by 220 carbon particles, a 20-MeV alpha ray or by 20-MeV protons in the AVF cyclotron in TIARA at the Takasaki Radiation Chemistry Research Establishment. The fluence of particles was varied between 10^{10} and 10^{13} 1/cm^{2}.

Current/voltage (I/V) and capacitance/voltage (C/V) characteristics of the diodes were measured before and after irradiation, with applied voltages ranging from -1 to 1 V and -1 to 0 V, respectively. The active boron concentration and the width of depletion layer in the Si$_{1-x}$Ge$_{x}$ epitaxial layer have been calculated from C/V measurements on the reverse biased diodes, assuming that the square of the capacitance is inversely proportional to the reverse voltage. The hole and electron capture levels in the Si$_{1-x}$Ge$_{x}$ epitaxial layer of the diodes were studied using deep level transient spectroscopy (DLTS) in the temperature range between 77 and 300 K. The emission rate window used in the DLTS measurements is chosen between 1.18 to 26.51 msec. The applied filling pulse was ranging from -1 to 0.5 V to observe electron capture levels, and from -1 to 0 V to observe hole capture levels.

RESULTS AND DISCUSSION

Degradation by carbon irradiation

Figures 1 (a) and (b) show the typical I/V and C/V characteristics for the x = 0.12 diodes, with different fluences of 220-MeV carbon. From Figure 1a it is noted that both the reverse and forward current increase with fluence, except for the largest value. The forward current is, however, lower after irradiation for a forward voltage (V$_{F}$) larger than 0.5 V. The reason for this is an increased resistivity of the Si substrate. Severe degradation is observed in Figure 1a for a fluence of 1 x 10^{13} 1/cm^{2}, resulting in a more or less symmetric characteristic and hence bad diode operation. It is also found from Figure 1b that the capacitance in the Si$_{1-x}$Ge$_{x}$ epitaxial layer decreases by irradiation due to the deactivation of the active boron atoms. The degradation of the performance increases with increasing carbon fluence, while the rate of increase decreases with increasing fluence. The deactivation of the boron is caused by the formation of interstitial boron as further discussed below. The associated trap levels are shown to be mainly responsible for the increase of the reverse current.

To investigate the radiation source dependence of the degradation of device performance, the damage coefficient of the reverse current and of the deactivated boron concentration at fixed fluence and reverse bias (K$_{I}$, K$_{B}$) are calculated by means of the following equations:

$$K_I = \frac{I_R(\phi)-I_R(0)}{\phi} \tag{1}$$

where ϕ is the fluence. I$_{R}$ (ϕ) and I$_{R}$ (0) are the reverse current at V$_{R}$ = - 0.8 V after and before irradiation, respectively, and

$$K_B = \frac{N_B(0) - N_B(\phi)}{\phi} \tag{2}$$

where N$_{B}$(0) and N$_{B}$(ϕ) are the boron concentration before and after irradiation, respectively.

Table 1 lists K_I and K_B for different radiation sources studied here and corresponding to a fluence of 10^{12} $1/\text{cm}^2$. Also shown are the calculated NIEL values. It is clear from the table that the carbon irradiation corresponds to the largest radiation damage. The difference in radiation damage can be clearly explained by the difference of energy transferred to the target atoms during irradiation.

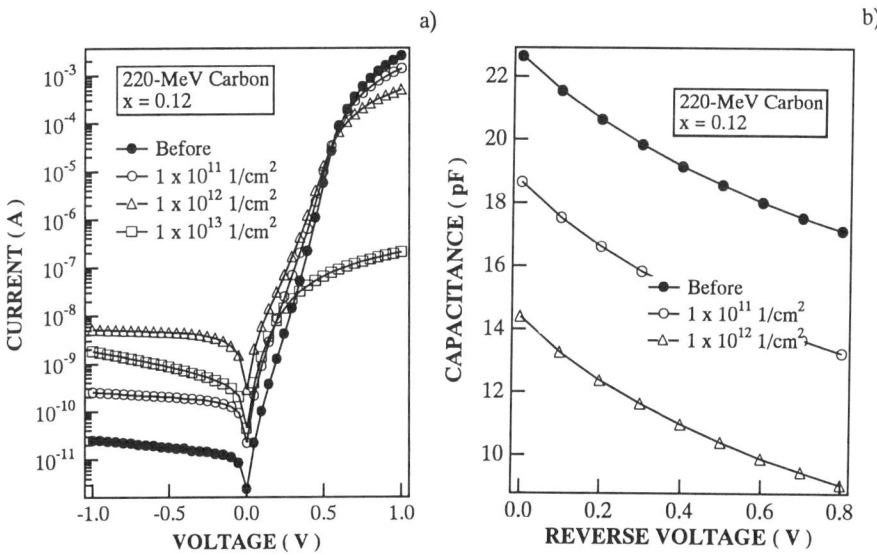

Figure 1 : Influence of 220-MeV carbon irradiation on I/V (a) and C/V (b) characteristics of x = 0.12 diodes.

Table 1: Damage coefficient of the reverse current and the deactivated boron concentration for 1 x 10^{12} $1/\text{cm}^2$ irradiations, and corresponding NIEL value for different radiation source.

	220-MeV carbon	20-MeV alpha rays	20-MeV protons
K_I (x 10^{-22} Acm^2)	40	6.6	0.95
K_B (x 10^5 1/cm)	20	1.6	0.47
NIEL (x 10^{-3} $\text{MeVcm}^2\text{g}^{-1}$)	169	93	8.5

Ge content dependence of radiation damage

It has been observed on several occasions that the Ge content plays a crucial role in the creation of radiation damage in SiGe layers [3-4]. Here, the relationship between both damage factors defined in Eqs (1) and (2) is investigated for 20-MeV alpha ray irradiation. Figure 2 (a) and (b) show the I/V and C/V characteristics of $Si_{1-x}Ge_x$ epitaxial diodes with $x = 0.12$, subjected to 20-MeV alpha rays. At first sight, a similar degradation behaviour as for the carbon particles, reported in Figure 1 is observed.

a)   b)

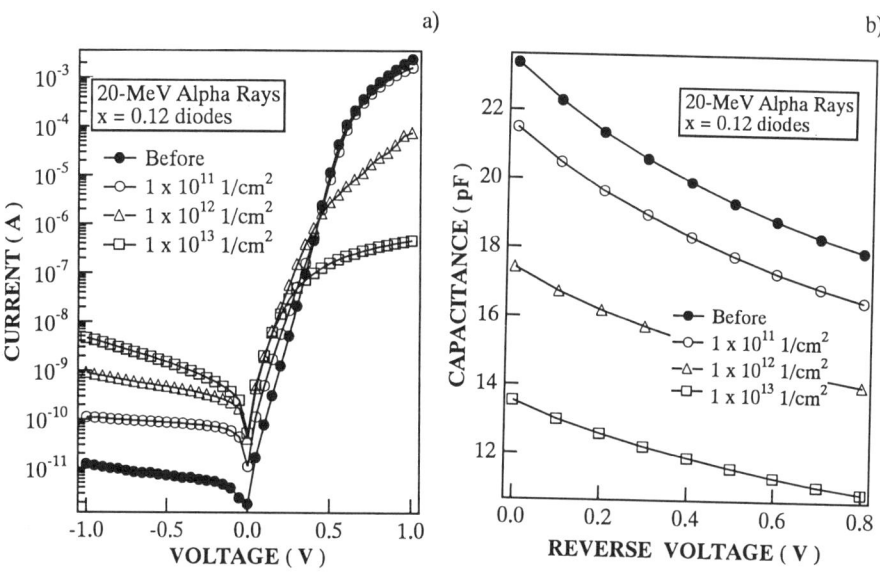

Figure 2 : I/V (a) and C/V (b) characteristics of $x = 0.12$ diodes after 20-MeV alpha ray irradiation.

Figures 3 (a) and (b) show K_I and K_B as a function of alpha ray fluence, respectively. From both figures it is clear that the radiation damage decreases with increasing Ge content for the same fluence. Combining all data points of both figures one can obtain the following empirical trend:

$$K_I = 1.58 \times 10^{-24} K_B^{0.49} \tag{3}$$

This square root curve suggests that the deactivated boron atoms are clearly related with the reverse current increase.

The DLTS spectra corresponding to a different Ge content for diodes subjected to 20-MeV alpha rays, for a fluence of 1×10^{13} $1/cm^2$, are shown in Figure 4. For $x = 0.08$, two hole capture levels and one electron capture level are induced, while for $x = 0.12$ and 0.16 only one hole capture level is induced. Following Monakhov et al. [3] one could ascribe the $H8_{13-1}$ level to the interstitial boron -substitutional carbon (B_iC_s) center. The electron capture level is probably associated with the double negative charge state of the divacancy center (V_2) [4]. The origin of the near mid-gap levels is not known at present. However, they could be related to the same defect center, whereby the activation energy shifts with increasing Ge content, similar as for other trap levels [3, 5]

a)   b)

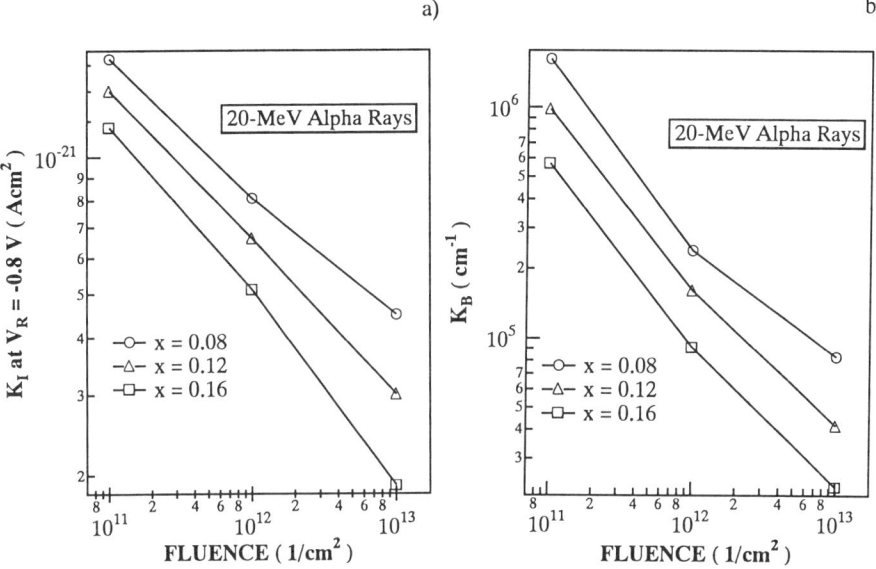

Figure 3 : K_I (a) and K_B (b) as a function of alpha ray fluence.

For the three different germanium contents shown in Fig. 4, the total defect density is estimated to be 2.4 x 10^{15}, 1.5 x 10^{15} and 0.8 x 10^{15} cm^{-3}, respectively. The corresponding global introduction rates R_T for a fluence of 10^{13} cm^{-2} are calculated to be about 100. Similar large introduction rates have been observed before, in the case of 1-MeV electron [1] and 2 MeV proton irradiation [6]. In the latter case, it was found that the trap introduction rate reduced from 90 1/cm for 5% Ge, down to 20 1/cm for 25% Ge. Taking into account the calculated number of knock on atoms in the range of 10^4 cm^{-3} per incident alpha particle, one arrives at about 1% of the primary vacancies and interstitials which escape initial recombination to form more or less stable irradiation defects. However, at lower fluences (10^{12} cm^{-2}), higher R_T values can be derived from the DLTS results, which are in the range 3000 1/cm (12% Ge) to 1500 (16% Ge), indicating that a significantly larger fraction of the primary damage participates in the formation of stable electrically active defect centres.

Combining the Fig. 3 and DLTS results, the B deactivation rate in the epilayer is approximately 300 times larger than the observed trap level concentrations. This result suggests that either some of the created B-related complexes give rise to deep levels occurring below 77 K or above 300 K, or that the displaced B atoms are in an electrically inactive clustered form. The fact that the amount of 'missing' B is increasing with increasing fluence supports the latter idea

CONCLUSIONS

The main conclusions which can be made from the present study:
1. The damage coefficients of electrical characteristics for carbon irradiation are larger than for alpha and proton irradiation. The radiation source dependence of the performance degradation is

thought to be attributed to the difference of mass and the probability of nuclear collision for the formation of lattice defects.

2. The degradation of the electrical performance of $Si_{1-x}Ge_x$ devices by 20-MeV alpha particles has been studied in a broad fluence range. A significant hardening effect has been observed by going from 8% Ge up to 16% Ge, which is related to a reduction of the introduction rate of the near midgap deep levels, or even the absence of deep levels, which were only observed in the lowest Ge content diodes. A close to square root dependence has been observed between the increase in the reverse diode current and the B deactivation in the $Si_{1-x}Ge_x$ epilayer. This points to the important role the interstitial B plays in the degradation of the device performance.

ACKNOWLEDGMENTS

Part of this work was supported by Giant-in-Aid for Scientific Research (No. 07555105 and No. 09045063) from the Japanese Ministry of Education for Science.

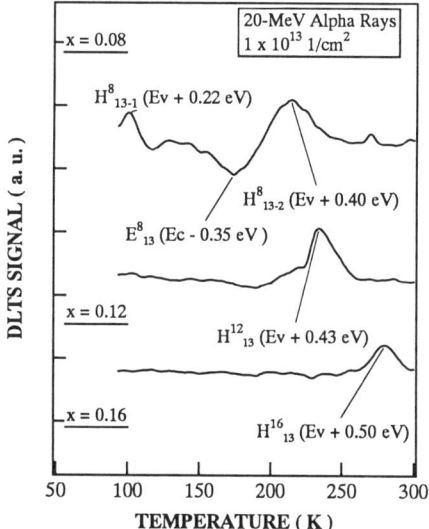

Figure 4 : DLTS spectra for different Ge contents and corresponding to a fluence of 10^{13} $1/cm^2$ 20-MeV alpha's.

REFERENCES

1. H. Ohyama et al., IEEE Trans. Nucl. Sci. **41**, 487 (1994).
2. H. Ohyama at al., Appl. Phys. Lett. **69**, 2429 (1996).
3. E. V. Monakhov et al., J. Appl. Phys. **81**, 1180 (1997).
4. J. J. Goubet et al., Appl. Phys. Lett. **66**, 1409 (1995).
5. S. A. Goodman et al., Journal of Electronics Materials **26**, 463 (1997).
6. P. Kringhøj et al., Phys. Rev. B. **52**, 16333 (1995).

EPITAXIALLY GROWN N+ PHOSPHORUS COLLECTOR PEAKS IN HIGH-FREQUENCY HBT'S WITH IMPLANTED EMITTERS

C.C.G. VISSER, L.K. NANVER, A. VAN DEN BOGAARD
Laboratory of ECTM, DIMES, Delft University of Technology, Feldmannweg 17, 2628 CB Delft, The Netherlands, visser@dimes.tudelft.nl

ABSTRACT

The boron out-diffusion into the Si collector of SiGe HBT's with implanted emitters is compensated by an epitaxially grown n+ phosphorus peak effectively positioned adjacent to the as-grown boron peak. Detrimental barrier formation is suppressed in devices with base sheet resistance 3.3 kΩ and cut-off frequency of 31 GHz.

INTRODUCTION

High-frequency SiGe HBT's with implanted phosphorus emitters can be produced by using high concentration SiGe (30%) in combination with 700 °C thermal annealing [1, 2]. At this temperature, the extremely low diffusivity of both boron and phosphorus in $Si_{0.7}Ge_{0.3}$ limits the implantation damage induced transient enhanced diffusion (TED) in the SiGe region. The resulting doping profiles have been used to produce devices with high current gain, an intrinsic base sheet resistance of 7 kΩ and $f_T > 40$ GHz. For devices with lower base sheet resistance, boron out-diffusion on the collector side must be compensated to avoid detrimental potential barriers. A phosphorus doped n+ peak, grown just before the SiGe base, is demonstrated to be suitable for this purpose.

PROCESSING

All layers are grown in a commercially available single wafer epitaxial reactor. The wafers were loaded and unloaded in a nitrogen-purged loadlock, and then placed on a SiC coated graphite susceptor. The quartz reaction chamber is lamp heated. Before each deposition the chamber was cleaned with HCl at 1150 °C. Wafers are prebaked in H_2 at 1120 °C for 2 min.

The basic layer structure that is examined here is shown in Fig. 1. A silicon buffer layer is grown at 1050 °C and 60 Torr. The phosphorus doped layers, the 325 Å strained SiGe layer containing the boron doped base, and the Si cap layer are all deposited at 700 °C and atmospheric pressure. The SiGe layers are grown from $SiCl_2H_2$ and GeH_4, with H_2 as carrier gas. Dopant sources are respectively B_2H_6 (200 ppm in H_2) and PH_3 (0.5 % in H_2). To eliminate the phosphorus in the system after phosphorus doping, an extra 1 min

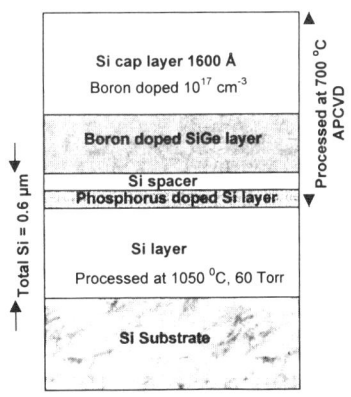

Fig. 1: Basic structure of the epitaxially grown layers.

purge step at high hydrogen flow is performed before continuing the epitaxy. Deposition times varied from 30 minutes for the silicon cap layers to 5 seconds for the SiGe layers.

A cross section of the SiGe HBT is shown in Fig. 2. The epitaxially grown Si collector is 0.6 μm thick. In the devices compared here, this collector is either uniformly doped or grown in two steps, so that a 180 keV phosphorus pedestal collector implantation can be placed in the first Si epi layer. The base is isolated by shallow trenches. A 3000 Å LPCVD TEOS oxide is deposited at 700 °C and all contact windows are plasma etched. The boron base contact doping and the phosphorus emitter are both implanted at 15 keV in their respective contact windows by resist masking. The emitter implantation dose is $5 \times 10^{15}/cm^2$. The dopants are activated by a single 30 min thermal anneal step at 700 °C, and the windows are contacted by sputtering Al/1%Si. The metal pitch is 3 μm and determines the size of the c-b junction.

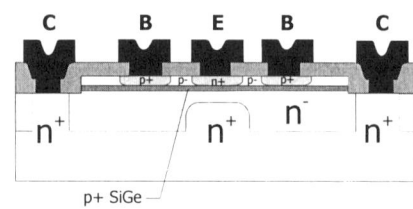

Fig. 2: Schematic of the SiGe HBT structure.

THE AS-GROWN DOPING PROFILES

Under the given growth conditions, the amount of boron that can be incorporated in the SiGe is limited by the B_6H_6 gas flow, which is at its maximum in these experiments. The largest possible peak width is achieved by boron doping the SiGe up to the Si emitter interface. The attainable B concentration increases with increasing GeH_4 flow. The relationship between the resulting concentrations of B and Ge is shown in Fig. 3. Also included is the boron concentration measured in samples where a phosphorus-doped layer was grown in the underlying Si. This gives a further enhancement of the boron incorporation.

Despite extra purging, there is a phosphorus auto-doping effect that is related to the incorporated phosphorus concentration. The level of auto-doping in Si is about a factor 10 lower than the peak phosphorus concentration, and dies out very slowly in Si. In earlier experiments [2], highly doped phosphorus peaks were grown, followed by a Si spacer of widths 0, 200 and 400 Å, respectively. Phosphorus is much more readily incorporated in SiGe than Si [3], which for the sample with no spacer layer resulted in doubling of the peak phosphorus concentration at the Si/SiGe interface. In the samples with a spacer layer this

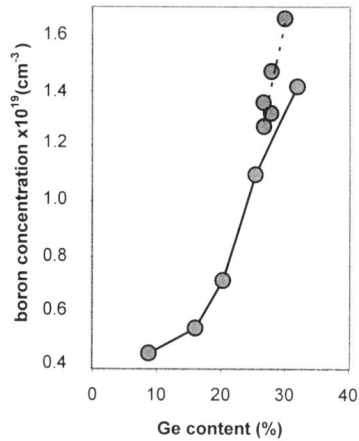

Fig. 3: Boron concentration in SiGe layers as a function of Ge content. The dashed line indicates the concentration obtained when a phosphorus doped layer is grown in the underlying Si.

effect did not increase the auto-doping level upon the onset of SiGe epitaxy and the peaks are well defined. The spacer layer, however, moves the n^+ peaks so far from the boron peak that the resulting n-type doping of the collector after thermal annealing extends undesirably far into the collector.

In the present experiments n^+ peaks without a Si spacer are examined. The targeted peak width was 100 Å and the concentrations were $3 \times 10^{17}/cm^3$, $6 \times 10^{17}/cm^3$, $8 \times 10^{17}/cm^3$, $10^{18}/cm^3$ and $2 \times 10^{18}/cm^3$ in samples N3x17, N6x17, N8x17, N1x18 and N2x18, respectively. The resulting peaks, shown in Fig. 4, are seen to be effectively placed adjacent to the boron peak due to the enhanced phosphorus incorporation in the SiGe. The total doping of the phosphorus peaks is about a factor 5 higher than targeted, but directly related to the PH$_3$ gasflow. Despite the strong enhancement of the auto-doping level at the onset of SiGe epitaxy, the further auto-doping dies out rapidly in the SiGe.

In principle it would be attractive to grow an even narrower phosphorus peak entirely in the SiGe. However, the very high P incorporation in the SiGe, together with the very short growth times, makes this a very difficult process to control with the present settings of our epi-reactor.

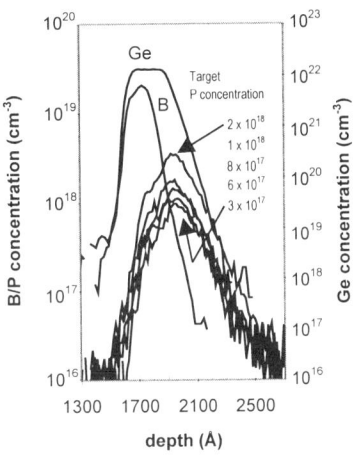

Fig. 4: SIMS profiles of the boron and phosphorus peaks grown in samples N3x17, N6x17, N8x17, N1x18 and N2x18.

DOPING PROFILES AFTER THERMAL ANNEALING

The normal thermal diffusivity of boron and phosphorus at 700 °C is extremely low, so outside the areas where processing damage induces TED, the as-grown doping profiles are barely changed by the emitter anneal. The implantation damage induced TED of these two dopants is on the other hand very large in silicon. This is seen in Fig. 5a, where the TED of the phosphorus emitter implantation diffuses the phosphorus away from the implantation damage region to the SiGe layer. Outside the SiGe, the diffusion of the boron and phosphorus peaks is also considerable.

A number of studies have shown that in strained SiGe, TED will be reduced due to clustering of the dopants with Ge [4]. When subjected to the emitter implantation and anneal, the boron peaks, will largely be retained in the SiGe. The amount of boron that diffuses into the Si collector, relative to the total boron doping, is shown in Fig. 6. Between 25% and 30% SiGe the clustering effect appears to become very strong and the out-diffusion is very effectively suppressed.

The diffusion of boron into the emitter is also limited by strong boron segregation at the Si/SiGe interface [5]. Phosphorus segregation at this interface from the epitaxially grown peak is also visible, particularly for the N2x18 sample. Nevertheless, the very sharp rise of the boron doping at the interface results in a positioning of the e-b junction at the Si/SiGe interface.

The TED of the phosphorus peak results in a phosphorus doping of the whole SiGe region. Table I compares a number of different samples with respect to electrical parameters and includes the sheet resistances, ρ_{bex} and ρ_{bin}, of respectively the boron layer in an unimplanted region, and of the intrinsic base formed after emitter implantation and anneal. In the sample N0 no n^+ peak has been grown, and ρ_{bex} and ρ_{bin} are 1600 and 2300 $\Omega/\square$, respectively. The higher ρ_{int} value is mainly due to the overdoping of the boron doped Si cap layer, since the penetration of the emitter implantation into the SiGe is very limited. The much higher phosphorus doping of the SiGe layer from the n^+ peaks results in 25 to 40% increase in ρ_{bin}. Although the boron and phosphorus levels are comparable at the Si/SiGe interface, the boron tail in the Si collector is overdoped by the

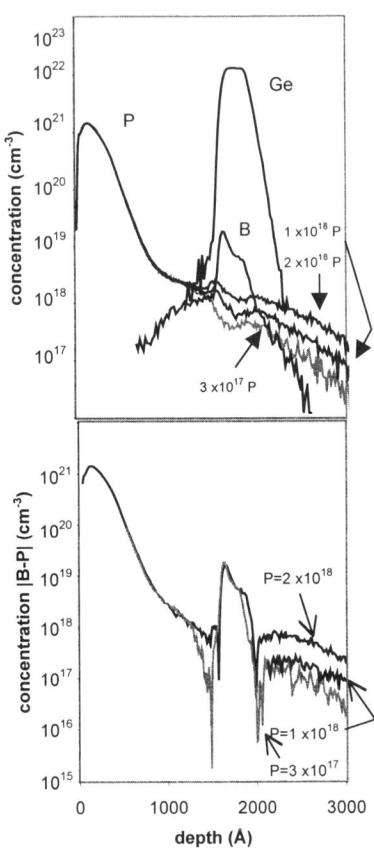

Fig. 5: (a) SIMS profiles of the samples N3x17, N1x18 and N2x18 after emitter implantation and 700 °C anneal. In (b) the absolute difference of the B and P concentrations is given.

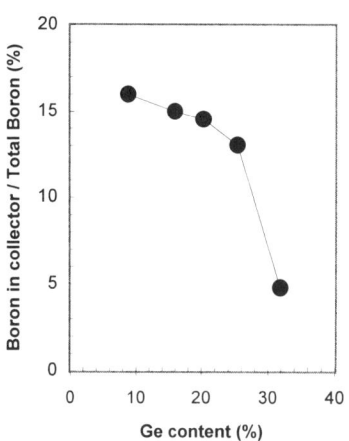

Fig. 6: The amount of boron that has diffused into the Si collector, normalized to the total boron concentration, versus the Ge concentration.

phosphorus because the phosphorus diffuses further into the Si than the boron. This is clearly seen in Fig. 5b. Nevertheless, the resulting n-doping drops to levels around $10^{17}/cm^2$ within 1000 Å of the c-b metallurgic junction.

Table I: Device parameters of SiGe HBT's fabricated in the samples N0LB, N0, N3x17, N8x17, N1x18 and N0HE, with and without pedestal collector. Device N07k has been described in references [1,2] and has no phosphorus peak but a collector epi doped to $3 \times 10^{16}/cm^2$. The emitter area is $20 \times 1 \ \mu m^2$.

Sample	Pedestal collector	Current Gain	ρ_{bex} ($\Omega/\square$)	ρ_{bin} ($\Omega/\square$)	BV_{CE0} (V)	C_{bc} (fF) ($V_{CB}=0$)	C_{bc} (fF) ($V_{CB}=3V$)	f_T (GHz)
N0LB	-	20	2600	4400	9	45	33	9
N0	-	5	1600	2300	11	44	32	7
N3x17	-	250	1900	3100	9	188	42	13
N8x17	-	300	2100	3300	5.5	370	45	15
N1x18	-	350	2300	3900	2.4	540	115	15
N0HE	-	250	1400	2400	2.2	430	240	30
N07k	-	350	5300	6800	6.5	110	63	27
N0LB	+	50	2600	4700	3.8	60	43	10
N0	+	8	1600	2400	5.0	59	41	8
N3x17	+	250	1900	3100	3.4	190	52	25
N8x17	+	400	2100	3300	2.6	373	54	31
N1x18	+	400	2300	3900	2.5	545	125	32
N07k	+	500	5300	7100	2.7	120	70	44

DEVICE CHARACTERISTICS

A number of devices were fabricated in samples both with and without n^+ peaks. The results are summarized in Table I. The N0 and N0LB are samples without n^+ peaks, the latter of which has a much lower base doping. Nevertheless, both device types suffer from collector barriers and have very low current gain and f_T. This was also the case for a device with a 50 Å wider SiGe collector spacer.

In sample N3x17, the barriers are compensated just enough to increase the current gain and f_T to 250 and 25 GHz, respectively. For N8x17 and N1x18, barriers are very effectively suppressed and f_T's above 30 GHz are achieved with a significantly lower ρ_{bin} than that of the N0 sample. Comparable results are achieved in sample N0HE, where the collector is uniformly doped to a value of $4 \times 10^{17}/cm^2$. The BV_{CE0} is, however, very low in this sample and the collector-base capacitance, C_{bc}, is very high. With respect to this parameter, the advantage of placing the bulk of the n^+-doping close to the

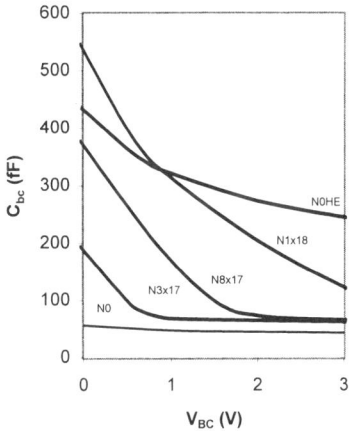

Fig. 7. The C_{bc} versus V_{CB} for SiGe HBT's with an emitter area of $20 \times 1 \ \mu m^2$ and a total collector-base area of $230 \ \mu m^2$.

base is seen in Fig. 7, where C_{bc} is compared for several samples. At $V_{CB} = 0$ the increasing n^+ peak doping results in a large increase in C_{bc}, but at higher voltages the values with and without n^+ peak become comparable. The results indicate that the best overall device characteristics are achieved with n^+ peak doping lying between that of the N3x17 and N8x17 samples. An intrinsic base sheet resistance of 3200 $\Omega/\square$ with $BV_{CE0} = 3$ V and $f_T = 30$ GHz appears feasible.

CONCLUSIONS

The advantages of using epitaxially grown phosphorus peaks in the collector to compensate boron base out-diffusion, have been demonstrated with respect to the trade-off between f_T and the parameters BV_{CE0} and C_{bc}. Cut-off frequencies in the 30 GHz range have been achieved for intrinsic base sheet resistances around 3 kΩ. The reasons for the much higher f_T obtained in samples fabricated earlier with $f_T = 44$ GHz and $\rho_{bin} = 7$ kΩ, are not understood at present. Discrepancies in the basic device structure and doping profiles prohibit a direct comparison of these devices.

Under the epitaxy conditions chosen here, the phosphorus peak is grown just before the SiGe layer. This gives a controllable but broadened peak profile, which excessively overdopes the boron that diffuses into the Si collector. A more narrow peak, grown exclusively in the SiGe collector spacer, would possibly give less phosphorus out-diffusion and a more precise compensation of the boron. It is not clear at present whether this can be realized in a reliable manner.

ACKNOWLEDGEMENTS

The authors are very indebted to J.W. Slotboom for many simulating discussions, and to E.J.G. Goudena, R. Mallee, J. Shi, K. Grimm and S. Beckers for their contributions to the experimental material.

REFERENCES

1. L.K. Nanver, E.J.G. Goudena, C.C.G. Visser, H.W. van Zeijl, and J.W. Slotboom, Proceedings 26th ESSDERC, Bologna, Italy, 9-11 Sept., pp. 469-472 (1996).

2. L.K. Nanver, C.C.G. Visser, A. van den Bogaard, to appear in J. Vac. Sci. and Techn. B, June 1998.

3. T.I. Kamins, and D. Lefforge, J. Electrochem. Soc., **144**, 2, 674 (1997).

4. P. Kuo, J.L. Hoyt, J.F. Gibbons, J.E. Turner, R.D. Jacowitz, and T.I. Kamins, Appl. Phys. Lett. **62**, 612 (1995).

5. R.F. Lever, J.M. Bonar and A.F.W. Willoughby, J. Appl. Phys. **83**, 4, 1988 (1988).

WET AND DRY OXIDATION OF POLYCRYSTALLINE Si_xGe_{1-x} FILMS

P.-E. Hellberg, S.-L. Zhang, F. M. d'Heurle *, and C. S. Petersson
Royal Institute of Technology, Department of Electronics, S-164 40 Kista, Stockholm, Sweden
*Also with IBM Thomas J. Watson Research Center, Yorktown Heights, NY 10598, USA

ABSTRACT

Wet and dry oxidations of polycrystalline Si_xGe_{1-x} with various compositions have been studied at different temperatures. The growth rate of SiO_2 is found to be enhanced by Ge, and the enhancement effect is more pronounced in H_2O than in O_2. A mathematical model, which assumes simultaneous oxidation of Si and Ge and reduction of GeO_2 by free Si available at the growing-oxide/Si_xGe_{1-x} interface, is found to give a quantitative description of the SiO_2 growth during thermal oxidation of Si_xGe_{1-x}. Kinetic parameters are extracted by comparing the model with experiments. The linear and parabolic rate constants for Si oxidation are determined on control Si (100) wafers and polycrystalline Si films. Simple expressions are used for the interdiffusion of Si and Ge in Si_xGe_{1-x}. For wet oxidation, the activation energy for the reaction rate constant of Ge oxidation is found to be smaller than that of Si oxidation.

INTRODUCTION

The importance of Si_xGe_{1-x}, single crystal or polycrystalline, in Si very-large-scale integration technology has called for numerous investigations of thermal oxidation of the material [1]. A mathematical model [2] has been proposed to describe the growth of SiO_2 during thermal oxidation of Si_xGe_{1-x}. The basic feature of the model is the assumption of three chemical reactions, i.e., (1) Si oxidation, (2) Ge oxidation and (3) the reduction of GeO_2 by free Si, taking place simultaneously at the growing-oxide/Si_xGe_{1-x} interface. The reduction reaction is further assumed to be infinitely fast at the temperatures used. The rate of SiO_2 growth is then:

$$\frac{dL_{ox}}{dt} = B/A \frac{\left[r + a_{iSi}(1-r)\right]}{1 + \frac{2L_{ox}\left[r + a_{iSi}(1-r)\right]B/A}{B}} \tag{1}$$

where, L_{ox} is the SiO_2 thickness, t is the oxidation time, B/A and B are, respectively, the linear and parabolic rate constant for oxidation of Si [3], a_{iSi} is the Si activity at the SiO_2/Si_xGe_{1-x} interface and r is the ratio of the reaction rate constant of Ge oxidation (k_{Ge}) to that of Si oxidation (k_{Si}). The silicon activity is determined by solving:

$$N\frac{\partial a}{\partial t} = N\frac{\partial}{\partial z}D(x)\frac{\partial a}{\partial z} \tag{2}$$

over the entire thickness of the remaining Si_xGe_{1-x}. Here, N is the atomic concentration of Si_xGe_{1-x}, a is the Si activity and z represents the coordinate (i.e., the depth in Si_xGe_{1-x}). Since the

111

self-diffusion of Ge in Si_xGe_{1-x} is known [4] to vary considerably with composition, the interdiffusion coefficient, $D(x)$, can be expected to be a strong function of the local concentration x in Si_xGe_{1-x}. A detailed description of the numerical solution of Eqs. (1) and (2) is given in [2].

More results are presented in this study to compare the model suggested in [2] with the experiment, for oxidation at various temperatures in both H_2O and O_2 atmosphere. Kinetic parameters, i.e., diffusion coefficient and reaction rate constant, are extracted from the comparison.

EXPERIMENT

Polycrystalline Si_xGe_{1-x} films, with various compositions Si, $Si_{0.75}Ge_{0.25}$, $Si_{0.52}Ge_{0.48}$ and $Si_{0.30}Ge_{0.70}$, were formed on oxidized Si wafers via chemical vapor deposited. On some wafers, a 20 nm Si cap layer was deposited subsequently, without breaking the vacuum, on top of Si_xGe_{1-x}. Native oxide was removed from the Si_xGe_{1-x} surface by hydrofluoric acid (HF) and then the wafers were immediately loaded in an oxidation furnace. Before the initiation of the oxidation process the samples were heated in N_2. The oxidation was performed in pyrogenic steam or O_2 (at atmospheric pressure) at 800, 850 and 900 °C for various lengths of time from 0.5 to 8 hours. A Si (100) wafer was always used as a control. After oxidation, the temperature was decreased to 625 °C before the wafers were unloaded. Rutherford backscattering spectroscopy (RBS) was used to measure the oxide thickness as well as the concentration of Si and Ge in the remaining Si_xGe_{1-x} layer. The RBS results were analyzed with the assistance of the RUMP simulation program [5]. The oxide thickness on the Si (100) control wafer was measured by ellipsometry.

RESULTS

Wet Oxidation

The RBS results are shown in Fig. 1 for $Si_{0.75}Ge_{0.25}$, $Si_{0.52}Ge_{0.48}$ and $Si_{0.30}Ge_{0.70}$ with (a) and without (b) the 20 nm Si cap layer, after oxidation in pyrogenic steam at 900 °C for 30 min. A

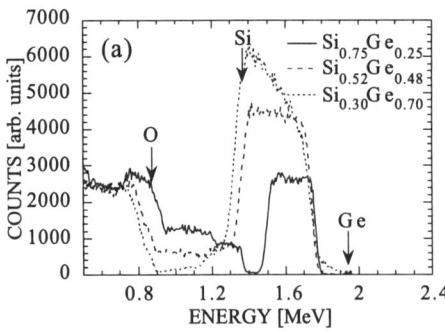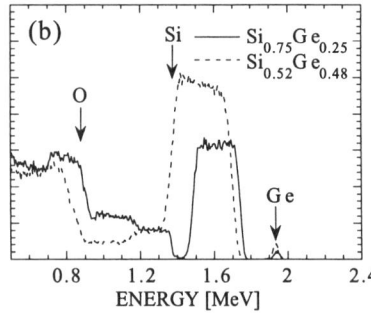

Fig. 1 - RBS results for wet oxidation of $Si_{0.75}Ge_{0.25}$, $Si_{0.52}Ge_{0.48}$ and $Si_{0.30}Ge_{0.70}$ at 900 °C for 60 min, with (a) and without (b) the 20 nm thick Si cap layer.

small Ge peak at the surface, that appears in all spectra, shall be discussed later. The presentation of the RBS results below does not include this Ge peak. With the Si cap (Fig. 1a), no Ge is detected at the surface in the case of oxidation of $Si_{0.75}Ge_{0.25}$ or $Si_{0.52}Ge_{0.48}$. Silicon and oxygen are found at the surface and their respective intensities indicate the formation of a SiO_2 layer on top of the Si_xGe_{1-x}. The thickness of the formed SiO_2 increases with increasing Ge content. For $Si_{0.30}Ge_{0.70}$, the Ge counts increase continuously below the Ge surface position, indicating the formation of a mixed oxide $(Si,Ge)O_2$ or SiO_2-GeO_2. Without the Si cap, Fig. 1b, basically the same features as with the cap are observed in the RBS spectra. But the SiO_2 layers grown are thicker.

The RBS spectra of the $Si_{0.52}Ge_{0.48}$ films with the cap are shown in Fig. 2, after oxidation at (a) 800, (b) 850 and (c) 900 °C for various lengths of time. A surface Ge peak, as in Fig. 1, is also detected in the RBS spectra and will again be excluded from the result presentation here. At all oxidation temperatures, a SiO_2 layer is initially formed on top of Si_xGe_{1-x}. However, at 800 °C a mixed oxide starts to grow after 60 min corresponding to a SiO_2 thickness of 115 nm. At 850 and 900 °C, only SiO_2 is found even after 2 hours of oxidation. Regions with high Ge concentrations are observed beneath the growing SiO_2 at 800 °C. The Ge pile-up is smaller at 850 °C and it is not observable at 900 °C.

Dry Oxidation

The RBS spectra are depicted in Fig. 3 for $Si_{0.75}Ge_{0.25}$, $Si_{0.52}Ge_{0.48}$ and $Si_{0.30}Ge_{0.70}$ films

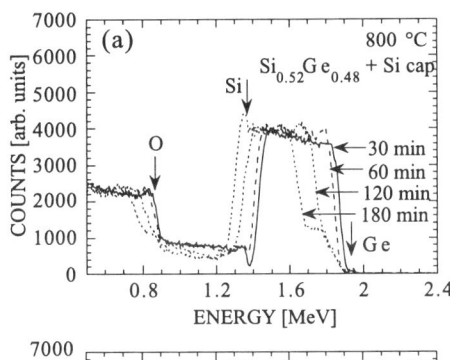

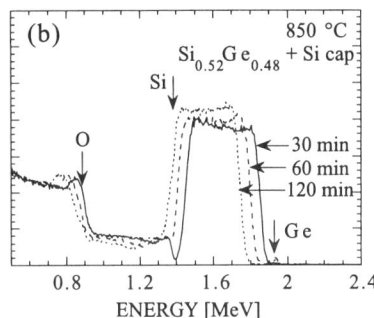

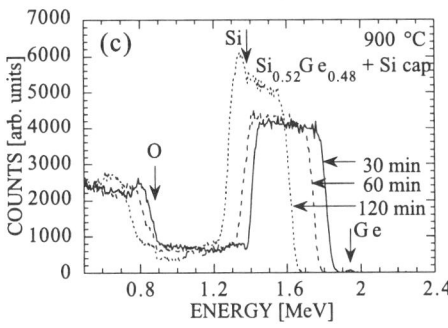

Fig. 2 - RBS results for the $Si_{0.52}Ge_{0.48}$ films oxidized at (a) 800, (b) 850 and (c) 900 °C for various lengths of time. The $Si_{0.52}Ge_{0.48}$ films were capped with 20 nm Si.

without the Si cap and after the 8-hour oxidation at 850 °C. A small surface peak of Ge is, again, found as in the case of wet oxidation (Figs. 1-2). Except for this surface Ge peak, a SiO_2 layer isfound on all Si_xGe_{1-x} films and the thickness of the formed SiO_2 increases with increasing Ge content. The RBS spectra displayed similar features at 900 °C (not shown), but with thicker SiO_2 formed.

Simulation

To facilitate the mathematical simulation of the growth of SiO_2, according to the model in [2], the linear and parabolic rate constant for Si oxidation were first determined by analyzing the SiO_2 growth on the control wafers as well as on the polycrystalline Si. Table I shows the results of wet oxidation together with the corresponding literature data. The value of r, the ratio of the reaction rate constant of Ge oxidation to that of Si oxidation, is also given.

In order to solve Eq. (2), information about the interdiffusion of Si and Ge in Si_xGe_{1-x} is required. Since, to the authors' knowledge, the interdiffusion coefficient is not known over the whole composition range in Si_xGe_{1-x}, simple exponential functions were used. Figure 4 shows the variation of the interdiffusion coefficient with film composition used in the simulation (continuous lines). For reference, the literature data for the self-diffusion of Ge in Si_xGe_{1-x} [4,7,8] are also shown as dots and triangles.

Table I - Rate constants for wet oxidation of Si_xGe_{1-x}.

| Temp. [°C] | Linear and parabolic rate constants for Si (100) | | | | Value of |
| | This work | | Ref. [6] | | |
	B [nm²/min]	B/A [nm/min]	B [nm²/min]	B/A [nm/min]	r = k_{Ge}/k_{Si}
800	382	0.75	800	0.50	7.5
850	1237	0.99	1500	0.83	6.5
900	2660	2.63	2500	2.50	5.5

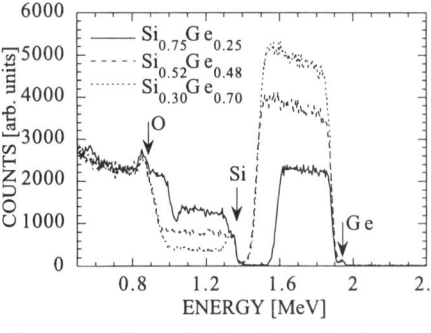

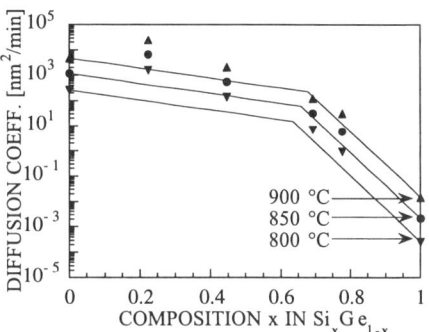

Fig. 3 - RBS results for dry oxidation of $Si_{0.75}Ge_{0.25}$, $Si_{0.52}Ge_{0.48}$ and $Si_{0.30}Ge_{0.70}$ without the Si cap, at 850 °C for 8 hrs.

Fig. 4 - Interdiffusion coefficient of Ge and Si in Si_xGe_{1-x} used in the simulation (continuous lines). The self-diffusion (extrapolated from higher temperatures) of Ge in Si_xGe_{1-x} [4,7,8] is also shown as dots and triangles.

Some examples are shown in Fig. 5 for the comparison between the simulated (lines) and experimental (round, square and triangular dots) oxide thicknesses versus time. The results obtained on the Si control wafers are also shown as reference. With the Si cap, the oxide growth rate on Si_xGe_{1-x} is, according to the simulation, the same as on polycrystalline Si in the beginning of oxidation (Fig. 5b and [2]). After a certain length of time, the interdiffusion of Si and Ge and the consumption of the surface Si by oxidation bring the Ge to the oxidizing interface. The growth rate is then enhanced, see, e.g., the simulated curve at 800 °C in Fig. 5b. Different values of r were chosen so as to fit the simulation results to the measured data at different temperatures (Table I). The simulation is terminated when the Si concentration at the SiO_2/Si_xGe_{1-x} interface decreases to zero. At this point, Ge begins to be incorporated in the formed oxide. This is the case for the wet oxidation of $Si_{0.52}Ge_{0.48}$ at 800 °C: a mixed oxide is found by RBS after about 60 min (Fig. 2a) and by simulation after 57 min (Fig. 5b) of oxidation.

DISCUSSION AND CONCLUSIONS

The observation of the surface Ge peak in Figs. 1-3 is challenging. One feasible interpretation would be that at the very beginning of the process when the rate of oxidation is at its maximum, it may exceed the reduction rate of GeO_2 by Si. A mixed oxide is then left at the sample surface. The same occurs in the presence of the Si cap layer, because the interdiffusion of Si and Ge brings Ge to the surface during the ramp-up (in N_2) period. As a continuous oxide forms the concentration of the oxidant in the oxide drops, and so does the oxidation rate. The

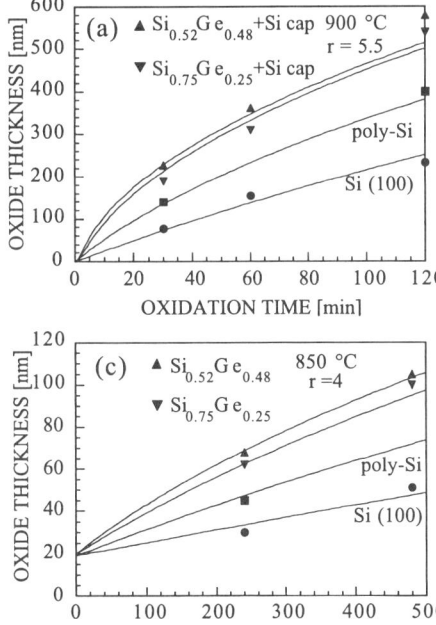

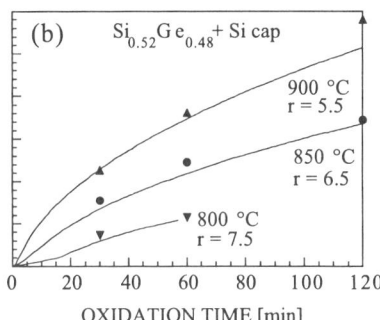

Fig. 5 - Comparison between simulation and experiment for (a) wet oxidation of Si_xGe_{1-x} with various compositions at 900 °C with the Si cap, (b) wet oxidation of $Si_{0.52}Ge_{0.48}$ with the cap at different temperatures, and (c) dry oxidation of Si_xGe_{1-x} with various compositions at 850 °C without the cap.

assumption of an infinite reduction rate becomes valid and the process now enters the steady-state where SiO_2 grows, as found on the majority of the samples examined. The steady-state growth is evidenced by the separation of the surface Ge peak from the main Ge signal in the RBS spectra (Figs. 1-3). The presence of the surface Ge peak may influence the accuracy of the (SiO_2) thickness measurement. However, it does not affect the validity of the model, since the enhanced growth of SiO_2 continues far beyond the surface region where the trace GeO_2 is found.

In order to fit the SiO_2 thickness measured on polycrystalline Si using the linear and parabolic rate constant extracted from the oxide grown on Si (100), the linear rate constant has to be scaled up by a factor of 2.25 for both dry and wet oxidation. Although this factor is considerably greater than the reported value for the ratio of the oxidation of polycrystalline Si to that of Si (100) [9], satisfactory simulation results are obtained for the SiO_2 growth on Si_xGe_{1-x} (including polycrystalline Si) as compared to the experimental data in this work. In order to simulate the SiO_2 growth during dry oxidation, an initial oxide thickness of 20 nm [3] was used for all Si_xGe_{1-x} films oxidized at different temperatures.

The reaction-rate-constant ratio r was used as a fitting parameter in the simulation. For wet oxidation, its value decreases with increasing oxidation temperature (Table I). Therefore, the activation energy for the reaction rate constant of Ge oxidation, k_{Ge}, is lower than that of Si oxidation, k_{Si}. For dry oxidation, the value of r does not vary significantly when changing the oxidation temperature from 850 to 900 °C.

In conclusion, the mathematical model proposed in [2] is found to give a good description of the oxidation process (Fig. 5); the simulation results reproduce the experimental ones for Si_xGe_{1-x} of various compositions, with and without a Si cap, oxidized at different temperatures and in pyrogenic steam or O_2.

ACKNOWLEDGMENTS

The authors are in debt to professor Gerald H. Meier and professor Frederick S. Pettit for their hospitality in Pittsburgh and their kind interest in our efforts, and to Martin Sparby for RBS data acquisition. This work was financially supported by the Swedish National Board for Technical and Industrial Development (NUTEK) and Mitel Semiconductor AB.

REFERENCES

1. P.-E Hellberg, S.-L. Zhang, F. M. d'Heurle, and C. S. Petersson, J. Appl. Phys. **82**, 5773 (1997), and the references therein.
2. P.-E Hellberg, S.-L. Zhang, F. M. d'Heurle, and C. S. Petersson, J. Appl. Phys. **82**, 5779 (1997).
3. B. E. Deal and A. S. Grove, J. Appl. Phys. **36**, 3770 (1965).
4. G. L. McVay and A. R. Ducharme, Phys. Res. **B 9**, 627 (1974).
5. L. R. Doolittle, Nucl. Instrum. Methods Phys. Res. **B 9**, 334 (1985).
6. R. R. Razouk, L. N. Lie, and B. E. Deal, J. Electrochem. Soc. **128**, 2214 (1981).
7. W. Frank, U. Gösele, H. Mehrer, and A. Seeger, Diffusion in crystalline solids (Academic, New York, 1984) pp. 74-88.
8. B. L. Sharma, Defect and Diffusion Forum **70&71**, 1 (1990).
9. H. Sunami, J. Electrochem. Soc. **125**, 892 (1978).

CHARACTERIZATION of SiGeC USING Pt(SiGeC) SILICIDE SCHOTTKY CONTACTS

Jeff J. Peterson[1], Charles E. Hunt[1], McDonald Robinson[2], Robin Scott[2]

[1] Dept. of Electrical and Computer Engineering, Univ. of California, Davis, CA 95616 USA

[2] Lawrence Semiconductor Research Laboratory, Inc.; Tempe, AZ 85282 USA

ABSTRACT

Material and electrical characterization of n-type and p-type $Si_{1-x-y}Ge_xC_y$ epitaxial layers on Si(100) was performed using silicided platinum Schottky contacts. XRD studies show Pt silcidation of SiGeC proceeds from non-reacted Pt to $Pt_2(SiGeC)$ and completes with the Pt(SiGeC) phase similar to Pt/Si silicides, but Pt silicide reactions with SiGeC are shown to require higher temperatures than Pt reactions with Si. Electrical characterization of Pt(SiGeC) contacts to n-type $Si_{1-x-y}Ge_xC_y/Si$ shows rectifying behavior with constant barrier heights of 0.67eV independent of composition, indicating Fermi level pinning relative to the SiGeC conduction band is occurring. Pt(SiGeC) contacts to p-type $Si_{1-x-y}Ge_xC_y/Si$ are also rectifying with barrier heights that track the variation of the SiGeC energy bandgap.

INTRODUCTION

SiGe epitaxial layers on Si substrates (SiGe/Si) show promise to be a column IV alternative to III-V materials used in heterojunction and bandgap engineered devices. The addition of C to SiGe epitaxial layers provides a solution for strain induced dislocation defects in SiGe on Si by using smaller C atoms to provide stress compensation in the SiGe layer [1]. The resulting SiGeC on Si layer has little known material and electrical properties, however.

The study of metal/SiGeC contacts and barrier heights is one area in which a thorough understanding does not yet exist. Among the metal/SiGeC contacts to have been studied are Al [2], Co [3], Ti [4], and W [5] contacts. It is reported that Schottky contacts to p-type SiGeC show compositional dependence in the barrier height, while Schottky contacts to n-type SiGeC show no compositional dependence, an indication of Fermi level pinning [6]. We report here the first phase morphology study of Pt silicidation of SiGeC layers as well as the electrical characteristics of silicided Pt(SiGeC) Schottky contacts to $Si_{1-x-y}Ge_xC_y$ epitaxial layers on Si.

EXPERIMENT

$Si_{1-x-y}Ge_xC_y$ epitaxial layers (x ranging from 0 to 0.27 and y ranging from 0 to 0.02) were grown to thicknesses of 300nm on Si(100) wafers using atmospheric-pressure CVD at 625°C. N^+ SiGeC layers were grown on N^- Si substrates, while P^+ SiGeC layers were grown on P^- Si substrates. Using RBS measurements, these epilayers have been analyzed to determine alloy composition and crystallinity, which are used in comparing the material and device characteristics of the Pt(SiGeC) Schottky diodes.

Elemental Pt was deposited to a thickness of 100nm onto the SiGeC epilayer surface using E-beam evaporation at pressures of 10^{-6} Torr or less. After deposition, the Pt was patterned using liftoff techniques and cleaned to remove contaminants which would affect the Pt(SiGeC) silicide. Next, the patterned Pt was heated to silicide reaction temperatures. The RTA (Rapid Thermal Anneal) reaction time and temperature were set by using XRD to determine which silicide phases were present in the reacted layer after a given RTA treatment; RTA took place in a 30mTorr N_2 ambient. Furnace reacted samples were annealed in atmospheric pressure N_2 for 30 minutes at 700°C. At this point, the Pt(SiGeC) silicides were characterized to determine the film quality discussed in this work.

Following the formation of the Pt(SiGeC) Schottky diodes, a 150nm thick oxide layer was deposited to act as an insulator between the bondpad metallization and epilayers. Contact holes were formed using liftoff techniques in preparation for deposition of the metallization. The metallization, a 10nm Ti adhesion layer and a 500nm Al layer, was sequentially deposited in the E-beam evaporator at a base pressure in the 10^{-6}Torr range. After metallization liftoff, a contact anneal was done for 30m at 400°C. For the final fabrication step, the silicide side of the sample was covered with photoresist while the oxide on the reverse of the wafer was removed using BOE etch. After photoresist removal, the Schottky diodes were measured for electrical characteristics.

Current-Voltage measurements were done at 25°C on a temperature controlled wafer probe station using a HP4145 Semiconductor Parameter Analyzer. Schottky diode I-V data was analyzed using either graphical or curve fitting methods. Both methods assume the diode operates under the TE (Thermionic Emission) model. The graphical analysis plots the data in a Log I vs. V form and uses the Log I intercept to estimate the J_s, from which the barrier height is calculated. The curve fitting method of data analysis uses a mathematical fit of the data to the TE Schottky diode model [7] to estimate J_s, ϕ_b, and n. The series resistance, R_s, between the diode and the backside wafer contact, was also incorporated into the curve fitting, but is typically unnecessary when the data is fitted to low current values.

RESULTS

Our results indicate the formation of Pt/SiGeC silicides - hereafter referred to as $Pt_x(SiGeC)_y$ - to be similar to that of Pt on Si, although the formation of $Pt_x(SiGeC)_y$ was seen to vary significantly in appearance and phase. Completely reacted $Pt_x(SiGeC)_y$ silicides are uniform in consistency and reflect the smoothness of the underlying SiGeC layers, having an average roughness of between 1nm and 2nm. Figure 1, below, shows an example of a completely reacted silicide, which we will later show to be the Pt(SiGeC) phase. Figure 2 shows an example of another silicide, which includes elemental Pt, $Pt_x(SiGeC)_y$ silicides, and the final Pt(SiGeC) phase. This mixed silicide, rough and granular in appearance, had an average roughness in excess of 20nm.

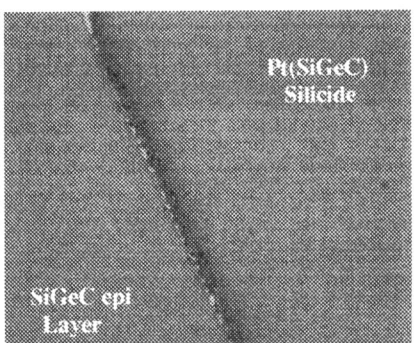

Figure 1 – Optical micrograph showing the phase morphology of a completely reacted Pt(SiGeC) silicide (1000x).

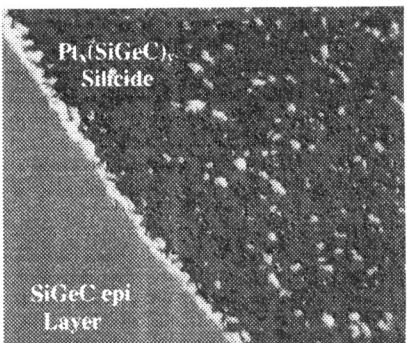

Figure 2 – Optical micrograph showing the phase morphology of an incompletely reacted silicide of Pt, $Pt_x(SiGeC)_y$, and Pt(SiGeC) mixed phases (1000x).

XRD and sheet resistance measurements were used to characterize the order of phase formation of the $Pt_x(SiGeC)_y$ silicides. Figure 3 shows XRD spectra of the $Pt_x(SiGeC)_y$ phase transformation for a 600°C SP (Set Point) RTA. Figure 3, curve (a), shows that after 20s of RTA the Pt film showed some evidence of $Pt_2(SiGeC)$ formation, but still shows a close match to the x-ray spectrum for Pt. Figure 3, curve (b), shows the XRD spectrum after 80s of 600°C SP RTA showing the reacted layer contains a mixture of $Pt_2(SiGeC)$ and Pt(SiGeC) with a slight Pt signal still present. Figure 3, curve (c), shows the XRD spectrum after 160s of 600°C SP RTA; in this

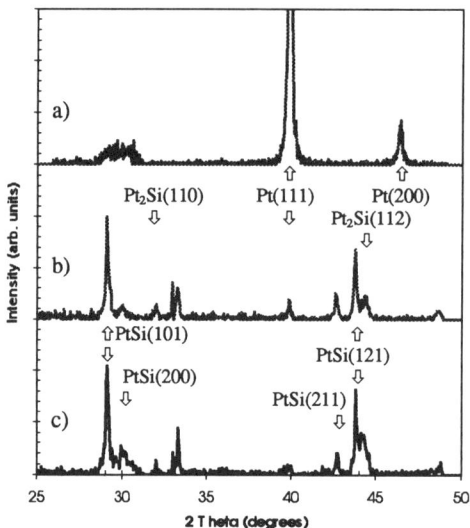

Figure 3 – XRD spectra of $Pt_x(SiGeC)_y$ phase with respect to RTA time. a) XRD spectra after 20s/600°C SP RTA showing elemental Pt, b) XRD spectra after 80s/600°C SP RTA showing mixture of Pt, $Pt_2(SiGeC)$ and $Pt(SiGeC)$ c) XRD spectra after 160s/600°C SP showing completely reacted $Pt(SiGeC)$.

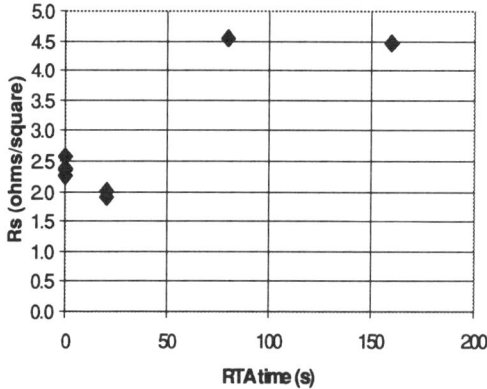

Figure 4 – Rs vs. 600°C SP RTA anneal time showing $Pt_x(SiGeC)_y$ phase formation beginning with Pt, continuing to form low resistive $Pt_2(SiGeC)$, and completing with the formation of higher resistance $Pt(SiGeC)$.

curve the silicide appears to be completely converted to the $Pt(SiGeC)$ final phase. Additional RTA time after this point shows little change in the silicide, indicating the $Pt(SiGeC)$ phase is the final phase, similar to the reaction between Pt and Si. Schottky diodes characterized in this work were annealed for at least 320s of 600°C SP RTA to complete the $Pt(SiGeC)$ silicidation.

Sheet resistance measurements were also used to characterize the $Pt_x(SiGeC)_y$ phase formation. For Pt/Si layers, [8] showed R_s lowers as Pt_2Si formation takes place and raises again as the reaction completes at the PtSi phase. Figure 4 shows a graph of R_s vs. 600°C SP RTA anneal time. Similar to the results obtained in [8], the R_s lowers from the initial value of the elemental Pt film as the reaction begins. From XRD analysis we know this intermediate phase to be $Pt_2(SiGeC)$. As the reaction proceeds, R_s begins to rise and finally levels out as the silicide reaches the final $Pt(SiGeC)$ phase. This is entirely consistant with the results of our XRD characterization and shows Pt silicide formation on $Si_{1-x-y}Ge_xC_y$ epitaxial layers to follow a path very similar to that for Pt silicide formation on Si.

Both RTA and standard furnace

anneals of Pt/SiGeC required higher reaction temperatures than would have been predicted from Pt/Si reactions. Standard Furnace anneals of Pt/SiGeC showed initial Pt(SiGeC) phase formation at temperatures of 600C, while for Pt/Si we would expect the reaction to begin at a temperature of about 300°C [9]. Higher silicide formation temperatures were also reported in the formation of Co silicides on SiGeC [3], where the effect was attributed to Ge/C pile-up at the silicide/SiGeC interface and to C precipitates at higher temperatures.

We now present the electrical properties of Schottky contacts to SiGeC/Si formed using the above silicides. Figure 5 plots Log I vs. V for a Pt(SiGeC) to n-type SiGeC/Si contact, demonstrating rectifying behavior. Forward bias turn-on voltages were about -0.45V with reverse bias breakdown voltages of roughly -8V. Graphical methods showed the reverse bias current density, J_s, to be in the $10^{-5} A/cm^2$ range, while curve fitting methods showed a larger bound of values. Ideality factors (n) for these diodes ranged from 1.02 to 2.02. Table I, page 6, shows a summary of the characterization results of these diodes. Note that the graphical ϕ_b values (which we believe to more accurate than the curve fitting ϕ_b values for these diodes) in Table I are independent of composition with an average value of 0.67eV. This is in agreement with other results for metal/SiGeC in the literature [6], which suggest Fermi level pinning relative to the conduction band in n-type SiGeC/Si removes the compositional dependence of

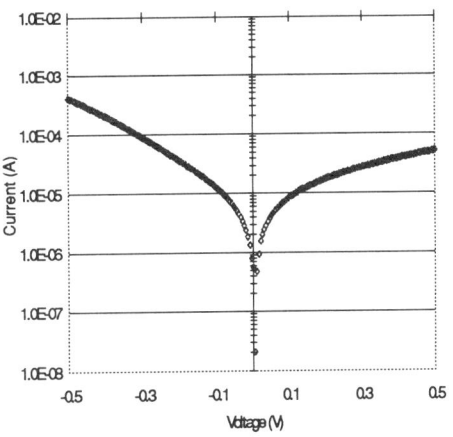

Figure 5 - Log I vs. V characteristic for Pt(SiGeC) contacts to N⁺ SiGeC/Si.

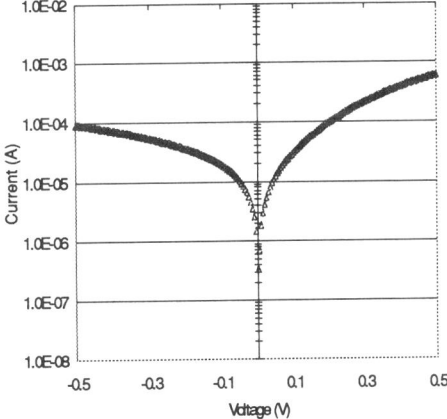

Figure 6 - Log I vs. V characteristic for Pt(SiGeC) contacts to P⁺ SiGeC/Si.

ϕ_b for Schottky contacts to n-type SiGeC/Si layers. Our results are also consistent with Pt silicide Schottky barrier contacts to n-type SiGe which show Fermi pinning at values of 0.68eV [10].

Figure 6 shows a Log I vs. V plot for a Pt(SiGeC) to p-type SiGeC/Si Schottky contact, also showing rectifying behavior. Forward bias turn-on voltages were about 0.45V. Reverse bias breakdown voltages ranged -8V to -16V. Diode ideality factors ranged from 1.05 to 2.03. Table II displays the electrical characteristics of Pt(SiGeC) contacts to p-type SiGeC/Si layers.

TABLE I - Pt(SiGeC) to n-type SiGeC/Si Schottky contact electrical results.

Sample	Anneal	Dopant	%Ge	%C	Graphical j_s (A/cm2)	Curve Fit j_s (A/cm2)	Curve Fit n (model)	Graphical ϕ_b (eV)	Curve Fit ϕ_b (eV)
D2	RTA	N(1E18)	27.0%	1.5%	4.8E-05	4.5E-04	1.56	0.67	0.61
G2	RTA	N(1E17)	20.0%	1.5%	7.5E-05	7.6E-05	1.05	0.66	0.66
H2	RTA	N(1E18)	20.0%	1.0%	4.8E-05	2.1E-03	2.02	0.67	0.57
D	Furnace	N(1E17)	27.0%	1.5%	4.9E-05	1.5E-05	1.05	0.67	0.70
H	Furnace	N(1E18)	20.0%	1.0%	7.2E-05	8.3E-05	1.02	0.66	0.65

TABLE II - Pt(SiGeC) to p-type SiGeC/Si Schottky contact electrical results.

Sample	Anneal	Dopant	%Ge	%C	Graphical j_s (A/cm2)	Curve Fit j_s (A/cm2)	Curve Fit n (model)	Graphical ϕ_b (eV)	Curve Fit ϕ_b (eV)
B2	RTA	P(1E18)	27.0%	2.0%	1.9E-05	1.6E-03	2.00	0.69	0.58
E2	RTA	P(1E17)	20.0%	1.5%	1.5E-05	2.5E-03	1.70	0.70	0.57
F2	RTA	P(1E18)	20.0%	1.5%	3.6E-05	3.8E-04	1.53	0.68	0.61
A	Furnace	P(1E17)	27.0%	2.0%	1.1E-07	7.1E-05	1.10	0.70	0.66
B	Furnace	P(1E18)	27.0%	2.0%	1.2E-09	2.0E-05	1.16	0.81	0.69

As shown in Tables I and II, graphical analysis showed the reverse bias current density, J_s, to be in the 10^{-5} A/cm^2 range for RTA annealed samples, while furnace annealed samples were at least 2 decades smaller. Our previous work has shown the magnitude of the saturation currents to increase as films become more relaxed, possibly due to the presence of dislocation defects at the heterojunction interface. We therefore conclude the RTA samples reached higher silicidation temperatures than the furnace anneal samples causing some relaxation in the RTA strained layers. Ideality factors may also show an indication of dislocations as well since the ideality is both a measure of the current conduction mechanism and recombination centers which may be present in the device [11]. Both tables I and II show that furnace anneal diodes tend to be more ideal than RTA diodes for Pt(SiGeC) contacts to both N^+ and P^+ SiGeC.

Figure 7 shows a plot of the Pt(SiGeC) to p-type SiGeC/Si measured barrier height versus calculated barrier height. Since the p-type barrier height calculates E_v, the valence band energy, this graph becomes a measure of how the p-type barrier height varies with E_v.

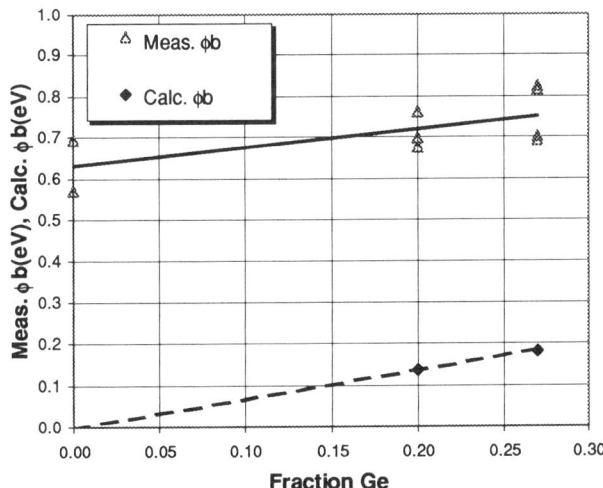

Figure 7 – Pt(SiGeC) to P^+ SiGeC/Si measured barrier height vs. composition. Shown also is a plot of the calculated barrier height variation for the compositional dependent SiGeC bandgap.

From Figure 7, we must conclude that the p-type SiGeC/Si barrier height does track the SiGeC valence band energy. This being the case, Pt(SiGeC) contacts to P^+ SiGeC alloy layers could prove to be a valuable tool in measuring the energy bandgap variation due to the composition of the SiGeC alloy layer.

CONCLUSIONS

We have presented the results of electrical and material characterization of Pt on SiGeC Schottky contacts. Pt silicidation of SiGeC follows a process similar to that of Pt/Si silicidation with Si in proceeding from non-reacted Pt to Pt_2(SiGeC) and completing with the Pt(SiGeC) phase. Pt silicide reactions with SiGeC are shown to require higher temperatures than Pt reactions with Si. It is not known at this time if segregation of the Ge/C species takes place during Pt(SiGeC) silicide formation.

Pt(SiGeC) Schottky contacts to n-type SiGeC/Si show rectifying behavior with a constant ϕ_b of 0.67eV, indicating Fermi level pinning relative to the SiGeC conduction band. Pt(SiGeC) Schottky contacts to p-type SiGeC/Si also show rectifying behavior with a barrier height that varies at the same rate as the valence band energy. These contacts should prove to be a useful tool in measuring the variation of the SiGeC bandgap energy with respect to composition.

ACKNOWLEDGMENTS

This project was supported by LSRL, MICRO, ONR#N00014-93-C-0114, and ONR#N00014-96-C-0219.

REFERENCES

1. A. R. Powell, K. Eberl, B. A. Ek and S. S. Iyer, J. Crystal Growth **127**, p. 425 – 429 (1993).

2. Jian Mi, Ashwant Gupta, Cary Y. Yang, Jintian Zhu, Paul K. L. Yu, Patricia Warren and Michel Dutoit, Appl. Phys. Lett. **69** (24), pp. 3743-3745 (9 December 1996).

3. R. A. Donaton, K. Maex, A. Vantomme, G. Langouche, Y. Morciaux, A. St. Amour and J. C. Sturm, Appl. Phys. Lett. **70** (10), pp. 1266-1268 (10 March 1997).

4. A. Eyal, R. Brener, R. Beserman, M. Eizenberg, Z. Atzmon, David J. Smith and J. W. Mayer, Appl. Phys. Lett. **69** (1), pp. 64-66 (1 July 1996).

5. M. Mamor, C. Guedj, P. Boucaud, F. Meyer, D. Bouchier, S. Bodnar and J. L. Regolini in Strained Layer Epitaxy – Materials, Processing, and Device Applications, edited by E. Fitzgerald, J. Hoyt, K. Cheng and J. Bean (Mater. Res. Soc. Proc. 379, Pittsburg, PA 1995), pp. 137-141.

6. F. Meyer, M. Mamor, V. Aubry-Fortuna, P. Warren, S. Bodnar , D. Dutartre and J. L. Regolini, J. Elec. Mater. **25** (11), pp.1748-1753 (1996).

7. M. Shur, Physics of Semiconductor Devices, Prentice Hall, Englewood Cliffs, New Jersey, 1990, p. 205.

8. E. G. Colgan, J. Mater. Res. **10** (8), pp. 1953-1957 (August 1995).

9. M.–A. Nicolet and S. S. Lau in VLSI Electronics – Materials and Process Characterization, edited by N. Einspruch and G. Larrabee (VLSI Electronics Microstructure Science, Vol. 6, Academic Press, New York, NY 1983), p. 346.

10. H. K. Liou, X. Wu, U. Gennser, V. P. Kesan, S. S. Iyer, K. N. Tu and E. S. Yang, Appl. Phys. Lett. **60** (5), pp. 577-579 (3 February 1992).

11. M. Shur, Physics of Semiconductor Devices, Prentice Hall, Englewood Cliffs, New Jersey, 1990, p. 169.

ELLIPSOMETRY STUDIES, OPTICAL PROPERTIES, AND BAND STRUCTURE OF Ge$_{1-y}$C$_y$, Ge-RICH Si$_{1-x-y}$Ge$_x$C$_y$, AND BORON-DOPED Si$_{1-x}$Ge$_x$ ALLOYS

Kelly E. Junge *, Rüdiger Lange *, Jennifer M. Dolan *, Stefan Zollner **, Josef Humlíček ***,
M.W. Dashiell ****, D.A. Hits ****, B.A. Orner ****, James Kolodzey ****
*Ames Laboratory and Dept. of Physics and Astronomy, Iowa State University, Ames, IA 50011
**Motorola Semiconductor Technology, Arizona Technology Laboratories, Technology Test and
Analysis Laboratory, MD M360, 2200 West Broadway Road, Mesa, AZ 85202
***Department of Solid State Physics, Faculty of Science, Masaryk University, Kotlárská 2,
61137 Brno, Czech Republic
****Department of Electrical and Computer Engineering, 140 Evans Hall, University of
Delaware, Newark, DE 19716

ABSTRACT

We measured the pseudodielectric function (PDF) of Ge$_{1-y}$C$_y$ and Ge-rich Si$_{1-x-y}$Ge$_x$C$_y$ alloys
from 1.1 to 5.2 eV using spectroscopic ellipsometry. These alloys were grown by molecular
beam epitaxy at 600°C on (001) Si substrates. Analytical lineshapes fitted to numerically
calculated derivatives of their PDFs determined the critical-point parameters of the E$_1$, E$_1$+Δ$_1$,
E$_0$', and E$_2$ transitions. The critical-point energies of the Ge$_{1-y}$C$_y$ alloys were found to be similar
to bulk Ge. This indicates that the presence of C in these alloys only has a small influence on the
band structure. For some samples, the amplitude of the PDF is much lower than in bulk Ge,
which can be attributed to surface roughness and explained within the framework of the
Kirchhoff theory of diffraction or using effective medium theory. The degree of surface
roughness indicated by optical measurements was confirmed by atomic force microscopy. We
also studied bulk Czochralski-grown Si$_{1-x}$Ge$_x$ alloys (0<x<0.28) doped with boron. Due to
doping, the critical points shift to lower energies as reported previously for bulk Si and Ge.

INTRODUCTION

Group-IV alloys and compounds promise improved performance of Si-based devices without
using smaller ultra large scale integration length scales. The introduction of C into Si$_{1-x}$Ge$_x$
reduces the strain (and associated defects) in pseudomorphic layers and also changes the band
gap, thus providing an additional parameter for band gap engineering. A detailed understanding
of the optical properties of such alloys is needed for monitoring epitaxial growth and processing
of devices, but our knowledge is still limited. See, for example, [1–4].

The E$_1$ and E$_1$+Δ$_1$ critical points (CPs) of pseudomorphically strained Si$_{1-x}$Ge$_x$ films on Si can
be understood using interpolation of bulk Si and Ge data, if the biaxial strain is taken into
account using deformation potentials. Surprisingly [3] (in contrast to theoretical predictions and
photoluminescence measurements of the indirect gap), the same is also true for Si$_{1-x}$C$_y$ and
Si$_{1-x-y}$Ge$_x$C$_y$ alloys (x<0.3, y<0.02). The CP parameters for the ternary alloys are difficult to
interpret, since the C and Ge fractions are not known with sufficient accuracy (0.1%). A
broadening of the CPs is observed with increasing disorder, particularly when adding C.

In this work, we report ellipsometry measurements of Ge$_{1-y}$C$_y$, Ge-rich Si$_{1-x-y}$Ge$_x$C$_y$, and bulk
Czochralski-grown boron-doped Si$_{1-x}$Ge$_x$ alloys. While the doping effects on the optical
properties have been discussed for bulk Si and Ge [5], there has been no such study for the
group-IV alloys.

Mat. Res. Soc. Symp. Proc. Vol. 533 © 1998 Materials Research Society

Table 1: Ge content x and hole concentration p for the boron-doped $Si_{1-x}Ge_x$ alloys studied.

Sample	x	p $(10^{20}\,cm^{-3})$
1	0.00	3.1
2	0.10	4.0
3	0.12	3.8
4	0.20	1.7
5	0.24	5.0
6	0.28	0.85

GERMANIUM-CARBON ALLOYS

Theory predicts that the binary compound SiC is stable relative to phase separation, whereas GeC is not [6,7]. Therefore, we expect that $Ge_{1-y}C_y$ alloys are even more difficult to produce than $Si_{1-y}C_y$ alloys. Nevertheless, the growth of such alloys has been reported by several groups [3,4,8], but their properties have not yet been well established. In Fig. 1, we show the PDF $<\varepsilon>$ [9] of a thick relaxed $Ge_{1-y}C_y$ alloy on Si grown by MBE ($y=0.01$). The optical constants are very similar to those expected for a relaxed epitaxial Ge film on Si covered with a native oxide [9]. The critical-point parameters determined from a lineshape analysis (see Fig. 2) are also similar to bulk Ge [3]. For substitutional C in Ge or Si, we expect to observe an increase of E_1 by about 40 to 60 meV for $y=0.01$ (after subtraction of strain effects, which do not exist here, since our films are relaxed). While this has been found for $Si_{1-y}C_y$, our $Ge_{1-y}C_y$ alloys show a much smaller blueshift (5 meV or less). We conclude that either the C content in our alloys is much less than 1% or that C in Ge behaves very different from C in Si (which should not be a surprise).

GE-RICH SILICON-GERMANIUM-CARBON ALLOYS

Figure 3 shows PDFs for several $Si_{1-x-y}Ge_xC_y$ alloys with different C concentrations. The amplitude of $<\varepsilon>$ is smaller than for $Ge_{1-y}C_y$ alloys (Fig. 1) because of surface roughness. This was confirmed using atomic force microscopy and has been discussed in more detail in [3]. The dielectric functions and the critical-point parameters are very similar for all four samples and do not seem to depend on C content. This is consistent with our observations for $Ge_{1-y}C_y$ alloys.

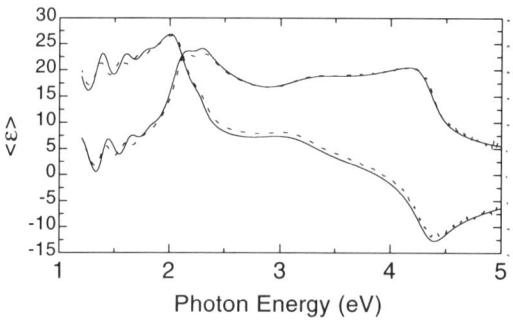

Figure 1: Pseudodielectric function $<\varepsilon>$ for a $Ge_{1-y}C_y$ alloy ($y\approx1\%$) on Si (dashed). Data for other samples are similar. The solid lines show theoretical data for a model [9]: 4400 Å Ge on Si covered with 32 Å of native oxide GeO_2. Apparently, the optical properties of this alloy are very similar to bulk Ge [9].

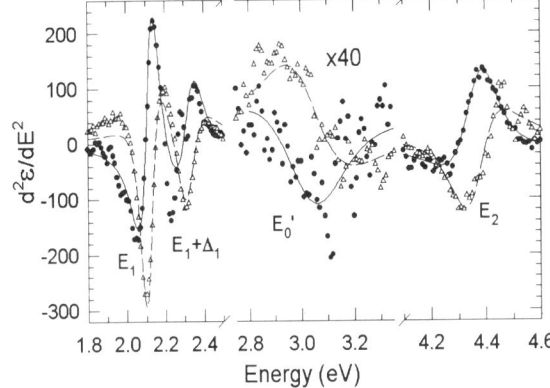

Figure 2: Numerically calculated second derivatives of the pseudo-dielectric function for a $Ge_{1-y}C_y$ alloy ($y \approx 1\%$) on Si as a function of photon energy (symbols). The lines give the best fit to analytical line-shapes allowing the accurate determination of the critical-point parameters (energy, broadening, amplitude, and phase angle). These parameters are very similar to those of bulk Ge.

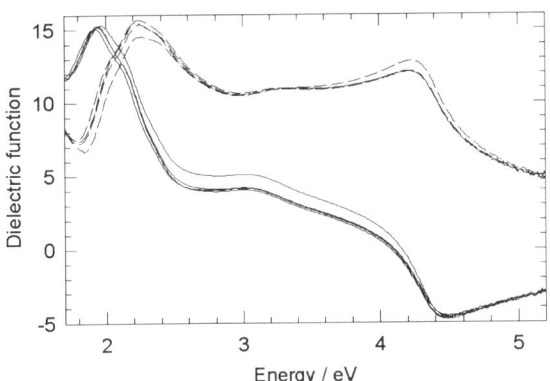

Figure 3: Real (solid) and imaginary (dashed) part of $\langle \varepsilon \rangle$ for four $Si_{1-x-y}Ge_xC_y$ alloys ($x \approx 0.9$, $y = 0..0.02$). The amplitude of $\langle \varepsilon \rangle$ is affected by surface roughness [3], but there are no obvious differences in peak positions between the four samples, although the C content changes. The critical-point parameters for the four samples are also very similar.

DOPED SILICON-GERMANIUM ALLOYS

The doped $Si_{1-x}Ge_x$ samples were grown by the Czochralski method. Crystals were pulled from a melt of Si, Ge, and B. Hydrostatic weighing (Archimedes' law) was used to obtain the sample composition, see Table I. The doping level (hole concentration p) of the samples was found using resistance and Hall effect measurements. For comparison, we also studied commercial bulk Si and Ge.

Figure 4 shows $\langle \varepsilon \rangle$ for doped Si and doped $Si_{0.72}Ge_{0.28}$ (sample 6) in comparison with undoped Si. The spectra of the doped samples resemble those of the undoped ones, but the peaks are broadened and shifted towards lower energies due to the scattering of the electron Bloch waves by the potential of the ionized impurities [4]. We determine the CP energies and broadenings accurately by fitting the derivatives of our spectra with analytical lineshapes. We find that E_1 and E_2 shift downward by about 65 and 35 meV, respectively, (see Fig. 5) relative to the critcal points in undoped $Si_{1-x}Ge_x$ [1]. This agrees qualitatively with the results of [4] for bulk Si. The observed broadenings in doped $Si_{1-x}Ge_x$ are larger than in bulk Si (see Fig. 6) for two reasons: (i) alloy broadening, (ii) broadening due to doping as described in [4]. Since the alloy broadenings in undoped $Si_{1-x}Ge_x$ have never been studied systematically, we are unable to separate the two contributions. A small E_1 redshift was also seen in boron-doped $Ge_{1-y}C_y$ alloys.

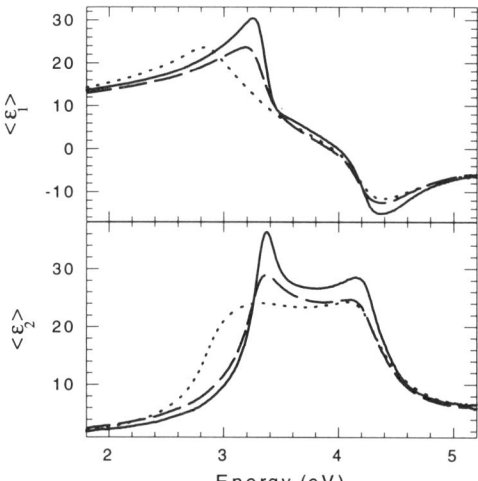

Figure 4: Pseudodielectric functions of undoped bulk Si (solid), doped Si (sample 1, dashed), and doped $Si_{1-x}Ge_x$ (sample 6, dotted). Spectra for undoped $Si_{1-x}Ge_x$ are given in [1].

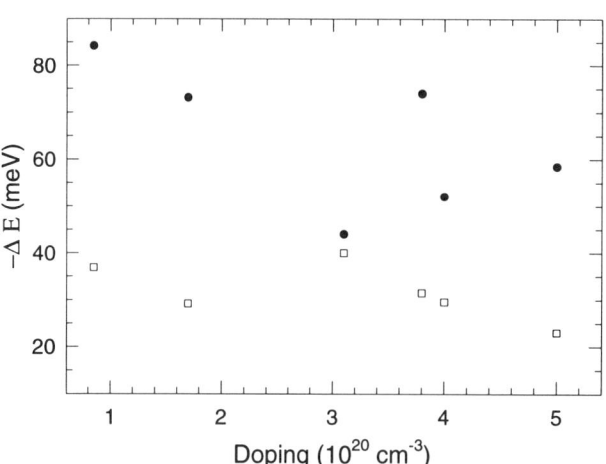

Figure 5: Energy shifts of the E_1 (●) and E_2 (□) critical point energies for $Si_{1-x}Ge_x$ alloys with different compositions as a function of doping concentration. The energies of the critical points in undoped $Si_{1-x}Ge_x$ have been subtracted. The error for ΔE is about 5 meV if the critical points are well separated, larger if they overlap.

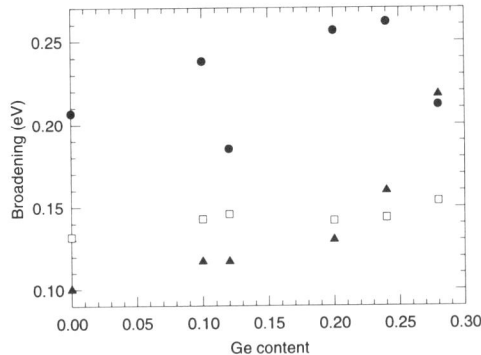

Figure 6: Broadenings of the E_1 (●), E_0' (▲), and E_2 (□) critical points of doped $Si_{1-x}Ge_x$ alloys as a function of composition. The doping concentrations are given in Table 1. It is not possible to subtract the broadenings of the undoped alloys, since these values are not well known. The error is about 0.01 eV if the critical points are well separated, larger if they overlap.

SUMMARY

The pseudodielectric functions and critical-point parameters of the $Ge_{1-y}C_y$ and Ge-rich $Si_{1-x-y}Ge_xC_y$ alloys studied here are not affected by small amounts of carbon. The critical points for boron-doped $Si_{1-x}Ge_x$ alloys shift to lower energies and broaden in a similar fashion as in bulk doped Si and Ge.

ACKNOWLEDGEMENTS

Ames Laboratory is operated for the U.S. Department of Energy by Iowa State University under Contract No. W-7405-Eng-82. The work at Ames was supported by the Director for Energy Research, Office of Basic Energy Sciences, and by the Iowa Space Grant Consortium. JMD acknowledges support through a Bernice Black Durant Undergraduate Research Scholarship. JH is grateful to the International Institute for Theoretical and Applied Physics and the Department of Physics and Astronomy at Iowa State University for supporting his stay in Ames. The work at Delaware was supported by AFOSR (F49620-95-0135), ARO (DAAH04-95-1-0625), and ONR (N00014-93-1-0393). The boron-doped $Si_{1-x}Ge_x$ alloys were grown by M.G. Kekoua and E.V. Khoutsishvili.

REFERENCES

1. J. Humlícek, M. Garriga, M.I. Alonso, and M. Cardona, J. Appl. Phys. **65**, p. 2827 (1989).
2. C. Pickering and R.T. Carline, J. Appl. Phys. **75**, p. 4642 (1994).
3. K.E. Junge, N.R. Voss, R. Lange, J.M. Dolan, S. Zollner, M. Dashiell, D.A. Hits, B.A. Orner, R. Jonczyk, and J. Kolodzey, Thin Solid Films (in print), and references therein.
4. M. Krishnamurthy, B.-K. Yang, and W.H. Weber, Appl. Phys. Lett. **69**, p. 2572 (1996).
5. L. Viña and M. Cardona, Phys. Rev. B **29**, p. 6739 (1984); **34**, p. 2586 (1986).
6. O.F. Sankey, A.A. Demkov, W.T. Petuskey, and P.F. McMillan, Modelling Simul. Mater. Sci. Eng. **1**, p. 741 (1993).
7. M.A. Berding, A. Sher, and M. van Schilfgaarde, Phys. Rev. B **56**, p. 3885 (1997).
8. H.J. Osten, E. Bugiel, and P. Zaumseil, J. Crystal Growth **142**, p. 322 (1994).
9. D.E. Aspnes and A.A. Studna, Phys. Rev. B **27**, p. 985 (1983).

Part III

Photonics and Optoelectronics

LIGHT EMISSION FROM ERBIUM DOPED Si$_{1-x}$Ge$_x$ HETEROSTRUCTURES

J H EVANS-FREEMAN, A T NAVEED, M Q HUDA*, A R PEAKER, D C HOUGHTON**, A C WRIGHT***
Centre for Electronic Materials, UMIST, PO Box 88, Manchester, M60 1QD, UK
*Dept EEE, Bangladesh University of Engineering and Technology, Dhaka-1000, Bangladesh
** SiGe Microsystems Inc, Ottawa, Canada
*** North East Wales Institute of Higher Education, Wrexham, UK

ABSTRACT

UHVCVD-grown Si$_{0.87}$Ge$_{0.13}$/Si heterostructures have been implanted with erbium, and photoluminescence and electroluminescence centred on 1.54μm have been studied. Implantation conditions were chosen so that the erbium concentration profile was flat over the spatial location of the SiGe quantum well region. We demonstrate that the technology of implantation and regrowth is feasible even when Si/SiGe interfaces are present. We have obtained more intense photoluminescence from erbium implanted SiGe heterostructures than that from a silicon layer implanted with a higher erbium dose. We report forward bias electroluminescence from the Er doped SiGe/Si heterostructures; the photoluminescence and electroluminescence from these structures demonstrates that the detailed mechanism of excitation is different from the Er:Si case.

INTRODUCTION

The use of rare earth doped semiconductors presents new opportunities for optoelectronic devices. Emission from the rare earth erbium is due to an intra-f-shell transition, and occurs at the technologically important wavelength of 1.54μm. Therefore, erbium doping of silicon is of particular interest because it provides a route to realising a silicon-based optical emitter at this wavelength. However, erbium doping of silicon-germanium/silicon heterostructures should offer several advantages when considering the design of an optical emitter, and is the preferred structure for an electrically excited laser. There have recently been several advances in the application of silicon-germanium alloys as waveguides [1,2]; the heterostructures will provide optical and carrier confinement, as in the case of III-V heterostructure lasers

Implantation of a heavy atom, such as erbium into silicon, amorphises the region of the implant, and care must be taken to ensure that regrowth does not result in further defect formation. Recrystallisation and optical optimisation is carried out by combinations of Solid Phase Epitaxial (SPE) regrowth and Rapid Thermal Annealing (RTA) [3,4]. Further self-implantations of silicon are often applied to ensure the layer is fully amorphous up to the surface [5]. In this work, we report implantation and optical activity of erbium implanted into silicon layers which contain strained SiGe quantum wells. There have been reports of regrowth of *relaxed* SiGe[6], but for good carrier injection, the SiGe layers in this work must remain strained. Paine[7] showed that epitaxial regrowth of strained Si$_{1-x}$Ge$_x$ alloy layers continues for a certain critical distance after the crystalline/amorphous (c/α) growth front crosses the heterointerface. After this limit, the SPE regrowth is accompanied by the formation of strain relieving misfit dislocations. Therefore, the annealing schedules which ensure good regrowth will be a critical factor in this technology, if both the strain and good

Mat. Res. Soc. Symp. Proc. Vol. 533 © 1998 Materials Research Society

optical performance are to be preserved. Defects in relaxed layers could provide non-radiative competition to both the excitation and de-excitation of the erbium in the layers.

In this paper we report a study of the structure and optical activity of Si/SiGe heterostructures that have implanted with erbium. The layers were studied by Transmission Electron Microscopy (TEM) and X-ray diffraction, to ensure that the correct annealing schedule was implemented to protect the interface quality. Photoluminescence (PL) and electroluminescence (EL) have been carried out as a function of temperature, and in the case of EL, drive current.

SAMPLE PREPARATION

Erbium was implanted into a UHVCVD-grown Si/SiGe layer (sample A), and also into crystalline silicon for comparison (sample B). Sample A consisted of ten SiGe layers, thickness 70Å with 13% Ge, each separated by 288Å silicon on a p-type silicon substrate, capped with 288Å silicon. The as-grown sample was of high quality and showed zero-phonon and TO-phonon PL related to the quantum wells. Implantation profiles were calculated by TRIM simulation, and the samples were implanted with erbium in the range of 0.3-1.0 MeV to produce a flat erbium profile throughout the SiGe structures. The peak erbium concentration was 10^{18} cm^{-3}. The residual oxygen level in the starting material was greater than this and therefore no oxygen was implanted [8]. Following the erbium implantation, the samples were implanted with silicon, to ensure that there was a continuous amorphous layer up to the surface. Sample A was then annealed at 550°C for 5 hours in a nitrogen ambient to facilitate Solid Phase Epitaxial (SPE) regrowth. The flat erbium concentration profile in the region of the SiGe low dimensional structures was verified by SIMS measurements. A subsequent Rapid Thermal Anneal (RTA) at 900°C was also applied to sample A to see if this increased the optical activity of the implanted erbium, as is observed by us for the case of pure silicon. Interestingly, the optical activity of the erbium was severely degraded by this step, which we attribute to degradation of the quantum well structures at these elevated temperatures.

Sample A was assumed to contain a rectifying p-n junction by virtue of the fact that the erbium introduces donors [9] and this was found, by I-V measurements, to be the case. Devices were designed in order that holes were injected into the active (erbium doped) region, to study the effect of the shape of the valence band upon the EL spectra. The SiGe quantum wells, being grown on silicon, exhibit the majority of the bandgap discontinuity in the valence band and it is predicted that hole trapping here will influence the excitation/de-excitation of the erbium.

P-type silicon was also implanted with multiple implants of erbium, oxygen, silicon (to post-amorphize) and finally with arsenic to provide extra carriers in the n-region, assuming the erbium acts as a donor. The peak erbium concentration was 4×10^{18} cm^{-3}. Table 1 lists all the implantation details and resulting peak concentrations. The peak concentrations of Er in sample A is only 25% of that in sample B, and the total dose was only 1/7th of that implanted into sample B. Sample B was designed in order that the erbium implant was located below the arsenic implant (to avoid any further defect formation between implanted erbium and donor), hence the higher implant energies. The most efficient luminescence from sample B was achieved after an anneal schedule that consisted only of an RTA at 1100°C for 90 seconds.

The EL samples were fabricated using 4mm^2 samples mounted on ceramic. Aluminium was used for the top and bottom contacts. The shape and size of the sample pieces is analogous to using 4mm^2 mesa structures, which were also fabricated.

TEM analysis was carried out in a Philips EM430ST machine operating at 300kV. The X-ray diffraction experiments were performed using a Bede 300 double crystal X-ray diffraction machine with the CuKα_1 line as the radiation source. For the PL and EL experiments the samples were placed in a Leybold refrigeration cryostat, at a minimum temperature of 5.5K. Excitation was with either with the 488nm line of an argon ion laser, or minority carrier injection under forward bias. The luminescence signal was dispersed with a spectrometer, detected by a cooled germanium detector and analysed with conventional lock-in techniques.

SAMPLE NO:	ERBIUM IMPLANTATION DETAILS			SILICON IMPLANT DETAILS		OXYGEN	
	Energy (MeV)	Dose (cm^{-2})	Peak Conc. (cm^{-3})	Energy (keV)	Dose (cm^{-2})	Energy (keV)	Dose (cm^{-2})
A	1.0	1.2×10^{13}	1.0×10^{18}	480	6.0×10^{15}	------	------
	0.5	0.5×10^{13}		200	2.5×10^{15}		
	0.3	0.3×10^{13}		80	1.5×10^{15}		
B	1.5	9.5×10^{13}	4.0×10^{18}	480	6.0×10^{15}	260	3×10^{14}
	1.0	4.7×10^{13}		200	2.5×10^{15}	200	7×10^{14}
				80	1.5×10^{15}	140	5×10^{14}

Table 1 *Details of the erbium, oxygen and silicon implants in samples A and B*

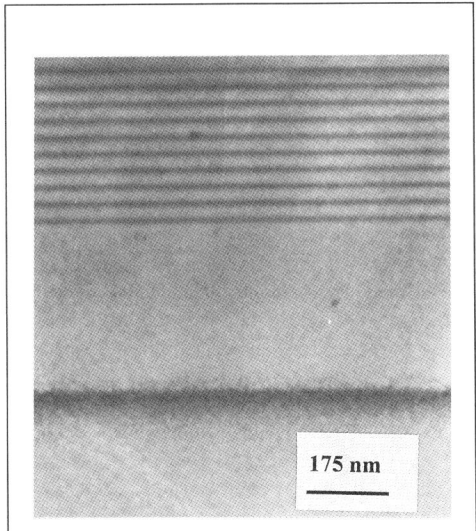

Figure 1. *TEM of Er-doped 10 period Si/Si$_{0.87}$Ge$_{0.13}$ quantum wells after Er and Si implant and regrowth*

RESULTS AND DISCUSSION

Figure 1 shows the cross-sectional TEM of sample A, after erbium and silicon implantation and the SPE regrowth. It can be clearly seen that there are no visible extended defects in the layer, which suggest that the strain in these layers has been retained. There is a line of small defects approximately 1μm below the surface. These are due to end-of-range defects, but will not be detrimental to device performance if they are some distance from the active region, i.e. the quantum wells.

Figure 2 shows the X-ray diffraction (XRD) rocking curve from the implanted and regrown sample A, and also the rocking curve from the as-grown sample A. This confirms the retention of

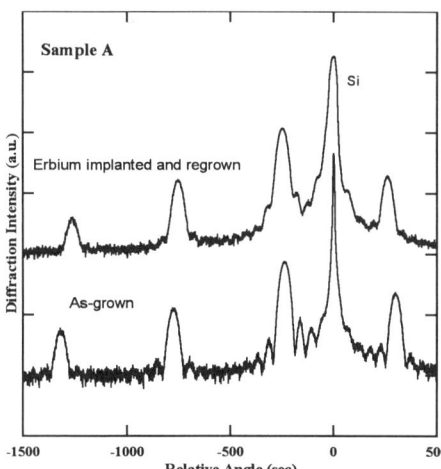

Figure 2. XRD from Er-implanted Si/SiGe after implant and regrowth

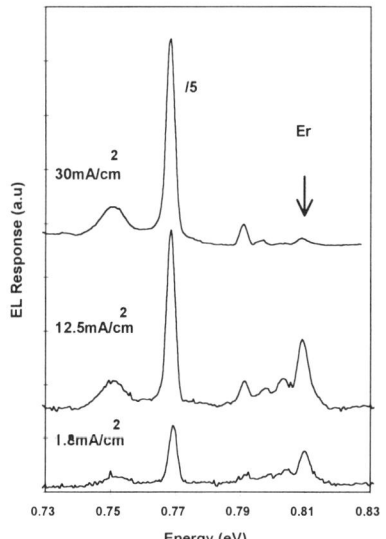

Figure 3. EL of Er-doped Si/SiGe structures as a function of current ,sample A.

strain in this sample after implantation and regrowth. The diffraction patterns both consist of a strong peak originating from the silicon substrate, together with satellite peaks originating from the quantum wells. The small ripples seen in the data are Pendelossung fringes, which arise because of interference between the incident wave and the interfaces. The presence of these fringes in the data from the implanted layer is a good indication that the interface quality has not degraded by out - diffusion of germanium, and the presence of the clear satellite peaks indicates that the strain is retained.

We have previously reported that PL from Er in these Si/SiGe quantum structures is more intense and efficient than PL originating from higher doses of erbium in silicon [5]. Figure 3 shows EL measurements on sample A as a function of current density. Weak but clear peaks in the spectra in the region of the erbium emission energy (maximum at 0.810eV) are observed at low current densities, but a defect peak at 0.76eV is also apparent. This is possibly due to end-of-range damage excited because minority carrier injection is occurring in both directions. This peak can be used as an internal calibration on the efficiency of the erbium excitation as a function of current. Interestingly, as the current increases, the Er-related peaks reduce in intensity, and the defect peaks increase in intensity. The current density could only be increased by increasing the voltage across the device, despite the fact that the diode was in forward bias, which we attribute to the high series resistance measured in this diode. The net effect of this is that the diode is subject to higher electric fields across the erbium-doped and p regions as the current is increased. As the field strength is increased, the holes in the SiGe quantum wells may be excited out of the wells by the field, and therefore free to recombine at other defects in the layer. PL data from the as-grown quantum wells suggest the confinement energy for the holes is less than 70meV.

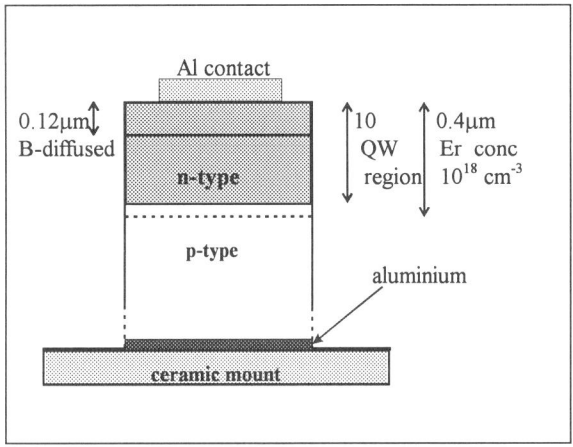

Figure 4. Structure of B-diffused EL sample A1

The weak erbium-related EL signal from sample A could be attributed to two effects. Firstly, the erbium concentration is lower than that generally employed for the fabrication of LEDs using the doping by the erbium[10]. However, in sample A the PL was more intense and efficient at higher temperatures than in erbium-doped crystalline silicon. The second effect, assumed to be the principal reason, is that sample A relied upon a junction between the n-type erbium region, and the p-type substrate. This is actually a very diffuse junction, because of the nature of the range and straggle of the erbium implant. The large area of the device also means that this device was quite leaky at reverse bias. Therefore, a piece of sample A (called sample A1) was diffused with boron from the front surface. This was carried out for two reasons, 1) to move the injecting junction nearer the surface and away from the end-of-range defects, and 2) to inject a higher concentration of holes into the erbium doped region.

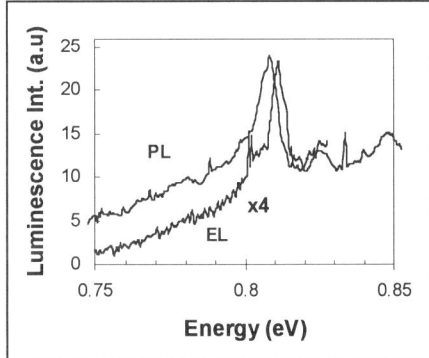

Figure 5. EL and PL of B-diffused Er-doped Si/SiGe structures.

Figure 5 shows PL and EL (at 50K) from sample A1. The electric field present in this diode should be largely across the p-type region because of the polarity of the connections. This is not in the region of erbium doping or the quantum wells. As the current (i.e. field) increases across the diode the defect peaks detected in sample A were not observed in sample A1, and the EL intensity at the erbium wavelength increased until limited by series resistance[10]. It can be seen that the PL and EL spectra in Fig 5 are also similar but the principal emission wavelength is lower in the EL case. This is not due to temperature variations: the EL peak energy from the sample did not vary in the temperature range 50-170K. It has been reported previously that forward-biased EL and PL occur at slightly different emission energies to reverse-biased EL[10]. This was attributed in ref 10 to either excitation of different Er sites, or excitation of a larger set of the Er population in the reverse bias EL case. In our case the confinement of the holes in the quantum wells may have contributed to the excitation of a larger set of the Er population. The EL and PL intensities are comparable to that of erbium-doped silicon at low temperatures, even though the Er dose and peak concentration in the SiGe/Si layer are lower. PL and forward bias EL from sample B at 6K (not shown) exhibit very similar emission characteristics however, with no change in emission wavelength. This is to be expected because there are no quantum wells in this layer. Sample B also exhibited room temperature EL.

It was not possible to carry out a detailed study of the temperature dependence of the EL from Si/SiGe samples as the current densities through the diode were small because of the low residual doping in the as-grown layer, and carrier freeze-out at low temperatures.

CONCLUSIONS

We have implanted erbium into Si/SiGe quantum well structures, and demonstrated that good regrowth can be achieved, despite the presence of heterointerfaces in the implanted region. We have compared the PL and forward-bias EL at 1.54μm from Er-doped crystalline silicon, and the Er-doped SiGe/Si heterostructures. In both cases excitation was by hole injection into the Er-doped region. The Er-doped silicon structure, with implants designed to fabricate an LED structure, exhibited EL at room temperature, and the low temperature PL and forward-bias EL exhibited the same spectral features. The Er-doped SiGe sample was not specially designed for electrical excitation, nevertheless we were able to detect forward-bias EL by using the self-doping effect of the erbium in the p⁻ substrate. A controlled boron diffusion was also carried out on the Si/SiGe samples to provide further hole injection. Forward-bias EL from the Si/SiGe sample yielded a lower wavelength emission than PL, which is attributed to excitation of a larger set of the Er population than in the Er-doped Si case. This is supported by the fact that the erbium concentration was lower in the Si/SiGe sample, but PL and EL intensities are comparable to the intensities from Er-doped crystalline silicon.

ACKNOWLEDGEMENTS

This work was funded by the Engineering and Physical Sciences Research Council, United Kingdom. ATN gratefully acknowledges the financial support of MoE, (Pak) and NIST (Pak).

REFERENCES

1. Z Yang, B L Weiss, G Shao and F Namavar, J Appl Phys **77**, 2254 (1995)
2. D-X Xu, S Janz, R Williams, E Allegretto, S Mailhot, J J He, J-M Baribeau, H Lafontaine, J Stapledon, J W Fraser, M Robillar, P Jessop ad S Kovacic, SPIE **3007**, 178 (1997)
3. A Polman, J S Custer, E Snoeks and G N van den Hoven, Nucl Inst and Methods in Phys Res, **B80/81**, 653 (1993)
4. B Zheng, J Michel, F Y G Ren, L C Kimerling, D C Jacobson and J M Poate, Appl Phys Lett **64**, 2842 (1994)
5. M Q Huda, A R Peaker, J H Evans-Freeman, D C Houghton and W P Gillin, Electron Lett **33**, 1182 (1997)
6. S Q Hong, Q Z Zong and J W Mayer, Appl Phys Lett **63**, 2053 (1993)
7. D C Paine, N D Evans and N G Stoffel J Appl Phys **70**, 4278 (1991)
8. G Franzo, F Priolo, S Coffa, A Polman and A Carnera, Appl Phys Lett **64**, 2235 (1994)
9. J L Benton, J Michel, L C Kimerling, D C Jacobson, Y-H Xie, D J Eaglesham, E A Fitzgerald and J M Poate, J Appl Phys **70** 2667 (1991)
10. G Franzo, S Coffa, F Priolo, C Spinella, J Appl Phys, **81**, 2784 (1997)

ERBIUM DOPED Si/SiGe WAVEGUIDE DIODES:
OPTICAL AND ELECTRICAL CHARACTERIZATION

E. NEUFELD, A. LUIGART, A. STICHT, K. BRUNNER and G. ABSTREITER
Walter Schottky Institut, Technische Universität München, Am Coulombwall,
D-85748 Garching, Germany

ABSTRACT

We have fabricated erbium- and oxygen-doped Si/SiGe waveguide diodes showing the characteristic 1.54 µm electroluminescence (EL) from incorporated Er^{3+} ions. All samples were grown by molecular beam epitaxy (MBE). The EL from the polished end facet of the waveguide was measured with a confocal microscope revealing a spatially narrow emission. Additional annealing was not necessary to improve the luminescence characteristics. Only a weak temperature dependence is found for the EL intensity between 4K and room temperature.

INTRODUCTION

The leading position of silicon for microelectronic applications raises a growing interest for optoelectronic devices compatible with silicon based technology. In particular, efficient light emitters are a basic requirement for optical data transfer. As has been shown by Ennen and coworkers in 1983 [1] and 1985 [2], both photo- and electroluminescence at 1.54 µm can be observed from erbium ions incorporated into a crystal silicon host. This infrared emission results from an intra-4f-transition of the excited Er^{3+} ions. The wavelength of 1.54µm lies in the minimum loss window of silica based fibers and is thus of technological importance. In the following years many efforts have been undertaken to achieve efficient electroluminescence at room temperature from the Si:Er material system [3-6]. Codoping with oxygen has turned out to play an important role to achieve incorporation of erbium to concentrations far above the solid solubility and also higher EL efficiency.

The SiGe alloy has a higher refractive index than silicon allowing the fabrication of waveguides and waveguide detectors which are compatible with existing silicon technology [7-12]. Thus, investigating the SiGe:Er system is a straight forward step towards optoelectronic devices like electrically pumped optical amplifiers etc. PL from erbium ions in a SiGe host has already been studied systematically for various germanium concentrations in a previous paper [13]. Here we present the realization of erbium-doped Si/SiGe waveguide diodes showing 1.54 µm electroluminescence at room temperature.

EXPERIMENT

All samples presented in this paper were grown by molecular beam epitaxy. In particular, erbium was evaporated by a standard Knudsen cell and molecular oxygen was supplied to the substrate through a quartz glass nozzle. The substrate temperature was kept at 500°C. Two dimensional layer growth was controlled by reflection high energy electron diffraction (RHEED).

First a 1 μm Si$_{1-x}$Ge$_x$:Er:O layer was deposited on top of a boron doped p$^+$ substrate ([B]=10^{19} at./cm^3) with erbium and oxygen concentrations of 4×10^{19} Er at./cm^3 and 2×10^{20} O at./cm^3, respectively. The germanium concentration x was chosen to be 12% and 30% for different samples. Finally, a 500 nm phosphorus doped silicon layer ([P]=10^{19} at./cm^3) has been deposited on top. Concentration profiles of the incorporated elements were investigated by secondary ion

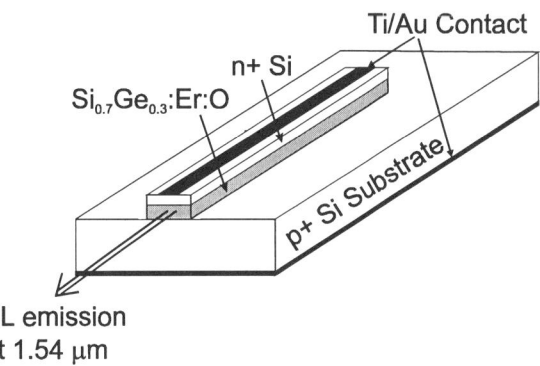

Fig. 1: Structure of a processed waveguide diode.

mass spectroscopy (SIMS). No sample annealing was required to optically activate the erbium ions in the MBE grown samples.

A schematic view of the processed waveguide diode is given in Fig. 1. Stripe mesas with a width of 150 μm and various lengths have been etched to a depth of 1.5 μm. The samples have been thinned from the back to a thickness of approximately 300 μm and were contacted by Ti/Au evaporation onto the top of the mesa waveguide and onto the backside of the thinned substrate. For the exit of the guided 1.54μm light wave the end facets of the waveguides have been polished mechanically using diamond paste. The metal contact on top of the mesa was chosen to have a width of only 17 μm in order to also allow EL detection from the surface.

The 1.54 μm emission from the waveguide was collected by a confocal microscope system and dispersed by a single grating monochromator. A CCD camera was used to control the microscope focus on the waveguide end facet. A liquid nitrogen cooled germanium detector was employed in conjunction with a lock-in amplifier to monitor the signal.

RESULTS

The I-V characteristics of the waveguide diodes are shown in Fig. 2. Due to the donor behavior of the erbium-oxygen-complexes [14] the pn-junction of the diode is given by the p$^+$-Si/SiGe:Er:O interface.

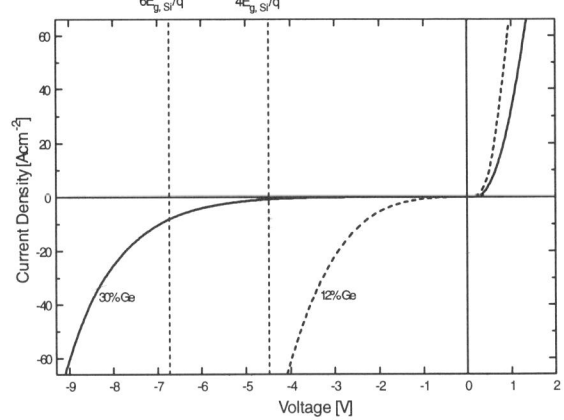

Fig. 2: Current density vs. applied voltage for waveguide diodes with different Ge concentrations in the SiGe:Er:O layer.

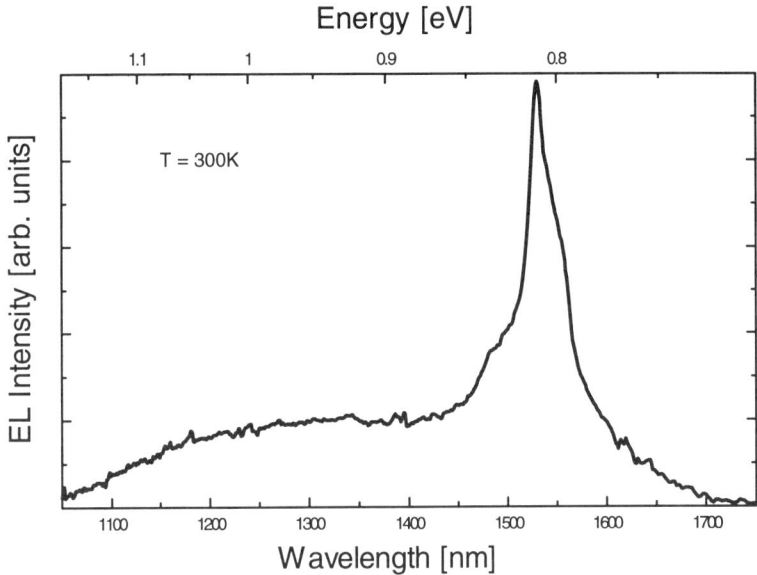

Fig. 3: Electroluminescence emission from the end facet of a $Si/Si_{0.7}Ge_{0.3}$:Er:O waveguide diode under reverse bias (-14 A/cm^2) recorded at room temperature.

The diodes are in the recombination regime under forward bias. Remarkable differences can be observed in the reverse bias regime. The I-V-characteristics of the diodes with 12% germanium in the active layer resemble those of Si:Er:O-diodes grown without germanium [5] showing an early and smooth onset of the current. The low breakdown voltage around $4E_g/q$ (E_g being the bandgap energy and q the electron charge) is characteristic for a tunneling current [15]. This interpretation is supported by the fact that the breakdown current increases with increasing temperature for a fixed voltage due to bandgap narrowing. Samples with 30% germanium in the active layer are also characterized by a smooth onset of the breakdown current, however at a significantly higher voltage around $6E_g/q$. This behavior indicates that part of the current is due to avalanche breakdown [15]. Again this interpretation is supported by the temperature dependence of the breakdown current, which in this case decreases with increasing temperature. A possible explanation for the different behavior under reverse bias could be a reduced carrier concentration at the p$^+$-Si/SiGe:Er:O interface due to dislocations and defects introduced by strain relaxation. This in turn would lower the probability for a tunneling current. It should be mentioned, that in either case the diodes show no degradation after operation for many hours under both biasing conditions.

As has already been reported for Si:Er:O light emitting diodes we also observe from these samples a much stronger electroluminescence under reverse bias than under forward bias. A reverse bias EL spectrum measured at room temperature from the end facet of the waveguide diode with 30% Ge is shown in Fig. 3. The characteristic erbium emission can be seen at wavelengths around 1.54μm. The excitation of erbium under reverse bias is achieved by impact of hot carriers. The flat EL signal at shorter wavelengths has been ascribed to radiative intraband transitions [16].

The strongest reverse bias electroluminescence at 1.54μm for a given current is observed from the waveguide diodes containing 30% germanium in the active layer. This might be understood by the significantly higher voltage as compared to the samples with 12% Ge (see Fig. 2). A higher voltage should result in a higher number of electrons with an energy greater than 0.81 eV. The latter energy value is required to excite an Er^{3+} ion by impact. It is interesting to note that the waveguide diode with 30% Ge is also characterized by a smaller reduction of EL intensity than the sample with 12% Ge when the temperature is raised from 4K to 300K. Keeping the current constant the intensity is reduced by a factor of 4 in the former case and by 20 in the latter case.

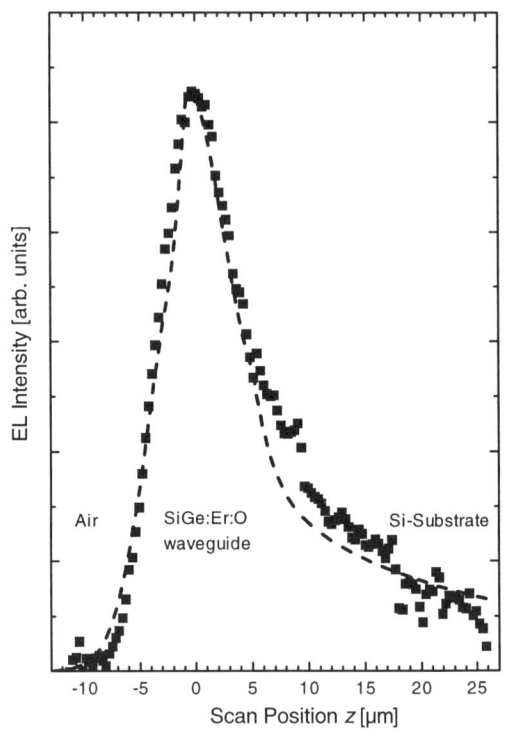

Fig. 4:EL scan at 1.54μm over the edge of the waveguide diode.

One should keep in mind that the reverse bias voltage increased slightly during this procedure for the sample with 30% Ge and decreased for the other one.

To characterize the 1.54μm emission out of the end facet of the waveguide we have performed an electroluminescence scan over the edge of the sample. The result is shown in Fig. 4 where the waveguide can be found at zero position. The top of the sample can be found on the left hand side and the p^+ Si substrate on the right. Towards the top of the sample (interface to air) the EL signal decreases to half of the maximum value on a length scale of approximately 4.5μm which corresponds to the spatial resolution of the experimental setup. An asymmetrically broadened intensity tail can however be observed in the direction to the substrate. The dashed curve in Fig. 4 is a theoretical calculation of the scan profile taking into account two different contributions. One component has a Gaussian profile representing the guided 1.54μm emission broadened by the detecting system. The other one takes into account that a big part of the EL generated at the p^+-Si/SiGe:Er:O interface is emitted into the substrate. The calculated curve shows good agreement with the experimental one. About 70% of the EL intensity at $z = 0$ is attributed to waveguide emission.

EL from erbium ions could also be measured from the top of the mesas. The spectral shape is the same as shown in Fig. 3 for the waveguide emission. By opening the slits of the monochromator and thus increasing the imaged sample area we observe a different behavior for the two detection geometries (Fig. 5). In surface geometry the corresponding EL signal at 1.54µm increases linearly as expected for a large emission area.

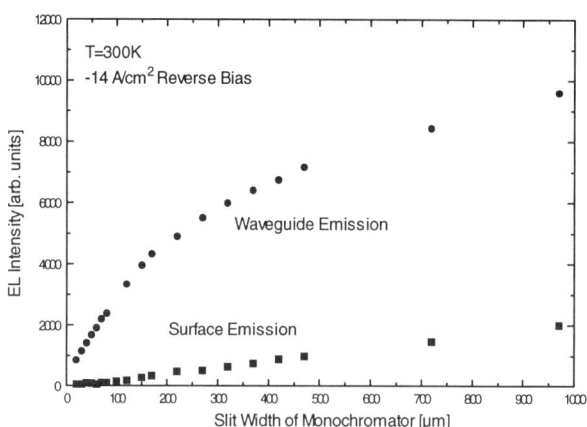

Fig. 5: EL intensity at 1.54µm for different monochromator slit widths in edge and surface geometry

When detecting the emission from the end facet of the waveguide the signal increases considerably faster initially and than starts to saturate for bigger slit widths. This confirms again the spatially narrow emission from the end facet with a higher power density.

During the EL measurements in surface geometry we have discovered an interesting feature when additionally focusing the 488nm line of a cw argon-ion laser onto the sample surface. This results in a strong increase of the EL signal. We attribute this to additional impact excitation of erbium ions by photogenerated carriers after acceleration in the pn-region. A further investigation of this phenomenon will be carried out in the near future.

CONCLUSIONS

We have investigated Si/SiGe:Er:O waveguide diodes grown completely by MBE. Samples with the high germanium concentration (30%) are characterized by a significantly higher breakdown voltage under reverse bias as compared to the samples with 12% Ge. The samples with 30% Ge also show the strongest electroluminescence at 1.54µm and a very weak temperature dependence of the EL intensity. Electroluminescence spectra were measured both in surface and in edge geometry, the latter showing spatially narrow emission from the end facet of the waveguide.

ACKNOWLEDGEMENTS

We thank U. Breuer for SIMS measurements. The samples were grown in a converted Riber MBE system which was provided by H. Kurz of RWTH Aachen, Germany. This project is supported financially by the Volkswagen Stiftung via Photonik programme and by the Bayerische Forschungsstiftung via FOROPTO.

REFERENCES

1. H. Ennen, J. Schneider, G. Pomrenke, and A. Axmann, Appl. Phys. Lett. **43**, 943 (1983).

2. H. Ennen, G. Pomrenke, A. Axmann, K. Eisele, W. Haydl, and J. Schneider, Appl. Phys. Lett. **46**, 381 (1985).

3. S. Coffa, G. Franzò, F. Priolo, A. Polman, and R. Serna, Phys. Rev. B **49**, 16313 (1994).

4. B. Zheng, J. Michel, F.Y.G. Ren, L.C. Kimerling, D.C. Jacobson, and J.M. Poate, Appl. Phys. Lett. **64**, 2842 (1994).

5. J. Stimmer, A. Reittinger, J. F. Nützel, G. Abstreiter, H. Holzbrecher, and Ch. Buchal, Appl. Phys. Lett. **68**, 3290 (1996).

6. J. Stimmer, A. Reittinger, E. Neufeld, G. Abstreiter, H. Holzbrecher, U. Breuer, and Ch. Buchal, Thin Solid Films **294**, 220 (1996).

7. F. Namavar, and R.A. Soref, J. Appl. Phys. **70**, 3370 (1991).

8. R.A. Soref, F. Namavar, and J.P. Lorenzo, Opt. Lett. **15**, 270 (1990).

9. K. Bernhard-Höfer, A. Zrenner, J. Brunner, G. Abstreiter, F. Wittmann, I. Eisele, Appl. Phys. Lett. **66**, 2226 (1995).

10. V.P. Kesan, P.G. May, E. Bassous, and S.S. Iyer, in *International Electron Devices Meeting Technical Digest* (IEEE, New York, 1990), p. 637.

11. A. Splett and K. Petermann, IEEE Photon. Technol. Lett. **6**, 425 (1994).

12. A. Splett, T. Zinke, K. Petermann, E. Kasper, H. Kibbel, H.-J. Herzog, and H. Presting, IEEE Photon. Technol. Lett. **6**, 425 (1994)

13. E. Neufeld, A. Sticht, K. Brunner, G. Abstreiter, H. Holzbrecher, H. Bay, Ch. Buchal, Appl. Phys. Lett. **71**, 3129 (1997).

14. F. Priolo, S. Coffa, G. Franzò, C. Spinella, A. Carnera, V. Bellani, J. Appl. Phys. **74**, 4936 (1993).

15. S. M. Sze, *Physics of Semiconductor Devices*, 2nd ed. (John Wiley & Sons, New York, 1981).

16. W. Haecker, phys. stat. sol. (a) **25**, 301 (1974).

EVIDENCE FOR HETEROJUNCTION EFFECTS IN POLYCRYSTALLINE Si$_{1-x}$Ge$_x$ THIN FILM TRANSISTORS WITH Si CAPS

ALBERT W. WANG and KRISHNA C. SARASWAT
Department of Electrical Engineering, Stanford University, Stanford, CA 94305

ABSTRACT

Polycrystalline silicon-germanium (poly-Si$_{1-x}$Ge$_x$) thin film transistors (TFTs) were fabricated with and without Si interlayers (caps) to buffer the SiO$_2$ gate dielectric and the Si$_{1-x}$Ge$_x$ channel. Both low temperature processes (≤ 550 °C) compatible with glass for flat panel displays and high temperature processes were used. NMOS TFTs show dramatic performance improvements up to moderate (5 to 10 nm) interlayer thicknesses. In contrast, PMOS TFTs show only small improvements at low interlayer thicknesses (< 5 nm), after which performance declines. Computer simulations using an effective medium model for polycrystalline materials suggest that in addition to interface improvement, a pseudomorphic heterojunction is formed from the strained Si cap and unstrained poly-Si$_{1-x}$Ge$_x$ channel with both conduction and valence band offsets. These offsets play a significant role in inversion layer formation.

INTRODUCTION

Current generation active matrix liquid crystal displays (AMLCDs) use amorphous silicon (α-Si) thin film transistors (TFTs) to control pixel charge and thus pixel light transmission. Although α-Si TFTs are fabricated using low process temperatures compatible with large glass substrates for AMLCDs, α-Si TFT performance is insufficient to allow driver circuit integration directly on the display. Consequently, there has been considerable interest in polycrystalline silicon (poly-Si) and polycrystalline silicon-germanium (poly-Si$_{1-x}$Ge$_x$) TFT technology. These technologies are capable of higher performance suitable for display drivers, as well as having potential for future 3-D integration applications. Poly-Si$_{1-x}$Ge$_x$ TFT technology requires lower thermal budgets than conventional poly-Si TFT technology [1], but to date, the performance of poly-Si$_{1-x}$Ge$_x$ TFTs has not matched that of poly-Si TFTs. An especially glaring deficiency is the difficulty in fabricating NMOS poly-Si$_{1-x}$Ge$_x$ TFTs, which are needed for CMOS circuits.

Previous work on interfaces between SiO$_2$ and single-crystal Si$_{1-x}$Ge$_x$ [2,3,4] indicates higher interface trap densities and negative fixed charge compared to SiO$_2$-Si interfaces. These interface problems can be reduced by inserting a silicon interlayer or cap between the Si$_{1-x}$Ge$_x$ and the SiO$_2$ to buffer the two materials [5]. For this reason, a handful of experimenters have explored the use of Si caps in poly-Si$_{1-x}$Ge$_x$ TFTs [6]. However, these experiments have tended to use high temperature processes incompatible with glass and only a single cap thickness.

However, the insertion of a silicon cap creates a heterojunction in the channel near the gate oxide interface. In single crystal Si$_{1-x}$Ge$_x$, this heterojunction is kept pseudomorphic to prevent dislocations and associated states at the heterojunction interface. Si caps have been used to enhance PMOS mobilities in strained Si$_{1-x}$Ge$_x$ on unstrained Si by strain enhancement and carrier confinement away from the gate oxide [7,8], and have also been used to enhance NMOS mobilities in strained Si on unstrained Si$_{1-x}$Ge$_x$ [9].

It must be considered if and how heterojunction effects will occur in poly-Si$_{1-x}$Ge$_x$ TFTs with Si caps. Consideration of strain and the possibility of a pseudomorphic heterojunction is therefore important. There is considerable evidence for strain in poly-Si and poly-Si$_{1-x}$Ge$_x$ films from experiments on gate workfunction [1], X-ray diffraction, and micromechanics applications. There is also the reasonable possibility of growth of pseudomorphic heterojunctions within individual grains in a TFT channel because of the fabrication method; the channel films of poly-Si and poly-Si$_{1-x}$Ge$_x$ TFTs are often deposited in amorphous form, then crystallized by annealing. The grains generally nucleate at the lower interface and grow upward and outward [10]. For an amorphous Si cap on top of a growing poly-Si$_{1-x}$Ge$_x$ film, it is likely that the Si will attempt to grow epitaxially on the Si$_{1-x}$Ge$_x$ within each grain. This would result in a Si cap in planar tension atop Si$_{1-x}$Ge$_x$, resulting in both a conduction and valence band offset. Strain can also cause mobility changes by removing the energy degeneracy of conduction valleys and hole bands and distorting the curvature of the hole bands, resulting in changes in conduction effective

145

mass and intervalley or interband scattering. However, these effects are directionally sensitive, especially in the case of electrons. Hence, electron mobility enhancement is weakened by the random orientation of the grains. For holes, mobility enhancement will occur in the strained Si cap, but the valence band offset will exclude holes from the cap. Ascertaining the mobility effects of this structure is therefore a complicated matter.

SIMULATION

In order to predict the effects of a Si cap in a poly-Si$_{1-x}$Ge$_x$ TFT, a set of 1-D Poisson equation computer simulations were carried out. The simulation treats a MOS capacitor and does not include a mobility model. The simulations assume an unstrained poly-Si$_{0.80}$Ge$_{0.20}$ channel covered with a poly-Si cap, psuedomorphic within each grain and under in-plane tensile strain. The grain boundary states and intragrain defect states are treated using the effective medium model [11], which spatially averages the effect of these states. The resulting distribution is U-shaped like that used to model gate oxide interfaces. The band gap is assumed also to be spatially averaged, with a value of 1 eV for both Si$_{0.80}$Ge$_{0.20}$ and Si and band offsets of about 0.125 eV on both band edges. The gate oxide is 1000 Å thick SiO$_2$, and its interface state density linearly decreases as the cap thickness is increased [5]. Because the parameter values are not well established, the simulation's value is in uncovering trends with changing cap thickness.

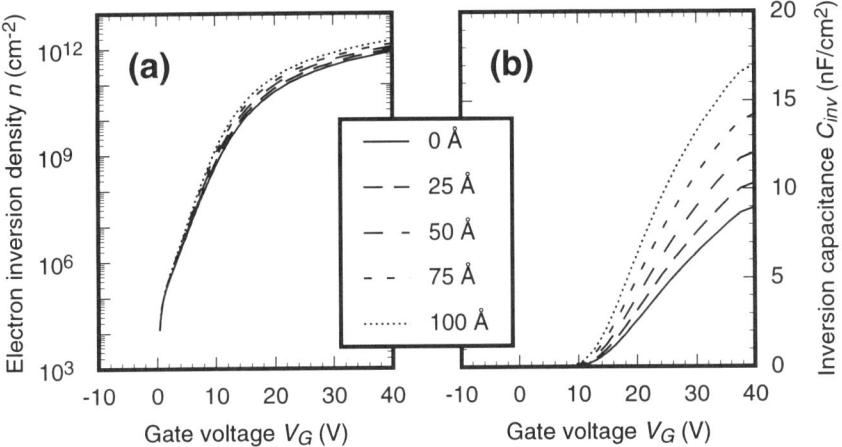

Figure 1. Results of 1-D Poisson simulation, assuming interface improvement but not heterojunction effects: (a) inversion electron area density; (b) inversion electron area capacitance.

Figure 1 shows the electron inversion charge and electron inversion capacitance if the heterojunction effects are ignored but the interface state improvement is treated. As the cap thickness increases, the carrier density increases and the subthreshold slope steepens, implying an improvement in TFT performance. The inversion capacitance also reveals another complication in determining mobility enhancement. In a single-crystal MOS field effect transistor, the gate voltage only induces inversion charge once the device is beyond threshold so that the oxide capacitance and inversion capacitance are equal. This is often assumed in mobility measurements for TFTs. But in this polycrystalline system, gate voltage also induces trapped charge even after the device is turned on, so that the inversion capacitance is lower than the oxide capacitance. Therefore, carrier population effects might be confused with mobility effects in TFTs. Hole effects would be similar in the absence of a heterojunction, but a valence band offset would be expected and the treatment is likely to be invalid.

Figure 2 show the hole and electron inversion charge, respectively, when both heterojunction and interface treatments are treated. The hole population initially increases when the cap is introduced because of the interface improvement. However, as the cap is made thicker, the hole population declines because the valence band offset causes the holes to be excluded from the cap and makes inversion more difficult. This contrasts sharply with the situation for electrons, where the introduction of the cap results in the presence of a potential well right under the gate oxide. The result is a large increase in the number of electrons, which can also be interpreted as a large threshold shift. Further thickening of the cap results in much smaller improvement.

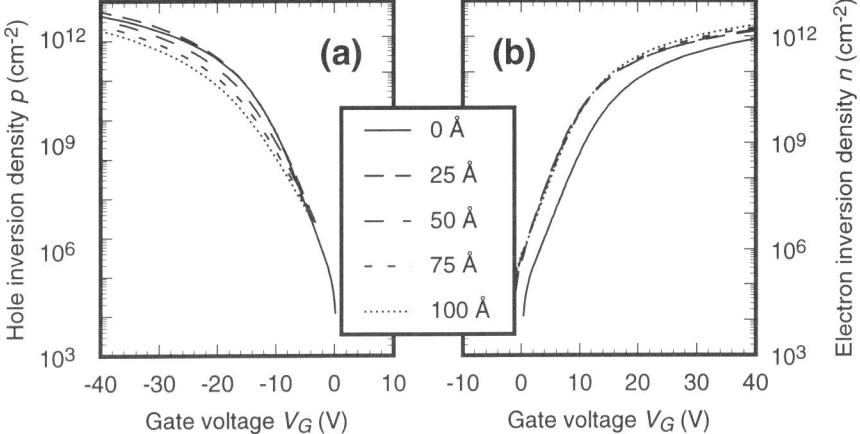

Figure 2. Results of 1-D Poisson simulation, assuming interface improvement and heterojunction effects: (a) inversion hole area density; (b) inversion electron area density.

EXPERIMENT

Poly-$Si_{0.80}Ge_{0.20}$ TFTs were fabricated with a variety of cap thicknesses from 0 to 400 Å. Although oxidized silicon wafers were used to simulate glass substrates, the maximum process temperature was kept to 550 °C for most wafers for compatibility with glass substrates used in AMLCDs. The $Si_{0.80}Ge_{0.20}$ channel layer was deposited in amorphous form at 475 °C. Without breaking vacuum, the temperature was raised to 500 °C and the Si cap is deposited in amorphous form. The sum of the $Si_{0.80}Ge_{0.20}$ and Si cap thicknesses was targeted to 1000 Å. This channel layer was crystallized by annealing at 550 °C for 24 hr, then patterned into device islands. A 1000 Å SiO_2 gate dielectric was then deposited at 450 °C and annealed in inert ambient at 550 °C. A limited selection of wafers was then annealed at 800 °C for 30 min to examine the effects of high temperature exposure. After this, gate electrode film was deposited and patterned, and the sources and drains implanted. Intermetal dielectric and aluminum were then sequentially deposited and patterned. Grain boundary passivation was carried out by immersion in hydrogen plasma at 300 °C for 8 hr. For further comparison, poly-Si devices were also fabricated. The process sequence was identical except that the entire channel film was deposited amorphous at 500 °C and crystallized at 550 °C for 48 hr.

Figure 3 shows the characteristics of PMOS TFTs with a maximum process temperature of 550 °C. It is seen that when the cap is first included, the device characteristics improve. However, the characteristics degrade as the cap is thickened. Some of these effects are easier to observe if the TFT parameter trends are examined These results are consistent with the initial interface improvement and the valence band offset effects observed in the simulation.

Figure 4 shows the characteristics of NMOS TFTs with a maximum process temperature of 550 °C. When the cap is included in the device, a performance improvement or threshold shift is observed, followed by small incremental improvements. Again, the trends are easier to observe from the TFT parameters. Although it is possible to conceive an interface state density consistent with these trends, it is found that these trends match very well with the simulation of Figure 2 based on the hypothesis of a conduction band offset originating from strained Si layer atop an unstrained $Si_{1-x}Ge_x$ layer.

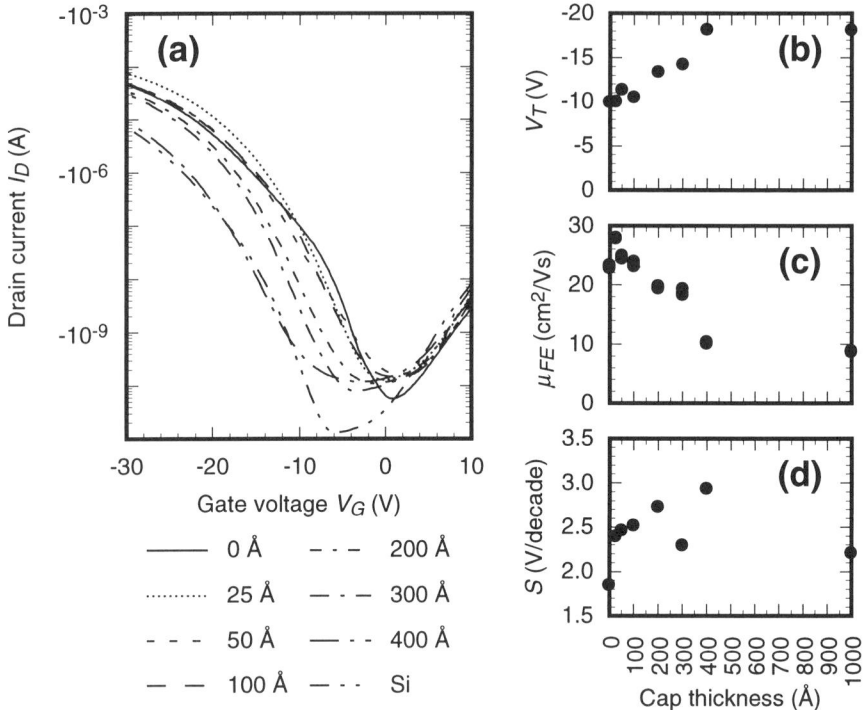

Figure 3. Measurements of 20 μm/20 μm PMOS TFTs fabricated with a maximum process temperature of 550 °C: (a) I_D-V_G characteristics, V_D = -10 V; (b) threshold voltage V_T at I_D = -100 nA; (c) field effect mobility μ_{FE} at V_D = -0.1 V; (d) subthreshold slope S at V_D = -10 V.

The only major disagreement with this explanation is that at high cap thicknesses, device performance begins to degrade. The likely reason is that at high cap thicknesses, the strain in the cap film changes so that the conduction band offset is no longer present. There are two methods by which this could occur. The first is that the cap thickness exceeds the critical thickness for strained layers, which is estimated to be on the order of 100 Å in thickness. Once this happens, dislocations appear between the Si cap and the $Si_{1-x}Ge_x$ base layer, relieving the strain and causing the conduction band offset to disappear. The second is that as the Si cap gets thicker, it begins to exert strain on the $Si_{1-x}Ge_x$ base layer to relieve its own strain. This will cause the conduction band offset to diminish. Although both of these effects affect the valence band offset as well, the effect is much weaker since a valence band offset is present whether or not the cap is strained.

It is known that strained layers can be grown beyond the critical thickness if the process temperature is kept low because of kinetic limitations on the strain relief. An anneal at high temperature will help to relieve the strain. Figure 5 shows the characteristics of NMOS TFTs with 800 °C, 30 min anneal. The trends are very similar to those observed in the simulations of Figure 2 and the measurements of Figure 4, except that the device parameters now degrade at lower cap thickness and drop off more sharply than before. This behavior is consistent with strain relaxation and the accompanying removal of the conduction band offset.

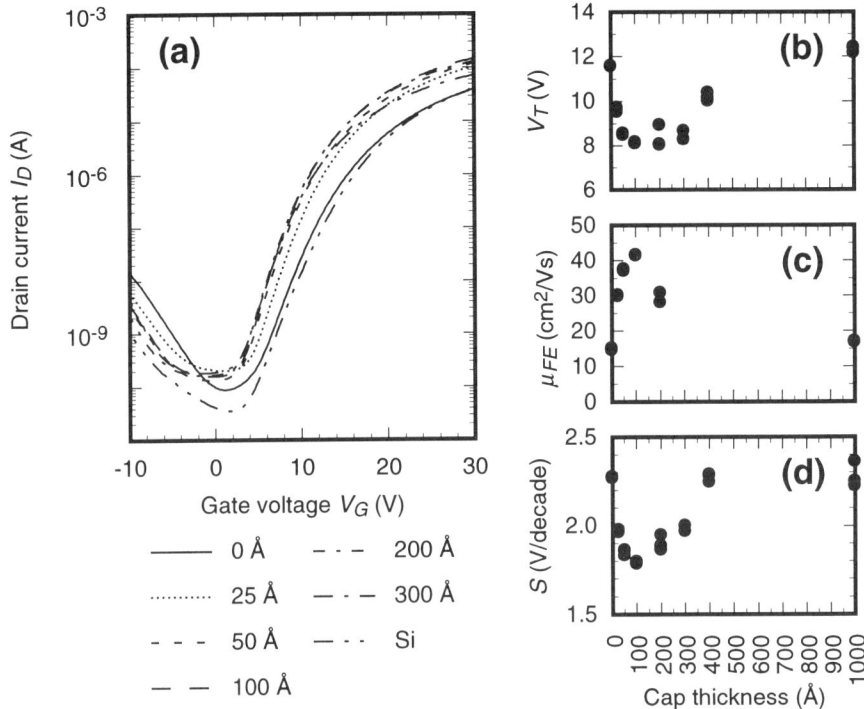

Figure 4. Measurements of 20 μm/20 μm NMOS TFTs fabricated with a maximum process temperature of 550 °C: (a) I_D-V_G characteristics, $V_D = 10$ V; (b) threshold voltage V_T at $I_D = 100$ nA; (c) field effect mobility μ_{FE} at $V_D = 0.1$ V; (d) subthreshold slope S at $V_D = 10$ V.

CONCLUSIONS

A set of poly-Si$_{1-x}$Ge$_x$ TFTs has been fabricated in a low temperature (≤ 550 °C) process with Si caps to buffer the SiO$_2$ and the channel. While previous workers have used the Si to improve the interface properties and thus TFT performance, computer simulations and experimental measurements strongly suggest that the Si cap can grow pseudomorphically on the Si$_{1-x}$Ge$_x$ channel. The resulting strain creates a conduction band offset in addition to the expected valence band offset. As a consequence, NMOS TFT performance improves dramatically because the conduction band offset favors inversion in the cap until the cap is thick enough that the strain effects diminish, while PMOS TFT performance improves for very thin caps because of interface improvement and declines for thicker caps because the valence band offset excludes holes from the cap.

ACKNOWLEDGMENTS

This work was funded by Defense Advanced Research Projects Agency grant MDA972-95-1-0004. The authors would also like to thank Dr. Judy Hoyt, Prof. Tsu-Jae King of UC Berkeley, Kern Rim, Vivek Subramanian, and Dan Connelly for discussions; the staff of the Center for Integrated Systems and Dr. Jackson Ho of the Xerox Palo Alto Research Center for assistance with device fabrication; and Prof. Michael Kelly, Brian Holloway, Ron Cicero, and Dr. Dimitry Kirillov for assistance with materials analysis.

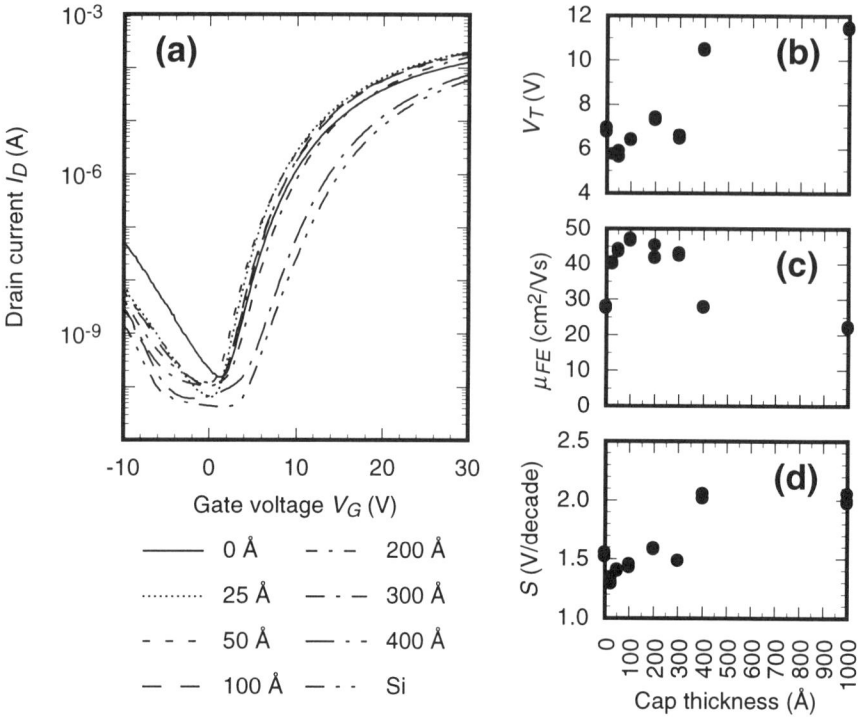

Figure 5. Measurements of 20 μm/20 μm NMOS TFTs annealed at 800 °C for 30 min: (a) I_D-V_G characteristics, V_D = 10 V; (b) threshold voltage V_T at I_D = 100 nA; (c) field effect mobility μ_{FE} at V_D = 0.1 V; (d) subthreshold slope S at V_D = 10 V.

REFERENCES

1. T.-J. King and K.C. Saraswat, IEEE Trans. Electron Devices **41**, 1581 (1994).
2. A. Neugroschel, S. Margalit, and A. Bar-Lev, J. Physics D **6**, 1606 (1973).
3. R. Turan and T.G. Finstad, Semicon. Sci. Tech. **7**, 75 (1992).
4. C. Caragianis, Y. Shigesato, and D.C Paine, J. Electronic Mat. **23**, 883 (1994).
5. S.S. Iyer, P.M. Solomon, V.P. Kesan, A.A. Bright, J.L. Freeouf, T.N. Nguyen, and A.C. Warren, IEEE Electron Device Lett. **12**, 246 (1991).
6. A.J. Tang, J.A. Tsai, R. Reif, and T.-J. King, Int'l Electron Devices Mtg., 513 (1995).
7. S. Verdonckt-Vandebroek, E.F. Crabbé, B.S. Meyerson, D.L. Harame, P.J. Restle, J.M.C. Stork, A.C. Megdanis, C.L. Stanis, A.A. Bright, G.M.W. Kroesen, and A.C. Warren, IEEE Electron Device Lett. **12**, 447 (1991).
8. P.M. Garone, V. Venkataraman, and J.C. Sturm, IEEE Electron Device Lett. **13**, 56 (1992).
9. J. Welser, J.L. Hoyt, and J.F. Gibbons, Int'l Electron Devices Mtg., 1000 (1992).
10. I.-W. Wu, A. Chiang, M. Fuse, L. Öveçoglu, and T.Y. Huang, J. Appl. Phys. **65**, 4036 (1989).
11. M. Hack, J.G. Shaw, P.G. LeComber, and M. Willums, Jap. J. Appl. Phys. **29**, L2360 (1990).

PHOTONIC CRYSTALS BASED ON MACROPOROUS SILICON

V. LEHMANN* U. GRÜNING*, A. BIRNER**
* Siemens AG, Dept. ZT ME 1, 81730 München, Germany
** MPI für Mikrostrukturphysik, Am Weinberg 2, 01620 Halle, Germany

ABSTRACT

The formation of macropore arrays with high aspect ratios by electrochemical etching of n-type silicon in hydrofluoric acid is a well established technique. By using standard photolithograpy the geometry of the array can be controlled with high precision. This enables us to fabricate two-dimensional photonic crystals for the infrared regime. The calculated photonic band structure of the crystal corresponds well with the transmission observed experimentally. Furthermore first defect structures like waveguides and cavities have been realized.

INTRODUCTION

In some sense a macropore array can be seen as a heterostructure composed of bulk silicon and void. The high contrast in refractive index between silicon and air together with the high geometric precision of the pore array etching process are exploited for the fabrication of photonic crystals. A photonic crystal is a dielectric or metallic structure that is periodic on a wavelength scale. The propagation of radiation in such dielectric heterostructures can be described similar to the case of electrons in the periodic potential of a semiconductor. If the wavelength is in the order of the dimensions of the dielectric lattice, a photonic band gap - a frequency range where photons are not allowed to propagate - can open up in two or three dimensions. Photonic crystals have been proposed in 1987 [1] and first been realized for the microwave regime [2]. Such structures serve, for example, as antenna reflectors in aircraft.

RESULTS AND DISCUSSION

The fabrication of macropore arrays by electrochemical etching (ECE) in optional patterns pre-determined by photolithography is known since 1990 [3]. Since this time new devices [4] and materials [5] based on this electrochemical growth process have been developed. The basic physics of the formation process and its limitations are well understood [6, 7]. For the manufacturing of photonic structures with a pitch in the micrometer regime, as shown in Figs. 1 and 3, moderately doped n-type silicon (1 Ωcm) is used as a substrate material.

In order to measure the transmittance of the macroporous structure small bars, as shown in Figs. 1 and 3 are required. This bars are formed by a micromechanical technique which employs

Fig. 1: a) SEM micrographs of a 200μm wide bar of a macropore array forming a two-dimensional photonic crystal. b) A tenfold magnification reveals the triangular lattice of the pores. c) A hundredfold magnification shows the lattice constant (2.3μm) and the thin remaining pore walls (0.17μm).

photolithography and subsequent plasma etching. This micromechanical technique is described in detail elsewhere [8]. After structuring the pore diameter was enlarged to the required dimension by thermal oxidation and stripping of the oxide.

The transmission of the photonic crystal shown in Fig. 1 is measured by a standard FT-IR spectrometer (Nicolet 740). By stacking several samples in z-direction, the use of glass fibers to couple the radiation in was avoided. The edges of the substrate were covered with silver paint in order to avoid transmission through the bulk substrate.

For the theoretical calculation of the band structure the plane wave method was employed, where both the electromagnetic field and the periodic dielectric structure are expanded in a Fourier series [5]. While orthogonal patterns are not sufficient to produce a full two-dimensional photonic band gap [9], a triangular pattern of pores as shown in Fig. 1 and Fig. 3, forms a two dimensional photonic crystal with a common gap to both polarizations. The band gap is centered at 5 μm wavelength for a lattice constant of 2.3 μm and pore diameters of 2.13 μm. The structure is etched 75 μm deep in an n-type silicon substrate by electrochemical pore formation. The measured IR-transmission of the photonic lattice showed an excellent agreement with its theoretically calculated bandstructure. This is shown for the TE-polarization for both the wavevector k parallel to the Γ-M direction (left side of Fig. 2) and k parallel to the Γ-K direction (right side). Note that the transmission is plotted in a logarithmic scale. The transmittance in the gap should drop to zero. However, the measured attenuation in the gap regions was limited both by the sensitivity of the detector and leakage paths around the photonic lattice, resulting in the observed noise in the forbidden regions.

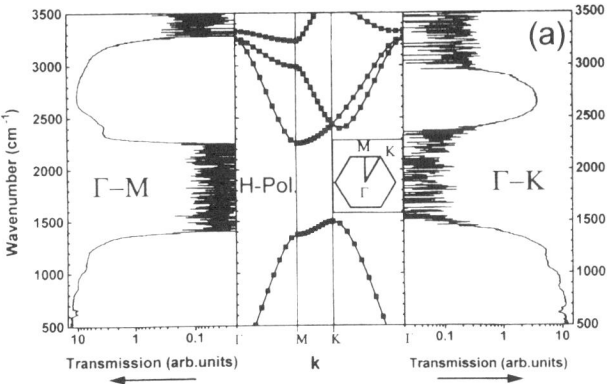

Fig. 2: In the center part the calculated photonic band structure of the two-dimensional photonic lattice shown in Fig. 1 is plotted, the inset shows the first Brillouin zone. The measured transmission spectra (TE-polarisation) for the Γ-M direction and the Γ-K direction are shown between 500 cm^{-1} and 3500 cm^{-1}.

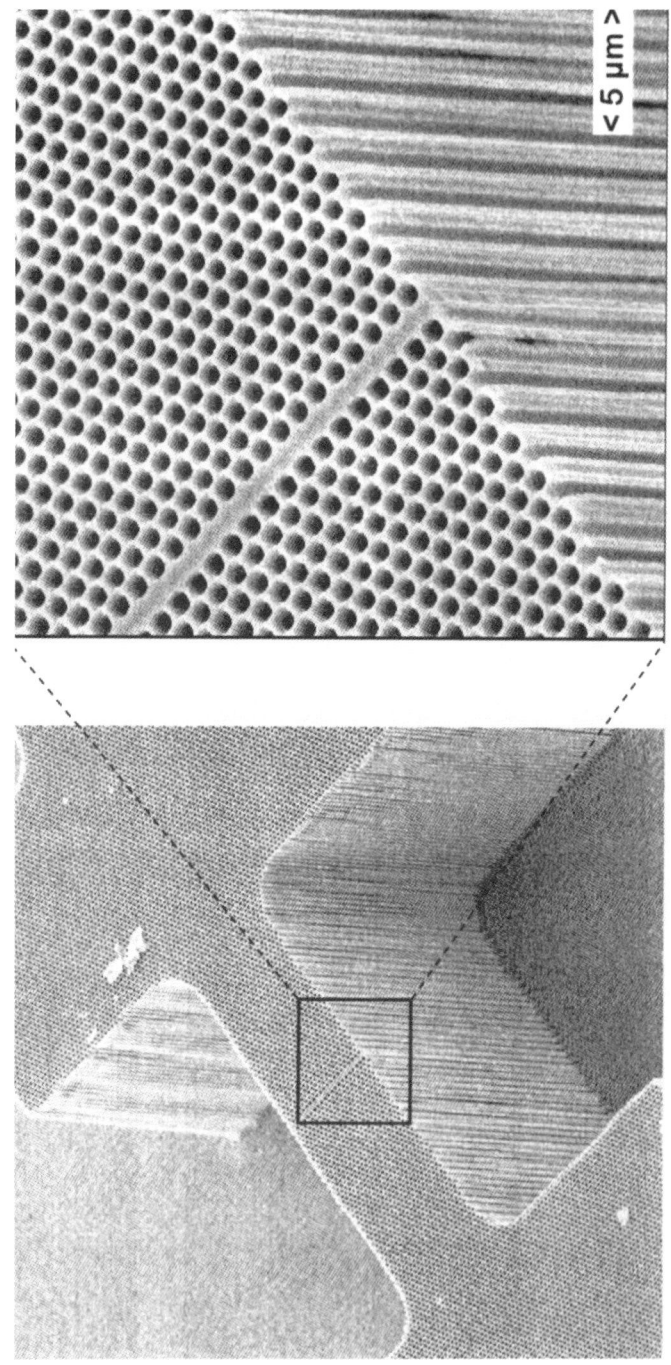

Fig. 3: By a row of missing pores a waveguide is established in the triangular photonic lattice. The photonic crystal is structured like a **H** for an easier positioning of glass fibers which are required to couple the radiation in and out.

The next step after having manufactured a perfect two-dimensional photonic crystal is the integration of defect structures. For example, a row of missing pores, as shown in Fig. 3, should establish a waveguide for frequencies in the bandgap. In a similar way optical microcavities can be defined in the photonic crystal, which exhibits a narrow transmission band within the gap [10].

For such defect strutures the radiation has to be coupled in and out locally at the defect. This can be done by a glass fiber, for example. However, it requires a refined setup for the transmission measurements. Therefore transmission measurements of the structure shown in Fig. 3 are not performed jet.

CONCLUSIONS

It has been shown that photonic crystals for the IR regime can be manufactured with high precision by electrochemical etching of bulk silicon. Optical defect structures like wave guides and cavities are feasible an their optical properties will be investigated in the near future. The study of interesting phenomena such as the inhibition of spontaneous emission from a luminescent center inside the gap and low threshold laser are topic for medium-term research. Improved and new optical components are expected to originate from this artificially tailored optical material.

REFERENCES

[1] E. Yablonovitch, Phys. Rev. Lett., **58**, 2059, (1987).

[2] E. Yablonovitch, T. Gmitter, and K. M. Leong, Phys. Rev. Lett., **67**, 2017, (1991).

[3] V. Lehmann and H. Föll, J. Electrochem. Soc., **137**, 653, (1990).

[4] V. Lehmann,W. Hönlein, H. Reisinger, A. Spitzer, H. Wendt, and J. Willer, Solid State Technol. **38**, 99, (1995).

[5] U. Grüning, V. Lehmann, S. Ottow, and K. Busch, Appl. Phys. Lett. **68**, 747, (1996).

[6] V. Lehmann, J. Electrochem. Soc.,**140**, 2836, (1993).

[7] V. Lehmann, Thin Solid Films, **140**, 2836, (1993).

[8] S. Ottow and V. Lehmann, J. Electrochem. Soc., **140**, 2836, (1993).

[9] U. Grüning, V. Lehmann, and C. M. Engelhardt, Appl. Phys. Lett., **66**, 3254, (1995).

[10] J. S. Foresi, P. R. Villeneuve, J. Ferrera, E. R. Thoen, G. Steinmeyer, S. Fan, J. D. Joannopoulos, L. C. Kimerling, H. I. Smith, and E. P. Ippen, Nature, **390**, 144, (1997).

EFFECT OF INTERFACE-RELATED DEEP LEVELS ON HIGH SENSITIVITY OF SCHOTTKY DIODE PHOTODETECTOR BASED ON ULTRATHIN InGaAs FILM ON Si

A.V.KVIT*, M.V.YAKIMOV*, P.L.KONSTANTINOV**, M.N.NAIDENKOV***
*P.N.Lebedev Physical Institute RAS, Moscow, Russia, kvit@sci.lebedev.ru
**NPO "Pulsar", Moscow, Russia
***PTI RAS, Moscow, Russia.

ABSTRACT

Recently, the Fermi level has been demonstrated to be pinned in GaAs semi-insulating ultrathin (<100 A) films grown at low temperatures [1]. This allows one to construct a detector with "internal" photocurrent amplification. The amplification effect compensates losses in sensitivity due to the small width of light-sensitive layer which absorbs 10 — 50 % of incident radiation. We fabricate photodetector structures with external quantum efficiency more than 1 for visible region. Photogenerated carriers in the GaAs layer are effectively separated by the built-in electric fields formed by the Schottky barrier and by the charge at the GaAs/Si interface. In this work, we show the relationships between spectral sensitivity of the metal-InGaAs/Si structures and In content. We observed the red shift in the photocurrent spectra with increasing In concentration, although photosensitivity of such structures dropped drastically. This shift demonstrates that the thin InGaAs film is actually responsible for photosensitivity. Despite the low photoluminescence intensity, the low-temperature PL spectra indicate that band gap decreases with indium flux rising during MEE growth. The surprise was that the decrease of the film thickness caused the increase of photosensitivity. The GaAs(20 A)-InGaAs (20 A)/Si structure was the most sensitive one. We also observed high quantum efficiency in near-UV region (up to 0.8). We determined activation energy of elctron and hole traps and their concentration profiles by DLTS. The centers localized on interface between polar and nonpolar semiconductors are responsible for Fermi-level pinning in III-V semi-insulating materials and act as an electron trap with activation energy 0.59 eV. The origin of deep levels is discussed.

INTRODUCTION

Semiconductor devices fabricated on the base of Schottky diode structures are widely applied in optoelectronics as surface barrier photosensors. Photodiodes with Schottky barrier have a number of advantages in comparison with p-n junction photodiodes. The separation of photogenerated carriers by built-in electric field in the surface area makes it promising to use Schottky diodes as photodetectors in the near UV region, as the separation takes place just in the region of high energy photon absorption.

The quality of the metal/semiconductor interface considerably affects the performance of Schottky diodes. First, the energy barrier height, which significantly influences the sensitivity of such photodetectors, strongly depends on the surface traps density. Second, the high density of the surface traps leads to the photosensitivity decrease in UV region due to effective surface recombination of non-equilibrium

Mat. Res. Soc. Symp. Proc. Vol. 533 © 1998 Materials Research Society

carriers. We note that the surface states in such semiconductors as Si and GaAs are believed to be located approximately at $E_v + 1/3E_g$.

Recently, we demonstrated that Schottky diode on the base of heterostructure Al/GaAs/Si with ultra-thin semi-insulating layer of GaAs (100 Å) grown by migration enhanced molecular beam epitaxy has a high photosensitivity

in visible light region (external quantum efficiency is ~0.9) and rather good kinetic and noise characteristics [1]. We proposed that in our case only surface states on the metal/GaAs interface would determine the height of the energy barrier. However, the value of the barrier height derived from experimental data appeared to be much higher than the expected value (~1.15 eV from the bottom of the GaAs conductivity band).

We clarified that deep states located on the GaAs/Si interface as well as in the volume of ultra-thin semi-insulating GaAs layer mainly determined the energy diagram of our photodiode. The previous studies of such structures by transmission electron microscopy [2] have revealed that ultra-thin epitaxial films of GaAs grown on Si substrate have high density of extended defects such as extended dislocations (5×10^{11}-2×10^{12} cm^{-2}) and microtwins.

In initial growth stages of GaAs films on Si the misfit dislocations do not appear. Relaxation of mechanical strain caused by mismatch of lattice parameters and the difference of thermal expansion coefficients of Si and GaAs goes through formation of the above mentioned extended defects. Additional yield stress takes place due to considerable density of point defects, as we have shown in the case of relatively thick (100 nm — 1000 nm) GaAs-on-Si films. At the same time, the point defects could create high concentration of deep states in epitaxial layer of GaAs.

Deep level states are studied in such structures rather weakly. In this work we undertake efforts to clarify the effect of deep levels on characteristics of the photodetector. However, we note that the features of ultra-thin strained layers bring on certain difficulties both for studies by optical methods and for interpretation of data derived from DLTS investigations.

EXPERIMENT

The structures investigated in this work are GaAs and InGaAs films (2-20 nm thickness) grown on Si (001) epi-ready substrate (B-doped p-type, 12 Ω•cm). We used the methods of Si substrate pre-growth thermal treatment in vacuum described in [3]. The growth temperature was T= 300 ± 25 °C. We have grown the films by migration enhanced molecular beam epitaxy with growth rate of 1 monolayer per cycle and dead time between corresponding cycles. Schottky contacts were fabricated by *ex-situ* deposition of Al layer through the photoresist mask. We have cleaned surface of the epitaxial films with solvents and without use of etching procedure. DLTS measurements were carried out with modernized computer controlled DLS-30 spectrometer, which had a mode of capacity measuring in series of time intervals.

To carry out photoluminescence (PL) measurements, we used Ar$^+$-laser (λ~514,5 nm) as an excitation source. The PL spectra were measured at excitation power of 1 W/cm^2. During PL measurements we mounted the samples on the holder and placed in the cryostat in the volume of liquid helium.

We performed measurements of photosensitivity spectral characteristics on samples mounted on foiled insulating plate with etched separating strips. The foil contacted to indium contact plate from the side of Si-substrate. The upper point contacts to the GaAs ultra-thin layer were made by a probe needle (probe tip radius was 8 μm). Spectrophotometer SF-46 with two calibrated light- emitting units - one in visible and another in UV region was used as a source of incident light. The spectral characteristics were measured in photovoltaic mode. Spectral characteristics and absolute values of photosensitivity have not changed for 2 years.

RESULTS

Fig. 1 shows the spectral characteristics of photodetectors on the base of ultra-thin photosensitive $In_xGa_{1-x}As$ layer (12 nm thickness) grown with several In/Ga flux ratios. The ratios of In flux to Ga flux were varied from 0.00 to 0.35. Photosensitivity spectral characteristic in the 500-1100 nm region shifts to the longer wave side with In/Ga flux ratio increase. We expected this trend, but to our regret the efficiency of photoresponse decreased with increase of In contents. This shift proves that the photon absorption in epitaxial layer of polar semiconductor rather than in the silicon substrate give the main contribution to the value of photocurrent. Otherwise, the spectral characteristic form would not change. Moreover, the heterostructures with nanometer-scaled InGaAs film demonstrate considerable photoresponse in UV region, which exceeded even the photoresponse in the visible and IR regions.

We also present the spectral characteristic of ultra-thin heterostructure GaAs(10 Å)-$In_{0.25}Ga_{0.75}As$(10 Å)/Si here. The photosensitivity is extremely high, as it is clearly seen from Fig. 1. Furthermore, we observe the stable increase of photosensitivity in UV region up to λ= 0.2 μm.

Figure 2 shows the low-temperature photoluminescence spectra of the investigated structures. Despite rather weak yield of photoluminescence, we suggest that a number of bands in the 1.35 - 1.50 eV region should be related to the luminescence from photosensitive thin layers.

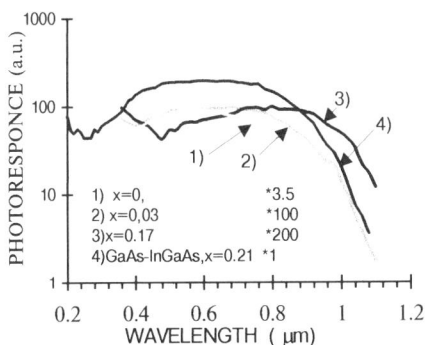

Fig 1 Photoresponce dependence
of $In_xGa_{1-x}As$/Si thin film on In content

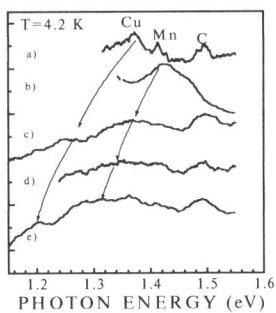

Fig.2. Low temperature photoluminescence spectra dependence on In contents a) GaAs (100 Å)/Si, b) InGaAs (100 Å)/Si, In/(Ga+In) flux ratio — 0.1 c) InGaAs (100 Å)/Si, In/(Ga+In) flux ratio 0.25 d) InGaAs (100 Å)/Si In/(Ga+In) flux ratio — 0.37, e) GaAs (10 Å) — InGaAs (10 Å)/Si In/(Ga+In) flux ratio — 0.21.

Using our previous experience of PL spectrum identification and estimation of biaxial strain in GaAs/Si with relatively thick GaAs layers [4], we also identified spectra of ultra-thin films. Three bands at ~1.49 eV, 1.41 eV, and 1.36 eV are associated with various impurities: a) shallow acceptors (C, Si); b) Mn_{Ga}; c) Cu_{Ga} respectively.

The comparison of spectral positions of bands related to these impurity states in ultra-thin films and in bulk GaAs gives a paradoxical result. Our estimation shows that the yield strain in the investigated structures are not considerable. The ultra-thin films have stress of less than 0.1%. Another unexpected result is that the In-flux increase gives only weakening and broadening of the 1.49 eV band rather than the red shift of this band. The presence of this band in the photoluminescence spectra indicates that there is GaAs phase in InGaAs ultra-thin films. However, in the 1.1-1.45 eV region the spectra successively change with In-flux increase. Evidently, the red-shift of the broad bands with increasing In fraction of InGaAs correlates with the decrease of band gap in these alloys. The bands widening can be a result of entropy rise with increasing In content of InGaAs alloy, which is typical for ternary compound semiconductors. On the other hand the random character of local strains in imperfect ultra-thin epitaxial film of photosensitive material can cause this effect. These effects limit the accuracy of In concentration determination from the photoluminescence spectra.

Figure 3(c) shows the capacity-voltage (C-V) characteristic of a diode based on Al/GaAs/Si structure. There are two regions with different density of centers emitting the carriers when the excitation light goes through the space charge region. The slope of the line in Fig. 3(c) in region >2 eV gives the carrier density in Si (in case of p-n junction this density would correspond to the density of electrically active impurities). The point of the line inter-section with voltage axis gives the energy barrier height (in this case ~1 eV). However, the estimation of the carrier concentration made according to the formula:

$$N_d = \frac{2}{q\varepsilon_s}\left[-\frac{1}{d(1/C^2)/dV}\right] \qquad (1)$$

where ε is dielectric constant, q is value of electron charge, C is capacity normalized to the junction area, V is voltage bias, gives the concentration of 9×10^{18} cm^{-3}, and 5.6×10^{18} cm^{-3}, respectively which is significantly higher than the impurity concentration in Si and GaAs. It will be shown that this discrepancy evidently attributed to change in the population of some trap, localized in the Si/GaAs interface, arising from the variation of applied bias. The presence of this trap is the only considerable feature of the DLTS spectra provided that the bias is rather small (U<1V; insert at the bottom of Fig. 4). Under relatively high voltage biases the traps localized in Si become active.

From dependence of the emission rate, p_n, on the temperature we calculated the activation enthalpy of the carrier emission from this defect, which equals 0.59 eV. To our regret, we can not identify the conductivity type of the film, so there is an uncertainty in identification of the trap type. We know that rather thick GaAs/Si films (0.2—1 μm) are p-type but the diffusion of Si from the substrate can invert the conductivity type of the ultrathin films. The activation energy from the DLTS spectra is close to that for the well-known electron trap EL2 in GaAs [5]. In this case the film at the initial stages of growth becomes enriched with As as EL2 is known to be a complex, including As_{Ga} anticite.

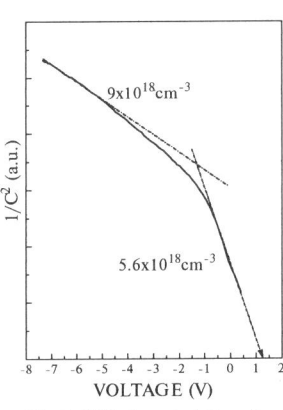

Fig.3. C-V characteristics of Al-GaAs/Si diode.

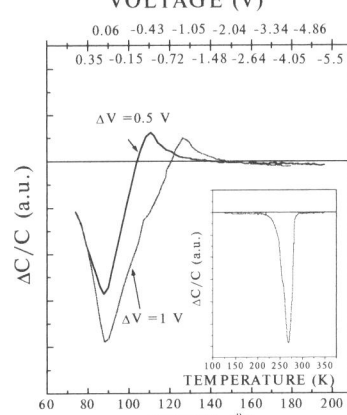

Fig.4. $\Delta C/C$ profile of Al-GaAs/Si diode measured at 291 K. Inset shows DLTS spectrum of Al-GaAs/Si diode.

The trap seems to be strongly thermalized at the room temperature. We detect the thermostimulated current of opposite polarity to the photo-induced current under heating of the sample without any change of the spectral sensitivity. We observe the spectral sensitivity changes after cooling, when the population of the trap increases. The photosensitivity decreases, mostly in the short- wavelength range. Under temperature of liquid nitrogen the sensitivity spectra closely resemble those of the Schottky diode Me-Si.

To show the region of the electron trap localization we plotted the profile of $\Delta C/C$ obtained from capacity-voltage dependencies measured at 18 °C (Fig. 4). In the vicinity of zero bias there was observed a strong minimum of the $\Delta C/C$ profile curve. The cause of the minimum is the trap localized in surface layer at the depth of <100 Å (which approximately corresponds to the thickness of GaAs film). This feature in the density profile becomes narrower with decrease of pulse amplitude. This effect indicates that the trap has δ-like spatial distribution. The dependencies of this sort are well known to represent the real profile of deep trap concentration. We give top scale in Angstroms. Depth is calculated from voltage bias according to formula of flat capacitor. This approximation is not quite correct in the case of such a short distance between Schottky contact and GaAs/Si interface. Even so, we observe the perfect coincidence of the trap localization depth with the GaAs film thickness.

The several factors lead to initiation of the effect of inner quantum multiplication of carriers in this structure: the Fermi level pinning in low temperature grown semi-insulating GaAs/Si, the high density of charge accumulated at the polar/covalent semiconductor interface, and tunnel transparency of the photosensitive layer. The decrease of photosensitive region width results in decline of the excitation light absorption to the level of 1-15% of the total incident intensity. The inner multiplication of carriers in the structure more than compensates for the decrease of photoresponse due to reduction of the fraction of photons absorbed in photosensitive layer. The carriers generated by photons absorbed in the semi-insulating GaAs layer are effectively separated by the built-in electric field at the GaAs/Si interface. The holes accelerated by this field obtain enough energy to cause the impact ionization in Si in the vicinity of the GaAs/Si interface. The deep levels described above capture the impact generated carriers.

The features of our photodetector are following: i) the high density of traps at the interface ($>4\mathrm{x}10^{13}$ cm^{-2}) is provided; ii) the energy levels attributed to these traps are very close to Si valence band edge (~ 0.1 eV), iii) these traps are mainly ionized at room temperature and zero bias voltage; iv) the trap relaxation time is rather short (<20 ns) to influence the photoresponse measured. On the other hand, the relaxation time is long enough for electrons to tunnel through the semi-insulating layer.

We suppose that the new carriers are generated due to transition from the Si valence band to the trap localized at the interface and then tunnel to the metal contact. In fact, the electrons use the trap as a "springboard" for tunneling to the contact through the film.

CONCLUSION

We discuss effect of traps on characteristics of photodetectors based on ultra-thin films of InGaAs-on-Si with various In contents. We observed that photosensitivity spectral characteristics shift to the longer wave side with In/Ga flux ratio increase that proves the photons absorption in thin epitaxial layer of polar semiconductor (20-200 Å). DLTS measurements show that the trap with activation enthalpy equaled to 0.59 eV determines a number of features of this photodetector. From the profile obtained from capacitance—voltage dependencies we demonstrate that the trap has δ-like spatial distribution and is localized at InGaAs/Si interface.

ACKNOLIGEMENT

The authors are grateful to the Russian Foundation for Basic Research for financial support under contract No.97-02-17747 and Russian scientific program "Physics of solid state low-dimension structures" under contract No. 97-2019.

REFERENCES

1. V.A.Joshkin, V.N.Pavlenko, A.V.Kvit and S.R.Oktyabrsky, J. Appl. Phys. **79**, 3774 (1996).
2. V.A.Joshkin, V.N.Pavlenko, M.N.Naidenkov, A.V.Kvit and A.A.Orlikovsky, Thesis of PTI RAS (edited by "Nauka", Moscow, Russia), **10**, 41 (1996).
3. V.A.Joshkin, S.R.Oktyabrsky, I.A.Bogonin and A.A.Orlikovsky, J. Crystal Growth, **132**, 209 (1993).
4. V.Joshkin, A.Orlikovsky, S.Oktyabrsky, K.Dovidenko, A.Kvit, I.Muhamedzhanov and E.Pashaev, J.Cryst.Growth **147**, 13 (1995).
5. Y.Takanashi, J.Appl.Phys. **80**, 4389 (1996).

Part IV

Epitaxy of Quantum Structures

LATERAL ORDERING OF SELF-ASSEMBLED Ge ISLANDS

JIAN-HONG ZHU, K. BRUNNER, G. ABSTREITER
Walter Schottky Institut, Technische Universität München, Am Coulombwall, D-85748 Garching, Germany

ABSTRACT

Two-dimensionally ordered arrays of Ge islands are realized by molecular beam epitaxy on vicinal Si(001) surfaces with regular ripples. Deposition of a 2.5 nm $Si_{0.55}Ge_{0.45}$/10 nm Si multilayer on vicinal Si(001) surfaces gives rise to the formation of regular ripples with a typical period of 100 nm, due to step-bunching. The ripples lead to the long-range line-up of the Ge islands along their direction, while the strong repulsive interaction between the dense Ge islands determines their relative arrangement on different step bunches of a ripple. The ordering pattern can be controlled by the Ge coverage as well as the direction of the ripples. The Ge islands show a narrow size distribution with the lateral size limited by the ripple period.

In contrast, when deposited directly on well-prepared biatomic-stepped vicinal Si(001) surfaces under the same growth conditions, only weak ordering of Ge islands along the step direction is achieved. No ordering of Ge islands has been observed, when a flat Si(001) surface is employed, where no obvious step-bunching occurs.

The results promise efficient control on the position and size of self-assembled and self-ordered Ge islands by the steps prepared on vicinal surfaces.

INTRODUCTION

Fabrication of semiconductor quantum dot structures with a regular in-plane spatial distribution and optimum size uniformity is highly desirable for applications of their novel optical and electronic properties. But this remains a challenging subject, especially on the nanometer scale. The strain-driven self-assembled formation of Ge islands on Si in the Stranski-Krastanov growth mode appears as a promising method. However, when grown directly on flat Si(001) substrates, the Ge islands are generally randomly distributed and have a broad size distribution. Several approaches of lateral ordering have been explored [1-4]. We have studied the growth of self-assembled Ge islands on vicinal Si(001) surfaces by molecular beam epitaxy (MBE). Two-dimensional (2D) ordering of self-assembled Ge islands has been realized on vicinal Si(001) surfaces with regular ripples. The ripples are realized via the growth of a SiGe/Si multilayer. Macroscopic 2D ordering is already achieved in a single layer of Ge islands.

EXPERIMENT

Samples were grown in a MBE system. The Si(001) substrates used were tilted by 1.5° towards $[1\bar{1}0]$ (substrate A) or by 2.0° towards [100] (substrate B). A multilayer of twenty periods of 2.5 nm $Si_{0.55}Ge_{0.45}$/10 nm Si bilayers was initially grown at 550 °C. Several monolayers (ML) pure Ge were subsequently deposited at 500 °C with a Ge deposition rate of 0.02 nm/s. Samples A3, A4, A5 and A6 using substrate A with 3, 4, 5 and 6 monolayers (ML) Ge, respectively, and a sample B5 using substrate B with 5 ML Ge were grown. A reference sample A0 using substrate A but without pure Ge layer was also grown. All samples were investigated by atomic force microscopy (AFM) in ambient air.

RESULTS

A regular one-dimensional ripple pattern with a period of about 60 nm is found to be formed on the multilayer surface of sample A0 (Fig. 1). The ripple has a direction parallel to the intrinsic substrate steps. The formation of the ripple is due to the instability of the SiGe vicinal surfaces under stress against step-bunching [5] as well as vertical correlation of the rippled interfaces, which was studied experimentally in detail in Ref. 6. The ripple pattern with a period comparable to the size of Ge islands will template the growth of Ge islands on it.

The Ge islands on sample A5 are found to be two-dimensionally well ordered [Fig. 2(a)]. They are characteristically aligned into rows parallel to the intrinsic substrate steps,

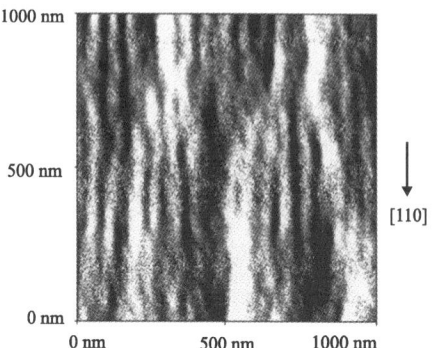

Fig. 1. AFM internal sensor image of sample A0. The height modulation is about 0.6 nm.

i.e. in the [110] direction. The rows are equally spaced with a period p of about 120 nm. Ge islands on neighboring rows are relatively shifted by half of their separation s (~140 nm) along the rows. Thus the Ge island lattice can be described by a centered rectangular unit mesh with sides s and $2p$, respectively, as indicated by the dotted frame in Fig. 2(a). The Ge islands on this sample seem to be closely packed, showing a 2D hexagon-like lattice. This 2D ordering of the Ge islands is also confirmed by the 2D fast Fourier transformation (2DFFT) from a 5x5 μm^2 AFM image of sample A5, which exhibits a clear sixfold pattern like a hexagon. The Ge island array turns out to be a stable self-organized state. The Ge islands do not ripen when kept at 500 °C for 5 min.

The Ge islands are found to form on the ridges of a ripple pattern on the Ge wetting layer. This ripple has a period of about 120 nm, twice the value of the ripple in Fig. 1. The mechanism for this is by far not clear. It might result from the bunching of every two neighboring step bunches on the SiGe/Si multilayer during deposition of the Ge wetting layer. The Ge islands have a

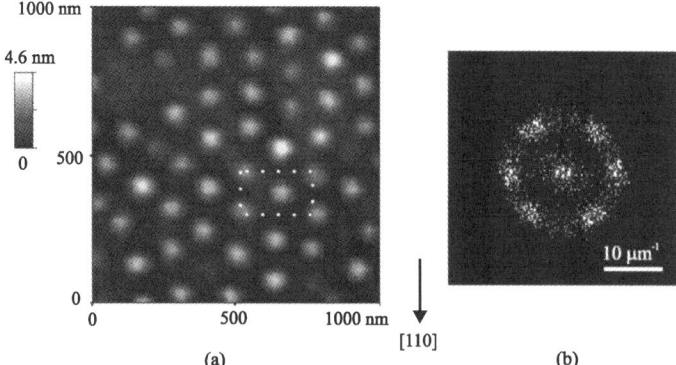

(a)   (b)

Fig. 2. (a) AFM image of the Ge islands on sample A5. (b) 2DFFT from a 5x5 μm^2 AFM image of sample A5. The center in (b) is the origin.

narrow size distribution, with both lateral size and height having a uniformity better than 10%. The lateral size distribution is found to be cutoff by the 120 nm ripple period. High resolution AFM measurements show that these Ge islands are of a compact pyramid shape, showing a square-like base [Fig. 3(a)]. Statistical distribution of the orientations of local surface normals [Fig. 3(b)], calculated from several AFM images following the procedure of Lutz *et al* [7], shows that these pyramids consist of {105} facets. Characteristic reflection high-energy electron diffraction patterns from {105} facets [8] have also been observed from this sample. The

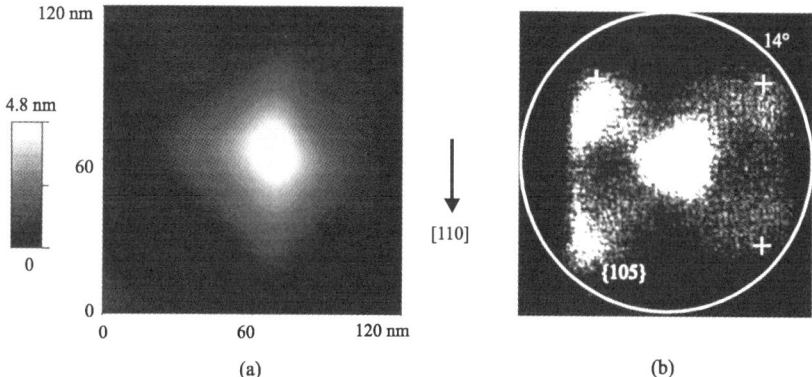

(a)   (b)

Fig. 3. (a) A typical pyramid-shaped Ge island on sample A5. (b) Statistical distribution of orientations of local surface normals on sample A5. It is displayed as a polar plot. Center in (b) corresponds to the normal of the vicinal surface. The white crosses mark the positions corresponding to four equivalent {105} facets.

inclination of pyramids with respect to the vicinal Si(001) surface results in the asymmetric base region observed by AFM in Fig. 3(a). No elongated 'hut' clusters observed on flat Si(001) [9] appear on this sample.

The Ge coverage is found to influence the final ordering pattern. Similar 2D ordering has also been found on samples A3, A4 and A6. The ordering patterns can also be described by a centered rectangular unit mesh like the one on sample A5 [see Fig. 2(a)]. Fig. 4 shows the separation s, period p and the height of the Ge islands versus the Ge coverage. Increasing the Ge coverage leads to an increase in the density as well as in the height of the Ge islands. The density increases mainly by a decrease in the separation s along the ripples. The period p keeps approximately constant. We ascribe this phenomenon to the ripple pattern with a fixed period due to identical multilayer and substrate used. The separation s saturates at $s = 2p / \sqrt{3}$ for large Ge coverage, corresponding to closely packed Ge islands.

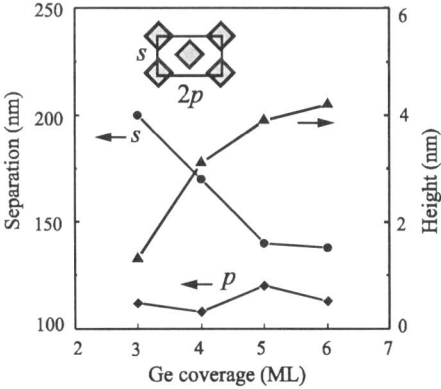

Fig. 4. The separation s, period p and the height of the Ge islands on samples using substrate A versus the Ge coverage.

When substrate B is used, the result is different. Figure 5 shows the AFM image of sample B5. In this case, the ripple also has a period of about 120 nm, but with a direction [010]. Although some Ge islands are a little elongated in the direction of the ripple, most of them tend to have a compact shape, showing a clear square base with edges parallel to the [100] or [010] direction. The Ge islands on this sample appear to align along these two perpendicular directions. The ordering pattern is different from that of sample A5 [see Fig. 2(a)], although the two samples share the same Ge coverage.

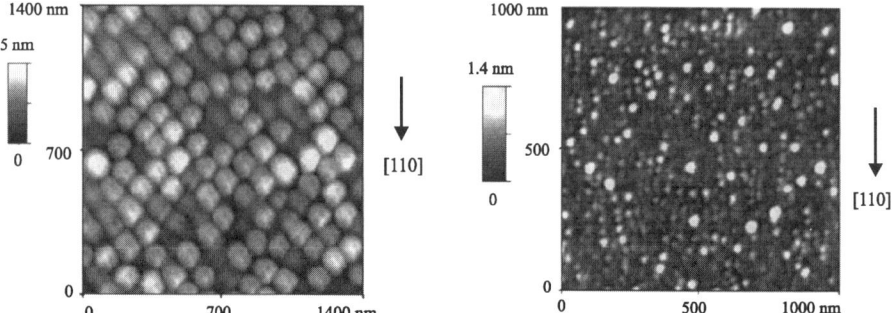

Fig. 5. AFM image of the Ge islands on sample B5.

Fig. 6. AFM image of the Ge islands on a vicinal Si(001) surface tilted by 2.5° towards [1$\bar{1}$0].

For comparison, 5 ML Ge are also deposited directly on vicinal as well as on nominally flat Si(001) surfaces under the same growth conditions. Fig. 6 shows that only weak ordering of Ge islands along the step direction is achieved on well-prepared biatomic-stepped vicinal Si(001) surfaces. The Ge islands, having a typical lateral size of about 40 nm, are not homogeneous in size. Some of them have developed into larger islands. No ordering of Ge islands has been observed, when a flat Si(001) surface is employed, where no obvious step-bunching occurs.

2D ordering of the Ge islands is driven by minimizing the strain energy of the whole system. There is a strong repulsive interaction between the dense Ge islands on our samples, due to the overlap of the strain fields induced by the Ge islands. It has been shown theoretically [10] that for square shaped islands on the (001)-surface of an elastically anisotropic cubic crystal, the total energy is minimum for the periodic square lattice with primitive lattice vectors along the elastically soft directions [100] and [010]. This has been experimentally observed for InAs islands on a GaAs(001) surface [11]. However, ordering of islands induced only by elastic interaction is not strong on a macroscopic scale and can be easily distorted in practice. In the presence of a periodic ripple like on our samples, the ripple provides preferential nucleation sites on their ridges and thus macroscopically aligns the Ge islands along its direction. In the case of sample B5, the ripple direction coincides with one of the two elastically soft directions. The ripple directly enhances the ordering of the square lattice potentially caused by the elastic interaction. In the case of the samples using substrate A, where the direction of the ripples is along [110], the ripples not only enhance the ordering degree on a large scale but also lead to the deviation of the ordering pattern from the periodic square lattice. The relative arrangement of the Ge islands on different step bunches of the ripple is determined by their elastic interaction. Two possible relative arrangements are schematically shown in Figs. 7(a) and 7(b). Each Ge island row along the step bunches creates a sawtooth contour of repulsive potential as indicated by thin solid lines in the two figures. Obviously, the arrangement in Fig. 7(a) has less overlap of the repulsive potentials than the one in Fig. 7(b) and is thus energetically favorable. The arrangement in Fig. 7(a) can be described by a centered rectangular unit mesh, as indicated by the dashed frame in the figure. This is what we have experimentally observed on substrate A. The final ordering pattern depends also on the ratio of $s/2p$. Decreasing in the separation s on a ripple leads to the variation of the ordering pattern. This variation will end up with a closely packed hexagonal lattice when $s/2p = 1/\sqrt{3}$, corresponding to a minimum value of s. This is nearly reached in sample A5 and A6.

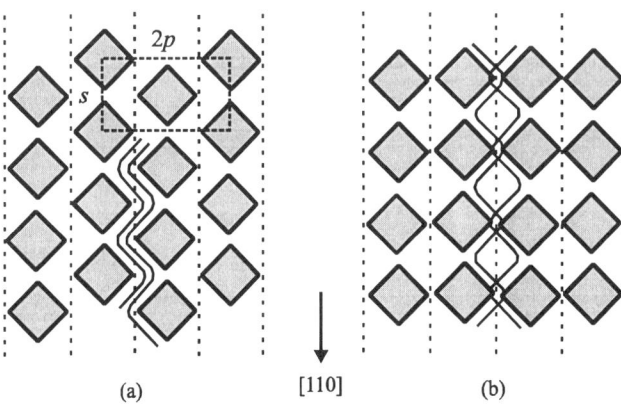

Fig. 7. Schematics of two possible arrangements of the Ge islands aligned along the step bunches of a ripple with [110] direction. Part of the repulsive potentials created by the Ge islands is also schematically shown by thin lines.

CONCLUSIONS

In summary, macroscopic 2D ordering is realized in a single layer of self-assembled Ge islands on vicinal Si(001) surfaces with regular ripples. The ordering pattern can be tuned by the Ge coverage and the direction of the ripples. The Ge islands have a compact shape of a {105} facetted pyramid. The ordering helps to improve the size distribution of the Ge islands. This provides an effective method for controlling the Ge islands in both ordering and size.

ACKNOWLEDGMENTS

This work was supported by the German Bundesministerium für Bildung and Forschung via project "Si-Nanoelektronik", Contract No. M2953 B2. One of us (J. Zhu) acknowledges financial support from Volkswagen-Stiftung.

REFERENCES

1. C. Teichert, M. G. Lagally, L. J. Peticolas, J. C. Bean, and J. Tersoff, Phys. Rev. **B 53**, 16334 (1996).

2. A. A. Darhuber, P. Schittenhelm, V. Holy, J. Stangl, G. Bauer, G. Abstreiter, Phys. Rev. **B 55**, 15652 (1997).

3. S. Yu. Shiryaev, F. Jensen, J. Lundsgaard Hansen, J. Wulff Petersen, and A. Nylandsted Larsen, Phys. Rev. Lett. **78**, 503 (1997).

4. T. I. Kamins, and R. Stanley Williams, Appl. Phys. Lett. **71**, 1201 (1997).

5. J. Tersoff, Y. H. Phang, Z. Zhang, and M. G. Lagally, Phys. Rev. Lett. **75**, 2730 (1995).

6. C. Teichert, Y. H. Phang, L. J. Peticolas, J. C. Bean, and M. G. Lagally, in Surface Diffusion: Atomistic and Collective Process, NATO-ASI Series, edited by M. C. Tringides (Plenum, New York, 1997), p. 297.

7. M. A. Lutz, R. M. Feenstra, P. M. Mooney, J. Tersoff, and J. O. Chu, Surf. Sci. **316**, L1075 (1994).

8. J. Zhu, K. Brunner, and G. Abstreiter, Appl. Phys. Lett. **72**, 424 (1998).

9. Y.-W. Mo, D. E. Savage, B. S. Swartzentruber, and M. G. Lagally, Phys. Rev. Lett. **65**, 1020 (1990).

10. V. A. Shchukin, N. N. Ledentsov, P. S. Kop'ev and D. Bimberg, Phys. Rev. Lett. **75**, 2968 (1995).

11. M. Grundmann, J. Christen, N. N. Ledentsov, J. Böhrer, D. Bimberg, S. S. Ruvimov, P. Werner, U. Richter, U. Gösele, J. Heydenreich, V. M. Ustinov, A. Yu. Egorov, A. E. Zhukov, P. S. Kop'ev, and Zh. I. Alferov, Phys. Rev. Lett. **74**, 4043 (1995).

STACKED LAYERS OF C-INDUCED Ge QUANTUM DOTS

O. G. Schmidt*, K. Eberl*, S. Schieker*, N.Y. Jin-Phillipp**, F. Phillipp**, J. Auerswald** and P. Lamperter**

*Max-Planck-Institut für Festkörperforschung, Heisenbergstraße 1, 70569 Stuttgart, Germany
**Max-Planck-Institut für Metallforschung, Heisenbergstraße 1, 70569 Stuttgart, Germany
and Seestraße 92, 70174 Stuttgart, Germany

ABSTRACT

Fifty layers of carbon-induced germanium dots, separated by 9.6 nm Si, are stacked by solid source molecular beam epitaxy. Each dot layer consists of 0.2 monolayers of pre-deposited carbon and 2.4 monolayers of post-grown Ge. These carbon-induced germanium dots are only 10 to 15 nm in diameter and 1 to 2 nm in height. Vertical alignment due to penetrating strain fields of underlying dot layers is not observed. Unlike to an identical structure without the pre-growth of carbon, a variety of advantageous aspects such as strain compensation, strongly enhanced no-phonon photoluminescence at a wavelength of around 1.3 μm and the possibility of effective waveguiding make this stack of C-induced Ge islands an attractive structure for Si based optoelectronic devices.

INTRODUCTION

In recent years there has been an increasing research effort dealing with self-assembling low dimensional structures [1-4]. However, thermodynamic limits impose restrictions to the structural design freedom. Taking the SiGe material system for example, it proves to be very difficult to form Ge islands with diameters smaller than about 50 nm [1]. Additionally, the required low growth temperatures together with only weak electron confinement prevent the small Ge islands from exhibiting intense photoluminescence (PL) signal. Consequently, new ways to control the formation of self-assembled islands such as growing on cross-hatch dislocation patterns [5,6] or pre-patterned substrates [7-9] have been proposed and developed. Although in these cases strong influence can be taken on the island distribution and sizes, to our knowledge only minor PL output has yet been reported on these structures. A promising way to enhance the PL intensity in Si-based structures is the use of spatially indirect electron-hole recombination in neighboring tensile and compressively strained quantum wells [10,11]. First demonstrated for Si/SiGe double quantum wells on relaxed virtual SiGe buffers[10] this type-II optical transition proved to be also very efficient in $Si_{1-x}Ge_x/Si_{1-y}C_y$ double quantum wells [11]. The approach of carbon-induced Ge dots combines the possibility to produce very small Ge islands with the exploitation of the efficient type-II recombination mechanism [12,13].

In this letter we use carbon-induced germanium (CGe) islands to fabricate a stack of 50 dot-layers, each layer separated by 9.6 nm Si. Dense packing of such a large number of self-assembled germanium quantum dots becomes possible because of the strain compensation effect of substitutional carbon in Si/SiGe. Together with considerably enhanced no-phonon PL at a wavelength of about 1.3 μm this structure might be an attractive candidate for future optoelectronic devices.

EXPERIMENTALS & RESULTS

Two samples, a 50 layer-stack of CGe-dots and a 50 period Si/Ge superlattice, are grown on a 4-inch (001) oriented Si wafer, using a modified Balzers ULS 400 solid source molecular beam epitaxy machine. After SiO_2 desorption from the Si substrate in the growth chamber at 900 °C, a 380 nm thick Si buffer is grown while decreasing the substrate temperature T_s from 900 °C to 460 °C. Subsequent growth steps all executed at $T_s = 460$ °C are as follows: 5 s growth interruption, 0.2 monolayer (ML) C deposition, 5 s growth interruption, 2.4 ML Ge, 5 s growth interruption, 150 nm Si cap layer. Naturally, the carbon layer is left out for the Si/Ge superlattice. Typical growth rates are 1.00, 0.20, 0.01 Å/s for Si, Ge and C, respectively. After growth the samples are annealed for 20 minutes at 500 °C in a rapid thermal annealing oven while flooded with forming

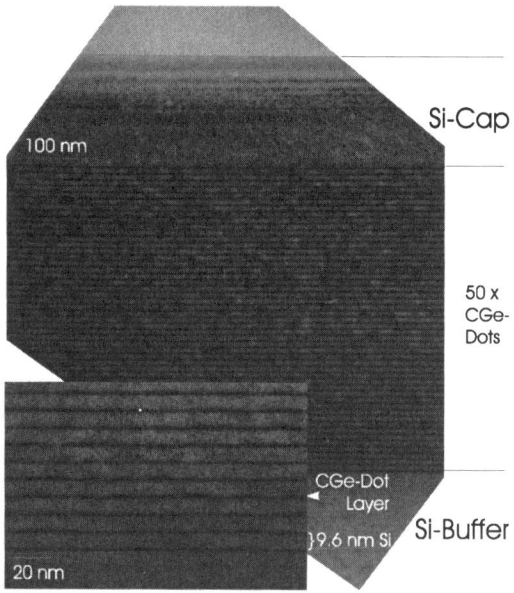

Figure. 1:
Cross-section TEM image of a stack of 50 CGe dot layers. Layers are separated by 10 nm thick Si spacer. No dislocations are observed. The inset shows a 3 times enlarged section of the image presenting well-developed but uncorrolated small CGe islands.

gas. It was checked that there is no PL energy shift due to material interdiffusion but instead a maximum gain in PL intensity. The structural properties of the samples are determined by transmission electron microscopy (TEM) and X-ray diffraction. PL is excited by a 488 nm Ar⁺ laser line with an excitation power of $P = 30$ mW and is analyzed by a single grating spectrometer and a Ge detector using standard lock-in technique. The beam is focused to a sample area of 2 mm². The samples are cooled down in a He-flow cryostat to a temperature of $T = 15$ K.

Figure 1 shows a cross-sectional TEM image of 50 layers of CGe dots. No dislocations can be found within the whole investigated sample area. The on average darker color of the dot-layer stack compared to the Si buffer and cap layers visualizes, what might be advantageous for future

Figure.2:
Comparison of X-ray diffraction (XRD) rocking curves of a 50 period Si/Ge superlattice with a 50 period CGe-dot superlattice. For the dot structure the superlattice (SL) peaks are shifted to the right hand side indicating that the pre-grown carbon partially compensates the compressive strain introduced by the Ge.

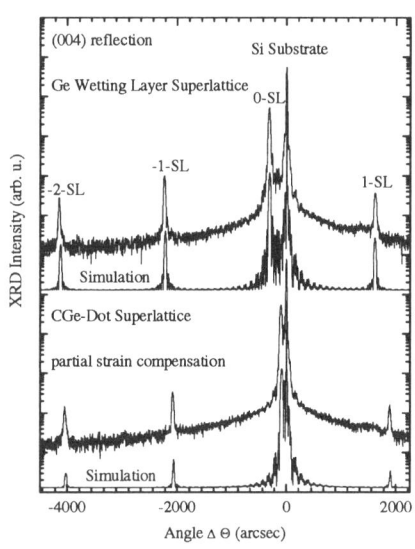

optolectronic applications: The stacked CGe dot layers are expected to exhibit a larger refractive index than Si because of the smaller average bandgap (note, that the darker color in TEM is of course not due to the larger refractive index). Thus, within the stack a light wave would on average experience a larger refractive index than in the surrounding Si cap and buffer layers. The CGe-dot stack could therefore be used as a light generating as well as light guiding structure.

The inset of figure 1 presents a three times enlarged image of the CGe dot layers. Well-developed islands are formed with diameters ranging from 10 to 15 nm and heights of around 1 to 2 nm. Surprisingly, there is practically no vertical alignment of the dots in contrast to what has been observed for stacked pure Ge islands in Si [14,15], InAs islands in GaAs [16] and InP dots in GaInP [17], where strain fields in the underlaying dot layers cause very pronounced vertical alignment of subsequently deposited dot layers. We appoint this effect to the pre-growth of carbon. The substitutionally incorporated carbon atoms partly compensate the compressive strain of Ge and introduce a rough surface with strong local strain fields. The surface diffusion length of the arriving Ge atoms is drastically reduced and vertical alignment of the Ge atoms on top of the underlying islands is suppressed. Reducing the thickness of the Si spacer layer to 2 nm still does not result in vertical dot alignment [18], which supports this explanation.

Figure.3:
Comparison of PL spectra of a 50 period 9.6 nm Si/2.4 ML Ge superlattice (SL, upper spectrum) with the 50 period 0.2 ML C/2.4 ML Ge dot superlattice (lower spectrum). The optical transition of the CGe-dots is red-shifted by 100 meV due to island formation. Efficient spatially indirect recombination channels result in strongly enhanced no-phonon PL signal for the CGe-dots. The insets show the band edge alignment of the two different structures. For the Si/Ge SL the main band-offset is given in the valence band (VB), whereas for the CGe dots a larger conduction band (CB) offset and spatially indirect recombination is assumed.

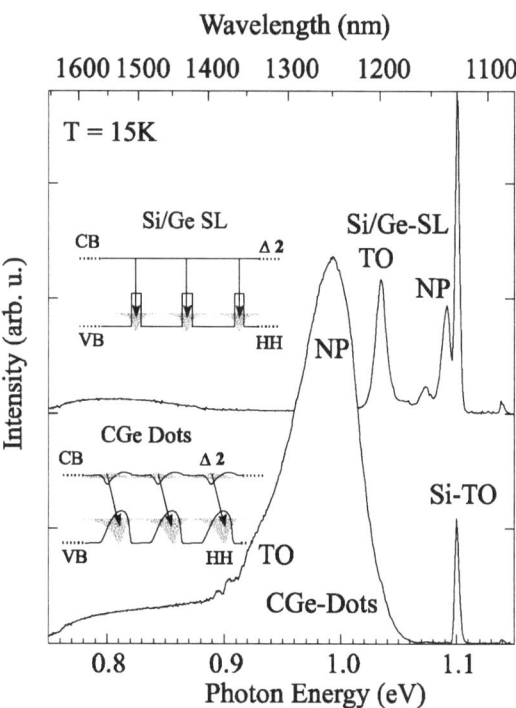

Figure 2 presents X-ray rocking curves of the 50 x (9.6 nm Si/2.4 ML Ge) reference sample in comparison to the CGe dot sample. The 2.4 ML Ge layers in the reference sample are planar with no island formation observed in TEM. The dominating peak is related to the (004) reflex of the Si substrate. Compared to the reference sample, it is evident that the zero-order superlattice related peak of the CGe-dot structure is shifted towards the substrate peak. This is explained by the tensile strain, introduced by the pre-grown carbon, which compensates compressive strain of the Ge layer. For the simulation of the upper spectrum we use 2.4 ML of pure Ge, whereas the lower spectrum is simulated taking 0.2 ML C followed by 2.4 ML Ge. Effects owing to island formation are neglected. In the latter case we assume that 1 carbon atom compensates compressive strain of 8.2 germanium atoms [19]. Note, that only 0.3 ML of pre-deposited carbon would be needed to achieve exact strain symmetrisation [20]. The results discussed in figure 1 and figure 2 yield the possibility to further reduce the Si spacer thickness and to increase the number of dot-layers without facing problems with nucleation of misfit dislocations.

In figure 3 PL spectra at $T = 15$ K of the two samples are presented. Besides the Si-TO line, PL peaks corresponding to the Si/Ge superlattice and the CGe-dot superlattice are observed at 1.090 eV (no-phonon) and 0.990 eV, respectively. Due to Ge island formation the optical transition of the CGe dots is red-shifted by 100 meV to a wavelength of 1252 nm, which is very close to the important 1.3 µm wavelength, needed for long-range optical fiber communication. Compared to the PL spectrum of a single layer of CGe dots [12], the transition energy is shifted by 30 meV to the low energy side. We attribute this red-shift to electronic coupling of carriers localized in the neighboring dot layers and to size increase in stacked dot layers [18]. For the Si/Ge superlattice the no-phonon (NP) and TO-phonon lines are well resolved and comparable in intensity, a much broader PL peak with a strongly enhanced NP-peak is observed for the CGe-dots. The broadness is attributed to the inhomogeneous dot size distribution. But, more important, the integrated intensity of the CGe islands is by a factor of 15 larger than that of the Si/Ge superlattice and is explained by efficient, spatially indirect recombination of electrons confined in an underlying carbon-rich SiGeC wetting layer with heavy holes confined in the Ge rich upper part of the island, which has been discussed in detail elsewhere [12, 21].

Figure 4 shows excitation dependent PL spectra of a sample with 50 times stacked dots with 0.2 ML C / 2.5 ML Ge. The excitation power was varied from 10 kWcm^{-2} to 100 MWcm^{-2}. The PL peaks above 1.03 eV are well known in the literature and are attributed to intrinsic energy transitions of the Si substrate and Si epilayers [22]. Two further PL peaks at 0.99 eV and 0.96 eV are observed. The energetically lower peak is attributed to the ground state of the dots. The second peak exhibits a similar line width and is related to the first excited state. The evolution of the spectra shows that the CGe dot ground state starts to saturate first, followed by the first excited state at about 10 MWcm^{-2}, whereas none of the Si-related peaks show such a saturation. The pronounced saturation behavior of the dot-related PL reflects the finite density of states associated with quantum dots.

Figure 4:
PL spectra of a stack of 50 CGe dot layers at excitation densities of 10, 100 kWcm^{-2}, 1, 10, 50 and 100 MWcm^{-2}. Saturation of the ground state (GS) and the first excited state (1. ES) at the low energy side is evident. PL peaks at higher energies are related to the Si substrate and Si epilayers. For clarity all spectra are normalized at the point indicated. Spectra below 1.05 eV are multiplied by a factor of 10.

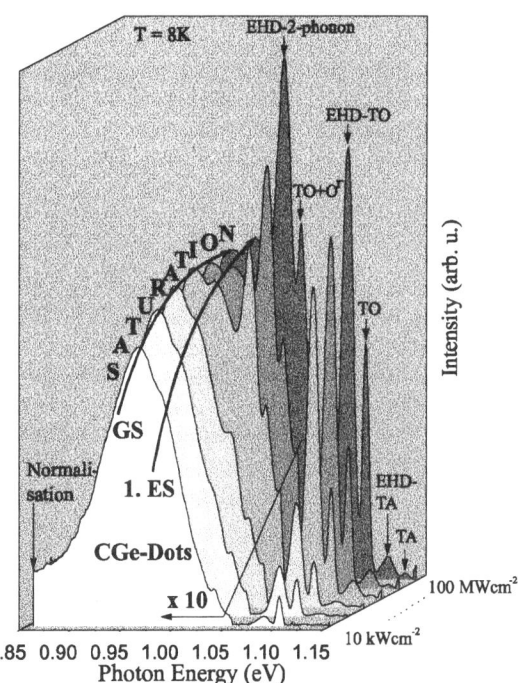

175

CONCLUSIONS

In conclusion, we have presented a stack of 50 layers of carbon-induced Ge dots. This structure exhibits very small, well-developed but vertically uncorrelated islands. Due to the pre-grown carbon, net compressive strain within the whole structure is kept much smaller than for a reference Si/Ge superlattice without C. Together with strongly enhanced photoluminescence signal this structure offers a large potential for future optoelectronic devices. There is significant design flexibility for further light output optimization by changing the Si spacer thickness, the C pre-deposition, the Ge thickness and the total number of stacked layers.

The technical assistance of W. Winter is gratefully acknowledged. We thank K. von Klitzing and E. Kasper for their continuous interest and support. This work was financially supported by the German Ministry of Education and Research within the Si-Nanoelektronik project.

REFERENCES

1 G. Abstreiter, P. Schittenhelm, C. Engel, E. Silveira, A. Zrenner, D. Meertens, and W. Jäger, Semicond. Sci. Technol. **11**, 1521 (1996).

2 H. Sunamura, N. Usami, Y. Shiraki, and S. Fukatsu, Appl. Phys. Lett. **66**, 3024 (1995).

3 D. Bimberg, et al., Phys. Stat. Sol. (b) **194**, 159 (1996).

4 K. Eberl, Physics World **10**, 47 (1997).

5 S. Yu. Shiryaev, F. Jensen, J. Lundsgaard Hansen, J. Wulff Petersen, and A. Nylandsted Larsen, Phys. Rev. Lett. **78**, 503 (1997).

6 Y. H. Xie, S.B. Samavedam, T.A. Langgo, and E.A. Fitzgerald, Thin Solid Films 1998 (in press).

7 N. Usami, T. Mine, S. Fukatsu, and Y. Shiraki, Appl. Phys. Lett. **64**, 539 (1994).

8 L. Vescan, Thin Solid Films **294**, 284 (1997).

9 T. I. Kamins and R. Stanley Williams, Appl. Phys. Lett. **71**, 1201 (1997).

10 N. Usami, F. Issiki, D. K. Nayak, Y. Shiraki, and F. Fukatsu, Appl. Phys. Lett. **68**, 2340 (1996).

11 K. Brunner, W. Winter, and K. Eberl, Appl. Phys. Lett. **69**, 1279 (1996).

12 O. G. Schmidt, C. Lange, K. Eberl, O. Kienzle, and F. Ernst, Appl. Phys.Lett. **71**, 2240, (1997).

13 O. G. Schmidt, C. Lange, K. Eberl, O. Kienzle, and F. Ernst, Thin Solid Films 1998 (in press).

14 L.Vescan, W. Jäger, C. Dieker, K. Schmidt, A. Hartmann, and H. Lüth, Mat. Res. Symp. Proc. **263**, 23 (1992).

15 P. Schittenhelm, G. Abstreiter, A. Darhuber, G. Bauer, A. Kosogov, and P. Werner, Thin Solid Films **294**, 291 (1997).

16 G. Solomon, J. Trezza, A. Marshall and J. Harris, Phys. Rev. Lett., **76**, 952 (1996).

17 M. K. Zundel, P. Specht, K. Eberl, N.Y. Jin Phillipp, and F. Phillipp, Appl. Phys. Lett. **71**, 2972, (1997).

18 O. G. Schmidt, C. Lange, K. Eberl, O. Kienzle, and F. Ernst, Appl. Phys. Lett. 1998 (in press).

19 K. Eberl, S.S. Iyer, S. Zollner, J.C. Tsang, and F.K. Legoues, Appl. Phys. Lett. **60**, 3033 (1992).

20 K. Eberl, S. S. Iyer, and F. K. Legoues, Appl. Phys. Lett. **64**, 739 (1994).

21 K. Eberl, O. G. Schmidt, S. Schieker, N.Y. Jin-Phillipp, and F. Phillipp, Solid-State Electronics (1998) (in press).

22 M. A. Vouk and E. C. Lightowlers, J. Phys. C. **8**, 3695 (1975).

USING GROWTH KINETICS FOR NANOENGINEERING OF Si-Ge SURFACES

I. GOLDFARB, G.A.D. BRIGGS
University of Oxford, Department of Materials, Parks Road, Oxford OX1 3PH, England,
ilan.goldfarb@materials.ox.ac.uk

ABSTRACT

In this work we explore how various growth characteristics of Ge on Si(001) can be used to fabricate structures for potential nanodevices. In the first example, the self-assembling tendency of germanium for three-dimensional islanding on Si(001) is considered, e.g. for application in devices based on quantum dots and wires. We aimed at achieving a detailed understanding of dot nucleation and growth mechanisms from germane. By controlling the deposition parameters, such as the germane pressure and substrate temperature, arrays of dots and antidots can be created on the grown surface, and further modified by post-deposition anneals. While lower temperature deposition leads to randomly distributed dots (i.e. small and coherent three-dimensional clusters with pyramidal shapes), a higher temperature deposition results in formation of antidots (i.e. pyramidal pits), which, in turn, are gradually replaced by the clusters, if the deposition is allowed to continue. The difference is caused by the different hydrogen behaviour at the respective temperature ranges. The germanium tendency to incorporate preferentially at the step and island edges is another beneficial property, which can be used to align the dots along step edges, creating wires rather than dots, or to fabricate ultrasmall Si-Ge heterojunctions, of a less than 10 nanometer size.

INTRODUCTION

The Si-Ge system plays an important role in the rapidly growing field of nanotechnology. The 4.2 % lattice mismatch between Ge and Si induces epitaxial strain in Si_xGe_{1-x}/Si layers (where $0 < x < 1$), leading to a band-gap narrowing [1] which can be exploited in microelectronic devices based on band-gap engineering (e.g. high-speed heterobipolar transistors [2]); strained-layer superlattices of alternating Si and Ge layers exhibit quasi-direct band-gaps [3,4] which are desirable in optoelectronics. The latter belong to a new family of devices based on reduced dimensionality, namely quantum wells, wires and dots [5], where electron confinement leads to enhancement of density of states and oscillator strength of the band-edge transition [6]. Although quantum wells of a sufficiently high quality can be grown, e.g. by molecular beam epitaxy (MBE), fabrication of quantum wires and dots remains a considerable challenge owing to the limited capability of microlithographic patterning to produce ultrasmall and defect-free structures. Therefore the ability of the small and coherent Ge "hut" clusters bounded by {501} facets to self-assemble on Si(001) [7] in a Stranski-Krastanow growth mode, is also very attractive from this point of view, if one can learn to control this self-assembly via the growth kinetics and thermodynamics [8-10]. This was one of the motivations behind our previous work, aimed at achieving a detailed understanding of the nucleation and growth mechanisms of the Ge/Si(001) [11-12] and Si_xGe_{1-x}/Si(001) [13] three-dimensional (3D) clusters, by monitoring the growth in real-time with near-atomic resolution in a scanning tunneling microscope (STM). (STM is a powerful tool which allows to investigate conducting surfaces in real-space with near-atomic resolution.) Another useful Ge/Si(001) tendency, when compared with Si/Si(001), is for an increased preferential incorporation at the step edges [14,15]. This tendency has been used to produce Ge quantum wires self-aligned at the Si(001) step edges [16], and by the present authors to produce ultrasmall, lateral Si-Ge heterojunctions by self-assembly [17].

In gas-source molecular beam epitaxy (GSMBE), a germane (GeH_4) molecule undergoes dissociative adsorption upon landing on a preheated substrate, to eventually yield clean epitaxial dimers and hydrogen chemisorbed on the surface [14,15], where the temperature/germane flux ratio determines not only diffusion lengths, but the hydrogen coverage as well. The latter can either hinder the surface diffusion at low temperatures, or serve as a surfactant, reducing segregation and element intermixing on one hand, and allowing for the growth of thicker 2D germanium layers on the other hand, at higher temperatures [11,14]. Such thick layers promote the formation of pyramidal pits, or antidots, which can be used e.g. to create quantum dots by filling them with different band-gap materials.

EXPERIMENT

The Si wafers used for this study were n-doped 0.1 Ω-cm, cut into 1×7 mm^2 pieces and chemically degreased *ex-vacuo*. The samples were handled with ceramic tweezers and clamped to the Ta support on the holder by Ta clamps. In UHV, the samples were degassed for several hours, repeatedly flashed at 1400 K, quenched below 800 K and slowly cooled to the desired temperature. During the sample flashes and anneals, the pressure was kept below 10^{-7} Pa. Such treatment has generally proved effective in producing well-ordered (2×1) Si surfaces. Sample heating was achieved by passing a direct current through it. Temperatures were measured by an optical pyrometer with an accuracy of 30 K. Polycrystalline 0.3 mm W wires were electrochemically etched in 2M KOH solution to produce atomically sharp tips.

A JEOL elevated-temperature STM, equipped with Low-Energy-Electron-Diffraction (LEED/Auger) and Reflection-High-Energy-Electron-Diffraction (RHEED), and capable of operation up to 1500 K was used to image the growing surface in real-time. The base pressure of the STM chamber prior to growth was 1×10^{-8} Pa. Growth movies were taken during the exposure to germane and disilane (separately) at the growth temperatures and in "constant current" mode, using currents around 0.1 nA and voltages between ±3V. GeH$_4$ and Si$_2$H$_6$ (99.99 %) were separately fed through a precision needle-valve onto the sample mounted in the STM stage, and the tip was allowed to scan while a desired constant pressure was maintained. Temperatures from room-temperature (RT) to 800 K, and gas pressures in the 10^{-7}-10^{-4} Pa range, were used for growth. The coverage was determined by computerised subtraction of images from one another, after fixing the same fiducial points in each successive pair of images.

RESULTS

Growth of Ge/Si(001) dots

For 3D islands to be used as quantum dots in optoelectronics, their size, i.e. the dimensions of the confining potential, must be smaller than the electron wavelength, and, as the luminescence of a quantum dot array is a convolution of spectra from every individual dot in that array, their size distribution must be narrow enough to preserve sharp photoluminescent transitions. The both parameters are intimately related to misfit strain and growth temperature, which constraint the smallest possible cluster size. As the highest lattice mismatch in Ge$_x$Si$_{1-x}$/Si(001) system is 4.2% for x = 1, i.e. pure Ge/Si(001), while it is 7% in InAs/GaAs, it is clear that the smaller dots can be obtained in the latter case. This is because the ultimate cluster size is inversely proportional to the square of strain, $w \propto \varepsilon^{-2}$ [18]. The effect of the growth temperature is apparent in Fig. 1: as the temperature increases the average cluster size increases as well, while the cluster number density decreases, because the cluster nucleation rate is exponentially dependent on the temperature, $R \propto \exp(T^{-1})$ [8]. Hence while at lower temperatures the dominating process at the growing surface is nucleation, at higher temperatures the growth of the cluster nuclei prevails. As the cluster growth is governed by the hut-shape instabilities [9,12,15], growth at higher temperatures (T ≤ 770 K) also leads to strikingly elongated shapes, as can be seen going from Fig. 1(a) to Fig. 1(d).

Being metastable, hut clusters only appear at Ge/Si(001) surfaces grown and/or annealed at T ≤ 800 K [7,12]. Above 800K large pyramidal nanocrystals, also bounded by {501} facets, appear, which upon further growth transform into even larger domes with complicated facet structure [10]. This transition gives considerable insight into the factors constraining the growth, but these nanocrystals and domes are too large for quantum applications. Thus the huts have been the most promising candidates for quantum dots so far. Fig. 2 shows that to achieve dense array

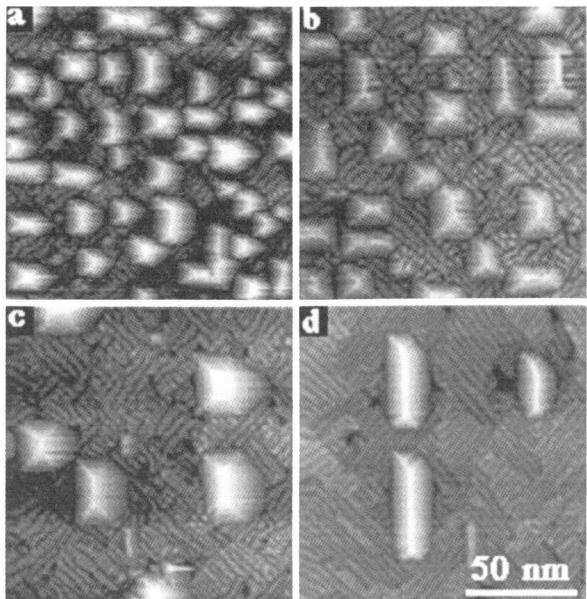

Figure 1 : Constant-current STM images of Ge/Si(001) hut-type dot arrays grown at different temperatures. (a) 630 K, (b) 690 K, (c) 720 K, and (d) 770 K.

of the smallest dots with a minimal scatter of the mean size, the growth must proceed at the lowest possible temperature. Therefore growth from solid sources can be preferred to that from gas sources, as the lowest growth temperature in the latter is limited to about 600 K by the hydrogen-blocking of surface diffusion [14]. As we have previously shown, mean cluster size and other statistical characteristics of the Ge/Si(001) dot arrays can be additionally modified by post-deposition anneals [12], but only to a limited extent.

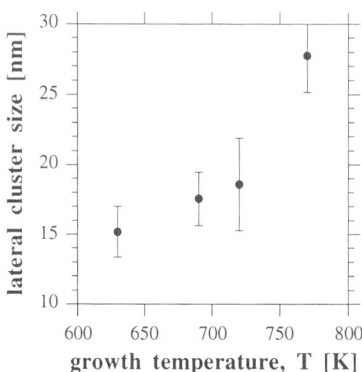

Figure 2 : Variation of the mean Ge/Si(001) hut cluster size with growth temperature.

Growth of Ge/Si(001) pits

The driving force behind the 2D-to-3D growth transition in strained-layer heteroepitaxial growth is the relaxation of mismatch strain. The strain can be relieved plastically by misfit dislocations, but frequently initial elastic relaxation is kinetically favoured [8]. Such an elastic relaxation is realised via exertion of stress on the surface by the thickness gradient of the pyramidal-shaped island, as well as by dilatation of lattice planes (which are compressed in the 2D layer) at the pyramid apex (see Fig. 3(a)). A similar, or even slightly higher, degree of relaxation can be achieved by forming a pyramidal pit (see Fig. 3(a)), provided the 2D wetting layer is sufficiently thick to accommodate the entire pit depth [8]. As the minimal thickness for pit

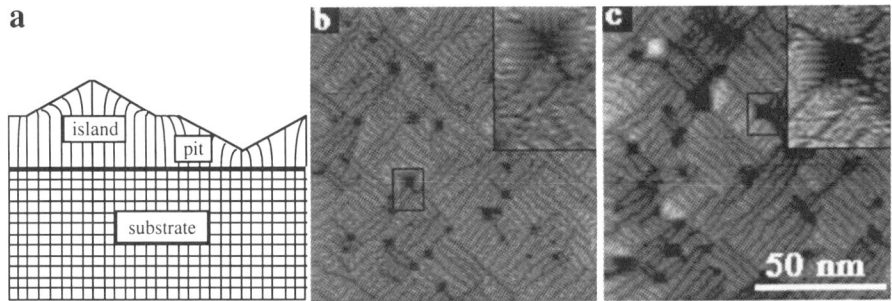

Figure 3 : Formation of small pyramidal pits at the Ge/Si(001) surface at 700 K. (a) Schematic drawing showing elastic strain-relaxation by means of pyramidal islands and pits, (b) constant-current STM image of 8.5 ML of Ge/Si(001), and (c) the same surface after additional ML (note the formation of huts, identifiable by their bright contrast, by the side of the pits). Note also the perfectly rectangular pit bases parallel to <100> directions and characteristic contrast of {501} facets, in the insets of (b) and (c), which are the blown-up images of the pits outlined in black.

formation is a rather strong function of the lattice mismatch, $h \propto \varepsilon^{-2}$ [19], in the Ge/Si(001) case, where the thickest 2D layers do not exceed 3-4 monolayers (ML) [20], pits cannot form without the use of surfactants. As the mismatch strain decreases the wetting layer thickness increases, causing sequential nucleation of pits and clusters in $Ge_{0.5}Si_{0.5}/Si(001)$ layers [19]. When growing germanium from hydride precursors, such as GeH_4, the hydrogen from dissociative adsorption of GeH_4 molecules exhibits threefold behaviour as a function of temperature; at T < 600 K it behaves as a diffusion-blocker and completely inhibits the growth, at 600 K < T < 700K it does not significantly influences the growth, and at T > 700 K it exhibits surfactant properties, e.g. delays the 2D-to-3D transition. Thus above 700 K we have been able to grow 2D layers up to about 10 ML [11,13-15], and hence to obtain the pits. Two examples of a typical pitted surface are given in Fig. 3(b)-(c). The pits exhibit precise geometrical characteristics of the hut clusters, i.e. {501} facets and <100>-oriented bases, but are even smaller than the huts. One idea how to transform these antidots into dots is to fill them with another semiconducting material, or with metals, creating embedded dots. Due to their very small dimensions, such embedded dots should exhibit even stronger electron confinement properties, close to those of isolated atoms. It should be pointed out that the range of the pit existence is very narrow; if more than 1 ML is added to the surface after the appearance of the pits, they are replaced by hut clusters [11,14]. By monitoring in real-time with STM, it is possible to catch the moment of formation of the pits.

Growth of lateral Ge-Si nanoheterojunctions

So far only the applications of 2D-to-3D transition to self-assembled quantum dots have been discussed. Yet other types of structure can be created by controlled self-assembly, e.g. lateral Si-Ge junctions of nanometer size, as demonstrated in Fig. 4. To achieve that we make use of the Ge/Si(001) tendency for an increased preferential incorporation at the step edges, when compared with Si/Si(001) [15]. As mentioned above, at low temperatures the surface hydrogen hinders the diffusion, but it also hinders the attachment rate of the adatoms and addimers to preferential sites at the island and step edges, by sticking to these edges and poisoning them. Above the hydrogen desorption temperatures the edges become vacant again, allowing edge-adsorption and step-flow growth. As the hydrogen desorption temperature from the Ge-covered surfaces is lower than from Si surfaces, even at temperatures as low as 600 K the hydrogen begins to desorb from the step and island edges [15], converting them (especially kinks) into preferential binding sites for the diffusing germanium and even GeH$_4$ precursors.

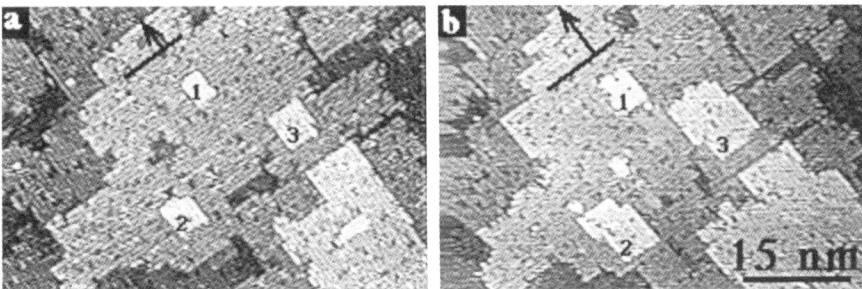

Figure 4 (a): Constant-current STM image of the Si/Si(001) surface grown during the first fabrication step, and (b) the same surface after submonolayer germanium deposition from germane at 600 K. Note the evolution of the initial Si islands in (a) due to the germanium addition in (b).

The first step towards the lateral heterostructures consisted of the formation of small Si islands on the surface by selecting the Si$_2$H$_6$ pressure in the order of 10^{-7} Pa and 600 K substrate temperature to promote island- rather than step-growth mode [15] (see Fig. 4(a)). At this point the disilane flux was stopped and the pressure was allowed to recover. We then deposited germane at the same temperature. It had to be deposited in such a way that all of it joins the existing Si-islands to form heterojunctions (i.e. suppressed nucleation), which, in principal, requires sufficiently high temperature. On the other hand using high temperatures can cause an undesired Si-Ge interdiffusion and thus rough interfaces. The hydrogen-desorption temperature from Ge-covered silicon steps is lower than from clean silicon steps, so that at the growth temperature at which the silicon forms islands the germanium adsorbs preferentially at the island edges rather than nucleating new germanium islands, at sufficiently low temperature where Si-Ge interdiffusion is negligible. As can be seen by comparing Fig. 4(a) to Fig. 4(b), indeed the vast majority of the deposited germanium joined the edges of Si islands increasing their size. Si-Ge-Si junctions can also be created by repeating step 1. The size of these Si-Ge "diodes" and Si-Ge-Si "transistors" can be precisely tuned simply by stopping the growth at each successive step when the islands have reached the desired size.

CONCLUSIONS

In this work we have explored the possibilities to create various nanostructures by precisely controlled self-assembly. In other words, in order to produce features on the scale which lies below the capabilities of even modern microlithography we had to use the growth kinetics itself to yield the desired effects. Three examples of such a "surface nanoengineering" have been given to illustrate the possible Ge-Si nanostructures. At present, due to size and nature of these structures, all three of them can only be obtained by precise real-time STM monitoring of the growth. (1) The first is a rather familiar example of self-assembled dots. We have shown that optimal dot type can be produced by low-temperature Ge/Si(001) growth. As lower-temperature growth is more viable using solid- rather than gas-sources, better (small and equally-sized) dots should be easier to grow by MBE. Statistical characteristics of the dot arrays, e.g. mean size, homogeneity, and especially number density, can be additionally "tailored" by post-deposition anneals. (2) On the other hand, even smaller and thus more strongly confined dots can be obtained in gas-source MBE without using additional surfactants, as the surface hydrogen inherent to this process plays a role of a natural surfactant by delaying the roughening transition. To obtain this kind of dots, sufficiently thick 2D layers have to be grown first, then tiny pyramidal pits are formed to relieve the mismatch strain (in the same way as dots do). These pits could be subsequently filled with a different band-gap material, producing embedded dots. (3) In the third example we have shown how ultrasmall Si-Ge heterojunctions can be self-assembled.

ACKNOWLEDGEMENTS

This work is supported by EPSRC (GR/K08161).

REFERENCES

1. J.C. Bean, A.T. Fiory, R. Hull, and R.T. Lynch, in Proceedings of the 1st International Symposium on Si-MBE, edited by J.C. Bean (Electrochemical Society, Remington, NY, 1985), Vol. 85-7, p. 161.
2. D.C. Houghton, Compound Semicond. 1, p.31 (1995).
3. E. Kasper, Surf. Sci. 174, p.630 (1986).
4. G. Abstreiter, Thin Solid Films 183, p.1 (1989).
5. C. Weisbuch and B. Vinter, Quantum Semiconductor Structures, Academic Press, San-Diego, 1991.
6. T. Ogawa and T. Takagawara, Phys. Rev. B 44, p.8138 (1991).
7. Y.-W. Mo, D.E. Savage, B.S. Swartzentruber, and M.G. Lagally, Phys. Rev. Lett. 65, p.1020 (1990).
8. J. Tersoff and F.K. LeGoues, Phys. Rev. Lett. 72, p.3570 (1994).
9. D.E. Jesson, K.M. Chen and S.J. Pennycook, MRS Bull. 21(4), p.31 (1996).
10. G. Medeiros-Ribeiro, A.M. Bratkowsky, T.I. Kamins, D.A.A. Ohlberg, R.S. Williams, Science 279, p.353 (1998).
11. I. Goldfarb, P.T. Hayden, J.H.G. Owen and G.A.D. Briggs, Phys. Rev. Lett. 78, p.3959 (1997).
12. I. Goldfarb, P.T. Hayden, J.H.G. Owen and G.A.D. Briggs, Phys. Rev. B 56, p.10459 (1997).
13. I. Goldfarb and G.A.D. Briggs, J. Cryst. Growth (submitted).
14. I. Goldfarb, J.H.G. Owen, P.T. Hayden, D.R. Bowler, K. Miki, and G.A.D. Briggs, Surf. Sci.. 394, p.105 (1997).
15. I. Goldfarb, J.H.G. Owen, D.R. Bowler, C.M. Goringe, P.T. Hayden, K. Miki, D.G. Pettifor, and G.A.D. Briggs, J. Vac. Sci. Technol. A (in print).
16. H. Sunamura, N. Usami, Y. Shiraki, S. Fukatsu, Appl. Phys. Lett. 68, p.1847 (1996).
17. I. Goldfarb and G.A.D. Briggs, Surf. Sci. Lett. (submitted).
18. W. Dorsch, H.P. Strunk, H. Wawra, G. Wagner, J. Groenen, and R. Carles, Appl. Phys. Lett. 72, p. 179 (1998)
19. D.E. Jesson, K.M. Chen, S.J. Pennycook, T. Thundat, R.J. Warmack, Phys. Rev. Lett. 77, p.1330 (1996).
20. M. Tomitori, K. Watanabe, M. Kobayashy, O. Nishikawa, Appl. Surf. Sci. 76/77, p.322 (1994).

GE-QUANTUM DOTS ON SI(001) TAILORED BY CARBON PREDEPOSITION

O. LEIFELD [*,**], D. GRÜTZMACHER[*], B. MÜLLER [***], K. KERN [**]

[*] Micro- and Nanostructures Laboratory, Paul-Scherrer-Institute, CH-5232 Villigen-PSI,
[**] Institut de Physique Expérimentale, EPFL, CH-1015 Lausanne, Switzerland
[***] Institute of Quantum Electronics, Nonlinear Optics Laboratory, ETHZ, CH-8093 Zürich, Switzerland

ABSTRACT

The morphology of Si(001) after carbon deposition of 0.05 to 0.11 monolayers (ML) was investigated in situ by ultra-high vacuum scanning tunneling microscopy (UHV-STM). The carbon induces a c(4x4)-reconstruction of the surface. In addition, carbon increases the surface roughness compared to clean Si(001) (2x1). In a second step, the influence of the carbon induced restructuring on Ge-island nucleation was investigated. The 3D-growth sets in at considerably lower Ge coverage compared to the clean Si(001) (2x1) surface. This leads to a high density of small though irregularly shaped dots, consisting of stepped terraces, already at 2.5 ML Ge. Increasing the Ge-coverage beyond the critical thickness for facet formation, the dots show {105}-facets well known from Ge-clusters on bare Si(001) (2x1). However, they are flat on top with a (001)-facet showing the typical buckled Ge rows and missing dimers. This indicates that the compressive strain is not fully relaxed in these hut clusters.

INTRODUCTION

Self organized quantum dots can be manufactured in many heteroepitaxial systems, driven by stress relaxation due to lattice mismatch. In the Si/Ge system, small optically active 3-D dots may be a viable route towards new Si-based optoelectronic devices. Recently small Ge dots (10 nm) showing strong photoluminescence were fabricated at temperatures as high as 550°C by MBE using Si(001) substrates with submonolayer C pre-deposition [1]. Their structural properties, however, were not revealed in detail. The aim of this paper is a detailed examination of the surface morphologies that occur in connection with the growth of these C-induced Ge-quantum dots by in-situ STM. This includes the investigation of the submonolayer C-coverages on Si(001) (2x1) as such, as this is a prerequisite for the understanding of the dot layer growth.

EXPERIMENT

Setup

For the investigations we have realized a combination of an STM with our MBE/UHV-CVD machine. The growth system is composed of a Balzers UMS 500 MBE-chamber combined with a UHV-CVD-reactor, both allowing processing of 4-inch samples. A schematic drawing is shown in Fig.1. Up to 5 wafers are introduced via one of the loadlock chambers and are distributed from there by means of the sample transfer system (sample handler) to the various locations. The MBE-chamber is equipped with electron beam evaporators for Si and Ge and resistively heated evaporation crucibles for Sb and B dopants. Carbon is sublimated from a well-shielded DC-heated pyrolytic graphite filament The substrate temperature of the 4-inch wafer can be varied from room temperature (RT) up to 1000°C. Substrate rotation provides homogeneous film thickness over the wafer surface. The 4-inch wafer STM is contained in a transportable ultra high vacuum (UHV) chamber, which is attached to the sample transfer chamber of the MBE/CVD-system via a small transfer lock between two CF150 gate valves, as seen in Fig.1. Processed wafers are

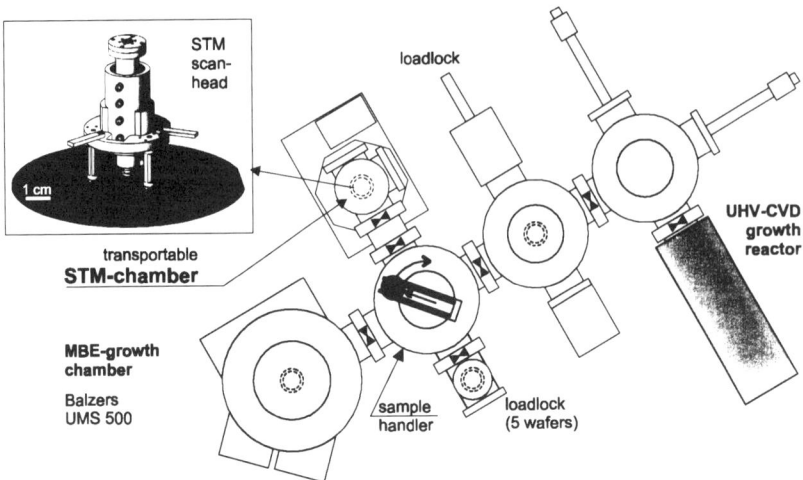

Fig.1: Scheme of the growth system for SiGe-based films, consisting of the MBE-chamber, the sample handler and the UHV-CVD reactor. The transportable STM-chamber is attached to the sample handler for in-situ sample transfer. The inset represents the STM head as described in the text.

transferred in-situ from the growth- to the STM-chamber by means of the sample handler. In order to avoid noise problems due to mechanical vibrations caused by the pumps of the growth system, the STM-chamber is detached from the growth system after the transfer lock has been vented. It is moved to another room for high resolution measurements.

The inset of Fig.1 displays the STM head of the home-built STM. It is based on the Beetle-type microscope originally proposed by Besocke [2] with three outer piezo tubes and a center piezo for scanning. The STM is lowered down directly onto the wafer surface at the desired position. This coarse positioning is done by a standard UHV xyz-manipulator with a range of ±25 mm in the xy-plane. Fine positioning is accomplished by inertial motion of the entire scan head on the wafer and by DC-offset voltages applied to the piezo tubes. Tip approach is effectuated by an Inchworm motor. The tunneling tips are in-situ exchangeable. The STM/sample-holder provides vibration isolation by a spring/eddy-current damping assembly. In addition, the whole chamber resides on laminar flow isolators during measurement. A more detailed description of the STM setup and performance is published elsewhere [3].

Growth procedure

Si-wafers are wet-chemically cleaned followed by an HF-dip for hydrogen passivation before they are loaded into UHV. In the MBE chamber they are heated to 600°C for 10 min for hydrogen desorption followed by a 30 min period at 950°C, resulting in a smooth Si(001) (2x1) surface, which has been confirmed by reflection high energy electron diffraction (RHEED) and STM. At a substrate temperature of 550°C a 200nm thick Si-buffer layer is grown. Onto this buffer carbon is deposited at the same temperature, followed by Ge for the dot layer. Typical deposition rates are 5×10^{-4}ML/s Carbon at a filament current of about 100A and 0.16 ML/s Ge from the e-gun evaporator. The layer thickness has been varied between 0 ML to 0.17 ML C and 2.5 - 5.8 ML Ge in the samples discussed in this paper. The base pressure of the MBE chamber is

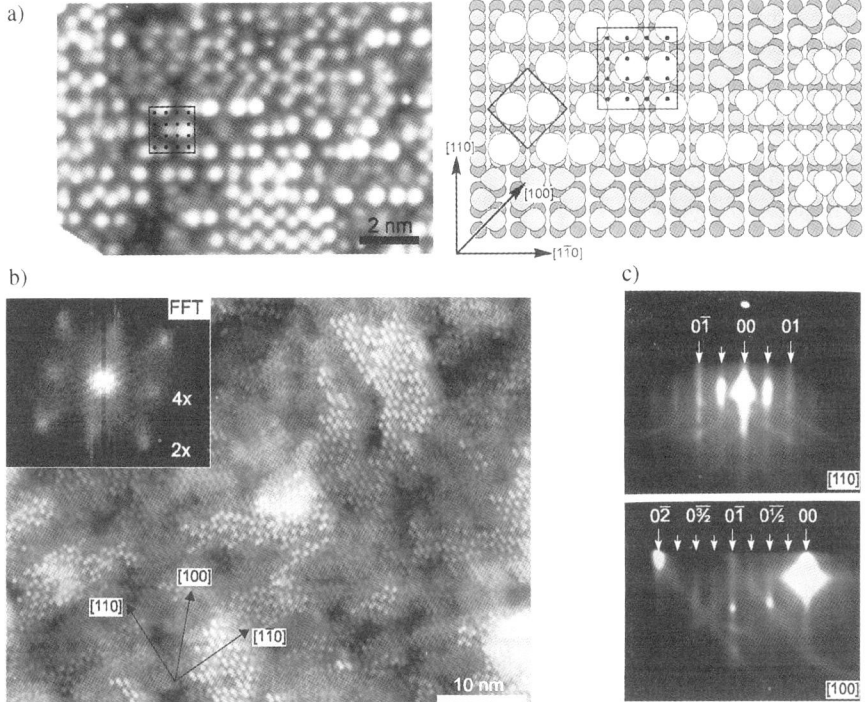

Fig. 2: Carbon induced c(4x4)-reconstruction. **a)** high resolution STM-micrograph showing the c(4x4) structure (bright double-spots) in coexistence with buckled Si-Dimers (smaller zigzag-pattern). A corresponding model is given aside, indicating the c(4x4) unit cell and a primitive one. **b)** larger scan area with c(4x4)-domains on different terraces (inset: Fast Fourier Transform of the image revealing the c(4x4) symmetry in reciprocal space. **c)** RHEED pattern proofing the long range order of the c(4x4) structure. 3 fractional order superstructure streaks can be observed in between the main streaks in [100]-azimuth.

lower than 1×10^{-10} mbar and raises to 1×10^{-8} mbar during carbon source operation. After deposition, the sample is cooled to RT and transferred to the STM with a chamber pressure of 1×10^{-10} mbar.

RESULTS

Carbon covered Si(001)

In a first step, the influence of the C pre-deposition on the clean Si(001) (2x1) surface has been studied in the range of 0.05 to 0.11 ML C. Filled state STM images of a surface covered with 0.11 ML C are displayed in Fig.2 a and b. Aside from an increase in surface roughness, they reveal areas of a surface reconstruction that differs from the well known (2x1) dimer reconstruction of Si(001). It has a unit cell of four by four Si(001) in-plane lattice constants ($a_0 = 3.84\text{Å}$) with a centered symmetry, thus called c(4x4) following the Wood-notation. In the large scale STM image, the c(4x4) patches are those covered with the elongated double-spots running parallel to the <110>-directions. In the RHEED pattern (Fig.2c) three fractional-order spots arising from the c(4x4) superstructure are observed between the main streaks in the [100]-azimuth. In a <110>-

azimuth, i.e. with the electron beam directed parallel or perpendicular to the dimer rows, only one half-order streak is obtained, caused by the centered symmetry. As RHEED is a non-local probe, this confirms the long-range order of the c(4x4) reconstruction. The high-resolution image and the schematic drawing in Fig.2a allow to determine the arrangement of the c(4x4) double-spots with respect to the Si-(2x1) buckled dimer rows (oval zigzag pattern). They are perfectly aligned with the Si-dimers in the same atomic layer. Neighboring rows are shifted to each other by two a_0. The center of the double-bumps and the voids between two of them are always aligned to the middle of the dimer rows in the underlying ML. A c(4x4) unit-cell is represented by the square. The black dots illustrate the non-reconstructed (bulk) positions of single adatoms in that layer. The rhombus indicates a primitive unit cell. A c(4x4)-reconstruction on Si(001), that looks similar to the one observed in this study, has been reported by various authors in the past. It has been prepared by exposure of Si to large amounts of hydrogen at temperatures as high as 700°C [4,5]. In these studies the reconstruction was always referred to as a metastable pure silicon surface, as H is not present at Si surfaces at these temperatures. Only recently it has been suggested that it might be related to C-contamination [6], derived from RHEED experiments. The fact, that the area covered by the c(4x4) pattern scales with the amount of deposited carbon, provides clear evidence that it stems from C-atoms.

Ge deposition onto C-pre-covered Si(001)

In a second step, Ge has been evaporated onto samples with 0.11 ML pre-deposited C at 550°C. This allows to investigate the influence of the C-induced restructuring on Ge island nucleation and to compare it to Ge on a bare Si(001) (2x1) surface. Ge coverages in the range of 2.5 ML to 5.8 ML were used. Fig.3 shows two such layers at Ge coverages of a) 2.5 ML and b) 4 ML. Already at nominally 2.5 ML three-dimensional growth is observed. For simplicity we call these islands Ge/C-dots. Island heights as high as 1.2 nm are observed already at this low coverage. The islands have an irregular shape and consist microscopically of stepped terraces with variable width. This can be seen in Fig.4a for a 3 ML Ge sample. The onset of 3D growth of Ge/C-dots is well below the critical thickness of island formation for Ge on bare Si(001) (2x1) [7], where Ge exhibits a Stranski-Krastanov (SK) growth mode. There it forms a 2-dimensional layer of up to 4 ML prior to island formation driven by elastic strain relaxation [8]. However, as

a)   b)

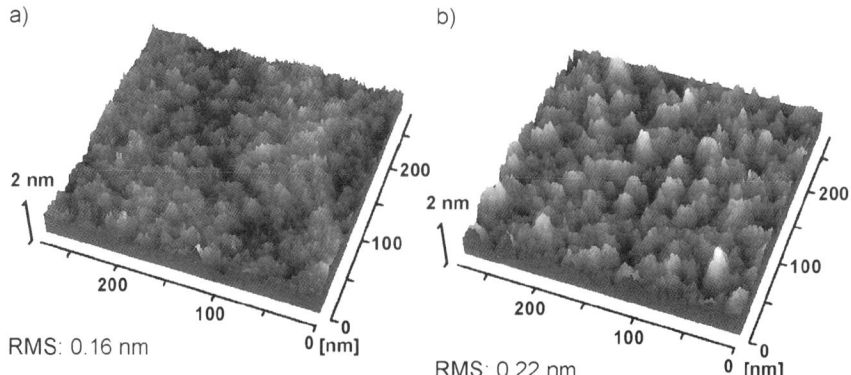

2 nm

RMS: 0.16 nm

RMS: 0.22 nm

Fig.3: STM-images of Ge-dot-layers deposited onto Si(001) at 550°C after pre-deposition of 0.11 ML C. **a)** 2.5 ML Ge, **b)** 4 ML Ge. Irregularly shaped islands have formed. An increase in island height with Ge-coverage is reflected by the increasing RMS-roughness. The island density remains unchanged at $10^{11} cm^{-2}$.

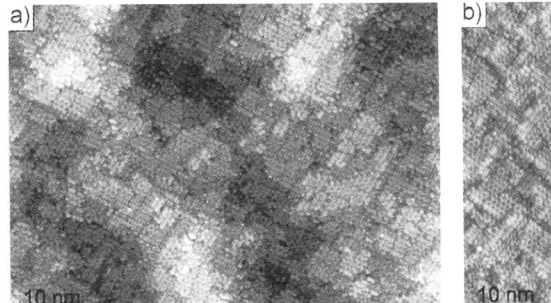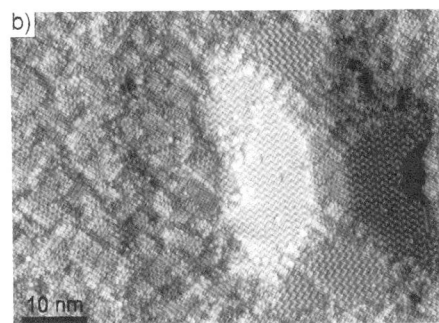

Fig.4: Atomic resolution STM-images of Ge/C-dots. At **a**) 3ML Ge stacks of irregularly stepped terraces with buckled Ge-dimer rows are clearly resolved. Height is 9 ML ~1.2 nm. One distinguishes four islands. At **b**) 5.8 ML Ge larger 'Hut-Cluster'-like islands occur exhibiting {105}-side facets and a (001)-top facet.

C reduces the overall amount of strain in the Ge/C-dot layer, strain is considered not to be the dominating driving force for the earlier onset of island formation. The additional substrate roughness introduced by C as compared to clean Si(001) (2x1) may be an explanation. Moreover, the formation of C-rich surface patches may have an impact on the island formation. On one hand the repulsive forces between Ge and C may lead to an agglomeration of Ge atoms in the areas between the patches. On the other hand, it might be favorable for Ge atoms to stay on the C-rich areas in order to relieve strain.

Increasing the Ge coverage further, the Ge/C-dots grow in height and width, which is obvious in Fig.3b for 4 ML Ge, where the stacks reach a height up to 14 ML ~1.9 nm. The measured root mean square roughness (RMS) rises monotonically from (0.07 ± 0.02)nm for the C-pre-covered Si to (0.22 ± 0.03)nm for the Ge/C-dot layer. The dots still reveal the same irregular shape. The dot density, although somewhat difficult to determine on this rugged surface, remains constant at 10^{11}cm^{-2} from 2.5 ML to 4 ML Ge, if only stacks are considered that are higher than the mean height in the STM-images plus the RMS-value. This can be explained by the changed nucleation kinetics on the C-covered surface compared to bare Si(001). We speculate, that the number of nucleation centers is determined by the amount of pre-deposited carbon, and that a variation of the C-coverage can influence the resulting island density.

Up to here, no sign of facet formation is observed, as it might be anticipated from the well known growth behavior of Ge on Si(001) [9]. At a coverage of 5.8 ML Ge, however, the critical thickness for facet formation is reached even for the Ge/C-dots. Dot geometries very similar to the well known 'hut clusters' with quadratic as well as rectangular shape are obtained, as shown in Fig.5a. The size range is 20 nm to 40 nm. At the same time, the surface between the islands smoothens and the terraced islands die out. Fig.4b demonstrates the facet geometry of such a faceted Ge/C-dot. {105}-side facets with their distinctive zigzag reconstruction are well resolved. In addition, all faceted Ge/C-dots observed exhibit a (001)-top facet. This facet shows the characteristic buckled Ge-dimer chains with rows of missing dimers across them. Such missing dimer rows are also observed in the 2D-layers of Ge on Si(001) [10] and are an effective way for relaxation of compressive strain. Thus we conclude, that this type of faceted Ge/C-dots may still contain a significant amount of strain. This is a difference to hut-clusters on bare Si that are completely relaxed towards their apex [11]. Nevertheless we assume, that the mechanism for facet formation is essentially the same as for pure Ge/Si hut-clusters, namely a reduction in sur-

 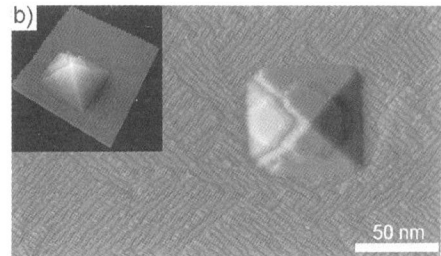

Fig.5: {105}-faceted dots ('hut-clusters') obtained by deposition of 5.8 ML Ge at 550°C **a)** on Si(001) with 0.11 ML pre-deposited C and **b)** on bare Si(001) (2x1). Dot density is at least one order of magnitude larger with C pre-deposition. Ge/C-dot sizes are significantly smaller. The ghostimage on the left slope of the large Ge-dot in b) is a doubletip artifact.

face free energy [12]. As the faceted dots grow at the expense of the terraced Ge/C-dots, their density is a factor of 4 smaller.

For comparison, a sample of 5.8 ML Ge was grown on a clean Si(001) (2x1) substrate at the same growth conditions (Fig. 5b). Here quadratic pyramidal {105}-faceted 'hut-clusters' with essentially uniform sizes about 65 - 70 nm are obtained on a smooth 2D-wetting layer with typical superstructure [10]. Their density is at least one order of magnitude lower than that of their Ge/C-equivalents ($\sim 10^9 cm^{-2}$). It is interesting to note, that this dots seem to grow layer by layer from top to bottom, since the collar-like structure was observed with STM on every dot.

CONCLUSIONS

We have studied submonolayer coverages of C and C-induced Ge dot layers on Si(001) with STM. A C-induced c(4x4)-reconstruction is found at 0.11 ML C on a generally roughened surface. Ge growth on this surface evolves by 3D-island formation already at coverages as low as 2.5 ML in contrast to Ge on bare Si(001) (2x1). Islands consist of irregularly stepped terraces. Their size increases with Ge coverage while their density remains constant. For 5.8 ML Ge the critical thickness for relaxation via {105}-facet formation is exceeded, for both, C-pre-covered and clean Si(001). Island density then reduces by a factor of 4.

REFERENCES

1 O.G. Schmidt, C. Lange, K. Eberl, O. Kienzle, and F. Ernst, Appl. Phys. Lett. 71, 2340 (1997)
2. K. Besocke, Surf. Sci. **181**, 145 (1987)
3. O. Leifeld, B. Müller, D.A. Grützmacher, and K. Kern, Appl. Phys. A **66**(3), S993, (1998)
4. T. Ide, and T. Mizutani,
5. R.I.G. Uhrberg, J.E. Northrup, D.K. Biegelsen, R.D. Bringans, and L.E. Swartz, Phys. Rev.B **46**(16), 10251 (1992)
6. K. Miki, K. Sakamoto, and T. Sakamoto, Appl. Phys. Lett. **71**(22), 3266 (1997)
7. Y.-W. Mo, D.E. Savage, B.S. Swartzentruber, and M.G. Lagally, Phys. Rev. Lett. **65**(8), 1020 (1990)

8. D.J. Eaglesham, and M. Cerullo, Phys. Rev. Lett. **64**(16), 1943 (1990)
9. M. Tomori, K. Watanabe, M. Kobayashi, O. Nishikawa, Appl. Surf. Sci. **76/77**, 322 (1994)
10. U. Köhler, O. Jusko, G. Pietsch, B. Müller, and M. Henzler, Surf. Sci. **248**, 321 (1991)
11. A.J. Steinfort, P.M.L.O. Scholte, A. Ettema, F. Tuinstra, M. Nielsen, E. Landmark, D.M. Smilgies, R. Feidenhans'l, G. Falkenberg, L. Seehofer, and R.L.Johnson, Phys. Rev. Lett. **77**(10), 2009 (1996)
12. K.E. Khor, and S. Das Sarma, J. Vac. Sci. Technol. B **15**(4), 1051 (1997)

MEASUREMENTS OF CONFINED ENERGY LEVELS AND COULOMB CHARGING EFFECT IN SELF-ASSEMBLED Ge QUANTUM DOTS BY ADMITTANCE SPECTROSCOPY

S.K.Zhang, Z.M.Jiang, H.J.Zhu and F.Lu
Surface Physics Laboratory, Fudan University, Shanghai 200433, CHINA

ABSTRACT

We perform measurements of the discrete quantum energy levels and Coulomb charging of self-assembled Ge quantum dots imbedded in Si barriers by using the admittance spectroscopy technique that was originally developed to measure the defect levels of bulk materials and the band offsets of heterojunctions. By varying the bias voltage, the population of carriers in the dot changes and the Coulomb charging effect could be clearly seen from the step-like change of the activation energy for hole emission in the admittance spectra. Up to five holes charged in a Ge dot with a lateral dimension of 13nm is observed. The energy levels of ground state and first excited state are determined. The advantages of this method are the relatively high measuring temperature, large signal to noise ratio, and its simple and straightforward use and interpretation. The requirement on the uniformity of the dot size distribution of the sample is quite tolerant.

INTRODUCTION

In recent years, semiconductor quantum dots have become the subject of great experimental and theoretical interest because they have potential device applications both as electronic memories and optoelectronic devices. Many methods have been developed to study the energy level structures and Coulomb charging effect in quantum dots, such as transport spectroscopy [1], far-infrared [2], conventional capacitance spectroscopy [3], single-electron capacitance spectroscopy [4] and so on. However, all these methods require well defined samples and extremely low temperature. Anand et al. [5] studied the Coulomb charging effect in InP dots using the deep level transient spectroscopy within the temperature range of 10~150K. However, the discrete energy levels were not obtained.

Admittance spectroscopy is a conventional method to measure the defect levels of semiconductor materials [6] and the band offsets of heterojunctions [7]. In this work it is employed to study the discrete quantum energy levels and Coulomb charging effect of self-assembled Ge quantum dots imbedded in Si barriers. The activation energies of the ground hole state and the first excited hole state are extracted and the Coulomb charging effect is clearly observed at measuring temperatures above 100K. The influence of the non-uniformity of the dot size distribution on the measurement results is discussed.

PRINCIPLE

The sample studied is a Si Schottky diode structure with Ge dots imbedded inside the Si depletion layer. Figure 1(a) shows the schematic valence band diagram of such a structure. In Fig. 1(b), the equivalent circuit of the sample structure with a single layer of Ge dots is a parallel circuit of the conductance G_{dot} and the capacitance C_{dot} of the dots, in series with a capacitance C_S of the Schottky barrier. According to the thermal equilibrium condition of hole emission and capturing, the conductance G_{dot} is deduced to be

$$G_{dot} = \alpha T \sigma_{\mathrm{I}} \exp(-\frac{E_I - E_V}{kT}) , \qquad (1)$$

where α is a temperature-independent constant, T is the temperature, σ_I and E_I are the capture cross section and the energy of the Ith discrete level respectively, E_V is the valance band maximum of Si, and k is the Boltzmann constant. At a measurement frequency ω, when the temperature scans through a certain value T_m which meets the following relation:

$$G_{dot}(T_m) = (C_S + C_{dot})\omega , \qquad (2)$$

the conductance of the sample reaches a maximum $G(T_m)$ [7]. The condition for the structure conductance reaching a maximum derived from Eqs. (1) and (2) is

$$\omega / T_m = \alpha_I \sigma_I \exp(\frac{E_I - E_V}{kT_m}) , \qquad (3)$$

where $\alpha_I = \dfrac{\alpha}{C_s + C_{dot}}$. The Arrhenius relation of above formula is

$$\ln(\omega / T_m) = \ln(\alpha_I \sigma_I) - E_{aI} / kT_m , \qquad (4)$$

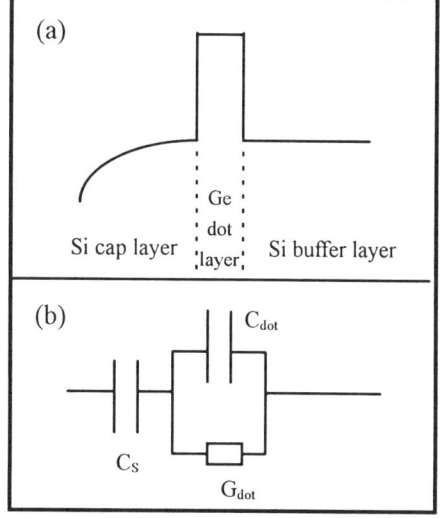

Fig.1 The schematic valence band diagram of a Si Schottky diode structure with Ge dots imbedded in Si barrier (a) and its equivalent circuit (b).

where $Ea_I = E_I - E_V$ is the activation energy of the Ith level.

Due to the small physical dimensions of the dots, the Coulomb charging effect cannot be ignored. The Coulomb charging energy required to charge the Nth holes into a disk-like dot is given by [8]

$$E_N = (N-1/2)e^2/4\varepsilon\varepsilon_0 D , \qquad (5)$$

where D is the typical diameter of dots. To account for the Coulomb charging energy, the activation energy of carriers should be expressed as

$$Ea_I = E_I - E_V - E_N . \qquad (6)$$

By measuring the peak temperatures T_m of G-T spectra at different frequencies, the activation energy Ea_I can be obtained from the slope of the Arrhenius plot $ln(\omega/T_m) \sim 1/kT_m$.

EXPERIMENT

Figure 2 is a schematic diagram of the sample structure used in this study which was grown by molecular beam epitaxy. Three periods of alternately-stacked Ge dot layers and Si spacer layers were deposited on a 100 nm thick Si buffer layer at 500°C and capped with a 400nm thick Si layer on the top. The thickness of the Si spacers is 50nm, which is thick enough to prevent the carrier

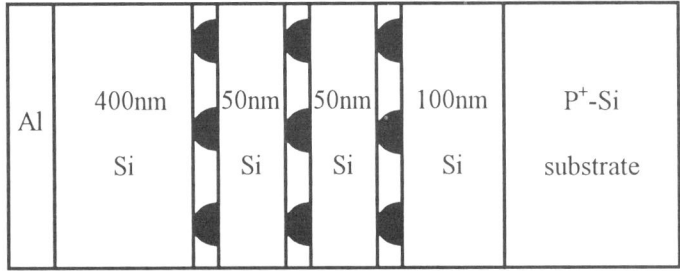

Fig. 2 The schematic diagram of the
sample structure.

coupling between neighboring Ge dot layers. All the Si layers were doped with boron to a concentration of about 1×10^{16}cm^{-3}. The nominal thickness of the Ge dot layers is 1.3nm. Under this growth condition, clustering of Ge occurs according to Stranski-Krastanow growth mode The cross-sectional transmission electron microscopic (TEM) observation verified the formation of Ge islands with their dimensions of typically 3nm in height and 13nm in diameter. The non-uniformity of the dot size is estimated to be ±10%. The areal density of the dots is about 2×10^{8}cm^{-2}

The sample was fabricated into a Schottky diode structure with an Al electrode at the top side and an ohmic contact at the back side. The diameter of the Al electrode is about 0.9 mm. Although the sample contains three dot layers, the basic formulas derived in previous section are still applicable. The equivalent circuit is the series of three circuits shown in Fig 1(b). Only the deepest dot layer contributes to the conductance signals (see below).

By using a Hewlett Packard 4275A LCR meter the capacitance-voltage measurement was carried out at the frequency of 1MHz and the admittance spectra were measured under different reverse bias voltages at frequencies of 1MHz, 500KHz, 300KHz, 100KHz, and 50KHz.

RESULTS AND DISCUSSION

Figure 3(a) shows the C-V curve of the sample measured at the frequency of 1MHz and at room temperature. The relation between the apparent carrier concentration and the voltage is derived and is shown in Fig.3(b). It is found that the three dot layers (the topmost, middle, and deepest ones) are

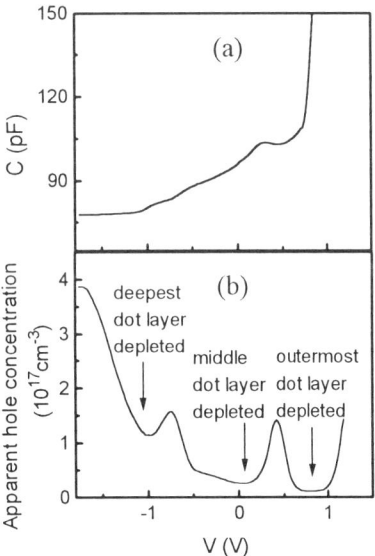

Fig.3 (a) The C-V curve of the sample measured at the frequency of 1MHz and at room temperature. (b) The relation between the apparent carrier concentration and the voltage.

unoccupied by holes at the bias voltages below +1.0, +0.2, and -1.2 V, respectively. At the bias voltage of -1.4V, the depletion region of the Schottky contact has extended over the Ge dot region. All the hole states in the quantum dots are empty. So no conductance peak will be observed in the admittance spectrum. In the bias region of -1.2V to +0.2V, the boundary of the depletion region is basically pinned near the location of the deepest dot layer. The variation of bias voltage moves the Fermi level with respect to the quantum levels in this dot layer. The hole population in the dot increases as the reverse bias decreases. For another two dot layers, the Fermi level is located well above the quantum levels so there is no hole emissions contributed by these two layers in the range of +0.2V to -1.4V. The upper two Ge dot layers do not participate the hole emission process in our measurements. The role played by these two layers is to block the extension of the Schottky barrier region over the third dot layer under zero bias.

Figure 4 (a) shows the admittance spectra at the frequency of 1MHz under different bias voltages in the range of +0.2V to -1.4V. The Arrhenius plots $ln(\omega/T_m)\sim1/kT_m$ obtained from the conductance spectra under different bias voltages are shown in Fig. 4(b). The linear correlation coefficients of all the lines are larger than 0.9998. From the slopes of these lines, the activation energies are obtained according to Eq.(4). The results are shown in Fig.5, where five discrete values are indicated at the energies of 417(Ea_1), 388(Ea_2), 263(Ea_3), 233(Ea_4), and 202(Ea_5) meV, respectively. The conductance peak at -1.2V is too weak, so the value of activation energy at -1.2 V was derived by another method, i.e., the capacitance differential method [9]. According to Eq.(5), the Coulomb charging energy per added hole in a disk-like dot with a diameter of 13nm is

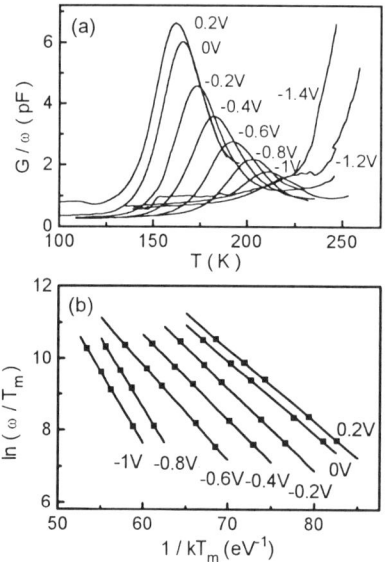

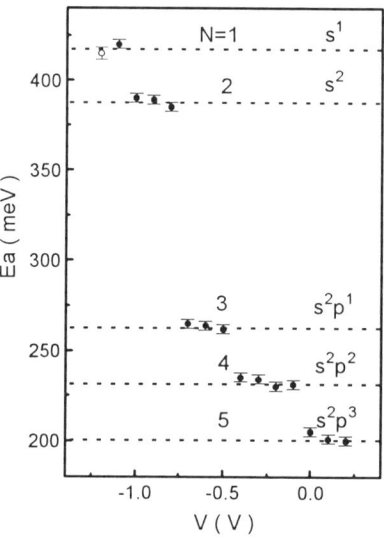

Fig. 4 The admittance spectra at the frequency of 1MHz under different bias voltages in the range of +0.2V to -1.4V (a) and the corresponding Arrhenius plots $ln(\omega/T_m)\sim1/kT_m$ (b).

Fig. 5 The activation energies under different bias voltages.

estimated to be 30meV, which is quite close to the energy differences between Ea_1 and Ea_2. At the reverse bias voltage larger than 0.8V, the holes occupy the lowest (ground) hole state in the deepest dot layer. It could be deduced that the activation energies Ea_1 and Ea_2 correspond to the hole emissions from the singly-charged and doubly-charged hole ground states of the dot. The s-like ground state can accommodate only two holes. With the decrease of reverse bias, the additional holes will occupy the p-like first excited state. Therefore, the activation energy changes to 263meV with the third hole occupying the excited state, and to 233meV and 202meV with additional Coulomb charging energies of the fourth and fifth holes. The charging of the sixth holes to the excited state is expected to occur at a larger forward bias voltage (>0.4V). In that case it is difficult to measure the admittance spectra since the signal will be smeared out by the large forward current. By using Eq.(6), the energy levels of ground state and the first excited state of the dots with respect to the Si valance band edge are derived to be 432meV and 338meV, respectively.

The influences of dot size distribution on the admittance spectra and peak temperatures are very limited. In our case, the sizes of the largest and smallest dots are about 10% deviated from that of the standard dots. The estimated activation energies for the first excited states are 361, 338 and 311 meV for the largest, standard and smallest dots, respectively. By adding the Coulumb charging energies, these excited states filled with 4 holes have the activation energies of 245, 233 and 216 meV, respectively. (The activation energy for the smallest dots charged with 3 holes is 249 meV). So a size non-uniformity of about ±10% leads to an activation energy fluctuation no larger than ±10%, although the energy distribution becomes asymmetric. At the reverse bias of 0.2 V, the standard dots are charged with 4 holes. The larger dots occupied by 4 holes and the smaller dots occupied by 3 holes will contribute to weak shoulders around the conductance peak since their densities are much smaller than that of the standard dots. (The activation energy of the largest dots occupied by 5 holes is 219 meV, so this level does not contribute to the conductance peak at that bias voltage.)

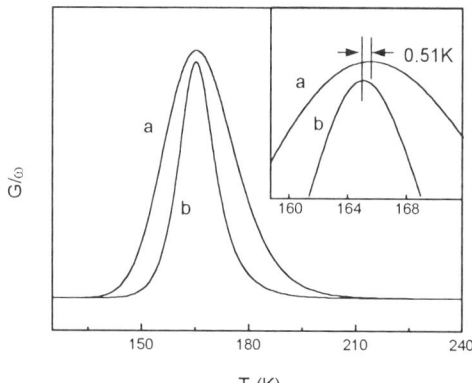

T (K)

Fig.6 The G/ω-T curve constructed by considering a Gaussian size distribution function with a standard deviation of ±10% in diameter (a) and the corresponding curve constructed for the standard dots (b).

To verify that the size distribution does not affect the conductance spectra very much, we have done a simulation of the conductance spectrum. Selecting the activation energy Ea of 200meV and considering a Gaussian size distribution function [10] with a standard deviation of ±10% in diameter, the G/ω-T curve is constructed as shown in Fig.6. The corresponding curve constructed for the standard dots is also included. It can be seen from Fig.6 that the width of the G/ω peak changes from 16K to 25K while the peak temperature shifts only 0.5K towards higher temperature. The shifts of T_m are all positive for the spectra measured at different frequencies, so the slope of the Arrhenius plot does not change too much and Ea is about the same as those without considering the broadening, as shown in Fig.7. The reason is quite easy to understand. The Ea measured from the admittance technique represents the energy level at the distribution maximum. The two wings away from the distribution maximum give only a small influence on the peak position of the spectrum. This is one of the advantages of admittance spectroscopy for application to in the self-assembled quantum dot samples.

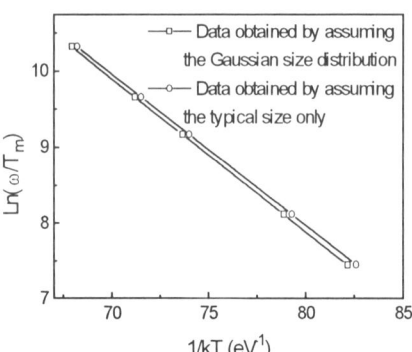

Fig.7 The Arrhenius plots $ln(\omega/T_m)\sim 1/kT_m$ obtained by considering a Gaussian size distribution and the typical size respectively.

CONCLUSION

The quantum confined energy levels and the Coulomb charging effect of holes in the self-assembled Ge dots embedded in Si barriers are studied by using admittance spectroscopy measured at temperatures above 100K. The loading of the ground state and the first excited state up to five holes is identified by varying the Fermi level position under different applied bias voltages in the admittance measurements.

ACKNOWLEDGMENTS

This work was supported by the State Commission of Science and Technology, State Education Commission and the National Natural Science Foundation of China.

REFERENCES

1. P. L. McMuen, E. B. Foxman, U. Meirav, M. A. Kastner, Yigal Meir, Ned S. Wingreen and S.J.Wind, Phys. Rev. Lett. **66**, 2252(1994).
2. Ch. Sikorski and U. Merkt, Phys. Rev. Lett. **62**, 2164(1989).
3. W. Hansen., T. P. Smith III, K. Y. Lee, J. A. Brum, C. M. Knoedler, J .M. Hong and D. P. Kern, Phys. Rev. Lett. **62**, 2168(1989).
4. R. C. Ashoori, H. L. Stormer, J. S. Weiner, L. N. Pfeiffer, S. J. Pearton, K. W. Baldwin and K. W. West, Phys. Rev. Lett. **68**, 3088(1992).
5. S. Anand, N. Carlsson, M. E. Pistol, L. Samuelson and W. Seifert, Appl. Phys. Lett. **67**, 3016(1995).
6. G. Vincent, D. Bois and P. Pinard, J. Appl. Phys. **46**, 5173(1975).
7. D. V. Lang, M. B. Panish, F. Capasso, J. Allam, R. A. Hamm, A. M. Sergent and W. T. Tsang, Appl. Phys. Lett. **50**, 736 (1987).
8. H. Drexler, D. Leonard, W. Hansen, J. P. Kotthaus and P. M. Petroff, Phys. Rev. Lett. **73**, 2252(1994).
9. F. Lu, S. Q. Wang, H. Jung, Z.. Q. Zhu and Takafumi Yao, J. Appl. Phys. **81**, 2425(1997).
10. M. Goryll, L. Vescan, K. Schmidt, S. Mesters, H. Luth and K. Szot, Appl. Phys. Lett. **71**, 410(1997).

REAL-TIME OBSERVATION OF TI SILICIDE EPITAXIAL ISLANDS GROWTH WITH THE PHOTOELECTRON EMISSION MICROSCOPY

W. Yang, H. Ade and R.J. Nemanich,
Department of Physics, North Carolina State University, Raleigh, NC 27695- 8202

ABSTRACT

The formation of nanoscale Ti silicide islands was observed by Photo-electron emission microscopy (PEEM). The islands were prepared by deposition of an ultrathin Ti (3-12ML) on Si(001) at room temperature and at an elevated temperature of $950^{\circ}C$. The island formation was initiated by *in situ* annealing to $1150^{\circ}C$. It was observed that initially Ti silicide islands form while longer annealing indicates some islands move and coalesce with other islands. Most of the islands are similar in size and have relatively uniform separation. Also, it was shown that for continued Ti deposition at a temperature of $950^{\circ}C$, the density of islands did not increase. However, islands grew together when their perimeter lines touch each other. The results are described in terms of island growth processes of coalescence and ripening.

INTRODUCTION

Photo-electron emission microscopy (PEEM) is an emission microscopy technique in which images of a solid surface are formed by photo excited electrons. Typically, ultra violet (UV) light above the photoelectron threshold will cause electrons to be emitted from a surface. The photo excited electrons originate from the near surface region (~10nm), and essentially reflect the electronic structure of the surface. These electrons may be accelerated and imaged, and the image will reflect the properties of the surface. PEEM has already been used to investigate surface chemistry of metals [1], epitaxial growth of materials [2] and characterization of semiconductor devices [3]. PEEM allows real time observation and direct imaging during processing with monolayer surface sensitivity and high resolution (~10nm). For this reason, PEEM is particularly suited for measurement of dynamical processes on semiconductor surfaces.

In this study, PEEM is used to obtain the growth of nanoscale epitaxial silicide islands on Si surface. The image contrast mechanism for a metal on a semiconductor system is the energy difference between the work function (WF) of a metal and the photo-threshold of the semiconductor (The semiconductor photo-threshold is the energy to excite an electron from the valence band to the vacuum level). Therefore, $TiSi_2$ islands on a Si surface would be imaged by this contrast mechanism. The photon energy of the employed UV-light should be below the photo-threshold of silicon and above the WF of the Ti silicide. In this case, electrons will be emitted from regions with Ti and no emission will occur from exposed Si surfaces.

Our previous studies have shown the tendency of $TiSi_2$ on Si to form epitaxial island structures. Furthermore, for ultrathin Ti (<1nm) deposited on Si substrate followed by high temperature annealing (~$1000^{\circ}C$), similarly sized islands of Ti silicide (~30nm) were obtained to be uniformly distributed over the Si (001) surface [4-6]. The Ti silicide island size and spatial distribution were very dependent on the growth conditions (i.e. Ti layer thickness, annealing temperature and surface roughness)[4,5]. It was suggested that the narrow island distribution is due to surface diffusion and the strain field effect induced in the substrate by the islands.

In this study, the formation of Ti silicide islands is explored with thin Ti deposition of 3-12 monolayers(ML) on Si(001) substrates. During in situ annealing at temperatures up to $1150^{\circ}C$, the island formation process and dynamics are observed in real-time with PEEM. After Ti silicide island formation, the surfaces are *ex-situ* analyzed with AFM and SEM to compare the surface morphology with the results from the PEEM. The size and distribution of islands are obtained, and the island growth is explained in terms of coalescence and ripening growth processes.

EXPERIMENT

The experiments were performed in an UHV PEEM system obtained from Elmitech. This system allowed heating of the substrates to > 1200°C, and the chamber is equipped with a Ti-filament deposition source. The base pressure in this system was < 2 x 10^{-10} Torr. The electric potential used for accelerating the imaging electrons is approximately 20kV across a gap of 2mm. The UV-light source is a Hg discharge lamp with upper cut-off energy near 5.1eV. We have demonstrated the resolution of our system to be ~12nm using a 100W Hg lamp as the UV excitation.

Sections of silicon (001) wafers (n-type, P-doped, resistivity 0.8-1.2 Ω-cm, 9x9mm^2) were used as substrates. The wafers were cleaned first by uv-ozone exposure and then by an HF based spin etch (HF:H$_2$O:ethanal=1:1:10). After *ex-situ* cleaning, the wafers were mounted to the sample holder and then introduced into the PEEM chamber. Before Ti deposition, the wafer was heated at a temperature of 800°C for 10 minutes by filament radiation and electron bombardment from the backside in a sample holder. The residual oxide and hydrogen were removed by the heat treatment. After cleaning, RHEED displayed a 2x1 pattern typical of the Si(001) reconstructed surface. Titanium was deposited *in situ* from the hot-filament titanium source in the PEEM chamber onto the cleaned sample which was held at room temperature or a temperature of 950°C. The titanium layer thickness was 3-12 monolayers (ML) with a deposition rate of 1ML/20S. To observe the Ti silicide island formation process, the substrates were annealed in UHV from 100°C up to 1200°C in 100°C increments for intervals of 10 minutes.

The PEEM images were displayed with a microchannel plate and phosphor screen installed in the PEEM which was monitored with a CCD camera. The images were simultaneously stored digitally with an image processor and on video tape. Image acquisition was obtained with a real-time image processor (DSP-2000) which is capable of integration of up to 128 single frames. For the data presented here, sixteen successive images were integrated. The resulting images correspond to a signal integrated over 16/30th of a second. After the substrates were unloaded from the PEEM, *ex-situ* AFM and SEM were performed to compare the surface morphology.

RESULTS

Initially, we monitored the surface morphology and the Ti silicide island formation processes while titanium is deposited on a clean Si(001) substrate followed by annealing to 1150°C. The PEEM images showed the overall surface of the Si as bright for the as-loaded wafers which became uniformly dark after the heat cleaning. The results suggest that the photo-threshold of the oxide terminated and the cleaned surfaces are below and above cut-off energy of Hg lamp(5.1eV), respectively. After Ti deposition of 10ML on the clean surface, the image is uniformly bright as the whole surface was covered uniformly by titanium (the WF of Ti is ~4.4eV). During annealing up to 700°C, the images of the surface did not change but grew darker. The darker surface indicates that the Si from the substrate has diffused into the overlayer of Ti, and a Ti silicide phase has formed. Our previous study showed Ti silicide island formation at this temperature [6]. One possibility to explain these inconsistent results may be that the photoelectron yield is too low to image the small islands. However, as shown in Fig. 1 separated TiSi$_2$ islands were observed after annealing at 1000°C. The image contrast originates from the photo-threshold difference between TiSi$_2$ (~4.6eV) and Si (>5.1eV).

Fig. 1 shows a sequential PEEM image of the TiSi$_2$ island formation sequence as the substrate is held at 1150°C for 15 minutes. Initially, similar sized islands are distributed uniformly and completely separated (Fig. 1-a). As the annealing time increases, the average size of the islands increases gradually, and the number of islands decreases (Fig. 1-f). Fig. 2 shows that the number of islands decreases for longer annealing time. The data is fitted with an

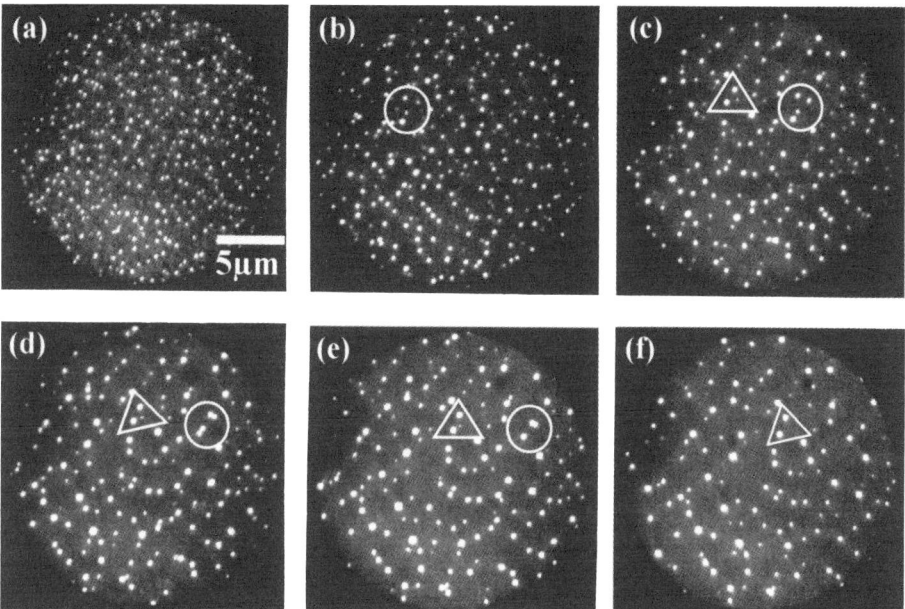

Fig. 1 PEEM images of 10ML Ti deposition followed by annealing at 1150°C for (a) 1min (b) 5min (c) 9 min (d) 10min (e) 11min (f) 15min. The islands (in the circle) coalesce and grow larger. The field of view of each image is 20μm. All images were obtained in situ, real time with the sample at the annealing temperature. The image shift is due to specimen drift at the elevated temperature.

exponential. Examination of islands highlighted by the circles in Fig. 1(b)-(e) indicates a coalescence growth process. The islands, with substantial thermal energy, diffuse on the surface. In this sequence, close islands are observed to combine with each other and transform into larger islands. The triangles in Fig. 1(c)-(f) display another island growth process. The large islands grow larger while the small islands disappear. This is characteristic of ripening integrated by surface diffusion.

Ex-situ AFM measurements of the same sample also display a uniform lateral and vertical size of the islands (Fig. 3). The average lateral size of the islands is ~500nm and the height is ~200nm. In comparison with our previous study (diameter of island: 50nm, height: 5nm) [4], the islands are larger and are separated by greater distances. This difference is attributed to different growth conditions (increased Ti thickness and higher annealing temperature). The SEM measurements also show a similar island distribution. The shape of the islands could be observed by tilting the sample be 60°[Fig. 4]. The larger islands are observed to be flatter than smaller ones. An interesting aspect of the shape of islands is the bump of Si under each island [Fig. 4-a]. This shape may be influenced by the high electric field of ~10^7 V/m between the sample and cathode lens of the PEEM.

To monitor the island formation during deposition at an elevated temperature, 3-12ML of Ti was deposited continuously at a temperature of 950°C (Fig. 5). It was observed that the growth mode of the Ti silicide at the high temperature is 3d-island growth (Volmer-Weber mode). Also, more dense and smaller islands were formed compared with Ti deposition followed by annealing at the same temperature. The density and average size of the islands did

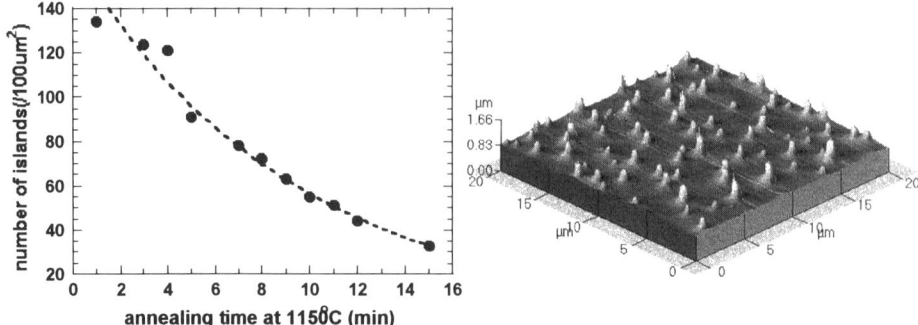

Fig. 2   The number density of islands observed at 1150°C as a function of annealing time.

Fig. 3   A 20 x 20 μm² AFM 3D-rending image of 10ML Ti deposition annealed at 1150°C.

Fig. 4 SEM micrographs of TiSi₂ islands showing the different shape and size of island. The sample was tilted 60° with respect to horizon.

not increase even though the Ti thickness increased.   However, islands next to each other coalesced as the deposition thickness of Ti increases.   In order to observe the motion and coalescence of these islands, it would be necessary to obtain images with higher magnification. Unfortunately, the low intensity of the electron emission from these samples limited the magnification of the PEEM.

We observed another image contrast mechanism with the sample at high temperature. During annealing at ~1100°C, the image of the Ti silicide islands could be obtained without the UV-light.   Furthermore, the overall brightness of the image was enhanced with increasing temperature (compare Fig. 1 with Fig. 5).   A sample heated to a sufficient temperature will emit thermionic electrons. These electrons will also be imaged in the PEEM. Thus, strictly speaking. PEEM over a range of temperatures utilizes the processes of both photoemission and thermionic electron emission.

DISCUSSION

Consider the growth processes of Ti silicide islands at high temperature (Fig. 1). Previous studies have explained the island formation in terms of the surface and interface energies of the structures [4-6].   Here we suggest two possibilities to account for the increasing size of the islands and the decreasing density obtained at the higher temperature annealing.   One growth

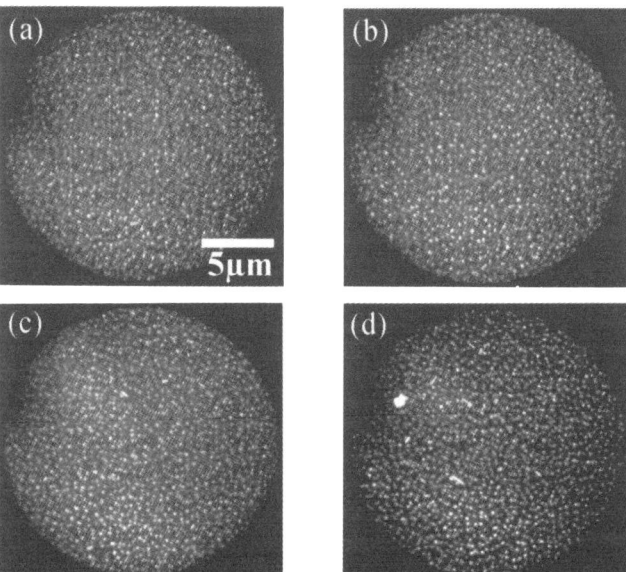

Fig. 5   PEEM images of Ti silicide islands formed with continuous deposition at 950°C. Ti deposition thickness is 3, 6, 9, 12ML for (a) through (d), respectively.   All images were obtained after cooling down to room temperature.   The field of view of each image is 20μm.

process is island coalescence (the circle in Fig. 1 (b)-(e)).   After the islands are grown epitaxially (facetted islands) by annealing at high temperature, substantial thermal energy could allow the islands to diffuse on the surface.   Finally, the islands grow together upon when they come into close proximity.   The other growth process is island ripening (the triangle in Fig. 1 (c)-(f)).   It is characterized by a local interaction between two neighboring islands of slightly different size. Thermodynamically each island grows and disintegrates with the same probability.   Due to the difference in island circumference, the larger islands grow at the expense of smaller islands. Both growth processes may reduce the surface-to-volume ratio of the employed system and result in a minimization of the total surface energy.

      For the case of continuous deposition of Ti at high temperature (Fig. 5), we observed another mechanism of island formation.   At the earliest stages, Ti silicide island initiation is a spatially random process.   The spatial distance between the islands is too far to interact by the island-induced strain fields.   Further deposition results in strain field overlapping caused by the dense and large islands.   The adatoms on surface between the islands are preferentially adsorbed by larger islands.   Furthermore, at the late stages, continued deposition allows the islands to grow together.   This is another coalescence mechanism of island growth.   The islands remain relatively in place because of the lower temperature.   However, with continuing deposition these islands grow larger at the perimeter of the islands.   The islands coalesce when their perimeter lines touch each other and grow together.   This process of coalescence can be referred to as static coalescence, compared with the dynamic coalescence process of the high temperature annealed samples [7].

Small islands (~30nm diameter) were not observed, even though our PEEM instrument has a high resolution capability of ~10nm. Regardless of the aberration of lens, the image resolution is limited by the emission current density. The emission current density depends upon the quantum efficiency (emitted electrons per incident photons) of the surface and the photon flux density of light source. We can estimate the photon flux and the current density required for imaging an island of ~ 10nm. A current density of > 1.6×10^{-4} A/cm^2 is required to observe the islands of ~10nm. Given a quantum efficiency of ~ 10^{-6} range as an average of many materials (pure metal : > 10^{-5}), the desired photon flux could be ~10^{21} photons/sec·cm². The photon flux of the Hg discharge lamp is several orders of magnitude lower than this(~10^{14} photons/ sec·cm²). Therefore, a higher intensity UV source will improve the imaging of smaller islands. Secondarily, chromatic aberration, which is caused from a spread in the velocity of the emission electron, blurs the image resolution. If a photon source with a small energy spread would be tuned to just above the work function of Ti silicide (~4.6eV), better image contrast of the Ti silicide islands should be achieved. If we can use an intense tunable light source with a narrow energy distribution, we may observe small Ti silicide island formation and dynamics. In the future, as we combine our PEEM with the UV-FEL at Duke University for a light source (the photon flux: >10^{25}/sec·cm², energy range: 3-10eV, energy resolution: $\Delta E/E > 10^{-4}$), we anticipate real time imaging of surface Dynamics at the 10nm resolution limit of the PEEM.

CONCLUSION

In this study, we have used PEEM to observe the formation and real-time dynamics of Ti silicide islands on Si (001). Similar sized and relatively uniformly separated islands were observed for Ti deposition followed by annealing or Ti deposition at high temperature. Both growth processes of coalescence and ripening contribute to large island growth for high temperature annealing. For continuous deposition of Ti, the island growth process is characterized as static coalescence. In the future, PEEM combined with a high intensity and tunable UV-FEL should promise observation of small island formation (~10nm) and real time surface dynamics.

ACKNOWLEDGEMENTS

We gratefully acknowledge the support of T. Franz and C. Koziol from Elmitech, and P. Goeller for helpful discussions. This work was supported in part by the NSF under grant DMR 9633547 and the ONR through grant N00014-95-1-1141.

REFERENCE

[1]   H. Rotermund, S. Nettesheim, A. von Oertzen and G. Ertl, Surf. Sci. Lett. **275**, L645 (1992)
[2]   M. Mundschau and E. Bauer, J. Appl. Phys. **65**(2), 581 (1988)

[3]   M. Giesen, R. Phaneuf, E. Williams, T. Einstein and H. Ibach, Appl. Phys. **A64**, 423 (1997)
[4]   W. Yang, F. Jedema, H. Ade, R.J. Nemanich, thin solid films **308-309**, 627(1997).
[5]   W. Yang, F. Jedema, H. Ade, R.J. Nemanich, Mat. Res. Soc. Symp. Proc. Vol **448**, 223(1997).
[6]   H. Jeon, C.A. Sukow, J.W. Honeycutt, G.A. Rozgonyi, and R.J. Nemanich, J. Appl. Phys. **71**, 4269(1992)
[7]   M. Zinke-Allmang, L. Feldman and M. Grabow, Surf. Sci. Rep. **16**, 377(1992)

MAGNETRON SPUTTER HETEROEPITAXY OF Si$_{1-x}$Ge$_x$/Si(001): THE EVOLUTION OF THE CROSS-HATCHED SURFACE

B. VÖGELI, M. KUMMER, AND H. VON KÄNEL
Laboratorium für Festkörperphysik, ETH Hönggerberg, CH-8093 Zürich

ABSTRACT

The surface step structure of step graded Si$_{1-x}$Ge$_x$ buffers was investigated by scanning tunneling microscopy (STM). On pseudomorphically grown samples, single atomic steps were found, with the 2x8 reconstruction indicating Ge segregation. Upon strain relaxation, surface slips and the development of the cross-hatched surface takes place. Here, close to the surface slips, the step structure was found to consist of double height D$_A$ steps. This feature is not simply kinematically driven step bunching but can be attributed to the anisotropic strain field on the surface and to the Ge segregation. Analysis of STM cross-sections were taken to estimate the relaxation of our samples. The obtained values agree well with x-ray measurements. The relaxation is found to depend exponentially on the temperature, as theoretically expected.

INTRODUCTION

Very high electron mobilities have been reported in modulation doped Si$_{1-x}$Ge$_x$/Si(100) structures, in which a strain-relaxed Si$_{1-x}$Ge$_x$ buffer has been used as a virtual substrate for the growth of a strained Si channel for the two dimensional electron gas (2DEG) [1]. The required strain in the quantum well is provided by growing it on the top of a nearly relaxed buffer. To keep the density of threading dislocations as low as possible, the relaxed buffers consist of a compositionally graded part and a buffer with constant Ge mole fraction. Much work has been done in the past years in order to understand the mechanisms for strain relaxation [2]. The nucleation, the glide and the multiplication of dislocations was investigated thoroughly mainly by transmission electron microscopy (TEM) [3]. Strain profiles as a function of the total layer thickness h_g and the final Ge fraction x_f were investigated using x-ray diffraction techniques [4]. The residual strain in the surface was found to decrease if an additional buffer with thickness h_{const} and x_f was deposited on the graded buffer. So far, the roughness of the cross-hatched surface has been monitored by atomic force microscopy (AFM). The roughness was shown to be influenced by the grading rate ($\frac{dx}{dz}$) and the final Ge fraction x_f [5]. In this work, UHV-STM investigations on Si$_{1-x}$Ge$_x$/Si(001) buffers are presented. Atomically resolved images give new insights into the growth mechansim. The startling feature of the biatomic (D_A) surface steps, present in most of the investigated samples, is being explained. Additionally, we analysed the STM-images of a set of structurally identical samples, grown at distinct temperatures and show that the residual strain in the samples can be estimated reliably.

EXPERIMENT

The films were grown by rf magnetron sputter epitaxy. Details of the growth system, the sample preparation and the growth conditions are described elsewhere [6, 7]. A 100 nm Si buffer, grown at 500°C, separated the substrate from the compositionally graded buffer.

Starting with a Ge mole fraction $x=0.12$, which was determined by the lowest power at which the Ge sputter gun could be operated, x was increased in increments of $\delta x=0.01$, up to the final Ge fraction $x_f=0.30$. Typical growth rates were between 0.15 (pure Si) and 0.21 nm/s. The step-graded buffers presented in this work consisted hence of 19 layers, each with a thickness of 26.3 nm. The total thickness of the graded buffers was therefore 500 nm, and the grading rate was $60\%/\mu m$. The growth temperatures in the first study was 485°C and, in the second one, between 370 and 570°C. The evolution of the cross-hatch was investigated by interrupting growth and transferring the samples in UHV into the STM chamber. The maximum scan-width of the Omicron UHV-STM was $4x4\mu m^2$. Relative to the sample, the tip could be positioned anywhere along a line of 5mm in the [110] direction, while the position in the [1$\bar{1}$0] direction is approximately fixed.

RESULTS

On the left hand side of Fig. 1 the surface is shown as it typically looks for x in the range between 0.12 and 0.16. The local misorientation can be determined. As can be deduced from

Figure 1: Left hand side: the surface after the deposition of the $Si_{1-x}Ge_x$ layers with x <0.16. No surface slips can be observed, indicating the film to be pseudomorphic. The inset emphasizes that Ge segregation occurs, leading to the 2x8 reconstruction. Right hand side: Upon further growth to $x=0.21$, strain relaxation by the formation of misfit dislocations takes place, leading to the surface slips. The inset is a magnification of the surface slip region. Though all kind of steps are present, D_A steps are dominant.

the inset, even in the case of a $Si_{0.84}Ge_{0.16}$ alloy, the surface reveals a 2x8 reconstruction with the dimer vacancy lines (VL) perpendicular to the dimer rows, identical to Ge/Si(001) [8]. Thus the top-most layer must be Ge-rich [7]. On a larger scale, no surface slips were found, indicating that the alloy has been grown pseudomorphically. This is in agreement with results found by other groups [9], if we consider an average composition of $x=0.14$ and the sample thickness of 131 nm. Upon further growth, the strain in the sample increases, and eventually the formation of misfit dislocations (MD) sets in. On the right hand side of Fig. 1, the surface is shown for $x=0.21$ and a total thickness of $h=263$ nm. Very few surface slips are present, implying a partial relaxation of the sample. An undulating surface can, in principal, involve steps of monolayer height, doublelayer height or a combination of the two. The inset is a magnification of the surface slip region. Double height D_A steps, with the dimer bond and the VL perpendicular to the step, are already dominant. During the growth of the last 9 layers, where x increases from 0.21 up to $x_f=0.30$, the nucleation of dislocations leads to the cross-hatched surface, as shown on the left hand side of Fig. 2. In

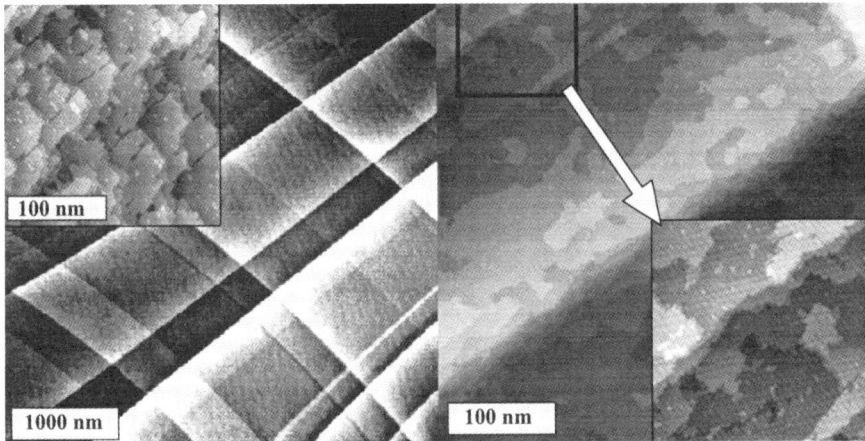

Figure 2: Left hand side: the $Si_{0.70}Ge_{0.30}$ buffer reveals the cross-hatched surface. Between the surface slips, the samples are planar. Growth proceeds by step flow and the formation of double atomic steps, see inset. Right hand side: the surface slip consists of D_A steps, with the dimer vacancy lines perpendicular to the steps. The mechanism leading to this feature is explained in the text.

between the surface slips, the samples are planar, as emphasized by the inset, where the local misorientation [10] can be determined. Note that the surface consists of double height steps. In contrast, the few islands are of single step height only. The right hand side of Fig. 2 shows the atomic structure of the surface slip region. This sample involved the growth of the linearly graded buffer as described above and additionally, a buffer maintaining the constant Ge mole fraction $x_f=0.30$ for another 100 nm. The buffer with x being constant lowers the residual strain in the surface [4]. Nevertheless, we observe the step structure in the surface slip region to consist of D_A steps, with the dimer VL perpendicular to the step. In the following, we show that the occurrance of these double height D_A steps is directly related to the residual strain in the surface. We should emphasize that we never observed this step structure on biaxially strained layers, such as thin $Si_{1-x}Ge_x/Si(001)$ films, e. g. in Fig. 1. The formation of the D_A steps is correlated with the formation of the cross-hatch pattern. In a recent work, Wu and Lagally [11] investigated *uniaxially compressively strained*

Si(001) surfaces as a function of the Ge coverage θ_{Ge} (ML). These experiments showed that the surface stress anisotropy $F = \sigma_{\parallel} - \sigma_{\perp}$ is a function of θ_{Ge} and changes sign at $\theta_{Ge} = 0.8$ ML (for $\theta_{Ge} = 0.8$ ML, the steps did not respond to the applied stress). The (tensile) stress $\sigma_{\parallel}$ along the dimer bond decreased dramatically with increasing Ge coverage. In contrast, the (tensile) stress $\sigma_{\perp}$ along the dimer rows remained almost uneffected if Ge was deposited. As a result, a uniaxially compressively strained Si(001) surface, with a Ge coverage $\theta_{Ge} > 1.6$ ML revealed a step structure, where the S_A steps (almost) reached the S_B steps, thus forming D_A steps. In our samples, as pointed out above, we assume a locally anisotropic strain field, which is more compressive along the direction of the surface slip line. The Ge induced reversal of the surface stress anisotropy, may hence explain why pure D_A steps occur.

We have also studied the evolution of strain relaxation in the $Si_{1-x}Ge_x/Si(001)$ buffers as a function of temperature. Most samples were 500 nm thick and had a final Ge mole fraction of $x_f = 0.30$; one sample had a 100 nm thick cap layer of constant composition $x_{const} = 0.30$ on top of the graded buffer. Our technique of determining the size of the relieved strain is based upon the relation between the surface cross-hatch pattern and the MD in the graded buffers [12]. The Burgers vector of the $\frac{a}{2}\langle 011 \rangle$ type dislocations prevalent in the diamond cubic lattice has a component of $\frac{a}{2}$ perpendicular to the interface, which leads to a measurable surface displacement. The density of MD was estimated by measuring the total height of the

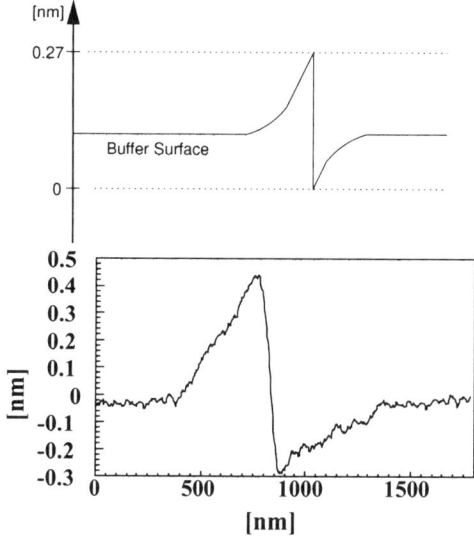

Figure 3: Surface displacement calculated for a single MD and measured for a pileup of 3 MDs, respectively. The lower image shows an average over 300 STM scanlines perpendicular to a single crosshatch line (the step appears slightly broadened due to piezo drift). The relaxation was measured by comparing the observed surface slip height with the one expected for complete relaxation.

surface slips within a certain length (typically 2 or 4 μm). The relieved strain in the samples was calculated by comparison of the measured height with the one expected for complete strain relaxation. The data are collected together in Fig. 4. The error bars were obtained by assuming a gaussian distribution over the readings on each sample. To assert the accuracy of

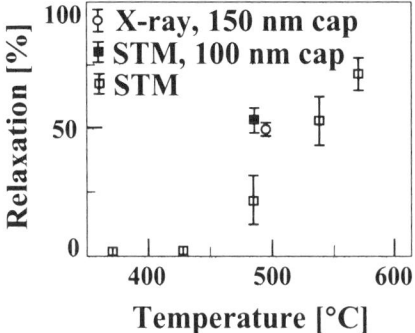

Figure 4: The strain relaxation, measured by STM, as a function of the substrate temperature.

the relaxation values obtained from the STM data, we included one measurement obtained from triple-axis x-ray diffraction. This sample had a 150 nm thick constant composition buffer with $x_f = 0.30$. As can easily be seen, below 450° C the buffer layers remain almost completely strained. For the two samples grown at the lowest temperatures, we found relaxations of about 1%. Due to the limited STM scan range of $4\text{x}4\mu\text{m}^2$, exact values cannot be given; one usually has to look at different places on these samples to find a surface slip. With increasing temperature, the strain relaxation in the buffers increases too. As expected [4], the two samples with compositionally uniform cap layers exhibit some additional relaxation. The overall behaviour is qualitatively consistent with theoretical models for dislocation nucleation, propagation and interaction [13, 14]. However, these theories are based on samples grown at much higher temperatures and, in some cases, on constant composition buffers. In turn, for temperatures below 550°C, they suggest a much lower relaxation as measured in our samples.

CONCLUSIONS

The surface of $Si_{1-x}Ge_x$ buffers, grown with a grading rate of 60%/μm at 485°C, was investigated by STM. For $x < 0.16$, the 2x8 reconstructed films were found to grow pseudomorphically. Upon further growth, strain relaxation takes place. Nucleation of misfit dislocations leads to surface slips. These were found to consist of D_A steps, with the VL perpendicular to the slip. The presence of this step structure can be attributed to the anisotropic (compressive) surface strain due to the cross-hatch, and the Ge induced reversal of the surface stress anisotropy F. Additionally we have shown, that STM images allow for an estimation of the relaxation in the buffers, provided that the Ge mole fraction x is known. The astonishingly high relaxation in our samples was found to depend exponentially on the temperature, as theoretically predicted.

ACKNOWLEDGMENT

This work was partially financed by the "Poly-Project" Nanorobotics 1 of the Swiss Federal Institute of Technology Zürich.

References

[1] For a review see, e. g. F. Schäffler, Semicond. Sci. Technol. **12**, 1515 (1997)

[2] R. Beanland, D. J. Dunstan, and P. J. Goodhew, Advances in Physics **45**, 87 (1996)

[3] F. K. LeGoues, B. S. Meyerson, J. F. Morar, and P. D. Kirchner, J. Appl. Phys. **71**, 396 (1993)

[4] J. H. Li, V. Holy, G. Bauer, and F. Schäffler, J. Appl. Phys. **82**, 2881 (1997)

[5] R. M. Feenstra *et al.*, J. Vac. Sci. Technol. **B 13**(4), 1608 (1995)

[6] B. Vögeli, S. Zimmermann, and H. von Känel, Thin Solid Films, in press

[7] B. Vögeli, M. Kummer, and H. von Känel, presented at the 1998 MRS Spring Meeting, San Francisco, CA, 1998 (to be published)

[8] J. Knall and J. B. Pethica, Surf. Sci. **265**, 156 (1992)

[9] R. People and J. C. Bean, Appl. Phys. Lett. **47**, 322 (1985)

[10] The misorientation of the substrates was 0.07°. We therefore would expect an average terrace width $\langle L \rangle$ of 110 nm. The measured terrace width of 16 nm indicates that the local misorientation is different.

[11] Fang Wu and M. G. Lagally, Phys. Rev. Lett. **75**, 2534 (1995)

[12] M. A. Lutz, R. M. Feenstra, F. K. LeGoues, P. M. Mooney, and J. O. Chu, Appl. Phys. Lett. **66**, 724 (1995)

[13] D. C. Houghton, J. Appl. Phys. **70**, 2136 (1991)

[14] B. W. Dodson and J. Y. Tsao, Appl. Phys. Lett. **51**, 1325 (1987)

Si/Ge/Si(001) MAGNETRON SPUTTER HETEROEPITAXY: THE INITIAL STAGES OF "HUT"- CLUSTER OVERGROWTH

B. VÖGELI, M. KUMMER, AND H. VON KÄNEL
Laboratorium für Festkörperphysik, ETH Hönggerberg, CH-8093 Zürich

ABSTRACT

We show that the capping of Ge/Si(001) hut clusters with epitaxial Si cannot be easily realised. Even for growth temperatures as low as 300°C, Ge segregation during the Si deposition leads to a $Si_{1-x}Ge_x$ alloy in both the wetting layer and the clusters which reduces the misfit of the system. In turn it is no longer beneficial to release the strain energy by forming small islands. On the contrary, there appears to be a metastable state for exhibiting a flat surface. Thus we find a 3D to 2D transition which at low Si coverages is probably thermally activated. At higher coverages, however, the transition becomes a barrier-less, resulting in the complete dissolution of the hut clusters. Annealing experiments prove that the resulting 2x8 reconstructed $Si_{1-x}Ge_x$ surface is not in thermodynamic equilibrium. During the melting of a hut cluster a number of new facets are formed, mainly at the edges of the clusters and at the top, where the apex becomes substituted by a {001} facet. Our results emphasize that the melting of the Ge/Si(001) hut clusters is a serious obstacle to their application, e. g. in electroluminescence devices. It has to be solved, e. g. by the use of surfactants.

INTRODUCTION

Ge on Si is known as a model Stranski-Krastanow growth system. The initial two-dimensional (2D) wetting layer grows pseudomorphically until the strain is relaxed by the formation of three-dimendional (3D) islands. So far, depending on the growth temperature, two distinct types of islands have been reported. Large clusters with facets changing from {118} to {113} during growth, and a diameter of more than 500 nm, were reported by Koide [1]. Mo et al. [2] observed much smaller clusters exhibiting {105} facets and a base size of typically 20 nm. According to their shape, they were henceforth referred to as hut clusters. Much work has been done in order to understand the physics of their nucleation [3, 4] and their self-limited size. A strain induced energy barrier at the edges of the hut clusters is expected to lead to the observed rather narrow size distribution [5, 6]. Apart from being an interesting growth system, the hut clusters could provide us with the possibility to create self-assembled quantum dot structures. Here, charge carriers would be confined in the zero-dimensional dots. On the other hand the self-assembling of Ge dots along the surface steps could be used to synthesize quantum wires [4]. To be technically applicable, however, the dots have to be buried in a matrix material. This is no obstacle in III-V systems, as the numerous work reported on buried InAs/GaAs quantum dots indicate. Conversely, for the Ge/Si system, only the capping of much larger islands than hut clusters was reported so far. In all cases, assessed by transmission electron microscopy (TEM) the initial pyramidal shape was found to be truncated, except when Si is deposited at room temperature [7]. The common flattening of the Ge islands upon Si deposition leads, in case of hut clusters, to their complete dissolution, as will be explained in the present study. In terms of technical applications, e. g. in electroluminescence devices, the described effect is a considerable drawback since lateral confinement requires typically islands with sizes in the range of hut clusters.

EXPERIMENTAL

The films were grown by rf magnetron sputter epitaxy in an UHV growth chamber equipped with three 2 inch magnetron sources. The targets were intrinsic Si and Ge. The epitaxial films were grown in a high purity Ar atmosphere, using working pressures at which the sputtered atoms are fully thermalized [8]. First, a 100 nm thick Si buffer was grown at 500°C. Then, the temperature was lowered to 430°C. The deposition of 5 ML Ge led to the formation of the desired hut clusters which were subsequently buried with Si at a rate of 0.5 Ås^{-1} in the temperature range of 300-450°C. The study of the films was carried out with an Omicron UHV-STM located in a separate chamber directly attached to the growth chamber.

RESULTS

It has been shown previously [9] that Ge grows layer by layer up to a coverage of 3 ML. Then, depending on the growth temperature the formation of islands or pits lowers the total free energy of the system. If the temperature is chosen to be 430°C, the so-called hut clusters are formed by depositing 5 ML Ge/Si(001), as shown on the left hand side of Fig. 1. The base size of typically 10 to 20 nm is similar to those of hut clusters prepared by other techniques. The shape is pyramidal, as emphasized by the inset of Fig. 1. Note that the wetting layer

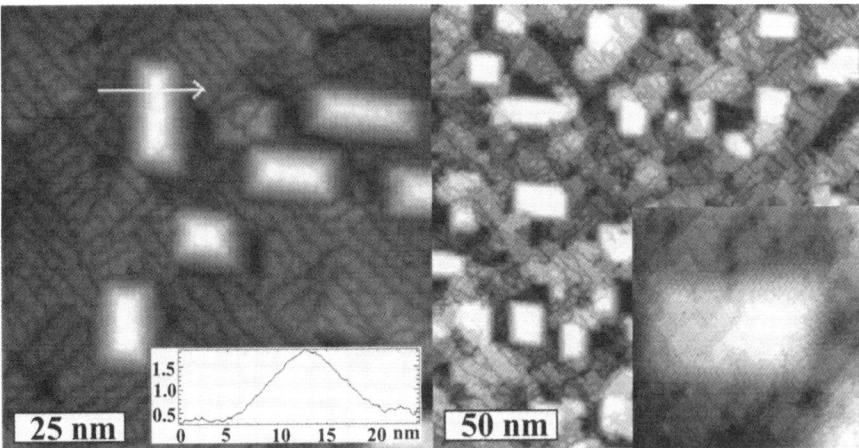

Figure 1: Left hand side: 5 ML Ge deposited on Si(001) at a substrate temperature of 430°C leads to hut clusters with a perfect pyramidal shape. Right hand side: the hut clusters after the deposition of 2.2 ML Si at 450°C. The inset shows a hut cluster with the apex substituted by a {001} facet. The facet and the wetting layer are both 2x8 reconstructed, indicating that Ge segregation takes place during the Si deposition.

is 2x8 reconstructed as expected in case of pseudomorphic Ge on Si(001). The right hand side of Fig. 1 shows the surface after depositing 2.2 ML of Si at 450°C. The hut clusters are still visible, but they have flattened considerably. The apex has been substituted by a

{001} facet, as can be seen from the inset. At slightly larger Si coverages the dots vanish completely as shown on the left hand side of Fig. 2, taken after depositing 3.3 ML Si. As can be deduced from Fig. 1 the flattened dots exhibit the same 2x8 surface reconstruction as the wetting layer. Since the Ge/Si(001) wetting layer grows pseudomorphically [10], its

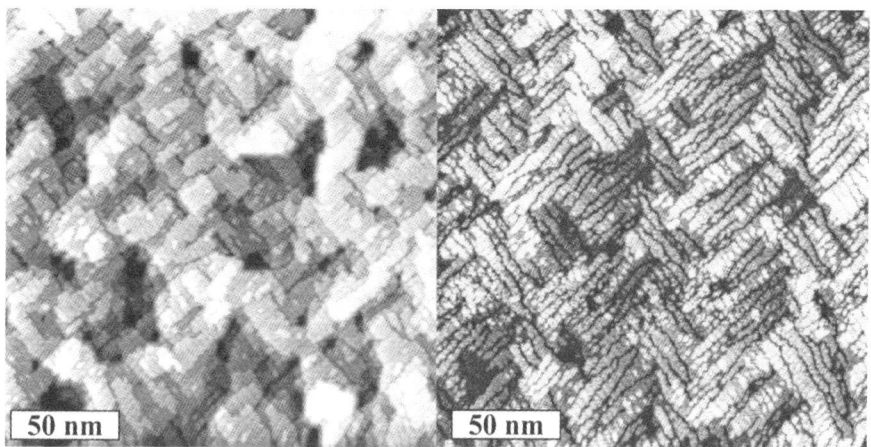

Figure 2: Left hand side: the hut clusters after the deposition of 3.3 ML Si. Despite the presence of some {105} facets, the exact position of the initial dots can no longer be determined. This is in qualitative agreement with the surface shown on the right hand side, where on a 10 ML thick $Si_{0.5}Ge_{0.5}$ layer grown at a substrate temperature of 450°C no islands can be found. Note the 2x8 reconstruction again indicating Ge segregation.

in-plane lattice constant is similar to the one of Si. Thus, there is no reason for a film of pure Si grown on the Ge/Si(001) wetting layer to exhibit a 2x8 reconstruction. From the fact that after Si deposition the reconstruction is observed we conclude that the surface must consist of pure Ge or a Ge-rich alloy under biaxial compressive strain. Considerable Ge segregation must hence be taking place during the Si deposition, leading to a $Si_{1-x}Ge_x$ alloy in the wetting layer. This is in qualitative agreement with the inhibited formation of hut clusters observed on a 10 ML thick $Si_{0.5}Ge_{0.5}$/Si(001) alloy grown at 450°C, as shown in Fig. 2 on the right hand side. Note that the surface is again 2x8 reconstructed. This does, of course, not necessarily mean that such a planar alloy layer is thermodynamically stable. Annealing the surface for 10 min at 560°C led to large {104} facetted islands with a diameter of more than 100 nm, as shown in Fig. 3. For comparison, a hut cluster is shown in the same Figure. Similar experiments were performed by Jesson *et al.* [5] who annealed a 14.5 ML thick $Si_{0.5}Ge_{0.5}$ layer for 5 min at 600°C and then observed clusters with {105} facets. These experiments indicate that the exact shape allowing for the maximum release of the elastic energy is very sensitive to distinct parameters, such as the Ge concentration x and the annealing (or growth) temperature.

The melting of the dots can be understood in the framework of a theory proposed by Tersoff for the inverse process [11]. According to this model, the formation of islands allows an initially compressed and planar film to reduce the elastic energy stored in it. On the other hand, the formation of an island costs surface energy. The total free energy of the surface is

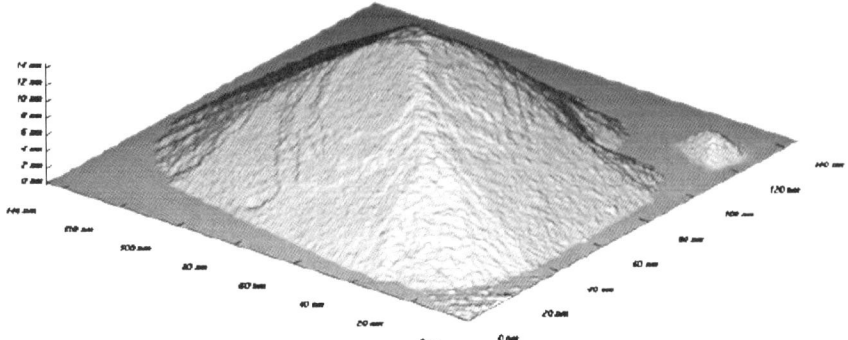

Figure 3: The surfaces shown in Fig. 2 are not in thermodynamic equilibrium. Annealing the samples for 10 min at a substrate temperature of 560°C leads to the formation of large clusters with a diameter of more than 100 nm and {104} facets. For comparison a hut cluster is shown in the same Figure.

therefore the sum of these two terms and can be written as

$$E_{tot}(V) = 4\Gamma V^{2/3} \tan^{1/3}\theta - 3V\frac{\sigma^2(1-\nu)}{\pi\mu}\tan\theta. \tag{1}$$

Here, V is the volume of an island, θ the contact angle between the island facets and the planar surface, σ the misfit stress, μ the shear modulus, ν Poisson's ratio, and $\Gamma = \gamma_f \csc\theta - \gamma_s \cot\theta$ with γ_f and γ_s being the facet and surface energy, respectively. According to that model, it would be beneficial to grow the islands to infinite size. The self-limitation of the volume V_{hut} of the hut clusters is thus not described by this model. The 2D to 3D transition for sufficiently large misfit does, however, apply to hut clusters, too. Upon Si deposition, as pointed out previously, the wetting layer becomes a $Si_{1-x}Ge_x$ alloy. This reduces the misfit, and thus the stress σ in the epilayer. If the volume V_{hut} of the islands is fixed and σ continuously decreased, the energy gained by elastic relaxation of the islands no longer counterbalances the surface energy term. While at the beginning of the resulting 3D to 2D transition an activation energy has to be overcome, the process will take place barrier-less for sufficiently small σ. A closer look at the melting of the hut clusters reveals an interesting feature, which was found for all temperatures investigated in this study. The melting appears not to proceed homogeneously. Rather the formation of new facets was observed, in particular a {001} facet at the top and {110} oriented facets at the edges of the hut clusters, as shown in Fig. 4. The facets at the edges were found to change from {117} to {113}, depending on the Si deposition and the growth temperature. Although we cannot give a detailed explanation of the mechanism, it appears that the most relaxed parts of the hut clusters, the apex [12] and the edges along the (hard) {110} directions [13], are the first to melt. In addition, one has to consider the anisotropy of the diffusion [14] which may selectively enhance the melting of the edges of the hut cluster. Regarding the melting of the hut clusters, we pointed out that Ge segregation during the Si deposition is responsible. This can be reduced by lowering the growth temperature or by the use of surfactants. Unfortunately, even at our lowest growth temperature of 300°C, the hut clusters were found to

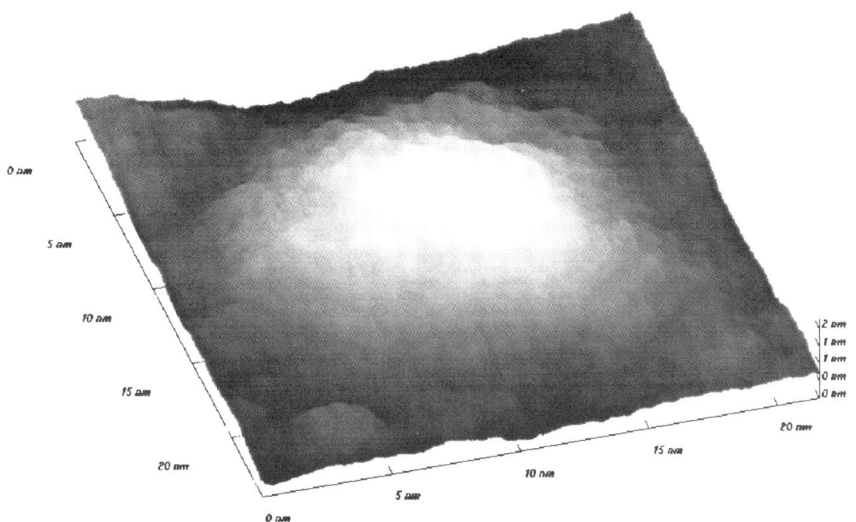

Figure 4: Upon further melting the formation of new facets can be observed: at the edges four {117} facets, which change to {113} facets, can be determined. The apex is substituted by a {001} facet which increases in size during the melting.

dissolve, indicating Ge segregation is still not sufficiently reduced. Due to the significantly reduced diffusion lengths rather rough, 2x8 reconstructed surfaces were, however, obtained even for Si coverages of a few ML. Indeed, since larger Ge clusters have been successfully buried [15], they appear so far to be the only pathway to exploit the Stranski-Krastanov growth mode for buried, self-assembled dot structures.

CONCLUSIONS

In conclusion, we have shown that the Ge hut clusters melt when covered by Si at temperatures between 300°C and 450°C. The observed melting is directly attributed to the Ge segregation during the Si deposition. The formation of an alloy in the initial wetting layer reduces the misfit of the system such that the presence of small islands is no longer energetically favorable. The subsequent 3D to 2D transition has first to overcome an energy barrier. With increasing alloying of the wetting layer a barrier-less transition takes place. Upon melting of the dots the formation of new facets, e.g. a {001} facet at the top and {110} oriented facets at the edges are observed. In terms of a possible application of the Ge hut clusters as quantum dots, the described melting is a serious obstacle which has to be overcome, e.g., by the use of surfactants such as As [16].

ACKNOWLEDGMENTS

This work was partially financed by the "Poly-Project" Nanorobotics 1 of the Swiss Federal Institute of Technology Zürich. We kindly thank Max Döbeli of the Institute of Particle Physics (ETH Zürich) for the RBS measurements.

References

[1] Y. Koide, S. Zaima, N. Ohshima and Y. Yasuda, Jpn. J. Appl. Phys. **28**, 690 (1989)

[2] Y.-W. Mo, D. E. Savage, B.S. Swartzentruber, and M. G. Lagally, Phys. Rev. Lett. **65**, 1020 (1990)

[3] D. E. Jesson, K. M. Chen, S. J. Pennycook, T. Thundat, and R. J. Warmack, Phys. Rev. Lett. **77**, 1330 (1996)

[4] I. Goldfarb, P. T. Hayden, J. H. G. Owen, and G. A. D. Briggs, Phys. Rev. Lett. **78**, 3959 (1997)

[5] D. E. Jesson, K. M. Chen, and S. J. Pennycook, Mater. Res. Soc. Symp. Proc. April 1996, 31

[6] I. Goldfarb, P. T. Hayden, J. H. G. Owen, and G. A. D. Briggs, Phys. Rev. B **56**, 10459 (1997)

[7] Y. Chen and J. Washburn, Phys. Rev. Lett. **77**, 4046 (1996)

[8] B. Vögeli, S. Zimmermann, and H. von Känel, Thin Solid Films, in press

[9] J. Tersoff and R. M. Tromp, Phys. Rev. Lett. **70**, 2782 (1993)

[10] J. Tersoff, Phys. Rev. B **43**, 9377 (1991)

[11] J. Tersoff and F. K. LeGoues, Phys. Rev. Lett. **72**, 3570 (1994)

[12] A. J. Steinfort *et al.*, Phys. Rev. Lett. **77**, 2009 (1996)

[13] G. Schiltges, private communication.

[14] Y. W. Mo and M. G. Lagally, Surf. Sci. **248**, 313 (1991)

[15] R. Apetz, L. Vescan, A. Hartmann, C. Dieker, and H. Lüth, Appl. Phys. Lett. **66**, 445 (1995)

[16] M. Copel, M. C. Reuter, E. Kaxiras, and R. M. Tromp, Phys. Rev. Lett. **63**, 632 (1989)

ROLE OF SURFACE INSTABILITY AND ANISOTROPY ON STRAIN RELAXATION OF SIGE ON SI(110)

X. DENG, M. KRISHNAMURTHY
Department of Metallurgical and Materials Engineering, Michigan Technological University, Houghton, MI 49931

ABSTRACT

SiGe alloys (< 1.5 % mismatch) were grown on Si (110) substrates at 700 °C by molecular beam epitaxy. The structure and surface morphology were studied as a function of SiGe coverage (1.5nm–9nm) using atomic force microscopy, transmission electron microscopy and RHEED. A very high density of nanoscale SiGe ledges (less than 1 nm high and 1 μm long) were observed to form on the step terraces created during buffer layer growth. An interesting phenomenon is that the step bunches generated by the buffer growth and SiGe ledges are aligned along the elastically hardest <111> direction on the Si (110) surface. With increasing alloy thickness, the nanoscale SiGe ledges become longer and straighter. The effects of the surface instability and elastic anisotropy on the elastic relaxation effects are discussed.

INTRODUCTION

Semiconductor nanostructures, due to their strong quantum confinement, have great potential in opto- and microelectronic applications [1]. Heteroepitaxial growth of high lattice-mismatched materials provides an attractive route for the process- and damage free fabrication of "self-assembled" nanostructures such as quantum wires and quantum dots [2,3]. For example, epitaxial growth of Ge/Si and InGaAs/GaAs has been used to fabricate quantum dot structures [4, 5]; Quantum wires for laser structures in the GaAs/AlAs/InAs system has been fabricated by self-ordered growth on V-groove patterned substrates [6]. However, most reports on the heteroepitaxial growth systems have concentrated on Si (100) surfaces. Recently, Omi et al reported on the growth of self-assembled Ge nanowires on Si (113) [7]. In our case, Si (110) surfaces are used to grow the self-assembled nanostructures. To our knowledge, very limited literature exists on the surface morphology evolution for Ge or SiGe growth on Si (110) surfaces [8, 9]. However, the structure and surface energy anisotropy of the (110) surface may significantly influence the nucleation and growth mechanisms of SiGe films and may provide new insights into the pathways for strain relaxation. In addition, the maximum film/substrate conduction-band offset is predicted to be larger for SiGe (110) than for SiGe (100) and to have higher electron confinement for the self-assembled SiGe nanostructures on Si (110) [10].

In this study, we have performed growth of SiGe on Si (110) surfaces with SiGe coverage between 1.5 nm and 9 nm at 700 °C substrate temperature. Step bunches were generated after the buffer layer growth. A dense array of self-assembled nanoscale ledges was observed to form on the step terraces rather than at step edges. These SiGe wires are coherent with the Si (110) substrates. Longer and straighter ledges were obtained with increasing SiGe coverage. One intriguing phenomenon is that the SiGe ledges are oriented along the elastically hardest <111> direction on the Si (110) surface. The surface morphology evolution is believed to be the continual contribution of the surface instability and the strong surface anisotropy in strain relaxation during the growth.

Mat. Res. Soc. Symp. Proc. Vol. 533 © 1998 Materials Research Society

EXPERIMENT

Experiments were performed in a Riber molecular beam epitaxy (MBE) system, with base pressure of about $\sim 5 \times 10^{-10}$ mbar. Si (110) substrates were degreased in acetone, methanol and de-ionized water. They were then passivated in a 10% HF solution, followed by a H_2O_2 : H_2SO_4 (30:70) oxidation. Before loading into the MBE chamber, the substrates were fully rinsed in de-ionized water and were blown dry with nitrogen gas. After the samples are degassed overnight at ~ 200 °C the oxide was desorbed by heating to ~ 925 °C. A low flux of Si was used during the oxide desorption in order to obtain a clean and smooth surface. A sharp (16x2) surface reconstruction pattern, which is consistent with a clean Si (110) surface [11], can be seen from RHEED when the substrate temperature is below ~ 800 °C. Samples were then cooled to 550 °C and a Si buffer layer (~ 30-45 nm) was grown. After the buffer growth, the samples were re-heated to 800 °C and annealed for 2-3 minutes to ensure a smooth surface. Then the samples were cooled to growth temperature of 700 °C ± 25 °C. Si from an electron beam source (calibrated at a rate of 0.2 ± 0.03 nm/min) and Ge from a pyrolitic boron nitride Knudsen source (calibrated at 0.15 nm /min) were co-deposited onto the Si (110) substrates, which were rotated during the growth. All samples were analyzed ex-situ using atomic force microscopy (AFM). Several samples were also analyzed by transmission electron microscopy (TEM).

RESULTS

The starting clean Si (110) surface shows a sharp (16x2) diffraction pattern (see figure 1(a)), which is consistent with the observations in the literature [11]. This diffraction pattern is taken slightly off the <112> direction. After about 3-monolayers of SiGe deposition, the RHEED pattern in figure 1 (b) was observed. This (2x1) pattern remained unchanged till the end of growth (highest coverage is about ~ 9 nm). We believe that the RHEED pattern shown in figure 1 (b) has evolved from the (16x2) pattern shown in figure 1(a) [11].

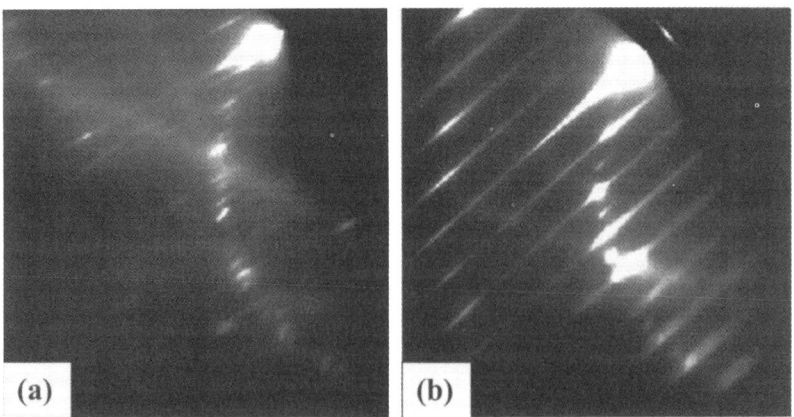

Figure 1. RHEED pattern in <112> direction (a) after the buffer layer growth, (b) with 3 nm SiGe coverage

Figure 2 (a) is the AFM image of a typical Si (110) surface after deposition of ~ 40 nm Si buffer layer. This morphology is typical for different areas of the same sample, and for different samples. The step bunches have an average ledge direction in one of the two <111> directions. In fact, the ledge direction along either [-111] or [-11-1] has been observed for different samples after the buffer layer growth. Figure 2 (b) shows the relationship between the two <111> directions on a Si (110) surface. The average step spacing in the <111> direction is partially dependent on the miscut, which is expected to be less than ±0.5°. The ledges themselves are faceted along <112> and <001> directions. A low density of pits is also observed on the surface; these are thought to form by the growth of a thin Si buffer over SiC contaminant particles formed on the original substrate [12]. Statistical measurements found the average ledge height to be about 4-5 Å. The number density of steps is about 10 ± 1 per micrometer and thus the width of the step terrace is about ~100 nm. We therefore estimate a local miscut of ~0.2°.

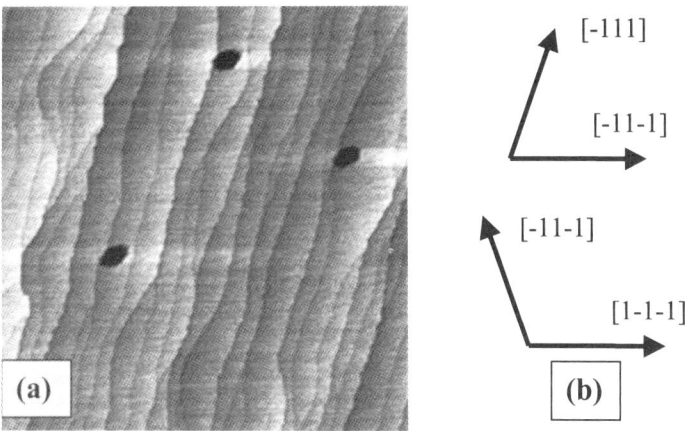

(a)

[-111]

[-11-1]

[-11-1]

[1-1-1]

(b)

Figure 2. (a). (2μ x 2μ) AFM image after 40 nm Si buffer layer growth on Si (110) surface at 550 °C, (b). Relationship between the two <111> directions on the Si (110) surface

Surface morphology of samples with 1.5 nm and 9 nm SiGe coverage on Si (110) surfaces are shown in figures 3 (a) and 3 (b) respectively. After about 1.5 nm SiGe growth, the traces of ledges from the buffer growth can still be seen. A very high density of SiGe ledges is observed to have formed on the terraces. After 9 nm SiGe alloy coverage, the surface morphology shows further dramatic changes. The SiGe ledges become longer and straighter with the increase of the SiGe coverage. In fact, the facets along the ledges, observed originally on the buffer ledges, seem to have disappeared after SiGe deposition. Comparison of the ledge height of the films shown in Figure 3(a) and 3(b) indicates that the ledge height of sample (b) is higher than that of sample (a). The SiGe ledge density, however, decreases from 27/μm in 3(a) to 22/μm in (b). Nonetheless, the SiGe ledge density of both samples is much higher than the buffer step density. Thus both SiGe films appear to have a much larger surface area than that of the initial buffer surface. Another interesting observation is that the SiGe ledges of both samples are aligned along the elastically hardest <111> direction on the Si (110) surface.

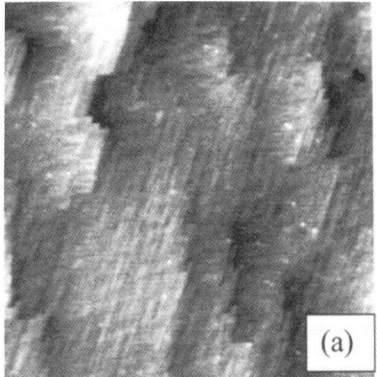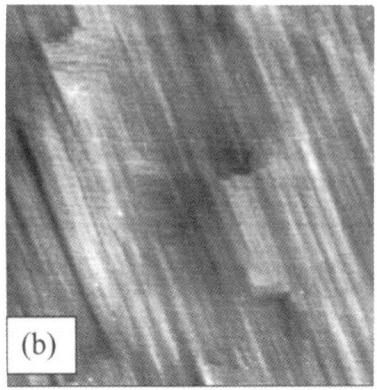

Figure 3. (2μ x 2μ) AFM image of (a) 1.5 nm SiGe coverage, and (b) 9 nm SiGe coverage grown on Si (110) surface at 700 °C

TEM observations indicate that the SiGe ledges are defect-free and are coherent with the Si (110) substrate. The alignment of the ledges along the <111> directions is also confirmed. However, for growths with SiGe coverage higher than 4.5 nm thick, a low density of misfit dislocations was observed in the thin film.

Statistical analyses of the ledge -density, -height and average ledge-length were performed. Figure 4(a) is the distribution of the step density with SiGe coverage. It can be seen that the SiGe ledge density is the largest at the initial stages of growth (1.5 nm) and significantly higher than the buffer ledge density. The ledge density begins to decrease gradually with increasing SiGe coverage, and remains almost unchanged after 3-nm deposition. The plots of the SiGe ledge -height and -length as a function of the alloy thickness are shown in figure 4(b). It seems that both the height and length of the ledges increase linearly with coverage until ~4.5 nm deposition. Beyond that the slope becomes smaller. It is also noted that there is an increased scatter in the values of the step height with increasing of SiGe coverage.

DISCUSSION

In this study, two interesting points need more discussion. First, a high density of elongated SiGe ledges were formed on the *terraces* instead of nucleation along *ledges*. The second interesting phenomenon is that the ledges on the original buffer surface, and the SiGe ledges are both aligned along the elastically hardest <111> direction on the Si (110) surface. In addition, the step bunches on the buffer surface have faceted kinks, while the SiGe ledges are extremely straight.

We start with a discussion of the energetics and kinetics of lattice mismatched epitaxy. During epitaxial growth, before the introduction of dislocations, strained thin films relax their excess strain energy by generating surface undulations (or roughness) at the cost of increased surface energy. The observed surface morphology is, however, controlled by the strain relief mechanisms operating under the specific kinetic conditions. The system reaches its lowest energy state by taking the most favorable configuration, such as, surface roughness, 3D islands, wires or ledges. Particular surface properties such as surface -energy and -anisotropy are likely to have a significant influence on the surface morphology of the SiGe films.

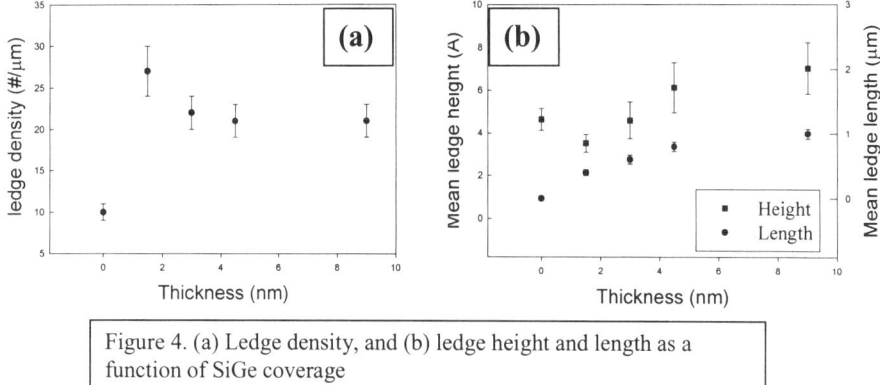

Figure 4. (a) Ledge density, and (b) ledge height and length as a function of SiGe coverage

As for the Si (110) surface, it is not surprising that step-bunches are formed during the buffer layer growth given the instability of the Si (110) surface relative to the (111) or (100) surface, especially in the presence of an unintentional miscut of the wafer. When the SiGe adatoms are deposited on the stepped surface, the adatoms should seek energetically favorable sites for attachment. The step edges should, in principle, serve as the low energy barrier nucleation sites for SiGe growth, given the sufficient diffusivity of SiGe adatoms at 700 °C . Such a mechanism has been reported for pure Ge growth on Si (110) surfaces [12], SiGe growth on (100) surfaces [13] and InGaAs growth on GaAs substrate [14]. However, in our case, a high density of SiGe ledges is formed on the terraces instead of nucleation of 3D islands along the ledges even after 9-nm deposition. (From our experimental results on Si(100) surfaces, SiGe islands are formed at a thickness significantly lower than 9 nm, under a similar set of growth conditions [15]). Moreover, it seems that the formation of the SiGe ledges may create an excessive amount of surface area in order to relieve strain energy.

To explain these intriguing results, we point to the unusual properties of the Si (110) surface. One distinguishing aspect of the Si (110) surface is that it has a higher surface energy than that of the (111), (100) or (311) surfaces [16]. It is also reasonable to assume that the surface energy of the (110) surface can be higher than some other high index planes. Therefore, the SiGe ledges formed on the step terraces may have the minimum excess surface *energy* even though the surface *area* created seems larger than that of forming islands along the ledges. With the increase of the SiGe coverage, the SiGe ledge density decreases or 'coarsens' after the initial 1.5 nm deposition in order to reduce the surface area. The SiGe ledges become longer and straighter due to the strain-induced growth.

In addition, we believe the strong anisotropy of the elastic constants on the Si (110) surface leads to the formation of SiGe ledges rather than islands. In particular, it is interesting that the SiGe ledges are aligned along the elastically hardest <111> direction, rather than the softer <100> direction. Thus the SiGe ledges are actually relaxed in the <112> direction perpendicular to the ledges. However, the Young's modulus along the <112> direction is identical to the <110> direction and appears to offer no elastic-energy advantage. We believe that some subtle surface energy/elastic anisotropy effects may be influencing the observed morphology. More detailed analysis is needed to explain our results.

CONCLUSIONS

In summary, step bunches with an average step direction along <111> were formed during Si buffer layer growth on the Si (110) surface. No 3D islands were formed even after 9 nm deposition of low mismatched (1.5 %) SiGe alloys. Instead, long and straight SiGe ledges appear to have nucleated on the step terraces. These SiGe ledges are oriented along the elastically hardest <111> direction rather than the softer <100> direction. Surface instability and anisotropy may be responsible for the observed morphology.

AKNOWLEDGEMENTS

We would like to thank Joe Mclaughlin for technical help and gratefully acknowledge financial support from NSF-DMR (9624456) and DARPA-ULTRA Nanoelectronics grant through AFOSR (F-49620-96-0313).

REFERENCES

1. L. Banyai and S. W. Koch, Semiconductor Quantum Dots, World Scientific, 1993, pp1-5.
2. E.Kapon, D.M.Hwang, and R.Bhat, Phys. Rev. Lett. **63**, 430 (1989)
3. D. Leonard, M. Krishnamurthy, C.M. Reaves, S.P. Denbaars, and P.M. Petroff, Appl. Phys. Lett. **63**, 3203 (1993)
4. R Notzel, Semicond. Sci. Technol. **11**, 1365 (1996)
5. N.N. Ledentsov, V.A. Schukin et al, Phys. Rev. **B 54**, 8743 (1996)
6. A. Hartmann, C. Dieker, R. Loo, L Vescan, and H. Luth, Appl. Phys. Lett. **67**, 1888 (1995)
7. Hiroo Omi and Toshio Ogino, Appl. Phys. Lett. **71**, 2163 (1997)
8. M, Krishnamurthy, Bi-ke Yang, J.D Weil, and C.G. Slough, Appl. Phys. Lett. **70**, 49 (1997)
9. J Arai, A. Ohga, and T. Hattori, Appl. Phys. Lett. **71**, 785 (1997)
10. C. W. Liu and J.C. Sturm, Appl. Phus. Lett. **65**, 76 (1994)
11. Youiti Yamamoto, Surf. Sci. **313**, 155(1994)
12. J.D. Weil, X. Deng and M. Krishnamurthy, J. Appl. Phys. **83**, 212 (1998)
13. K.M. Chen, D.E. Jesson, S.J.Pennycook, M. Mostoller, and T. Kaplan, Phys. Rev. Lett. **71**, 1409 (1995)
14. R. Leon, T.J.Senden, Yong Kim, C. Jagadish, and A. Clark, Phys.Rev. Lett. **78**, 4942 (1997)
15. X. Deng, J.D. Weil, and M. Krishnamurthy, Phys.Rev. Lett. in press (1998)
16. D. J. Eaglesham, A.E.White, L.C. Feldman, N. Moriya, and D.C. Jacobson, Phys. Rev. Lett. **70**, 1643 (1993)

FABRICATION OF SUB 30 NANOMETER SHEETS OF SINGLE CRYSTALLINE SILICON

J.G.FLEMING
Sandia National Laboratories,
MS 1084,
P.O. Box 5800,
Albuquerque NM 87185.
fleminjg@Sandia.gov

ABSTRACT

A relatively simple technique for the fabrication of sub 30 nm thick sheets of perfect single crystalline silicon is described. The thinnest sheets formed were 15 nm thick. The width of the sheets is 300-1000 nm and the length of the sheets can be tens of microns. The sheets are crystallographically well defined with the thin dimension being bound by atomically smooth {111} planes. The sheets are fabricated using a combination of reactive ion etching and wet KOH etching steps.

INTRODUCTION

One of the limitations of silicon technology is that silicon's indirect band structure makes it an inefficient emitter and detector of light. However, the advantages of silicon technology: readily available, large, high purity substrates; an extensive technology base; high temperature stability; a stable oxide; low toxicity; and ease of formation of ohmic contacts; fuel continuing efforts to find ways of overcoming this limitation. This is evident by the interest generated by the discovery of strong room temperature photoluminescence (PL) in porous silicon [1-2]. Other approaches to changing the electronic properties of silicon include ion implantation [3-4] and the deposition of thin films [5]. However, in the vast majority of approaches, the resulting materials are either heavily defective or suffer from some level of contamination, both of which may cloud the interpretation of the results obtained. In this work, a process is described which enables the fabrication of high quality, pure sheets of single crystalline silicon. These sheets are formed using a combination of dry reactive ion, and wet KOH, etching.

FABRICATION

The starting material for this study was single crystalline, monitor grade <100> and <110> 6 inch wafers. A schematic of the fabrication process for <100> type wafers is given in Fig. 1. The first step involved the fabrication of a fillet of silicon nitride. This was done by first growing silicon dioxide which is subsequently patterned in strips parallel to a <110> direction. A layer of silicon nitride was then deposited. The fillet was formed by anisotropic reactive ion etching. Since the silicon nitride running along the edge of the oxide step is effectively thicker than the rest of the silicon nitride thin film, a fillet remains along this edge after the silicon nitride has been cleared from planar regions of the sample. Following fillet formation, another anisotropic process was used to etch into silicon substrate ~2000 nm. In this etch the silicon nitride and silicon dioxide acted as etch masks. The silicon dioxide was then removed using hydrofluoric acid. Since the etch selectively between the silicon dioxide and silicon nitride is high, the silicon nitride fillet remained after the removal of the silicon dioxide. The wafers were then etched for 30 to 60 seconds in a 6M, 65°C KOH solution. This wet etch has been demonstrated to have very high selectivity to the silicon {111} planes [6]. The etch process occurring on the regions of the sample initially covered by oxide proceeded as usual, with the etch effectively stopping when the etch front intersects the {111} plane running along the edge of the base of the fillet. However, the silicon on the etched silicon sidewall was etched to undercut the fillet, Fig 2. Again, the etch front stops when it intersects the edge of the fillet. Due to the presence of the self aligned step in the silicon, the sides of the two {111} stopping surfaces are parallel to each other and atomically smooth as a result of the KOH etch.

Mat. Res. Soc. Symp. Proc. Vol. 533 © 1998 Materials Research Society

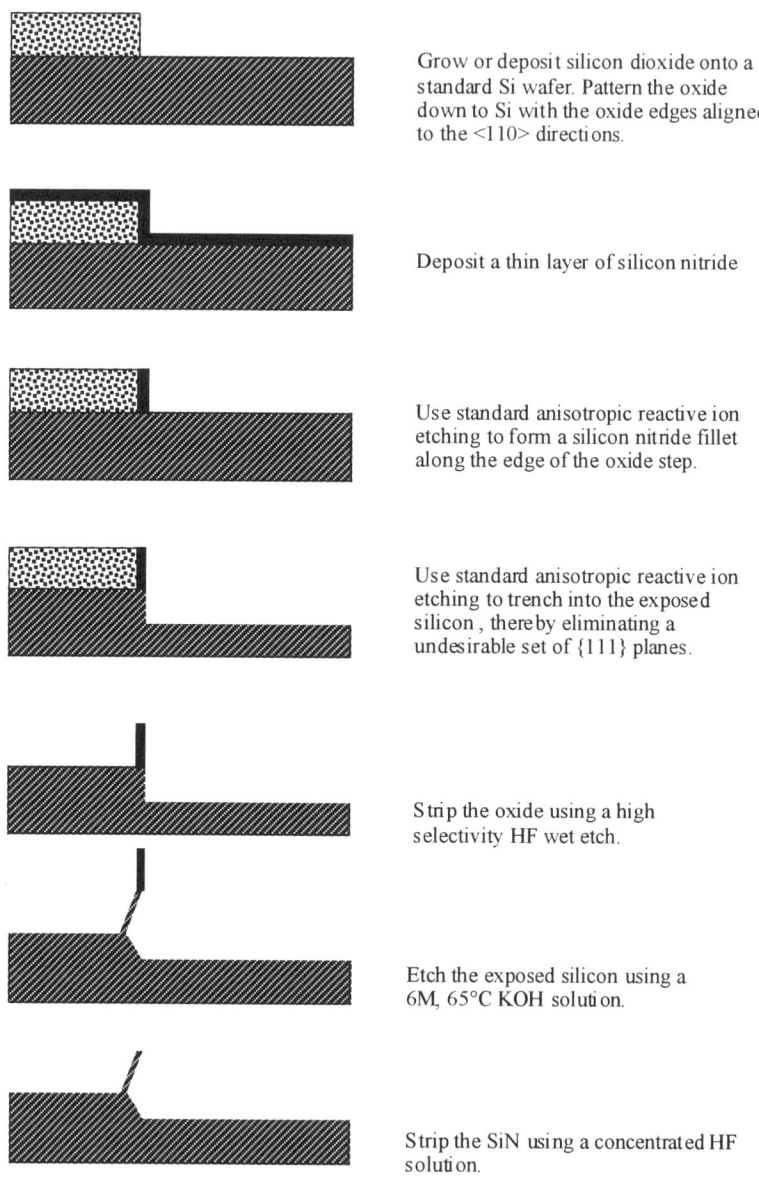

Grow or deposit silicon dioxide onto a standard Si wafer. Pattern the oxide down to Si with the oxide edges aligned to the <110> directions.

Deposit a thin layer of silicon nitride

Use standard anisotropic reactive ion etching to form a silicon nitride fillet along the edge of the oxide step.

Use standard anisotropic reactive ion etching to trench into the exposed silicon, thereby eliminating a undesirable set of {111} planes.

Strip the oxide using a high selectivity HF wet etch.

Etch the exposed silicon using a 6M, 65°C KOH solution.

Strip the SiN using a concentrated HF solution.

Fig. 1
Cross sections schematically outlining the fabrication process for <100> substrates.

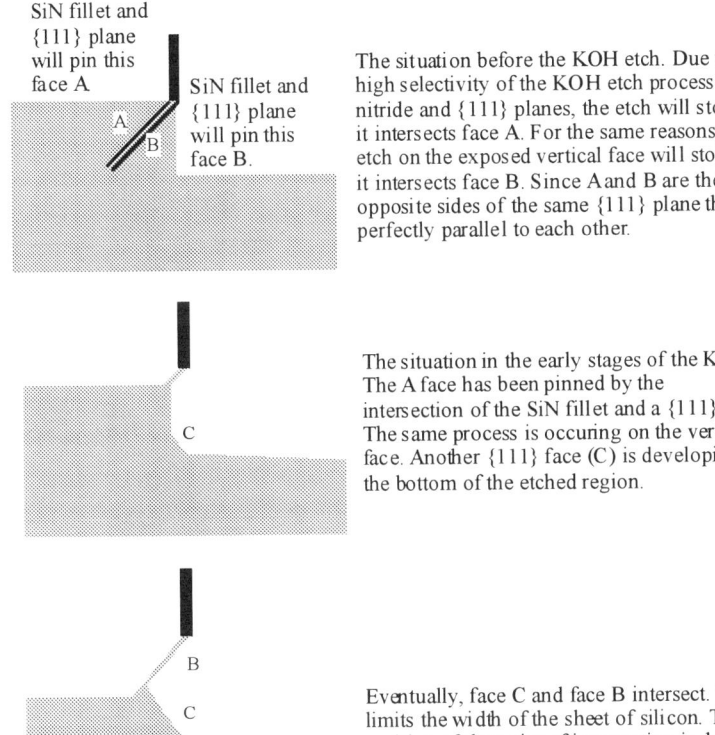

SiN fillet and
{111} plane
will pin this
face A

SiN fillet and
{111} plane
will pin this
face B.

The situation before the KOH etch. Due to the high selectivity of the KOH etch process to silicon nitride and {111} planes, the etch will stop when it intersects face A. For the same reasons, the etch on the exposed vertical face will stop when it intersects face B. Since A and B are the opposite sides of the same {111} plane they are perfectly parallel to each other.

The situation in the early stages of the KOH etch. The A face has been pinned by the intersection of the SiN fillet and a {111} plane. The same process is occuring on the vertical face. Another {111} face (C) is developing at the bottom of the etched region.

Eventually, face C and face B intersect. This limits the width of the sheet of silicon. The position of the point of intersection is determined by the depth of the silicon etch.

Fig. 2

Cross-section schematic showing the evolution of the sheet during the KOH etch.

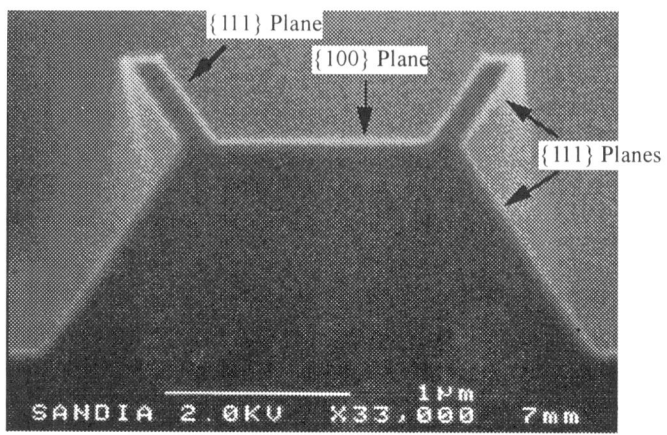

3a

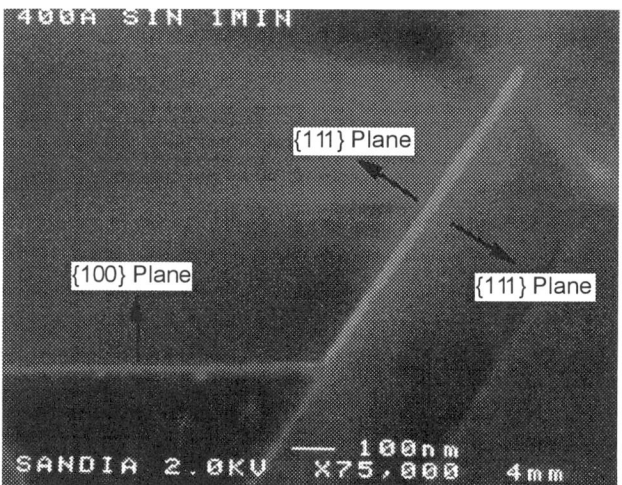

3b

Fig 3a,b.

Figure 3a shows a cross section through a relatively thick sheet.
The fillet used in this case was 100 nm thick. Figure 3b shows a
close-up of a ~25nm thick sheet.

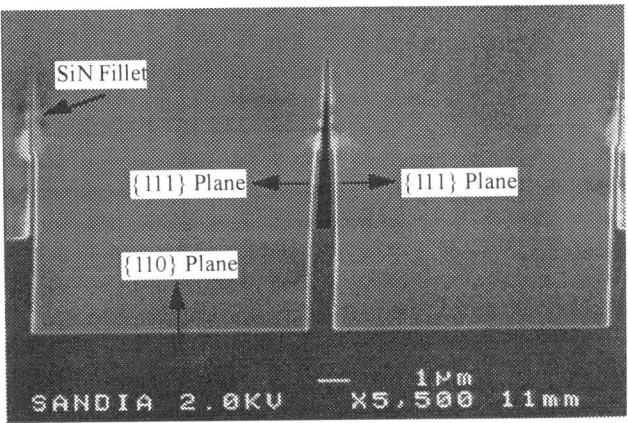

SiN Fillet

{111} Plane ←——→ {111} Plane

{110} Plane

SANDIA 2.0KV   X5,500   11mm

Fig. 4

Cross sections of samples obtained using <110> substrates. In
these cases, the ~200 nm thick sheets are oriented perpendicular
to the surface of the substrate. The SiN fillet has not yet been stripped.
In the center of the micrograph, there are two sheets separated by
less than a micron with touching SiN fillets.

RESULTS AND DISCUSSION

Examples of parts fabricated using this process are given in Fig.3. The thickness of the sheets is dependent
upon the thickness of the silicon nitride layer. Since this is a standard integrated circuit process, it can be
controlled with a high degree of precision. The width of the silicon sheet is determined by the depth of the silicon
etch. There is some undercut of the silicon nitride fillet during the KOH etch. If the amount of undercut is
excessive, the fillet of silicon nitride can be completely undercut. There are a number of sources of this undercut.
The first is roughness of the etched face of the silicon dioxide. This roughening arises during both the
photolithography and etch processes. The second major source of undercut is misorientation of the edge of the
oxide cut to the <110> direction. In this case alignment was made to the major wafer flat of the wafer, which can
be off orientation by a couple of degrees. These sources of undercut set a lower limit of the minimal silicon nitride
thickness of ~20 nm for sheets that are ~500nm wide. Due to the undercut and geometric effects, the final
thickness of the sheets is thinner than that of the fillet. Fillets which are too thin are completely undercut by the
end of the KOH etch. The minimum layer thickness obtained to date in this work is ~15 nm. This undercut
limitation can be eliminated by a better alignment to the <110>, for example using a rosette structure [7] and by
modifications to the fillet fabrication process. Further thinning of the sheets can be performed by oxidation
processing [8].

This process has also been applied to <110> oriented substrates. In this case, the orientation of the sheets to
the surface is vertical, Fig 4. A modification of the process can also be applied to <111> substrates.

In summary, a fabrication process has been developed for the formation of well defined thin layers of single
crystalline silicon. The processes used are commonly employed in the integrated circuit and micromachining
industries.

ACKNOWLEDGMENTS

This work was supported by the United States Department of Energy under contract DE-AC04-94AL85000. Sandia is a multiprogram laboratory operated by Sandia Corporation, a Lockheed Martin Company, for the United States Department of Energy. The author gratefully acknowledges the support of the MDL silicon processing team.

REFERENCES

1. M.H. Ludwig, Critical Reviews in Solid State and Materials Sciences **21** p265 (1996).

2. S.S. Iyer and Y.-H. Xie, Science **260** p40 (1993).

3. S. Guha, M.D. Pace, D.N. Dunn and I.L. Singer, Appl. Phys. Lett. **70** p1207 (1997).

4. G. Ghislotti, B. Nielsen, P. Asoka-Kumar, K.G. Lynn, L.F. DiMauro, C.E. Bottani, F. Corni, R. Tonini and G.P. Ottaviani, J. Electrochem. Soc. **144** p2196 (1997).

5. G. Allan, C. Delerue and M. Lannoo, Appl. Phys. Lett. **71** p1189 (1997).

6. K. E. Petersen, Proceedings of the IEEE **70**, p420 (1982).

7. G. Ensell, Proceedings of Transducers '95 p186 (1995).

8. H. Namastu, S. Horiguchi, M. Nagase and K. Kurihara, J. Vac. Sci Technol. B **15** p1688 (1997).

9. H.I. Liu, N.I. Maluf, R.F.W. Pease, D.K. Biegelsen, N.M. Johnson and F.A. Ponce, J. Vac. Sci Technol. B **10** p2846 (1992).

FEASIBILITY OF NOVEL Si-BASED INTERMINIBAND LASERS

Gregory Sun
Physics Department, University of Massachusetts at Boston
Boston, Massachusetts, 02125
gsun@cruiser.engin.umb.edu

Lionel Friedman and Richard A. Soref
Sensors Directorate, Air Force Research Laboratory
Hanscom AFB, Massachusetts, 01731-2909

ABSTRACT

We have designed a parallel interminiband lasing in superlattice structures of co-herently strained $Si_{0.5}Ge_{0.5}/Si$ quantum wells (QWs). Population inversion is achieved between the non-parabolic heavy-hole valence minibands locally in-k-space. Lasing transition is at $5.4\mu m$. Our analysis indicates that an optical gain of 134/cm can be obtained when the laser structure is pumped with a current density of $5kA/cm^2$.

I. INTRODUCTION

The quantum cascade laser has operated at mid-to-far infrared wavelengths[1]-[4]. For fiber-optic communications, it would be advantages to operate at $1 - 3\mu m$ wavelengths. However, in the near infrared, the QCL encounters a major problem, namely the large bias voltage making it difficult to be compatible with the integrated electronic devices, which also produces a strong electric field that could exceed the breakdown strength of the active material. To deal with this issue, we recently proposed the design of a new quantum-parallel laser (QPL)[5] where lasing transitions occur between the minibands of superlattice which allows a low-voltage operation under flat-band. The QPL is capable of operating at near-infrared communications wavelengths, as well as middle-infrared wavelengths. The interminiband lasing mechanism is employed in Si-based material systems since the indirect nature of the Si-bandgap can be avoided which opens up the possibility of monolithic integration of Si-photonics with Si-microelectronics. Also because of the absence of polar optical phonon scattering, the miniband lifetimes in Si-based structures are much longer than those in III-Vs. We shall consider a $Ge_{0.5}Si_{0.5}/Si$ QPL comprised of identical, square quantum wells. The lasing wavelength is at $5.4\mu m$. Population inversion is achieved between the valence minibands locally in k-space due to the band nonparabolicity. An optical gain of 134/cm is calculated for a current density of $5kA/cm^2$.

II. INTERMINIBAND LASER OPERATION

The bandedge diagram of our proposed, electrically pumped, unipolar, Si-based QPL is shown in Fig. 1. We consider a $Ge_{0.5}Si_{0.5}/Si$ superlattice to obtain a large valence band offset needed to increase the lasing energy. As shown in Fig. 1, the $Ge_{0.5}Si_{1-x}$ well width is $25\mathring{A}$ and the Si barrier $15\mathring{A}$. The SL is grown on a (100)

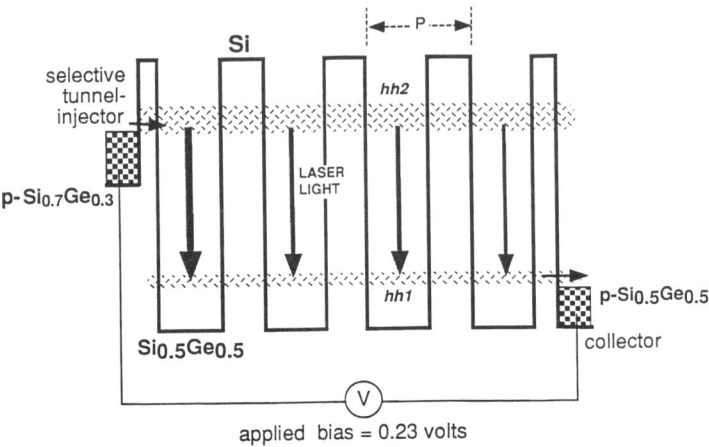

Figure 1: The quantum-parallel laser: p$^+$-i-p diode showing lasing between two valence minibands of Si$_{0.5}$Ge$_{0.5}$/Si superlattice. The heavy line is the bandedge for heavy-holes (HH).

Si substrate. Hole injection is from a larger bandgap p-SiGe emitter through a thin Si injection barrier; the larger bandgap of the emitter insures selective injection into the bottom of the upper HH2 miniband. Holes are coherently transported through the superlattice on the upper miniband. Holes that fall into the lower miniband HH1 following a lasing transition are collected via tunneling through a thin Si barrier for depletion. Fig.2 shows a schematic of hole energy E as a function of in-plane wavevector k_x and superlattice wavevector k_z. Holes are selectively injected near the very bottom of the HH2 miniband at $k_z = \pi/P$, where P is the width of a superlattice period. Thus, the initial state for the radiative and nonradiative phonon intersubband transitions is at the energy minimum of HH2 at $k_z = \pi/P$ and $k_x = 0$. For the radiative processes, photon emission is vertical at $k_x = 0$ as shown by the wavy line. Radiation is from $E_{HH2}(k_z = \pi/P)$ to $E_{HH1}(k_z = \pi/P)$ with a wavelength of 5.4μm and p-polarization along the z-direction. A ridge-waveguide laser cavity with a TM fundamental mode is most suitable for the z-polarized radiation. A bulk Si substrate could be employed for the resonator because the spatially averaged index of refraction for the superlattice is larger than that of silicon ($\Delta n \approx 0.2$). However, a silicon-on-insulator (SOI) platform is preferred because the buried SiO$_2$ layer gives a much larger index difference ($\Delta n \approx 2$) between the superlattice and the substrate. The SOI provides high Q, good mode confinement, and a high mode overlap factor. Fig.3 illustrates the proposed rib laser configuration with vertical end-facets in a Fabry-Perot cavity.

For the non-radiative HH2-HH1 transition in Fig.2, emission of an optical phonon

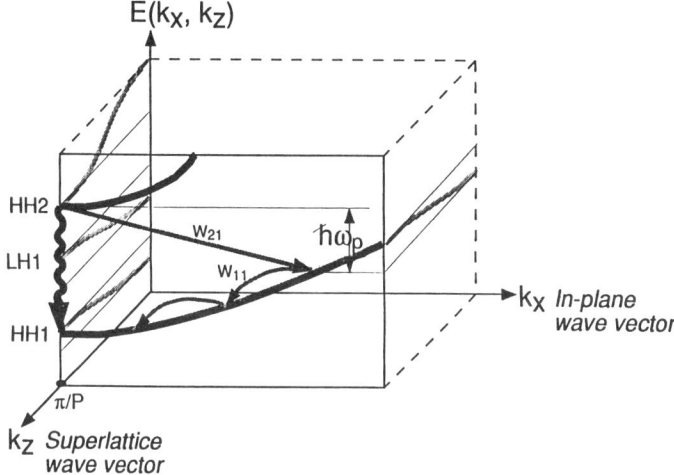

Figure 2: Dispersion of lowest-three valence minibands of $Si_{0.5}Ge_{0.5}/Si$ superlattice as a function of k_x and k_z. W_{21} is the intersuband transition rate and W_{11} is the intrasuband rate.

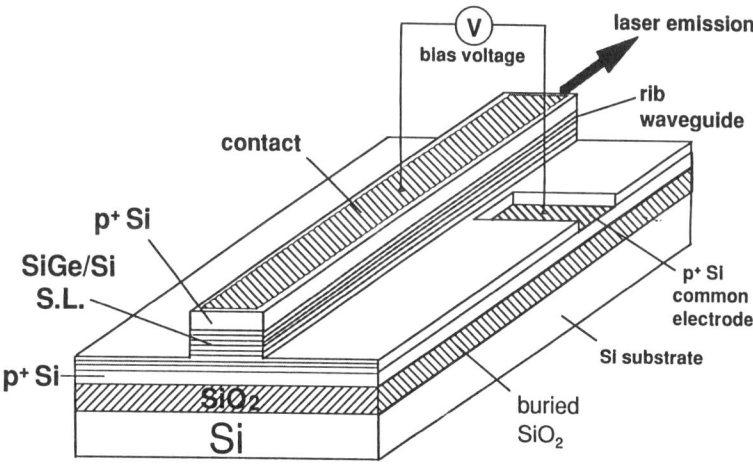

Figure 3: Waveguided unipolar SiGe/Si superlattice interminiband laser in silicon-on-insulator. Fabry-Perot resonator is shown.

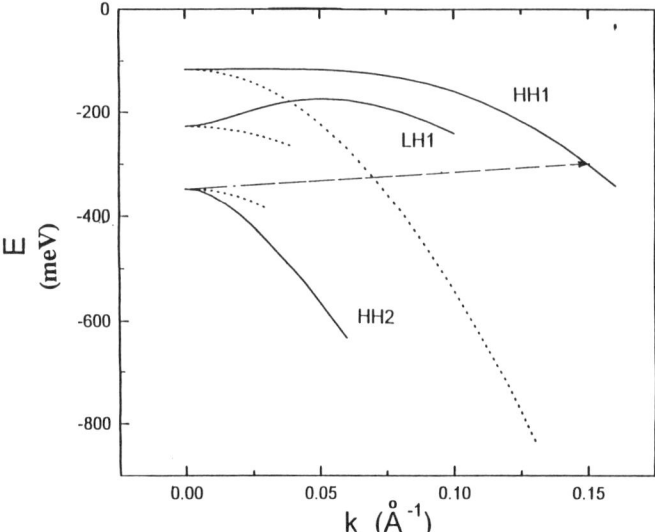

Figure 4: In-plane dispersion $E = E(k_x)$ of HH1, HH2, and LH1 bands at minizone boundary $k_z = \pi/P$ of $Si_{0.5}Ge_{0.5}/Si$ superlattice, solid lines. Dashed lines are in the absence of interaction (b=c=0). The long dashed straight line shows the intersubband optical phonon emission. Decreasing electron energy (increasing hole energy) is in the downward direction.

from HH2 occurs to superlattice states of finite in-plane wavevector, k_x, from which the holes cascade down via intrasubband phonon emission. We will assume that the holes relaxing into HH1 are distributed over the lower miniband and transit the superlattice at the mean miniband velocity in calculating the tunnel time. However, in the calculation of optical phonon scattering rates, we shall consider only the in-plane transition at $k_z = \pi/P$ for simplicity. Due to the interactions between the HH and LH states, the in-plane valence band dispersion is highly nonparabolic. As shown in Fig.4, $E_{HH1}(k_x)$ is forced to lower hole energies, $E_{HH2}(k_x)$ is forced to higher hole energies, and $E_{LH1}(k_x)$ becomes electron-like at $k_x = 0$. Thus the energy difference $E_{HH1}(k_x) - E_{HH2}(k_x)$ increases with increasing k_x and the photon emitted at $k_x = 0$ cannot be re-absorbed as the hole cascades down the $E_{HH1}(k_x)$ dispersion curve, until it reaches the minimum. More exactly, reabsorption will occur when the absorbed photon is within the linewidth of the upper level. We therefore have a local in k-space inversion as had been described for the InGaAs/InAlAs system by Faist et al [6], except that here the intersubband energy difference increases rather than decreases with k_x.

III. LASER GAIN

The gain can be written as

$$G_L = \sigma_{21}(N_2 - N_1), \tag{1}$$

where the lasing cross-section $\sigma_{21} = (4\pi\alpha_0/n_0)(E_L/\Gamma)|z_{21}|^2$, with the fine structure constant $\alpha_0 = 1/137$, refractive index $n_0 = 3.5$, the lasing energy $E_L = 230.5$meV, the linewidth of $E_{HH2}(k_x)$ at $k_z = \pi/P$ is chosen to be $\Gamma = 10$meV, and the dipole matrix element $|z_{21}| = 3.7\overset{\circ}{A}$. The population inversion difference is found from a rate equation for the coupled two level system:

$$(N_2 - N_1) = (J/e)\tau_{int}[1 - (\tau_{tun})^{eff}/\tau_{int}], \tag{2}$$

where J is the charge current density and e is the electron charge. We have estimated the effective tunneling time $\tau_{tun} = 0.43$ps and the intersubband lifetime $\tau_{int} = 2.8$ps, then $N_2 - N_1 = 7.4 \times 10^{10}/$cm^2. We obtain an optical gain $G_L =134/$cm for $J = 5$kA/cm^2. Since the emitter and collector Fermi levels are at the bottom of the upper and lower minibands respectively, the applied voltage required for the flat-band condition is $V_a = 0.23$V.

IV. CONCLUSIONS

We have theoretically studied a p-i-p coherently strained $Si_{0.5}Ge_{0.5}/Si$ superlattice quantum-parallel laser. We proposed a waveguided SOI resonator structure. Calculations have been made of the local in k-space population inversion between the non-parabolic HH2 and HH1 valence minibands at the 5.4μm laser wavelength. Analysis of radiative-and-phonon scattering between the mixed bands indicates a laser lifetime enhancement. The non-radiative intersubband optical phonon scattering is slower than the in-plane intrasubband scattering. The effective lifetime of the lower state is further decreased by the probability of sequential tunneling to the collector. At a current density of 5kA/cm^2, a laser gain of 134/cm is achieved.

References

[1] J. Faist, F. capasso, D. L. Sivco, A. L. Hutchinson, C. Sirtori, and A. Y. Cho, Science **264**, 553 (1994)

[2] C. Sirtori, J. Faist, F. Capasso, D. L. Sivco, A. L. Hutchinson, and A. Y. Cho, Appl. phys. lett. **69**, 2810 (1996)

[3] S. Slivken, C. Jelen, A. Rybaltowski, J. Diaz, and M. Razeghi, Appl. Phys. Lett. **71**, 2593 (1997)

[4] O. Gauthier-Lafaye, P. Boucaud, F. H. Julien, S. Sauvage, S. Cabaret, J.-M. Lourtioz, V. Thierry-Mieg, and R. Planel, Appl. Phys. lett. **71**, 3619 (1997)

[5] L. Friedman, R.A. Soref, and G. Sun, IEEE Photonics Technol. Lett. **9**, 593 (1997)

[6] J. Faist, F. Capasso, C. Sirtori, D.L. Sivco, A.L. Huchinson, M.S. Hybertsen, and A.Y. Cho, Phys. Rev. Lett. **76**, 411 (1996)

Part V

SiGeC Alloys

BAND ALIGNMENT OF Si$_{1-x}$Ge$_x$ AND Si$_{1-x-y}$Ge$_x$C$_y$ QUANTUM WELLS ON Si (001)

N.L. Rowell[*], R.L. Williams[*], G.C. Aers[*], H. Lafontaine[**], D.C. Houghton[**], K. Brunner[***], K. Eberl[***], O. Schmidt[***], W. Winter[***]
[*]National Research Council, Ottawa, Canada K1A 0R6
[**]SiGe Microsystems Inc., 1500 Montreal Road, Ottawa, Canada K1A 0R6
[***]Max-Planck-Institut für Festkörperforschung, Heisenbergstrasse 1, D-70569 Stuttgart, Germany

ABSTRACT

Recent low-temperature photoluminescence (PL) studies will be discussed for coherent Si$_{1-x}$Ge$_x$ and Si$_{1-x-y}$Ge$_x$C$_y$ alloy multiple quantum wells on Si (001) substrates grown by either ultra-high vacuum chemical vapour deposition or solid source molecular beam epitaxy. An in-plane applied-stress technique will be described which removes systematically band edge degeneracies revealing the lower, PL-active CB. Applied-stress data taken with this technique at ultra-low excitation intensity proved intrinsic type II CB alignment in SiGe on Si (001). Apparent type I alignment observed at higher intensity will also be discussed. New applied stress PL results are presented for Si$_{1-x-y}$Ge$_x$C$_y$ quantum wells under various grown-in stress condition

INTRODUCTION

Recently, strained layers of the semiconductor alloy, Si$_{1-x}$Ge$_x$, have been integrated within Si microelectronic technology to take advantage of this material's substantial performance advantage in low cost, high speed, Si-compatible devices for wireless communication and high speed digital applications. The band gap of such low defect density, compressively strained Si$_{1-x}$Ge$_x$ layers on Si(001) is smaller than Si [1] with nearly all the gap difference in the valence band (VB) leading to the use of Si$_{1-x}$Ge$_x$ in p-channel electronic devices. The related small conduction band (CB) offset precludes complementary Si$_{1-x}$Ge$_x$ n-channel applications except when higher defect density, relaxed Si$_{1-x}$Ge$_x$ buffer layers are employed as virtual substrates. However, tensile-strained Si$_{1-y}$C$_y$ layers [2] on Si - also with band gaps smaller than Si - have CB dominated band gap differences [3] suggesting this material to be suitable for low defect, n-channel electronic devices. For both alloy types it follows that proper characterization of conduction and valence band offsets at heterostructure interfaces is necessary to enable device applications. Although conventional low temperature photoluminescence (PL) provides accurate alloy band gap energies [4, 5], this method does not establish directly the alloy/Si interface band alignment. Yet with our wafer bending method [6], we can evaluate band alignment for Si$_{1-x}$Ge$_x$, Si$_{1-y}$C$_y$, Si$_{1-x}$Ge$_x$/Si$_{1-y}$Ge$_y$, and Si$_{1-x-y}$Ge$_x$C$_y$ quantum wells on Si by applying uniaxial stress to samples while observing the associated change in their low temperature PL spectra. For Si$_{1-x}$Ge$_x$ quantum well samples, an intrinsic type II band alignment (holes only confined in the wells) was obtained although this property is observable only at ultra-low excitation density [7,8]. At higher excitation density, type I alignment (both electrons and holes confined in the wells) is induced by excitation-related charge accumulation which shifts the SiGe CB below the Si CB via electrostatic interaction. For $Si_{1-y}C_y/Si$ heterostructures, we confirmed the previously suspected, type I band alignment and saw for the first time an elastic strain induced type I to type II transition in $Si_{1-y}C_y$ quantum wells [3]. Results for a $Si_{1-x}Ge_x/Si_{1-y}C_y/Si$ double-well structure showed a type II alignment with the observed PL transition from the SiC CB to the SiGe VB. A CB offset ~70% of the band gap difference was deduced for $Si_{1-y}C_y/Si$ [9]. Ternary pseudomorphic $Si_{1-x-y}Ge_xC_y$ layers may provide even further growth and design flexibility as exactly strain-compensated $Si_{1-x-y}Ge_xC_y$ layers [10] of arbitrary thickness can be grown on Si. New applied stress PL results will be presented for this material under various grown-in stress conditions.

EXPERIMENT

Photoluminescence

The PL experiments were conducted with a PL-optimized apparatus [11] consisting of a Fourier transform spectrometer, argon or krypton ion laser (514.5 or 647 nm), and variable-temperature cryostat (2 - 100 K) using a resolution of 0.5 meV from 700 to 1200 meV. Samples, approximately 15 mm x 5 mm and of wafer thickness ~0.5 mm, were bent lengthwise along a specific crystallograpic direction under remote control with the sample remaining at the measuring temperature in the helium cryostat. This method was used to introduce in-plane uniaxial tension or compression along [110] or [100] directions. For thin wafer substrates with large bending radii, the stress is compressive when the QW surface of the wafer is concave and tensile when convex. The PL was excited normal to the QW surface with an expanded parallel laser beam of uniform irradiance over a 3 to 5 mm diameter area. The stress was calibrated by comparing induced shifts of the substrate's Si transverse optical (TO) and no phonon (NP) lines with those predicted from the measured radius of curvature combined with deformation potential theory. Good correlation between PL line shifts and theory was observed. Furthermore PL with this bending apparatus provided reproducible data (±0.1 meV) with the necessary stress resolution ($\pm$ 10 Mpa).

Materials

Most $Si_{1-x}Ge_x$ layers discussed here were undoped and grown in a mixture of mass-flow controlled silane and germane at a pressure of 10^{-3} mbar, a rate of 0.1 to 1 Å/sec, and a temperature of 525 °C using a hot wall multiple 6" wafer UHV-CVD reactor with a base pressure below 1.5 x 10^{-9} mbar. Other relevant growth and preparation procedures associated with the production-ready, high throughput growth system are described elsewhere [12]. $Si_{1-y}C_y$ and $Si_{1-x-y}Ge_xC_y$ and related $Si_{1-x}Ge_x$ layers were grown by solid source molecular beam epitaxy at a substrate temperature of 550 °C as described elsewhere [2]. From various structural characterization methods and PL, the QW's were observed to be fully pseudomorphic with flat, regular, dislocation-free interfaces of atomic layer abruptness. For example, $Si_{1-x}Ge_x$ QW samples with single or multiple (typically ten - silicon spaced) wells have been grown with germanium fractions from 2 to 50% and layer thicknesses from ~10 Å to the metastable critical thickness. MBE grown material was of comparable quality [2].

RESULTS

$Si_{1-x}Ge_x$ / Si (001) Band Alignment

Baier *et al* [13] and Thewalt *et al* [7] have shown the positive charge (hole) accumulation in $Si_{1-x}Ge_x$ quantum wells introduced by photo-excitation during PL can lead to a accumulation of hitherto unconfined electrons in the same wells through the electrostatic interaction of electrons and holes. With this interaction included the charged particles no longer move in simply the quantum well potential but see a superposition of this interaction (Hartree) potential and the square well potential. Thus the overall potential can be bent down significantly (band bending) in both the CB and VB as shown in Fig. 1 where it is apparent that the recombination energy increases with increasing excitation density especially when the band bending overcomes the small original CB offset. As the photo-induced carrier density affects the size and shape of the Hartree potential and vice versa, the problem, described by Schrödinger and Poisson equations [13], must be solved self-consistently, *e.g* in a two band formulation. From Fig. 1 we expect a significant increase in the PL peak energy with increasing excitation density due to band bending at zero applied stress - an effect we observed for a variety of CVD-grown QW samples for Ge-concentrations up to 0.5. However suggestive, observation of PL shifts alone does not determine the particular intrinsic CB alignment.

For $Si_{1-x}Ge_x$ matched to Si(001), the application of in-plane uniaxial stress has the most significant effect if it serves to compensate the built-in SiGe stress [14] which is compressive in the plane and tensile along [001]. Thus the application of tension along [110] shifts the SiGe's E_g upwards towards its unstrained value while simultaneously shifting the bandgap energy of the neighbouring silicon downward. Even more significantly for PL, however, is the fact that the SiGe CB shifts upward at a rate of approximately 14.7 meV/GPa while the Si CB shifts downward by 28.3 meV/GPa and the SiGe VB shifts down by 11.2 meV/GPa [6]. Including hydrostatic effects for [110] applied tension, we expect the PL energy to shift upward by 31.8 meV/GPa for type I band alignment and to shift downward by 17.1 meV/GPa for type II band alignment - a clear differentiation between the two types. To study this effect we have used the four QW structure consisting of 4 nm thick wells each separated by undoped Si (30 nm thick) with Ge concentrations of 0.03, 0.05, 0.07, and 0.09 in ascending order from the lightly doped n⁻ substrate. With moderate excitation density (5 mW/cm^2 at 514.5 nm), the low applied stress PL spectrum (labelled 17 MPa in Fig. 2) consists of four sets of QW NP/TO peaks energy shifted from each other because of the different Ge fractions. Additionally, a set of Si NP/TO peaks is apparent in Fig. 2. As the applied [110] tensile stress is increased the peaks from all four QW's shift to higher energy indicative of the photo-induced type I band alignment which occurs at this excitation level. However near a 140 MPa [110] tension the 9% QW's signal strength decreases and for higher applied stress this well's PL lines move to lower energy. What has occured is that the applied stress has increased the QW intrinsic CB offset to where it is greater than the Hartree potential and the QW has reverted to its intrinsic type II alignment. The accompanying decrease in line strength is associated with the electron-hole wavefunction overlap which is considerably smaller for type II band alignment than for type I band alignment in the case of this well. Similarly, but at higher applied [110] tension the 7% well makes the transition to type II behavior as depicted in Fig. 2 at a higher tension level (near 200 MPa) since the intrinsic CB offset is smaller. As the lowest concentration well, 3% Ge fraction, becomes an electron barrier (type II CB alignment) at large applied stress its PL intensity is maintained probably since this barrier, being the closest to the n$^-$ substrate, can impede a large fraction of the electrons diffusing from the electron-rich substrate. We also observe an enhanced-strength transverse acoustic (TA) phonon replica PL line with type II band alignment, consistent with previous results [7].

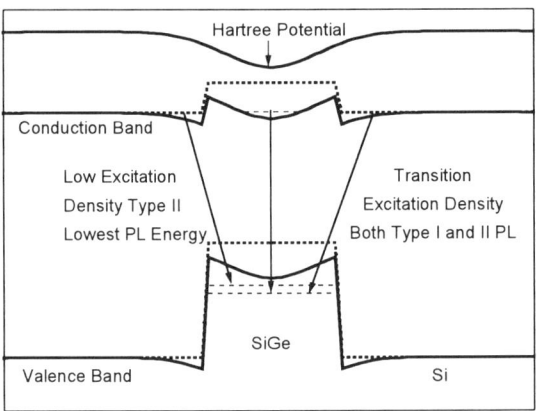

Figure 1 Photo-induced charge buildup leading to a band bending through superposition of self-consistent (Hartree) and quantum well potentials.

In Fig. 3 we show NP peak energies versus applied [110] tensile stress for a number of excitation intensities for three samples - each containing a single QW - with x = 0.5, 0.3, and 0.15 with QW thicknesses of 2.3, 3.0, and 3.5 nm respectively on p$^-$ - substrates. For a given applied stress the maximum excitation intensity for which type II behavior can be observed decreases with decreasing x because the wells of lower Ge concentrations have smaller type IICB offsets and thus have lower threshold photo-injected charge to switch to type I alignment. A simple self-consistent model predicts that the total QW charge at which a type II/I transition occurs, Q_{min}, increases linearly with CB offset. For example, a CB offset of 5 meV implies the value of Q_{min} must be ~1.8×10^{11} cm^{-2} for the transition to occur assuming that carrier lifetimes are much shorter for type I than for type II alignment. Activated p-dopants in the QW or adjoining Si may force type I behavior in samples with low x since such dopants can lead to an appreciable QW positive charge. A background QW charge estimated [7] at 2×10^{10} cm^{-2} corresponds to a CB offset of ~1 meV in the model. Given unintentional doping in the 10^{15} cm^{-3} range, this background charge has to be captured by the QW from adjacent Si several tens of nm thick. The opposite negative charging in the QW for n-doping does not occur since the SiGe QW presents a barrier, albeit small, to electrons. Nonetheless, due to this background positive charge and possible surface charge effects, the true intrinsic CB offsets are difficult to obtain from data such as Fig. 3 with the simple self-consistent model.

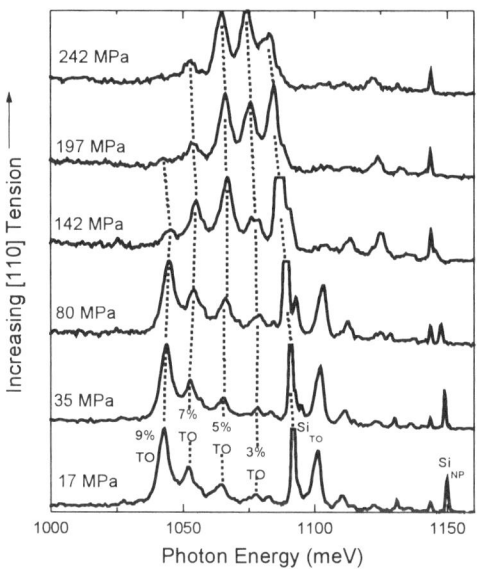

Figure 2 Photoluminescence spectra versus applied [110] tension at 4.5 K with 5 mW/cm^2 at 514.5nm for a four quantum well sample consisting of 3, 4, 7, and 9% Ge-fractions 4 nm thick.

Si$_{1-x-y}$Ge$_x$C$_y$ / Si (001) Band Alignment

In Fig. 4 we display the PL no-phonon peak energy for Si$_{1-x-y}$Ge$_x$C$_y$ quantum wells [15, 16] 4 nm thick, for six carbon concentrations versus applied [110] tensile stress. The carbon fraction varies from 0 to 1.7 %. At the lower three carbon concentrations, y = 0, 0.66, and 0.78 %, the behavior with applied [110] tensile stress is similar to Si$_{1-x}$Ge$_x$ of the same Ge fraction (see Fig. 3); i.e. at low applied stress the band alignment is photo-induced type I and at higher stress (>~50 MPa) the upshift of the quantum well CB with stress has exceeded the photo-induced band bending restoring the structure to type II CB alignment. However as the carbon concentration is increased the energy difference

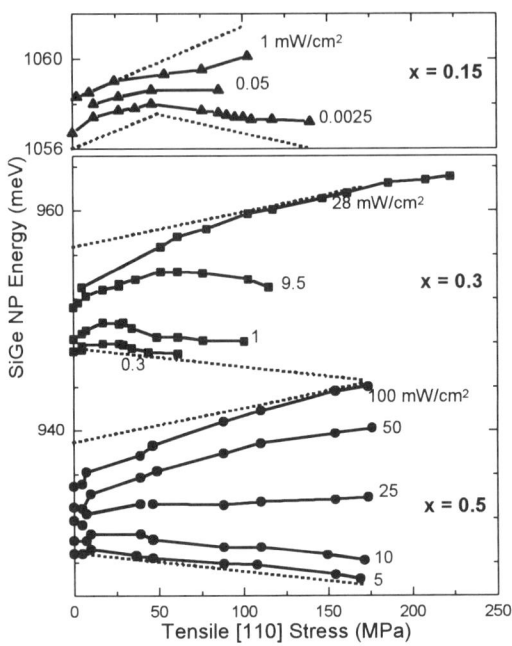

Figure 3 Photoluminescence peak energy versus stress for three germanium concentrations as a function of excitation density. As the concentration increases the minimum excitation density for observation of type II behavior increases consistent with increasing intrinsic conduction band offset.

between the $\Delta(4)$ $Si_{1-x-y}Ge_xC_y$ CB and the Δ_{001} Si CB increases as shown in Fig. 5 to a point where only type II behavior should be predominant even at low applied stress. This trend is apparent in Fig. 4 for the y = 1.04, 1.48, and 1.7% QW samples for which only type II behavior was observed at all intensity levels up to the maximum employed, 1 W/cm^2 (647 nm). It is also possible that the added carbon has reduced the exciton lifetime contributing further to the increase of the type I threshold. The 1.7% sample is slightly below in carbon concentration the strain compensated point (x = 8.2y) where the light hole (LH) and heavy hole (HH) valence bands are degenerate. However the observed slope with strain for this sample suggests the HH band is still the ground state at y = 1.7%. Nonetheless, with a lower concentration (x and y) exactly strain compensated sample [10], the observed slope with [110] tensile stress of nearly 40 meV/GPa indicated that the LH band can replace the HH band as the ground state valence band for such samples.

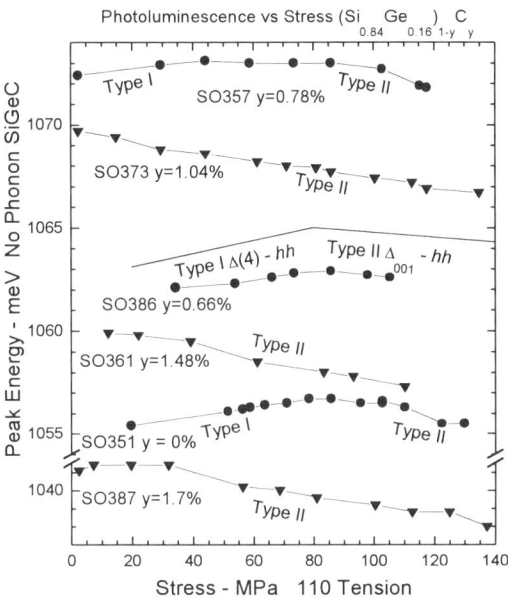

Figure 4 Photoluminescence peak energy versus applied [110] tensile stress of $(Si_{0.84}Ge_{0.16})_{1-y}C_y$ quantum wells 4 nm thick for six carbon concentrations. As the concentration increases the structure exhibit stronger of type II behavior consistent with increasing intrinsic conduction band offset.

CONCLUSIONS

We have presented results from PL results with the application of [110] tensile stress for SiGe QW's on Si(001) which support the conclusion that the CB alignment is intrinsically type II in SiGe QW's for a range of Ge-concentrations from 0.03 to 0.5 with the CB offset increasing with x. However direct evidence for type II behavior at zero applied stress was observed only for SiGe QW samples with Ge-concentrations between 0.3 and 0.5 since only in that range of concentrations was the CB offset was large enough to allow type II PL spectra to be observed at zero applied stress. For SiGe QW samples of lower x, although the transition to type II from type I behavior was observed at a finite applied stress, intrinsically type II band alignment was not proven but likely. Our PL results with [110] tensile stress for SiGeC QW's on Si(001) for a Ge fraction of 0.16 and variable C fraction from 0 to 0.017 indicate a type II band alignment which becomes more pronounced with increasing C fraction consistent with the deformation potential model. With an exactly strain compensated sample we concluded from the applied stress PL results that the LH band can replace the HH band as the ground state valence band for such samples.

In general determining actual CB offsets is affected by background dopant type and level and by possible surface charging since either effect can contribute to intrinsic QW charging and, hence, band bending. Such complications which do not depend directly on photo-excitation are variable from sample to sample and preclude a simple determination of intrinsic CB offset.

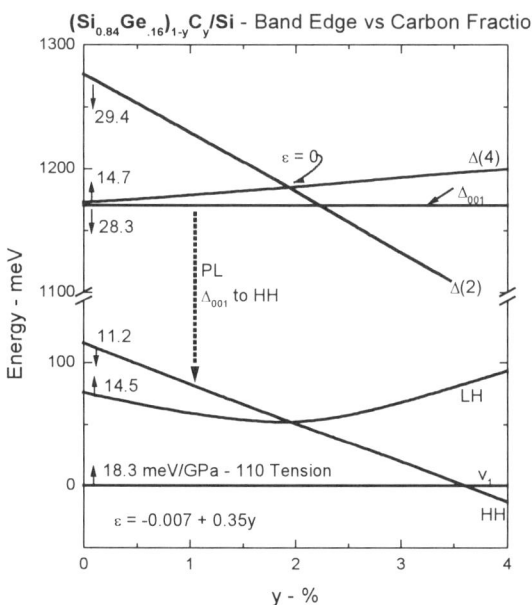

Figure 5 Schematic band edge energies for $(Si_{0.84}Ge_{0.16})_{1-y}C_y$ lattice matched to Si (001). Deformation potential theory is the basis for the energy shift. C intrinsic chemical effects on bandgap have not been included. Slopes with applied [110] tensile stress are indicated at y = 0.

REFERENCES

[1]   J.C. Sturm, H. Manoharan, L.C. Lenchyshyn, M.L.W. Thewalt, N.L. Rowell, J.-P. Noël, and D.C. Houghton, Phys. Rev. Lett. 66, 1362 (1991).
[2]   K. Brunner, K. Eberl, and W. Winter, Phys. Rev. Lett. 76, 303 (1996).
[3]   D.C. Houghton, G.C. Aers, N.L. Rowell, K. Brunner, W. Winter, and K. Eberl, Phys. Rev. Lett. 78, 2441, 1997.

[4]   D.J. Robbins, L.T. Canham, S.J. Barnett, A.D. Pitt, and P. Calcott, J. Appl. Phys. 71, 1407 (1992).

[5]   J. Weber and M.I. Alonzo, Phys. Rev. B 40, 5683 (1991).

[6]   D.J. Houghton, G.C. Aers, S.-R. Eric Yang, E. Wang, and N.L. Rowell, Phys. Rev. Lett. 75, 866 (1995).

[7]   M.L.W. Thewalt, D.A. Harrison, C.F. Reinhart, and J.A. Wolk, Phys. Rev. Lett. 79, 269 (1997).

[8]   N.L. Rowell, G.C. Aers, H. Lafontaine, and R.L. Williams, Thin Solid Films, in press (1998).

[9]   R.L. Williams, G.C. Aers, N.L. Rowell, K. Brunner, W. Winter, and K. Eberl, Appl. Phys. Lett. 72, 1320 (1998).

[10]   O.G. Schmidt and K. Eberl, Phys. Rev. Lett, in press (1998).

[11]   N.L. Rowell, J.-P. Noël, D.C. Houghton, A. Wang, L.C. Lenchyshyn, M.L.W. Thewalt, and D.D. Perovic, Appl. Phys. Lett. 74, 2790 (1993).

[12]   H. Lafontaine, D.C. Houghton, N.L. Rowell, G.I. Sproule, S.J. Rolfe, and R.L. Williams, *Physics in Canada*, 241 (1996).

[13]   T. Baier, U. Mantz, K. Thonke, R. Sauer, F. Schäffler, and H.-J. Herzog, Phys. Rev. B 50, 15191 (1994).

[14]   L.D. Laude, Fred H. Pollak, and M. Cardona, Phys. Rev. B 3, 2623 (1971).

[15]   K. Brunner, K. Eberl, O.G. Schmidt, W. Winter, M. Glück, and U. König, J. Vac. Sci. and Technol. B., in press (1998).

[16]   K. Brunner, W. Winter, and K. Eberl, Appl. Phys. Lett. 69, 1279 (1996).

Direct Optical Measurement of the valence band offset of p^+ $Si_{1-x-y}Ge_xC_y$ / p^- Si (100) by Heterojunction Internal Photoemission

C.L. Chang, L.P. Rokhinson, and J.C. Sturm, Department of Electrical Engineering, Center for Photonics and Optoelectronic Materials, Princeton University, Princeton, NJ 08544 USA, clchang@ee.princeton.edu

ABSTRACT

Optical absorption measurements have been performed to study the effect of carbon on the valence band offset of compressively strained p^+ $Si_{1-x-y}Ge_xC_y/(100)$ p^- Si heterojunction internal photoemission structures grown by Rapid Thermal Chemical Vapor Deposition (RTCVD) with substitutional carbon levels up to 2.5%. Results indicated that carbon decreased the valence band offset by 26 ± 1 meV/ %C. Results from optical measurement in this study agreed with previous data from capacitance-voltage measurements. Based on previous reports of carbon effect on the bandgap of compressively strained $Si_{1-x-y}Ge_xC_y$, our work suggests that the effect of carbon incorporation on the band alignment of $Si_{1-x-y}Ge_xC_y/Si$ is to reduce the valence band offset, with a negligible effect on the conduction band alignment.

INTRODUCTION

Strained $Si_{1-x}Ge_x/Si$ heterostructures have been extensively studied and have led to many device applications. The advantage of using strained $Si_{1-x}Ge_x/Si$ heterostructures results from the flexibility in bandgap engineering by controlling the amount of incorporated Ge into Si matrix. However, due to the 4% larger atomic size of Ge than that of Si, strain involved in $Si_{1-x}Ge_x$ prevents one from growing unlimited pseudomorphic $Si_{1-x}Ge_x$ layer on Si substrate without introducing misfit dislocations. Recently $Si_{1-x-y}Ge_xC_y$ has attracted a strong interest due to the ability of substitutional C to compensate the strain caused by Ge atoms, with 1% substitutional C compensating the strain caused by 8-10 % Ge [1-6].

Photoluminescence (PL) measurements on $Si_{1-x-y}Ge_xC_y$ as well as transport studies of heterojunction bipolar transistors (HBT's) with $Si_{1-x-y}Ge_xC_y$ as the base showed that the addition of 1% C increases the bandgap of $Si_{1-x}Ge_x$ by 21-26 meV [3-5]. However, reducing the strain in $Si_{1-x}Ge_x$ by adding C increases the bandgap less than does reducing the strain by merely removing Ge. These results imply that, for a given bandgap, $Si_{1-x-y}Ge_xC_y$ has less misfit strain and therefore allows a greater critical thickness than does $Si_{1-x}Ge_x$.

Although it is generally agreed that 1% C increases bandgap by 21~26 meV, it is still under debate regarding how the bandgap increase is allocated in the band alignment of $Si_{1-x-y}Ge_xC_y$ /Si heterostructures. Several electrical and optical methods have been used to determine the band alignment. A temperature-dependent leakage current study on p^+ $Si_{1-x-y}Ge_xC_y$ / p^- Si unipolar diodes indicated that C decreased the valence band offset (ΔE_v) of the resulting $Si_{1-x-y}Ge_xC_y$ /Si heterostructure [13]. However, no accurate quantitative number was extracted due to scatter in data among devices caused by strong dependence of leakage current on local defects. Capacitance-Voltage (C-V) measurements, on the other hand, are theoretically insensitive to anomalous sources of leakage current. They have demonstrated a clear downward trend of ΔE_v of $Si_{1-x-y}Ge_xC_y$ /Si by C incorporation with minimal scatter of data among devices and indicated that

Mat. Res. Soc. Symp. Proc. Vol. 533 © 1998 Materials Research Society

the increase in bandgap by C is fully accommodated in the valence band[7]. Similar results were also obtained from C-V analysis of $Si_{1-x-y}Ge_xC_y$ based metal-oxide-semiconductor structures[14]. X-ray photoelectron spectroscopy (XPS) evaluation on the $Si_{1-x-y}Ge_xC_y$ /Si valence band offset did not show significant change with carbon incorporation (with accuracy limit ±30 meV), consistent with the finding that the effect of carbon is small[18].

Conflicting results have also been reported. An indirect evidence from PL study on the $Si_{1-x-y}Ge_xC_y$ / $Si_{1-y}C_y$ quantum wells suggested that C increases ΔE_V of $Si_{1-x-y}Ge_xC_y$ /Si by 10 meV/ %C[15]. XPS measurements on Ge-rich $Si_{1-x-y}Ge_xC_y$ indicated an increase in ΔE_V by ~50 meV/%C[19]. Moreover, recent results from admittance spectroscopy on $Si_{1-x-y}Ge_xC_y$/Si multi-quantum wells suggested a large effect by C (~ 80 meV/ %C) on both the conduction and valence band offset of $Si_{1-x-y}Ge_xC_y$ /Si[16]. Given conflicting reports, it is therefore necessary to have a direct optical measurement on the $Si_{1-x-y}Ge_xC_y$ /Si heterostructures. In this study, we report such a measurement of the valence band offset by heterojunction internal photoemission (HIP) of $Si_{1-x-y}Ge_xC_y$/ Si (100) from the onset of photocurrent.

EXPERIMENT

The samples in this study contain 39% Ge and up to 2.5% substitutional carbon and were grown by RTCVD[7]. They contain a p^+ Si buffer for substrate contact, followed by 0.2 μm p^- Si, 2 nm undoped $Si_{1-x-y}Ge_xC_y$ spacer and 18 nm p^+ $Si_{1-x-y}Ge_xC_y$. Finally, a 20 nm heavily doped (~ 10^{20}/cm^3) Si layer was grown for a top contact. Substitutional carbon fractions were measured by X-ray diffraction, assuming 8.3 Ge/C strain compensation ratio. For the rest of this paper, all carbon levels refer to the substitutional levels measured by this method.

Device were fabricated by a simple mesa etching in CF_4/O_2 plasma and Al metallization by lift-off. A device structure is shown in figure 1. Good rectifying characteristics were observed at low temperatures (~ 77K), indicating a significant valence band offset between p^+ $Si_{1-x-y}Ge_xC_y$ / p^- Si. Samples were further cooled down to ~ 4 K to minimize thermionic leakage current for infrared photocurrent measurements and a good ohmic contact was still observed. Optical absorption measurements were performed at 4K using a calibrated glowbar IR source, a spectrometer and phase-sensitive detection.

Figure 2 shows the band diagram of the p^+ $Si_{1-x-y}Ge_xC_y$ / p^- Si HIP structure. Under a reverse bias, holes current is mostly blocked by the valence band offset and the ideal leakage current comes from thermionically emitted holes from p^+ $Si_{1-x-y}Ge_xC_y$ layer. When infrared light is incident on the p^+ $Si_{1-x-y}Ge_xC_y$ layer, holes will be excited to higher energy states, and if the photon energy is large enough for hole to overcome the barrier posed by the valence band offset, a photocurrent will result. From the band diagram, ΔE_V can be expressed as

$$\Delta E_V = E_{F(SiGeC)} + qV_{bi} + E_{F(Si)} \qquad (1)$$

where $E_{F(SiGeC)}$ is the distance between Fermi level and the valence band of $Si_{1-x-y}Ge_xC_y$, qV_{bi} is the built-in voltage of the junction, $E_{F(Si)}$ is the distance of the valence band of Si and the Fermi level. Since the $Si_{1-x-y}Ge_xC_y$ is heavily doped, the threshold energy for the onset of photocurrent is E_V - $E_{F(SiGeC)}$. To extract ΔE_V, one also needs to know the doping concentrations in $Si_{1-x-y}Ge_xC_y$ to know $E_{F(SiGeC)}$. Doping concentrations were obtained by SIMS measurement on similarly grown samples and SIMS data show no dependence of dopants (boron) incorporation on the carbon level. We thus assume the onset of photocurrent tracks accurately with ΔE_V.

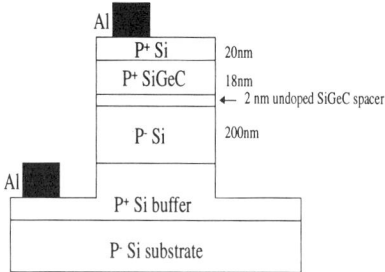

Figure 1: A heterojunction internal photoemission device structure.

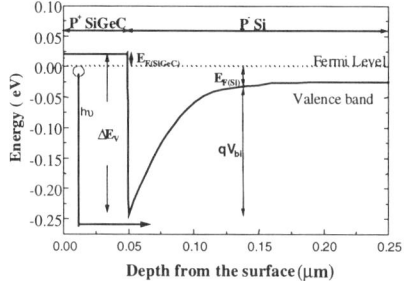

Figure 2: Zero-biased valence band diagram of p^+ $Si_{1-x-y}Ge_xC_y$ /p^- Si.

RESULTS

Figure 3 shows plots of the square root of photoresponse curves as a function of photon energy (Fowler plot) of $Si_{1-x-y}Ge_xC_y$/Si with different carbon concentrations. The onset of photocurrent decreases as carbon level increases, indicating a decreasing ΔE_V with carbon concentrations. Carbon decreases the ΔE_V of $Si_{1-x-y}Ge_xC_y$ / Si by 26±1 meV/%C, as shown in figure 4. This is consistent with previously reported values measured by C-V measurements[7,14], and similar to the increase in bandgap with carbon. We conclude that the increase in bandgap is reflected in the valence band of $Si_{1-x-y}Ge_xC_y$, with very little or no change in the conduction band. Thus little ΔE_c in $Si_{1-x-y}Ge_xC_y$ /Si (100) is expected as in $Si_{1-x}Ge_x$/Si.

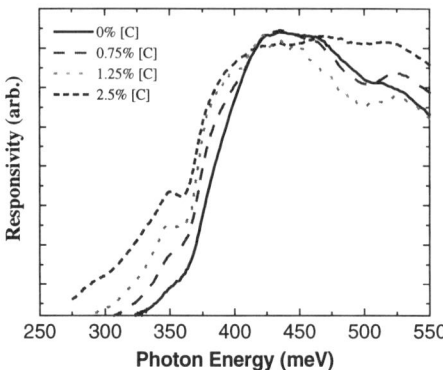

Figure 3: Photoresponse curves of p^+ $Si_{1-x-y}Ge_xC_y$ /p^- Si as well as p^+ $Si_{1-x}Ge_x$/ p^- Si. Samples were measured at 4K.

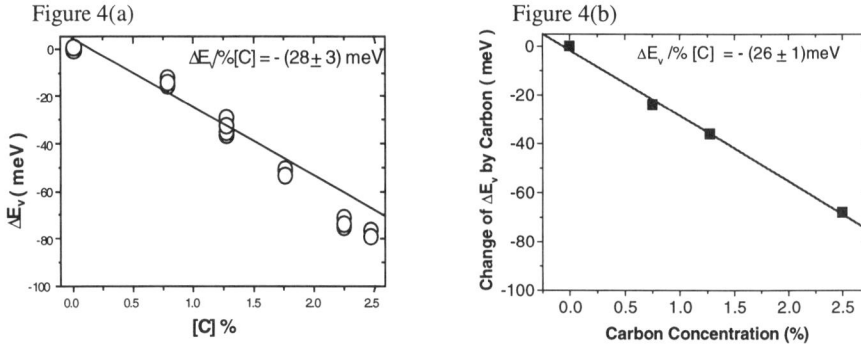

Figure 4: A comparison of change of ΔE_v of $Si_{1-x-y}Ge_xC_y$ /Si as a function of C concentrations. Figure 4(a) is measured by C-V method[7] and figure 4(b) is the results for HIP measurement, as extracted from figure 2.

Figure 5 shows the valence band offset of $Si_{1-x}Ge_x$/Si[18] and $Si_{1-x-y}Ge_xC_y$/Si as a function of lattice mismatch and equivalent Ge levels for $Si_{1-x}Ge_x$ of the given strain. Adding carbon to $Si_{1-x}Ge_x$ to form $Si_{1-x-y}Ge_xC_y$ on silicon (100) reduces both ΔE_v and the compressive strain. But compared to strain reduction by reducing the Ge fraction alone, the ΔE_v reduction by adding carbon is small. For example, the valence band offset of $Si_{0.585}Ge_{0.39}C_{0.025}$ / Si is ~100 meV larger than that of an equally strained $Si_{0.82}Ge_{0.18}$/Si heterostructure. Figure 5 also predicts that, by extrapolating the dashed line to the vertical axis, a strain-free $Si_{.563}Ge_{.39}C_{.047}$/Si heterostructure will have ~ 200 meV valence band offset.

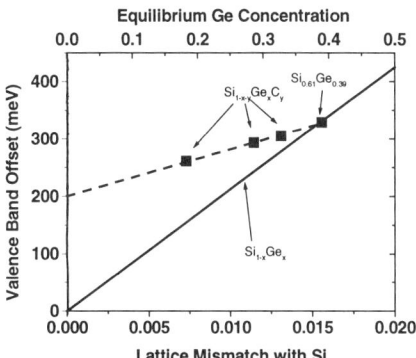

Figure 5: Summary of valence band offsets obtained from optical absorption measurement as a function of lattice mismatch with Si and equivalent Ge concentration. The solid line represents the valence band offsets of $Si_{1-x}Ge_x$/Si.

CONCLUSIONS

In summary, we have studied the valence band offset of compressively strained pseudomorphic $Si_{1-x-y}Ge_xC_y$ /Si (100) by heterojunction internal photomission. Carbon decreased the valence band offset of $Si_{1-x-y}Ge_xC_y$/Si by 26 ± 1 meV/% carbon. Combining this number with previously reported similar increases in the bandgap caused by carbon, we conclude that the band structure of $Si_{1-x-y}Ge_xC_y$ /Si exhibits a large valence band offset and a negligible conduction band offset, similar to that of $Si_{1-x}Ge_x$/Si heterostructures.

ACKNOLEDGEMENTS

The authors like to acknowledge the support by ONR, USAF Rome Lab, and Sandia National Laboratory.

REFERENCES

1. K. Eberl, S.S. Iyer, S. Zollner, J.C. Tsang, and F.K. LeGous, *Appl. Phys. Lett* **60**, 3033 (1992).
2. J.L. Regolini, F. Gisbert, G. Dolino, and P. Boucaud, *Mat. Lett* **18**, 57 (1993).
3. H.J. Osten, H. Rücker, M. Methfessel, E. Bugiel, S. Ruminov, and G. Lippert, *J. Cryst. Growth* **157**, 405 (1995).
4. P. Boucaud, C. Guedj, F. H. Julien, E. Finkman, S. Bodnar, and J.L. Regolini, *Thin Solid Films* **278**, 114 (1996).
5. J. Kolodzey, P. R. Berger, B.A. Orner, D. Hits, F. Chen, A. Khan, X. Shao, M.M. Waite, S. Ismat Shah, C.P. Swann, and K.M. Unruh, *J. Cryst. Growth* **157**, 386 (1995).

6. C.W. Liu, A. St. Amour, J.C. Sturm, Y.R.J. Lacroix, M.L.W. Thewalt, C.W. Magee, and D. Eaglesham, *J. Appl. Phys.* **80**, 3043 (1996).

7. C.L. Chang, A. St. Amour, and J.C. Sturm, *Appl. Phys. Lett* **70**, 1557 (1997).

8. K. Brunner K. Eberl, and W. Winter*, Phys. Rev. Lett.* **76**, 303, (1996).

9. A. St.Amour, C.W. Liu, J.C. Sturm, Y. Lacroix, and M.L.W. Thewalt, *Appl. Phys. Lett* **67**, 3915 (1995).

10. L.D. Lanzerotti, A. St.Amour, C.W. Liu, and J.C. Sturm, *Elec. Dev. Lett*, **17**, 334 (1996).

11. P.Boucaud, C. Francis, F. Julien, J. Lourtioz, D. Bouchier, S. Bodnar, B. Lambert, and J. Regolini, *Appl. Phys. Lett* **64**, 875(1994).

12. K. Brunner, W. Winter, and K. Eberl, *Appl. Phys. Lett* **69**, 1279 (1996).

13. C.L. Chang, A. St. Amour, L. Lanzerotti, and J.C. Sturm, *Mat. Res. Soc. Sym. Proc.* **402**, 437 (1995)

14. K. Rim, S. Takagi, J.J. Welser, J.L. Hoyt, and J.F. Gibbons, Mat. Res. Soc. Sym. Proc., **379**, 327, (1995).

15. Eberl, K. Brunner, and W. Winter, *European Mat. Res. Soc. Spr Mtg*, Strasbourg, France (1996)

16 B.L. Stein, E.T. Yu, E.T. Croke, A.T. Hunter, T. Laursen, A.E. Bair, J.W. Mayer, C.C. Ahn, *Appl. Phys. Lett.*, **70**, 3413 (1997).

17 C.L.Chang, S.P. Shukla, W. Pan, V. Venkataraman, J.C. Sturm, and M. Shayegan, 7th International MBE Symposium, Banff, Canada, (1997). Also to appear in *Thin Sloid Films*.

18 M.Kim and H.J. Osten, *APL*, **70**, 2702 (1997).

19. J. Kolodzey et al, *European Mat. Res. Soc. Spr Mtg*, Strasbourg, France (1996)

PHOTOLUMINESCENCE IN STRAIN COMPENSATED SI/SIGEC MULTIPLE QUANTUM WELLS

R. Hartmann*, U. Gennser*, D. Grützmacher*, H. Sigg*, E. Müller*, K. Ensslin* **
*Paul Scherrer Institut, CH-5232 Villigen-PSI, Switzerland
**Eidgenössische Technische Hochschule, CH-8093 Zürich, Switzerland

ABSTRACT

The effect of strain compensation on the band gap and band alignment of Si/SiGeC MQWs is studied by photoluminescence (PL) spectroscopy. Evidence for type-I band alignment of strain reduced SiGeC MQWs is found. Values for the conduction and valence band offsets are given. A band gap reduction for exactly strain compensated SiGeC compared to compressive SiGeC is observed. This behavior is interpreted in terms of strain induced splitting and confinement shifts of the quantum well states. A good agreement between the model and the PL data is obtained.

INTRODUCTION

Ternary $Si_{1-x-y}Ge_xC_y$ is an interesting material system both in terms of growth and physics issues. The incorporation of Ge atoms and C atoms at a ratio of 8.2:1 allows the realization of Si based heterostructures completely lattice matched to the Si substrate [1]. Growth restrictions such as the formation of misfit dislocations and thermal instability for layer thicknesses above certain critical values, which are typically faced in binary SiGe and SiC heteroepitaxy, are relieved. A band gap lowering of exactly strain compensated SiGeC compared to pure Si has been proposed by considering strain induced and intrinsic band gap shifts [2,3] and verified by PL measurements [4]. PL spectroscopy also indicates a type-I band alignment for Si/SiGeC MQWs with small compressive strain of $\varepsilon = 0.002$ [5]. However, the magnitude of the carrier confinement in the conduction and valence band has not been investigated. This work focuses on the gap energies and band alignments of nearly strain free Si/SiGeC MQWs. PL spectroscopy is used as the experimental tool to extract values for the band discontinuities. Using these band offsets the experimentally observed band gap reduction for exactly strain compensated SiGeC compared to slightly compressive SiGeC is described by a model, which takes the confinement shifts of the populated bands and the strain induced splitting of the quantum well states into account. For SiGeC layers with a fixed Ge concentration of 6% and C-contents up to 0.8% good agreement between the PL energies and the simulation results is found.

EXPERIMENTAL DETAILS

The Si/SiGeC MQWs were grown by solid source MBE using e-beam evaporators for Si and Ge, and a pyrolithic graphite filament for C sublimation. The growth was performed on undoped Si(100) substrates at a temperature of 500°C. The deposition rate was chosen 1Å/s for the Si layers and reduced to 0.3Å/s for the SiGeC layers in order to achieve a significant C incorporation [5]. During growth a bias of +600V was applied to the substrate that helps to reduce bombardment defects in the samples due to e-beam evaporation [6,7]. Each sample consists of 6 periods with period lengths between 135Å and 360Å. Ge concentrations between 0% and 10% and C concentrations of up to 1% allow the adjustment of different strain conditions

Mat. Res. Soc. Symp. Proc. Vol. 533 © 1998 Materials Research Society

in the SiGeC layers. The structural properties are determined by high resolution X-ray diffraction assuming Vegard's law and linearly interpolated elastic constants. After the growth the samples were treated by 10 min anneals at 800°C in a $N_2/6\%H_2$ gas mixture. The PL spectra were recorded at a probe temperature of 2.2K using a flow cryostat. The luminescence was excited by the 488nm line of an Ar^+ laser (excitation power: 3mW), dispersed by a grating monochromator and detected by a nitrogen cooled Ge photoconductor.

RESULTS AND DISCUSSION

Fig. 1 shows low temperature PL spectra taken on several 180Å Si / 45Å $Si_{1-x-y}Ge_xC_y$ MQWs of different compositions. The lower spectra in the figure correspond to MQWs with a Ge concentration fixed at 6% and C concentrations varying between 0% and 1%. The upper measurements are taken on samples where the Ge concentrations are increased up to 10%, and the C concentration is kept constant at 0.4%. Besides the Si related free exciton NP and TO peaks at 1.15 and 1.09 eV (indicated by × in Fig. 1), the main PL signals originate from the $SiGe_{0.06}C_y$ and $SiGe_xC_{0.004}$ layers, respectively. The SiGeC signals are found to shift in energy when C is added to $SiGe_{0.06}C_y$ and Ge to $SiGe_xC_{0.004}$. In both cases the initial blueshift of the PL signal is followed by a shift to lower energies.

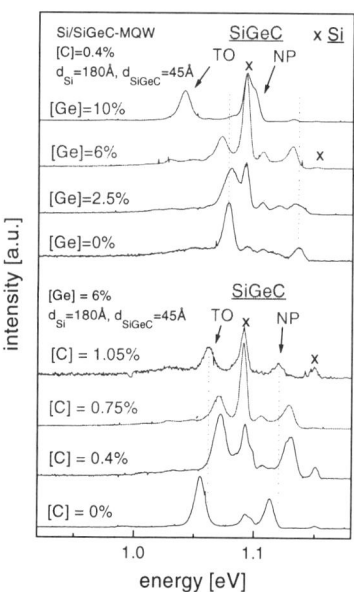

Fig. 1: PL spectra for 180Å Si / 45Å $SiGe_{0.06}C_y$ MQWs (lower part) and 180Å Si / 45Å $SiGe_xC_{0.004}$ MQWs (upper part) with varying C, respectively Ge concentrations. The dotted lines are the guides to the eye to follow the shifts of the SiGeC signals.

The addition of C to pseudomorphic SiGe reduces the compressive strain and therefore the strain induced energy splitting of the hole and electron states. A similar reduction of the band splitting occurs when Ge is mixed to pseudomorphic SiC, and decreases the tensile strain. Considering strain induced band splitting alone, and neglecting intrinsic effects, strain compensation is met with a band gap increase, thus shifting the PL signal to higher energies. In Fig. 2 the SiGeC NP luminescence energy is plotted versus lattice mismatch for both $SiGe_{0.06}C_y$ and $SiGe_xC_{0.004}$ MQWs. It is seen that the luminescence peak with highest transition energy does not occur for the samples with zero mismatch as would be expected from the strain considerations only. In agreement to previous PL measurements [4] a redshift of the PL signals occurs before full strain compensation is achieved, meaning that a band gap reduction occurs for strainless SiGeC. For $SiGe_{0.06}C_y$ the band gap is expanded under the conditions of compressive strain. For the samples with 0.4% C the measurements indicate that the transition of maximum energy occurs under slightly tensile strain. The band gap reduction for strainless SiGeC differs from the behavior observed for binary SiGe and SiC, where the fundamental band gap always shrinks by the presence of strain [8,9].

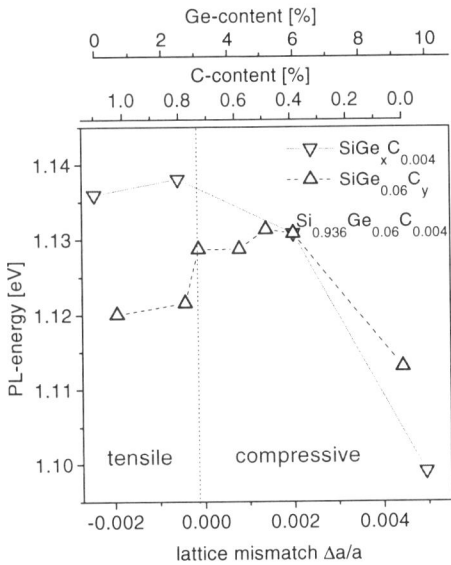

Fig. 2: SiGeC NP PL energy versus lattice mismatch. The PL signals with highest transition energies occur for tensile $SiGe_xC_{0.004}$ and compressive $SiGe_{0.06}C_y$. The upper axes give the corresponding Ge content x for $SiGe_xC_{0.004}$ and C content y for $SiGe_{0.06}C_y$ in percent.

Exact strain compensation in the $Si/SiGe_{0.06}C_y$ MQW is achieved for $\approx 0.7\%$ C substitutionally incorporated, i.e. for a C content higher than in the $SiGe_xC_{0.004}$ samples. The indicated PL blueshift for strainless $SiGe_xC_{0.004}$ compared to strainless $SiGe_{0.06}C_y$ agrees with PL measurements on strain compensated $Si_{1-x-y}Ge_xC_y$ which exhibit a linear decrease of the intrinsic SiGeC band gap with $x \cong y \times 8.2$ [4].

The overcompensation of the strain by adding larger amounts of C, respectively Ge into the layers leads to a further reduction of the band gap compared to exactly strain compensated SiGeC. This is attributed to the uniaxial strain induced band splitting and the intrinsic band gap lowering of Ge-rich, respectively C-rich SiGeC [3].

Because of the larger amount of data available on the $SiGe_{0.06}C_y$ samples, we restrict ourselves in the following to the samples with 6% Ge. The band gap reduction for zero mismatch SiGeC previously reported for $SiGe_{0.037}C_y$ [4] and $SiGe_{0.16}C_y$ [10] MQWs has been interpreted qualitatively by an energy crossing of the $\Delta(2)$ and $\Delta(4)$ electron states [4]. Such a crossing can occur only if there is a conduction band offset available that can account for the different confinement shifts of the electron states. PL measurements on MQWs with constant compositions but different layer widths allow us to extract the band alignments. For C concentrations of 0.4% and 0.6% the widths of the SiGeC and Si layers have been varied, while keeping either the Si or the SiGeC layer widths constant [5]. In Fig. 3 the NP energy of the SiGe$_{0.06}$C$_{0.004}$ PL is plotted versus d$_{SiGeC}$ and d$_{Si}$. Keeping the SiGeC thickness constant at 45Å and changing the Si layer width does not affect the energetic position of the NP line. Instead, the luminescence signal shifts monotonically up in energy when the SiGeC width is reduced from 180Å to 27Å. These results show that the PL line originates from quantum confined subband levels with a type I band alignment, where the ternary layer is acting as the quantum well. No shifts were observed with excitation intensity indicating a fairly strong confinement in contrast to Si/SiGe [11].

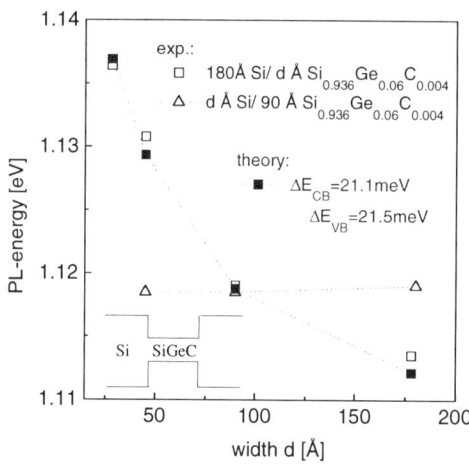

Fig. 3: Dependence of the SiGe$_{0.06}$C$_{0.004}$ NP energy as a function of d$_{SiGeC}$, respectively d$_{Si}$. The best fit is obtained for type-I band alignment with SiGeC acting as the quantum well (insert) and band offsets ΔE_{CB}=21.1meV and ΔE_{VB}= 21.5meV.

A comparison of the experimental data to model calculations allows us to deduce values for the band offsets. For the data in Fig. 3 we use the effective $\Delta(4)$ electron mass of Si bulk material $m_{\Delta(4)}$=0.19m_0 and the effective SiGe$_{0.06}$ heavy hole mass m_{hh}=0.27 for the SiGeC quantum well taking the band edge splitting due to strain into account. For the barrier the effective mass of the transverse $\Delta(2)$ Si electron state $m_{\Delta(2)}$=0.92m_0 was used. The use of the $\Delta(2)$ mass seems reasonable, since one can expect significant interface scattering of the carriers, as well as alloy scattering due to the large local strain in the vicinity of the C atoms. Furthermore, fitting the data assuming a conservation of the $\Delta(4)$ nature of the electrons in the barrier did not give physically reasonable results. Assuming type I band character, the Kronig-Penney model yields the best fit for the SiGe$_{0.06}$C$_{0.004}$ PL data for ΔE_{CB}=21.1meV and ΔE_{VB}=21.5meV. For the samples with 0.6% C the behavior of the PL energy is described best by assuming a type I band alignment with band offsets ΔE_{CB}=33.2meV in the conduction band and ΔE_{VB}=9.0meV in the valence band.

The interaction of the light hole states with spin orbit split holes makes the valence band splitting non-linear, and requires the extrapolation of the band offsets within 0%<[C]<1% to be done for unstrained materials. This has been done for the calculation of the band discontinuities of pseudomorphic Si/SiGe$_{0.06}$C$_y$ MQWs which are plotted in Fig. 4 [12]. The incorporation of C into pseudomorphic SiGe reduces the compressive strain and thus the energy splittings of the conduction and valence bands. At 0.7% C exact strain compensation is achieved, and the electron and hole states regain degeneracy. The tensile strain for higher C contents reverses the uniaxial band splittings, i.e. the band edges are formed by the $\Delta(2)$ and lh states. Strong confinement simultaneously for electrons and holes is not achieved for the C and Ge concentrations studied

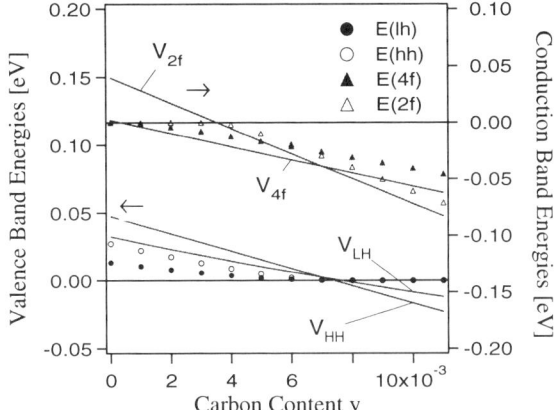

Fig. 4: Variation of the band offsets of strained Si/SiGe$_{0.06}$C$_y$ with C concentration. The confinement shifts are calculated for 45Å wide quantum wells. Because of the mass dependence of the $\Delta(2)$ and $\Delta(4)$ confinement energies the degeneracy of the electron states shifts to the compressive region.

here. The maximum energy separation between the band edges occurs for exact strain compensation as expected in terms of strain splitting. This contradicts to the PL measurements of Fig. 2 revealing the highest energy gap for compressive SiGe$_{0.06}$C$_{0.004}$. However, PL spectroscopy probes the energy gap between the band edges being shifted by quantum confinement. The conduction band offset of ≈40meV for strain compensated SiGe$_{0.06}$C$_{0.007}$ and the different masses for $\Delta(2)$ (0.92m$_0$) and $\Delta(4)$ (0.19m$_0$) cause confinement shifts which are different in energy for the $\Delta(2)$ and $\Delta(4)$ electron states, thus lifting the degeneracy for zero lattice mismatch. Instead the degeneracy of the electron states is found for compressive SiGeC with ≈0.65% C. A similar crossing of the valence bands can not occur for compressive SiGeC due to the strain splitting and the masses of the hh and lh states.

The SiGeC PL energy as a function of [C] can be calculated from the energy separation between the $\Delta(4)$ and hh, respectively $\Delta(2)$ and hh states. This is done in Fig. 5 including the Si band gap and an exciton binding energy of 15meV. The simulation matches the experimental data for C concentrations up to 0.8%. For higher C concentration the PL energy is not described

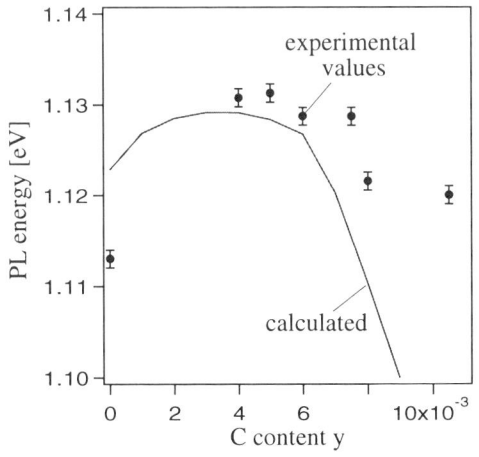

Fig. 5: Simulation of the experimental PL energies of Si/SiGe$_{0.06}$C$_y$ as a function of [C]. The model results agree to the PL data within the range of 0% and 0.8% C content. The occurence of the energy maximum under compressive strain is well reflected.

by the model which might be due to a non-linear dependency of the unstrained band offsets for this C content. The maximum energy for SiGeC under compressive strain is well reflected. The negligible conduction band offset for low C concentrations makes the energy of the electron states rather insensitive to the C content below $\approx 0.3\%$. The hh energy decreases with [C], thus increasing the band gap. At higher C concentrations the energy of the electron states decreases faster than of the hole states. This leads to a shrinkage of the fundamental band gap. The position of the PL signal as a function of [C] results from a combination of band crossing and changes in the band alignment of the MQWs.

CONCLUSION

PL measurements on strain reduced $Si/SiGe_{0.06}C_y$ MQWs with y=0.004 and y=0.006 give evidence for a type I band alignment with SiGeC forming the quantum well. The confinement shift of the PL signals for varying SiGeC layer widths can be well described by assuming rather big conduction band offsets of ΔE_{CB}=21.1meV and ΔE_{CB}=33.2meV, respectively. Strong confinements for both electrons and holes appear possible for SiGeC MQWs with similar strain conditions but higher C concentrations. Band offsets for C concentrations in the range of 0% and 0.8% are calculated by linearly extrapolating the unstrained band discontinuities. These values allow a quantitative description of the observed band gap reduction of strainless SiGeC layers relative to compressive SiGeC. The confinement shifts which in size depend on the type of band in the conduction and valence band that is populated and the influence of the strain on the band splitting are included into the model. The influence of the strain on the band alignment of nearly strain free Si/SiGeC MQWs will be discussed in more detail elsewhere [12]. The validity of our model in the tensile strain region including an energy crossing of the hole states remains to be verified.

REFERENCES

1. K. Eberl, S. S. Iyer, S. Zollner, J. C. Tsang, and F. K. LeGoues, Appl. Phys. Lett. **60**, 3033 (1992)
2. K. Brunner, W. Winter, K. Eberl, N. Y. Jin-Philipp, and F. Philipp, J. Cryst. Growth **175/176**, 451 (1997)
3. A. St. Amour, C. W. Liu, J. C. Sturm, Y. Lacroix, and M. L. W. Thewalt, Appl. Phys. Lett. **67**, 3915 (1995)
4. O. G. Schmidt and K. Eberl, submitted to Phys. Rev. Lett.
5. R. Hartmann, D. Grützmacher, E. Müller, U. Gennser, A. Dommann, P. Schröter, and P. Warren, accepted for publication in Thin Solid Films
6. R. Hartmann, D. Grützmacher, E. Müller, U. Gennser, and A. Dommann, Thin Solid Films **294**, 50 (1997)
7. D. Grützmacher, R. Hartmann, P. Schnappauf, U. Gennser, E. Müller, D. Bächle, and A. Dommann, accepted for publication in Thin Solid Films
8. D. Dutartre, G. Brémond, A. Souifi, and T. Benyattou, Phys. Rev. B **44**, 11525 (1991)
9. K. Brunner, K. Eberl, and W. Winter, Phys. Rev. Lett. **76**, 303 (1996)
10. K. Eberl, K. Brunner, and W. Winter, Thin Solid Films **294**, 98 (1997)
11. M. L. W. Thewalt, D. A. Harrison, C. F. Reinhart, J. A. Wolk, and H. Lafontaine, Phys. Rev. Lett. **79**, 269 (1997)
12. R. Hartmann, U. Gennser, H. Sigg, D. Grützmacher, and K. Ensslin, to be published

STRONG DEVIATION OF THE LATTICE PARAMETER IN $Si_{1-x-y}Ge_xC_y$ EPILAYERS FROM VEGARD'S RULE

J. STANGL, S. ZERLAUTH, F. SCHÄFFLER, G. BAUER
Inst. for Semiconductor Physics, Kepler-University Linz, Altenbersgerstr. 69, A-4040 Linz, Austria
M. BERTI, D. DE SALVADOR, A.V. DRIGO, F. ROMANATO
INFM at the Physics Dept. University of Padova, Via Marzolo 8, I-35131 Padova, Italy

ABSTRACT

From the comparison of precise determinations of the Ge and C contents of a series of $Si_{1-x-y}Ge_xC_y$ epilayer samples ($x < 0.18$, $y < 0.02$) by Rutherford and resonant backscattering experiments and x-ray diffraction, the variation of the $Si_{1-x-y}Ge_xC_y$ lattice spacing as a function of C content is determined. A significant negative deviation from Vegard's rule is observed, in agreement with theoretical predictions by Kelires.

INTRODUCTION

Si–based heterostructures have gained technological importance during the past years, and devices based on $Si_{1-x}Ge_x$ alloy layers are already available. The band alignment in such heterostructures depends on the composition as well as on the strain state of the individual layers. To achieve more freedom in device design, the effect of carbon as an additional alloy material has been extensively investigated. In contrast to pseudomorphic $Si/Si_{1-x}Ge_x$ structures in which the band offset occurs mainly in the valence band, $Si/Si_{1-y}C_y$ structures have a type–I band alignment with the offset mainly in the conduction band [1, 2]. As the lattice mismatch between Si and C (diamond) or β–SiC (cubic silicon carbide) is very large (about 34% and 20%, respectively), even substitutional C concentrations of a few percent lead to a significant tensile strain in pseudomorphic $Si_{1-y}C_y$ layers on Si [3, 4]. This effect can be exploited to engineer the strain state of $Si_{1-x-y}Ge_xC_y$ epilayers and thus to tailor also the band offsets. Choosing a ratio of $x/y \approx 11$, lattice matched layers can be grown pseudomorphically on Si substrates with virtually no theoretical limitation of thickness.

The Ge and C contents of $Si_{1-x}Ge_x$, $Si_{1-y}C_y$ and $Si_{1-x-y}Ge_xC_y$ epilayers are usually determined from x-ray diffraction (XRD). XRD, however, measures the lattice spacings rather than the C and Ge contents of the epilayers. The determination of the latter is only indirect and relies on assumptions on the variation of the lattice parameters and of the elastic constants with x and y. For most compounds the simplest assumption of a linear variation of these parameters with concentration (Vegard's rule) is used. For the $Si_{1-x}Ge_x$ system small deviations of a linear behavior have been detected [5]. In the $Si_{1-y}C_y$ system, only a small region of C content ($y < 0.02$ typically) is accessible experimentally as the solubility of C in Si is very small. The lattice parameter of the stoichiometric compound β–SiC, however, shows a strong deviation from Vegard's rule. Although β–SiC is not a disordered solid solution, this might already serve as an indication that deviations from Vegard's rule have to be expected in $Si_{1-y}C_y$ and also in $Si_{1-x-y}Ge_xC_y$ compounds. Recent theoretical

257

Table 1: Structural parameters of the investigated $Si_{1-x-y}Ge_xC_y$ epilayers as obtained from rBS: (total) Ge and C contents and fraction f of substitutionally incorporated C.

Sample No.	Z453	Z454	Z455	Z459	Z460	Z462
x_{Ge}	9.5	8.4	10.5	17.9	16.0	15.9
y_C	0.56	0.93	0.36	1.28	1.00	1.81
f_{subst}	1.0	1.0	1.0	0.95	0.94	0.79

predictions by P.C. Kelires indeed propose a strong deviation of the lattice parameter of $Si_{1-y}C_y$ and $Si_{1-x-y}Ge_xC_y$ from Vegard's rule [6]. The theoretically predicted values are close to a quadratic interpolation between those of Si, β-SiC and C, at least for the region of small C content.

For the absolute determination of Ge contents, Rutherford backscattering (RBS) is generally used. Due to the low scattering cross section σ_C of carbon, RBS cannot be applied for the absolute measurement of C contents in $Si_{1-x-y}Ge_xC_y$ layers. Only the use of resonances in σ_C (resonant backscattering; rBS) allows the direct measurement of C concentrations. In this letter the carbon content of pseudomorphic $Si_{1-x-y}Ge_xC_y$ epilayers ($x < 0.18$, $y < 0.02$) is determined by rBS. In contrast to former attempts at the 4.265MeV resonance [7, 8], in our investigations we employed the $^{12}C(\alpha,\alpha)^{12}C$ resonance at a He^+ beam energy of about 5.72MeV [9, 10], which allows much more reliable C concentration profiles to be extracted. Furthermore, the amount of carbon incorporated at substitutional sites in the Si matrix can be determined much less ambiguously using the 5.72MeV resonance. The comparison of these experiments with the lattice constants of the epilayers measured by XRD rocking scans revealed a negative deviation from Vegard's rule which is in good agreement with the theoretical predictions.

EXPERIMENT

The $Si_{1-x-y}Ge_xC_y$ samples were grown in a RIBER SIVA 45 MBE chamber. The substrate preparation is described in detail in Ref. [11]. A Si buffer layer of approx. 100 nm was grown while the substrate temperature was decreased from 600°C to the growth temperature of the $Si_{1-x-y}Ge_xC_y$ epilayers of 415°C. The $Si_{1-x-y}Ge_xC_y$ layers with a nominal thickness of 100nm were deposited at a growth rate of approximately 0.1 nm/s. To be able to separate the rBS signal from the surface carbon contamination and the C signal from the $Si_{1-x-y}Ge_xC_y$ epilayers, a Si cap layer of the order of 100 nm was grown at 470°C. The sample parameters obtained from RBS and rBS are listed in Table 1.

The RBS and rBS experiments have been performed at the Laboratori Nazionali di Legnaro using $^4He^+$ beams delivered by the 7 MV CN Van der Graaff accelerator. The Ge contents were determined by conventional RBS at 2 MeV He^+ beam energy (see Fig. 1a). For the C content determination a beam energy of 5.72 MeV has been used, as σ_C has a resonance around this energy, leading to a scattering yield enhanced by a factor of 120–130 with respect to the Rutherford value [9, 10]. The width of this resonance is much larger ($\pm 5\%$ variation within 170 keV) than for the one at 4.265 MeV ($\pm 5\%$ variation within 12 keV), so that a nearly constant σ_C is obtained over Si thicknesses of several hundreds of nm. The advantages of the use of the 5.72 MeV resonance as well as details on the experimental setup are discussed in Refs. [10, 11]. In this energy region, however, the Si signal is no longer Rutherford–like, and for the extraction of the C signal the measurement of a pure Si

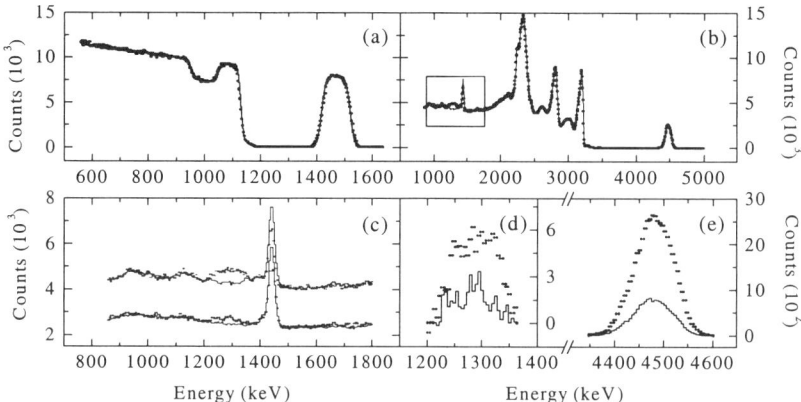

Figure 1: (a) RBS spectrum of sample Z-SGC 459, taken at a He$^+$ beam energy of 2 MeV in random orientation (dots), together with a computer simulation (thin line). (b) rBS spectrum of sample Z-SGC 460 (dots) and a virgin Si sample (thin line) taken at 5.72 MeV beam energy in random orientation. The signal of the C–containing layer is marked. (c) enlargement of the marked region of the rBS spectra of sample Z-SGC 460 and the Si reference sample (taken in random orientation; upper curves) and according spectra with the beam aligned in the (110) azimuth (lower curves). All the spectra are taken with the sample tilted 60 deg from the surface normal. From the intensity ratios of this curves (minimum yield) the substitutional C fraction f has been determined. The difference between SiGeC and Si reference spectra in random and channeling sample alignement is also shown: expanded view of the C (panel d) and Ge (panel e) signals.

reference sample in exactly the same scattering conditions as for the Si$_{1-x-y}$Ge$_x$C$_y$ sample, and a subtraction of both spectra is required (Fig. 1b,c; see also Refs. [10, 11] for details of analysis).

In order to determine the substitutional C fraction f, rBS spectra were taken both in random sample orientation as well as with the beam aligned in the sample's (110) azimuth (channelling configuration; see Figure 1c). The superpositions of the signals from C and Ge in random and aligned orientation are shown in Figure 1d, e, respectively. The ratio between random and aligned yields gives the minimum yields χ_C and χ_{Ge} for C and Ge, respectively. The substitutional fraction of C can then simply be calculated by $f = \frac{1-\chi_C}{1-\chi_{Ge}}$, the obtained values are listed in Tab.1.

XRD ω–2θ–scans around the (004) Bragg reflections were recorded to measure the lattice constants of the epilayers *in growth direction* (Fig.2; dots represent the measurement, thin solid lines are simulations based on dynamical scattering theory). Reciprocal space mappings around the (004) and (224) reciprocal lattice points were recorded to ensure that all samples have been grown fully pseudomorphic in the sense that the (average) in–plane lattice constant

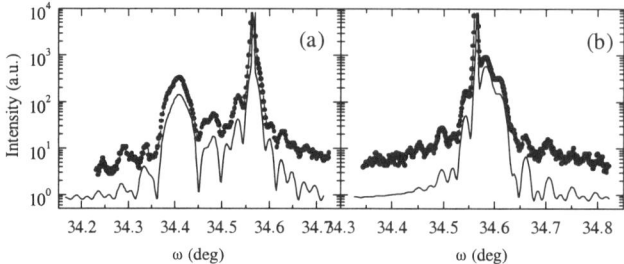

Figure 2: XRD spectra around the (004) Bragg reflection of samples Z–SGC 455 and 462 (dots), together with calculations (thin lines; shifted in intensity).

of the epilayer exactly matches the substrate's lattice parameter.

For tetragonally distorted $Si_{1-x-y}Ge_xC_y$ alloys the lattice constant in growth direction, $a_\perp$, for given Ge and C contents x and y is given by

$$a_\perp = \left[a_{rel}\left(x, y\right) - a_{\parallel} \right] \cdot \left(1 + 2\frac{C_{12}\left(x, y\right)}{C_{11}\left(x, y\right)}\right) + a_{\parallel}, \tag{1}$$

with $a_{\parallel} = a_{Si} = 5.431\text{Å}$. $a_{rel}\left(x, y\right)$ is the bulk lattice parameter of the compound and C_{11} and C_{12} are its elastic constants. For the latter we always use a linear interpolation between the values of Si, Ge and C. For the calculation of $a_{rel}\left(x, y\right)$ with the Ge and C concentrations we employ the following method: We "split" the alloy in a SiGe–"part" and a SiC–"part". For the SiGe–part we calculate $a_{rel}\left(x\right)$ according to Dismukes' model [5]. Using a linear interpolation between Si and Ge instead would result in an overestimation of the C content by approx. 6% relative at a Ge content of 10%. For the SiC–part we use three different models, namely a linear interpolation between Si and C (diamond), a linear interpolation between Si and β–SiC, and the model presented by P.C. Kelires [6]. Finally the lattice parameters of the SiGe–part and the SiC–part are linearly interpolated to obtain $a_{rel}\left(x, y\right)$. XRD measures only the *average strain* in the layers, and hence for a single $Si_{1-x-y}Ge_xC_y$ layer is sensitive mainly to the ratio of Ge and C contents. Thus the Ge concentration in the calculations was fixed at the value obtained from RBS, and the C content was determined by the comparison of measured (004)–ω–2θ–scans with simulations based on dynamical scattering theory.

RESULTS

Figure 3 shows the C content obtained from XRD with the three models for $a_{rel}\left(x, y\right)$ described above, as a function of the *substitutional* C content obtained from rBS experiments. It is obvious from Figure 3 that the rBS and XRD measurements are in good agreement if for $a_{rel}\left(x, y\right)$ the theoretical predictions of Kelires are used. The C contents obtained from XRD using linear interpolation models for the relaxed lattice parameters are *significantly higher*. The deviations from Vegard's rule between Si and C or Si and β–SiC amount to 25%

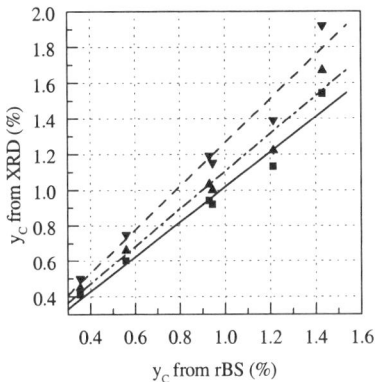

Figure 3: substitutional C content obtained from rBS measurements versus the C content of $Si_{1-x-y}Ge_xC_y$ epilayers obtained from XRD measurements and three different assumptions on the variation of lattice parameter with C content : i) down triangles/dashed line: Vegard's rule between Si and C (diamond). ii) up triangles/dot–dashed line: Vegard's rule between Si and β–SiC. iii) squares/solid line: Kelires' model (the lines are linear regressions to the respective symbols).

and 8%, respectively.

An important point in the comparison of rBS and XRD measurements is the fraction of C that is introduced in the Si lattice at *substitutional* sites. Mainly the substitutional C fraction is responsible for the lattice strain obtained by XRD, whereas rBS in a random beam orientation measures the *total* (substitutional and interstitial) C concentration in the epilayers. From previous experiments on pure $Si_{1-y}C_y$ layers a good agreement between the C contents obtained by XRD using Kelires' model and the substitutional fraction of C determined by rBS was found [11]. If we use the total instead of the substitutional C content as obtained from rBS in Figure 3, no good correspondence between XRD and rBS is found for any of the models. This further confirms that only the substitutionally incorporated carbon seems to have significant impact on the lattice strain, which has already been assumed [15].

Beside the only small principal uncertainties of the experimental methods (see Refs. [10, 11] for details), the main reason for the scattering of the data points is the lateral inhomogeneity of our samples, due to the fact that the substrates have not been rotated during epilayer growth, and that the subsequent XRD and rBS measurements might not have been performed at exactly the same position on the samples.

A physical explanation of the deviation from Vegard's rule has been given by Martins and Zunger. A charge transfer from Ge to Si or from Si to C, respectively, due to different electronegativities of the elements, leads to an additional attractive force between the atoms and hence to a smaller lattice parameter [14]. Whereas in the SiGe system this is only a minor effect, it is much more pronounced in the SiC system, giving rise to the comparatively large deviations from linearity as predicted by Kelires.

SUMMARY

We have studied the dependence of the lattice parameter in $Si_{1-x-y}Ge_xC_y$ epilayers on the C content by a comparison of XRD and rBS experiments. The absolute C concentration and the substitutional fraction of C were measured by rBS using the 5.72MeV $^{12}C(\alpha, \alpha)^{12}C$ resonance. The investigations revealed a strong deviation of the lattice parameter from Vegard's rule, in good agreement with recent theoretical predictions. The C contents obtained from XRD measurements taking into account this deviation are lower by about 25% and 8%, respectively, than those obtained previously under the assumption of a linear variation of the lattice constants between either Si and diamond or Si and β–SiC.

This project was partially supported by INFM PRO-FESR-RIM, FWF and GMe.

REFERENCES

[1] K. Brunner, K. Eberl, W. Winter, Phys.Rev.Lett. **76**, 303 (1996).

[2] D.C. Houghton, G.C. Aers, N.L. Rowell, K. Brunner, W. Winter, K. Eberl, Phys.Rev.Lett. **78**, 2441 (1997).

[3] H.J. Osten, M. Kim, K. Pressel, P. Zaumseil, J. Appl. Phys. **80**, 6711 (1996).

[4] E.T. Croke, A.T. Hunter, C.C. Ahn, T. Laursen, D. Chandrasekhar, A.E. Bair, D.J. Smith, J.W. Mayer, J. Cryst. Growth **175–176**, 486 (1997).

[5] J.P. Dismukes, L. Ekstrom, R.J. Paff, J. Phys. Chem. **68**, 3021 (1964).

[6] P.C. Kelires, Phys. Rev. **B75**, 8785 (1997), Phys. Rev. Lett. **75**, 1114, (1995).

[7] J.A. Leavitt, L.C. McIntyre Jr., P.Stoss, J.G. Oder, M.D. Ashbaugh, B. Dezfouly-Arjomandy, Z.M. Yang, Z. Lin, Nucl. Instr. Meth. in Phys. Res. **B40/41**, 776 (1989).

[8] D. Endisch, H.J. Osten, P. Zaumseil, M. Zinke–Allmang, Nucl. Inst. Meth. in Phys. Res. **B100**, 125 (1995).

[9] Y. Feng, Z. Zhou, Y. Zhou, G. Zhao, Nucl. Instr. Meth. in Phys. Res. **B86**, 255 (1995).

[10] M. Berti et al., subm. to Nucl. Inst. Met. in Phys. Res.

[11] M. Berti, D. De Salvador, A.V. Drigo, F. Romanato, J. Stangl, S. Zerlauth, F. Schäffler, G. Bauer, Appl. Phys. Lett. **72**, 1 (1998)

[12] S. Zerlauth, C. Penn, H. Seyringer, F. Schäffler, Appl. Phys. Lett. **71**, 3826 (1997)

[13] Handbook of Modern Ion Beam Materials Analysis, J.R. Tesmer and M.Nastasi editors. Materials Research Society, Pittsburg (USA), 1995

[14] J.Martins , A.Zunger, Phys.Rev.Lett. **56**, 1400 (1986).

[15] G.G. Fischer, P. Zaumseil, J. Phys. D: Appl. Phys. **28**, A109 (1995).

EPITAXIAL GROWTH AND ELECTRONIC CHARACTERIZATION OF CARBON-CONTAINING SILICON-BASED HETEROSTRUCTURES

J.L. Hoyt, T.O. Mitchell, K. Rim, D.V. Singh, and J.F. Gibbons, Solid State Electronics Laboratory, Stanford University, Stanford, CA, 94305.

ABSTRACT

Epitaxial $Si_{1-x-y}Ge_xC_y$ and $Si_{1-y}C_y$ layers grown on Si are opening up new possibilities for bandstructure engineering of electronic devices. Thin $Si_{1-y}C_y$ layers containing a few atomic percent substitutional carbon, grown on Si substrates, experience biaxial tensile strain, which produces a conduction band energy splitting that is expected to be favorable for in-plane electron transport. For other applications, C may be useful as a means of compensating the compressive strain of Ge in ternary $Si_{1-x-y}Ge_xC_y$ alloys. Although the understanding of the electronic properties of these materials is still at an early stage, interesting trends are emerging.

A key issue for synthesis of these alloys is the low equilibrium solubility of carbon in silicon. However, a number of non-equilibrium methods have been employed to grow these materials. This work focuses on the properties of $Si_{1-y}C_y$ and $Si_{1-x-y}Ge_xC_y$ grown by chemical vapor deposition. There is a strong influence of the growth conditions on the fraction of the total carbon concentration which is substitutional on the silicon lattice. Using low temperatures (e.g. 550°C) and very high silane partial pressures for $Si_{1-y}C_y$ growth, good agreement is obtained between the carbon contents determined by x-ray diffraction and secondary ion mass spectrometry, for carbon concentrations up to about 1.8 atomic percent. Metal-oxide-semiconductor capacitors fabricated on $Si/Si_{1-x-y}Ge_xC_y$ and $Si/Si_{1-y}C_y$ epitaxial layers show well-behaved electrical characteristics. Temperature dependent capacitance-voltage analysis is used to extract the band offsets, and indicates that the conduction band energy is lowered as carbon is added to Si. Complementary to the case of strained $Si_{1-x}Ge_x$ grown on Si, for which most of the energy offset is in the valence band, the band offset appears primarily in the conduction band for $Si_{1-y}C_y/Si$ heterojunctions.

INTRODUCTION

The addition of C to Si and $Si_{1-x}Ge_x$ epitaxial layers has attracted considerable attention in electronic materials research recently [1-5]. Carbon introduces tensile strain, which by analogy to strained Si grown pseudomorphically on relaxed $Si_{1-x}Ge_x$, has a conduction band structure that should be favorable for in-plane electron transport. In addition, Demkov et al. predicted that the conduction band energy of Si should be lowered for a few percent C in Si [6]. Since a conduction band offset is lacking for strained $Si_{1-x}Ge_x$ on Si, this opportunity is particularly attractive. Calculations by Berding, et al. indicate that certain ordered phases of SiGeC may exist and may exhibit metallic properties [7]. Finally, C-Ge strain compensation enables growth of $Si_{1-x-y}Ge_xC_y$ layers that are lattice-matched to Si, for those applications that require thick layers and/or high Ge contents. This work focuses on the properties of $Si_{1-y}C_y$ and $Si_{1-x-y}Ge_xC_y$ layers grown by chemical vapor deposition (CVD). In particular, we examine the fraction of C that is substitutional on the lattice, and its dependence on growth conditions. Measurements of some of the basic electronic properties such as band offsets for $Si/Si_{1-y}C_y$ and $Si/Si_{1-x-y}Ge_xC_y$ heterojunctions, which are key to evaluating the usefulness of these materials in devices, are discussed.

Mat. Res. Soc. Symp. Proc. Vol. 533 © 1998 Materials Research Society

CVD GROWTH OF C-CONTAINING Si-BASED HETEROSTRUCTURES

Epitaxial layers in this work were grown on 4" Si (100) substrates by rapid thermal chemical vapor deposition in a load-locked, lamp-heated reactor which has been described previously [8]. The growth pressure is 12 Torr, consisting of 8 slpm of palladium-purified hydrogen, and source gases of germane (GeH_4) for Ge, either silane (SiH_4) or dichlorosilane (SiH_2Cl_2) for Si, and either ethylene (C_2H_4) or methylsilane ($SiCH_6$) as the C precursor. The $Si_{1-x-y}Ge_xC_y$ layers were grown at temperatures in the range of 550 to 600°C, and the $Si_{1-y}C_y$ growth temperature was varied from 550 to 700°C. Growth conditions for the alloy layers are discussed in more detail in References [9-11]. A typical sample structure, illustrated in the inset of Fig. 1, consists of a 20 to 40 nm-thick alloy layer, with a 30 to 50 nm-thick epitaxial Si cap. The Si cap is grown at 600 or 700°C, and serves to separate the alloy layer from the near-surface region which is subject to contamination and transient effects during secondary ion mass spectrometry (SIMS). The total C concentrations were measured by SIMS depth profiling using Cs^+ bombardment. In order to accurately quantify the carbon concentration, SIMS standards with peak C concentration of about 0.7 at. % were created by implanting 60 KeV C^+ to a dose of 5×10^{15} cm^{-2} into Si wafers and thick epitaxial $Si_{1-x}Ge_x$ samples with appropriate Ge concentrations. For the $Si_{1-x-y}Ge_xC_y$ samples, the layer thicknesses and Ge contents were measured by grazing exit-angle Rutherford backscattering (RBS) using 2.2 MeV He^+.

X-ray Diffraction Analysis

The Philips materials research diffractometer was used in the high resolution, five-crystal mode to obtain (004) x-ray rocking curves (XRD) on all samples. These measurements yield the out-of-plane lattice spacing. On some samples, (224) rocking curves were also acquired, to obtain information about the in-plane lattice spacing. This analysis indicates that in the thickness range studied, the alloy layers are fully strained to within the accuracy of the measurement.

The Philips High Resolution Simulation program [12] was used to fit numerical simulations, based upon dynamical x-ray diffraction theory, to the measured rocking curves. In order to extract a C concentration from the measured rocking curves, assumptions must be made about the influence of C and Ge on the elastic constants (e.g. Poisson's ratio) and equilibrium lattice parameters of the alloys. In this work, linear interpolation was used in the program to obtain Poisson's ratio for the alloys, based upon values of 0.278 for Si, 0.271 for Ge, and 0.285 for β-SiC, which were calculated from the elastic moduli [13]. We have discussed in detail in Ref. [11] the impact of various assumptions concerning the equilibrium lattice parameters on the C concentration extracted from this type of analysis. A number of authors assume Vegard's law (linear interpolation) for the $Si_{1-x}Ge_x$ lattice parameter. This can have a significant impact on the relative error in the extracted carbon fraction, particularly for small amounts of carbon, where the carbon-induced mismatch relief is on the order of the deviation from Vegard's law for $Si_{1-x}Ge_x$. In the simulations of the $Si_{1-x-y}Ge_xC_y$ samples, the Ge fraction x was fixed according to the value determined by RBS. The negative deviation from Vegard's law for $Si_{1-x}Ge_x$, measured by Dismukes, [14] was verified in this work and used in the calculations for the Ge contribution to the equilibrium lattice parameter for the alloys. For the C contribution to the mismatch, some authors use linear interpolation between Si and C (diamond), while others interpolate to β-SiC. In this work, we have employed linear interpolation between Si and β-SiC, which yields a value of -0.395% mismatch per atomic % C, which is roughly 15% higher than the interpolation to diamond. An example of the impact of these assumptions on the extracted C concentration in $Si_{1-x-y}Ge_xC_y$ is given in Ref. [11]. For a sample with 20% Ge, Dismukes' lattice parameter for $Si_{0.8}Ge_{0.2}$ combined with the β-SiC interpolation produces a C fraction of 1 atomic %, whereas assuming linear $Si_{1-x}Ge_x$ lattice parameters and interpolating to diamond yields a C concentration of 1.38 atomic % [11]. Theoretical calculations suggest that the interpolation to β-SiC is more accurate for $Si_{1-y}C_y$ alloys [15]. In addition, recent measurements indicate that our assumptions contribute at most a relative error of 12% to the extracted C fraction for $Si_{1-y}C_y$ [16].

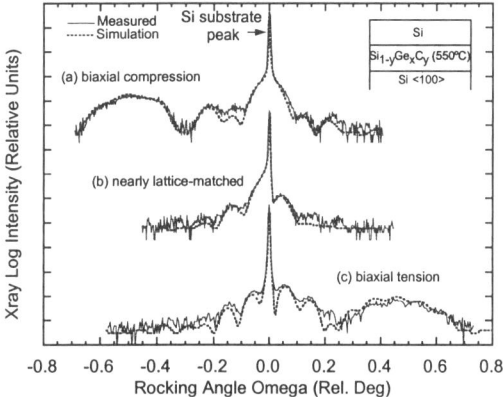

Fig. 1 High resolution measured (solid lines) and simulated (dashed lines) x-ray (004) rocking-curves for (a) 27 nm $Si_{0.8}Ge_{0.2}$, (b) 24 nm $Si_{0.783}Ge_{0.20}C_{0.017}$, and (c) 22 nm $Si_{0.982}C_{0.018}$. Epitaxial layers were grown at 550°C by RTCVD. The inset shows the sample structure.

The ability to vary the strain in the epitaxial layer from biaxial compression to tension is illustrated by the x-ray diffraction data shown in Fig. 1. The plot shows the rocking curves for three heterostructure samples with alloy layer thicknesses on the order of 24 nm: (a) $Si/Si_{0.8}Ge_{0.2}$ (b) $Si/Si_{0.783}Ge_{0.20}C_{0.017}$, and (c) $Si/Si_{0.982}C_{0.018}$. For these three samples, the measured in-plane strain varies from 0.76% for 20% Ge in $Si_{1-x}Ge_x$, to -0.71% for the $Si_{0.982}C_{0.018}$ layer. The $Si_{0.783}Ge_{0.20}C_{0.017}$ sample, shown in (b), is nearly lattice-matched, with a residual in-plane strain of only 0.1%. For these samples, the alloy layers were grown at 550°C using silane, germane, and methylsilane as the precursors. The $Si_{1-x-y}Ge_xC_y$ layers were grown with germane and silane partial pressures of 0.12 and 0.3 Torr respectively, and the $Si_{0.982}C_{0.018}$ was grown with a silane partial pressure of 1.2 Torr, the significance of which is discussed below. The Si capping layers were grown at 700°C.

Also shown in Fig. 1 (dashed lines) are the simulations of the rocking curves. The agreement between the simulations and measured data is in general excellent for C concentrations up to about 1.8 at %. Above that concentration, the measured data does not generally reproduce all of the interference fringes, indicating the onset of some degradation in the crystal quality.

Dependence of Substitutional Carbon Incorporation on Growth Conditions

For device applications, the goal is to produce material in which all of the C is incorporated on substitutional sites, since the substitutional C concentration will likely determine electronic properties including band offsets, and non-substitutional C may influence the presence of deep levels. For example, degradation of the electrical characteristics of $Si/Si_{1-x-y}Ge_xC_y$ capacitors has been reported for samples which were believed to contain significant fractions of non-substitutional C [17]. Using a dichlorosilane silicon source, and typical CVD growth conditions for $Si_{1-x}Ge_x$ at 600°C, we find good agreement between the XRD- and SIMS-measured C concentrations in $Si_{1-x-y}Ge_xC_y$ for concentrations up to about 1.3 at %, as illustrated in Fig. 2. Above 1.3 at. %, there appears to be less substitutional C measured by XRD than total C determined by SIMS. This is true for either methylsilane or ethylene source gases. When the growth temperature is lowered to 550°C, and silane is substituted for dichlorosilane (open squares), there is some improvement in C substitutionality. This is consistent with the observation that the substitutional incorporation of C in $Si_{1-y}C_y$ layers grown at 700°C is improved when silane was used in place of dichlorosilane [9].

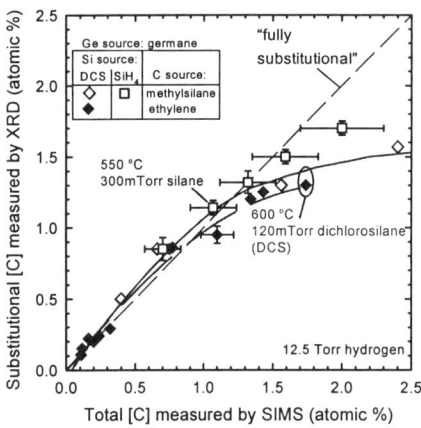

Fig. 2   Comparison of XRD- and SIMS-measured C fractions for $Si_{1-x-y}Ge_xC_y$ alloys grown at 600°C using a dichlorosilane ("DCS") silicon source (diamonds), and at 550°C using a silane silicon source (squares). The samples represented by the solid symbols have Ge contents varying from 10 up to 36%. All other samples have Ge fractions of 20 atomic %.

The dependence of the substitutional C fraction on the growth conditions for $Si_{1-y}C_y$ layers grown by CVD is discussed in Ref. [10]. The substitutional C concentration increases for higher silane partial pressures and lower growth temperatures. This is illustrated in Fig. 3. At 550°C, it is possible to obtain essentially fully substitutional C incorporation for concentrations up to about 1.8 at. %. Above that concentration, the material quality as judged by x-ray diffraction is degraded. In Fig. 3, focusing on total C concentrations above 1 atomic percent, we see that carbon substitutionality decreases as the growth temperature increases from 550 to 600°C. This is significantly higher than the temperature range of some earlier reports for the onset of non-substitutional C incorporation during MBE growth [18], but in the same temperature range reported more recently by Zerlauth, et al. for MBE growth of $Si_{1-y}C_y$ [19]. The latter group also report a correlation of non-substitutional C incorporation and three-dimensional growth [19]. However, cross-section transmission electron microscopy (XTEM) analysis of CVD-grown samples indicates that samples with non-substitutional C do not necessarily display islanding or evidence of three-dimensional growth. Fig. 4 shows an XTEM micrograph of a sample with 1.26 at. % C by SIMS and 1 at. % C by XRD. There is no evidence of extended defects, and the interfaces are quite smooth. It seems likely that the onset of non-substitutional C incorporation is associated with the formation of point defects and small clusters of C which are not detectable by TEM analysis.

Thermal Stability

Fig. 5 shows the XRD data and simulations for a strained $Si_{1-y}C_y$ sample grown at 550°C with 1.3 at % C, subject to a series of 20 minute anneals. Annealing at temperatures from 600 to 800°C reveals no measurable impact on the XRD data. After annealing for 20 minutes at 860°C, however, the extracted substitutional C concentration is reduced from 1.3 to 1.0 at. %. Comparison of the data to the simulation shows a degradation in the material quality after annealing at 860°C. Analysis of the (224) rocking curve indicates that the strain reduction is not associated with a decrease in the in-plane lattice constant (e.g. due to misfit dislocation formation). Instead, it appears that C is moving off of substitutional lattice sites.

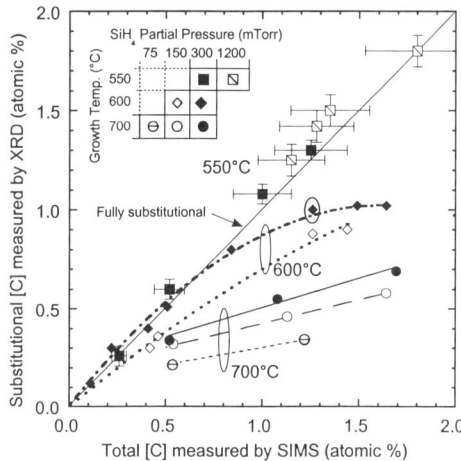

Fig. 3 Comparison of XRD- and SIMS-measured C fractions for $Si_{1-y}C_y$ alloys, as a function of growth conditions. The diamond symbol which is circled corresponds to the sample shown in Fig. 4. From Mitchell, *et al.* [10].

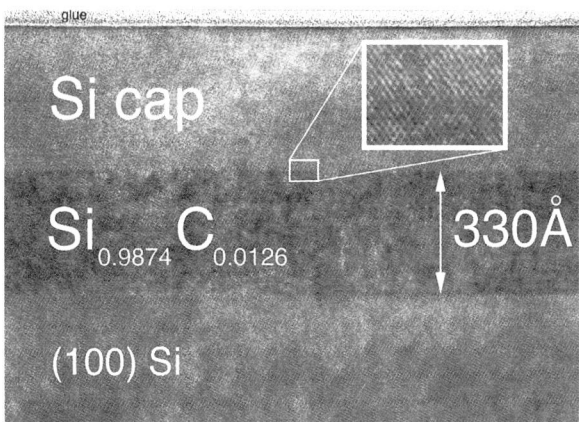

Fig. 4 XTEM micrograph of a $Si_{1-y}C_y$ layer with 1.26 at. % C measured by SIMS and 1 at. % C by XRD. Photo courtesy A. Marshall and T. Mitchell.

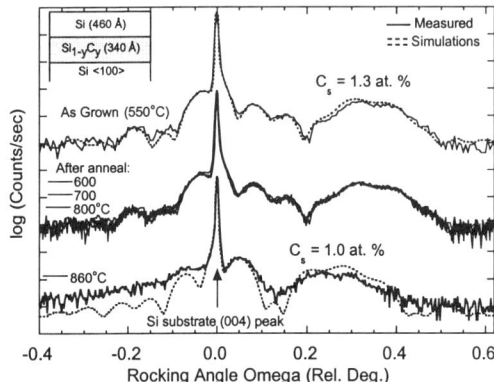

Fig. 5 High resolution (004) XRD rocking curves for a $Si/Si_{0.987}C_{0.013}$ sample annealed at various temperatures for 20 minutes. No measurable change is noted in the XRD data for annealing temperatures up to 800°C. The sample structure is shown in the inset. From Mitchell [20].

ELECTRONIC DEVICE APPLICATION EXAMPLES

Most devices reported to date utilizing these new materials have been fabricated for the purpose of extracting electronic properties such as bandgaps and band offsets. Lanzerotti, *et al.* used *npn* $Si/Si_{1-x-y}Ge_xC_y/Si$ heterojunction bipolar transistors (HBTs) to extract the bandgap difference between Si and $Si_{1-x-y}Ge_xC_y$, and found an increase in the $Si_{1-x-y}Ge_xC_y$ bandgap with the addition of C [21-22]. The measured bandgap widening of 26 meV/at. % C is less than the magnitude of the effect expected from the C strain compensation, i.e. for an equivalent mismatch to Si, the bandgap of $Si_{1-x-y}Ge_xC_y$ is smaller than that of $Si_{1-x}Ge_x$, indicating that there is some intrinsic impact of C on the bandgap. These authors also reported a reduction in the post-implant boron transient enhanced diffusion (TED), which indicates that C interacts with Si interstitials. Osten, *et al.*, used small amounts of C (e.g. 0.1 at. %) in the base of the transistor, and demonstrated complete suppression of boron TED for *npn* HBTs, with no significant degradation in the carrier lifetime [23]. Although this result is not directed towards bandgap engineering, from a technological and manufacturing point of view, the suppression of B transient enhanced diffusion during device processing is quite significant.

For Field-effect Transistors (FETs), enhanced thermal stability during processing has been demonstrated in *p*-type $Si/Si_{1-x-y}Ge_xC_y$ MOSFETs [24]. There is evidence, however, that either Coulomb or alloy scattering may be issues for FET applications, and these topics require further investigation. For example, the expected tensile-strain-induced electron mobility enhancement was not measured in a recent experiment on surface-channel $Si_{1-y}C_y$ MOSFETs [25].

The band alignment is important for determining the usefulness of these materials in device applications. In strained $Si_{1-x}Ge_x$ grown on Si, virtually all of the band offset is in the valence band, which limits the application to certain types of devices. Hence, we have focused on extracting the conduction band offset for $Si/Si_{1-x-y}Ge_xC_y$ and $Si/Si_{1-y}C_y$ heterojunctions. The remainder of this section discusses these band offset measurements.

Band Offsets

The extraction of band offsets from MOS capacitors with buried heterostructure layers is discussed in Refs. [17],[26-27]. In the MOS C-V technique, carrier confinement in a buried potential well which is separated from the Si/silicon dioxide interface produces a capacitance plateau region in the C-V curve. The magnitude of the conduction (valence) band offset can be extracted by fitting the simulated high frequency C-V curves to measurements of n-type (p-type) capacitors at various temperatures. Fig. 6 shows an example of C-V data for an n-type MOS capacitor with a buried $Si_{1-y}C_y$ layer. As the gate bias is swept from negative to positive, electrons are first accumulated in the buried potential well formed by the conduction band discontinuity at the $Si/Si_{1-y}C_y$ interface. This results in the formation of a plateau-like region in the C-V curve, and the shape of this feature depends upon the band offset. The solid lines in Fig. 6 are 1D Poisson solutions from a device simulator [28]. The best fit yields an extracted conduction band offset of 80 meV for 1.2 at. % C. Calculations with band offsets of 80 ± 30 meV are illustrated to indicate the sensitivity of the technique. Note that this technique has been successfully applied to measure the band offsets between strained Si and relaxed $Si_{1-x}Ge_x$ [29].

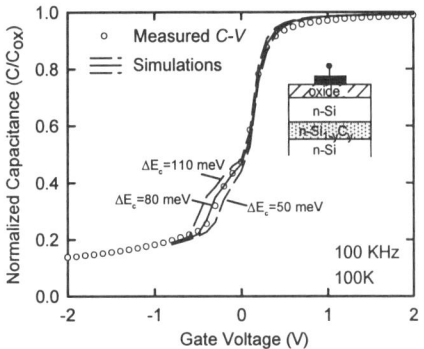

Fig. 6 Comparison of measured (symbols) and calculated (lines) MOS high-frequency C-V characteristics for an n-type $Si/Si_{1-y}C_y$ capacitor. The C concentration is 1.2 at. %. The gate oxide was grown by wet oxidation at 750°C for 40 min. The best fit to the extracted band offset is 80 ± 30 meV. In the simulations, the doping profile from SIMS measurements is assumed. From Singh, *et al.*, [30].

Table 1 summarizes the results of MOS C-V analysis on both n- and p-type $Si/Si_{1-x-y}Ge_xC_y$ and $Si/Si_{1-y}C_y$ capacitors. From the capacitor experiments, Rim *et al.* have shown that most of the band offset is in the conduction band for $Si/Si_{1-y}C_y$ heterojunctions [26]. For $Si/Si_{1-x-y}Ge_xC_y$ heterojunctions, most of the band offset appears in the valence band, and an electron potential well is not observed for n-type $Si/Si_{1-x-y}Ge_xC_y$ capacitors. The physical reason for a smaller conduction band offset in $Si/Si_{1-x-y}Ge_xC_y$ compared to $Si/Si_{1-y}C_y$ heterojunctions is discussed in the next section.

	Si/SiGeC	$Si/Si_{1-y}C_y$
n-type (ΔE_C)	ΔE_C < 30 meV	plateau observed
p-type (ΔE_V)	plateau observed	$|\Delta E_V|$ < 30 meV

Table 1 Summary of observations from MOS C-V analysis of capacitors with C contents up to approximately 1.5 at. %. From the results, we see that the band offset is primarily in the valence band for $Si/Si_{1-x-y}Ge_xC_y$ and in the conduction band for $Si/Si_{1-y}C_y$ heterojunctions.

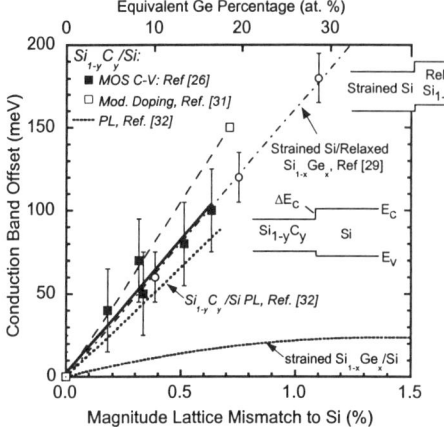

Fig. 7 Summary of conduction band offsets extracted from MOS C-V measurements of Si/Si$_{1-y}$C$_y$ heterojunctions (solid squares). The solid line is a least-squares fit to the MOS C-V data and has a slope of ~ 65 meV/% C. The dotted line is from PL measurements on MBE-grown samples [32]. For comparison, the conduction band offsets measured between strained Si and relaxed Si$_{1-x}$Ge$_x$ (open circles) [29] are also shown.

Fig. 7 summarizes the measured conduction band offsets from MOS C-V analysis of Si/Si$_{1-y}$C$_y$ capacitors (solid squares) as a function of lattice mismatch (or in-plane strain) to Si. For comparison, the theoretical conduction band offset for strained Si$_{1-x}$Ge$_x$/Si is also shown. A least-squares fit to the MOS C-V data (solid line) indicates a conduction band offset of approximately 65 meV/% C (using -0.395% mismatch per at. % C in the XRD analysis). Results for two other reports of the conduction band offset in this system are shown. The open square is an estimate from Faschinger, *et al.*, using an MBE-grown modulation-doped Hall effect sample [31]. The dotted line is from Williams, *et al.*, which is based upon their extraction of the band alignment from stress dependent photoluminescence (PL), which indicates that 70% of the total bandgap difference between Si and strained Si$_{1-y}$C$_y$ is in the conduction band [32]. In deriving this line in the figure, we have used a value of 65 meV per 0.345% mismatch for the total bandgap difference between Si and Si$_{1-y}$C$_y$, which is the value reported by this same group using photoluminescence [33]. From the figure, we see that the MOS C-V-measured band offsets are in good agreement with those extracted from the PL measurements. One other comparison is made in this figure. The open circles show the results for the conduction band offset for strained Si/relaxed Si$_{1-x}$Ge$_x$ heterojunctions, extracted using a similar technique [29]. For a given mismatch to Si, the Si/Si$_{1-y}$C$_y$ conduction band offset appears to be similar in magnitude to that measured for strained Si on relaxed Si$_{1-x}$Ge$_x$. The technological advantage for strained Si$_{1-y}$C$_y$ is that growth of a thick, relaxed Si$_{1-x}$Ge$_x$ buffer layer is not required.

Strain and Carbon Effects on the Conduction Band Offset

We estimate the relative contributions of biaxial tensile strain and the intrinsic contribution of the C itself to the conduction band offset for Si/Si$_{1-y}$C$_y$ heterojunctions. Theoretical considerations indicate that for biaxial tensile strain, the six-fold degenerate Si conduction band edge splits into a lower-energy two-fold degenerate band (Δ_2) and a higher-energy four-fold degenerate band (Δ_4), as show schematically in Fig. 8. The total energy splitting between these two levels is given by $\Delta E = D_u \cdot |e_T|$, where the symbol D_u is used here to denote the uniaxial deformation potential, and e_T denotes the tetragonal distortion of the lattice, and is given by [34]:

$$e_T = e_{zz} - e_{xx} = (1+v)/(1-v) \cdot m = g \cdot m \qquad (1)$$

where m is the mismatch defined by $m = (a_{alloy} - a_{Si})/a_{Si}$, g is the distortion factor, and v is Poisson's ratio. Using linear interpolation on the values of $v=0.278$ for Si and 0.285 for β-SiC, we obtain a distortion factor g of 1.77 for 1 at. % C. Substituting $g = 1.77$ and $m = -0.395$ into Eq. (1) yields $e_T = -0.007$ for 1 at. % C. Using a value of 9.2 eV for D_u [35], we obtain a total strain splitting of the conduction band, ΔE of roughly 65 meV for 1 at. % C, as illustrated in Fig. 8 (b). Because the degeneracy of the Δ_2 valley is half that of the Δ_4 valley, the energy lowering of the Δ_2 level due to the strain splitting is expected to be [34] $2/3 \cdot \Delta E = 43$ meV, as shown in Fig. 8 (b). Taking the measured conduction band offset to be 65 ± 15 meV, and subtracting the strain contribution, we obtain an intrinsic contribution to the conduction band energy due to the carbon to be approximately -22 meV for 1 at. % C. This estimate indicates that the C itself does appear to intrinsically lower the conduction band energy, which is consistent with the trends predicted theoretically [6].

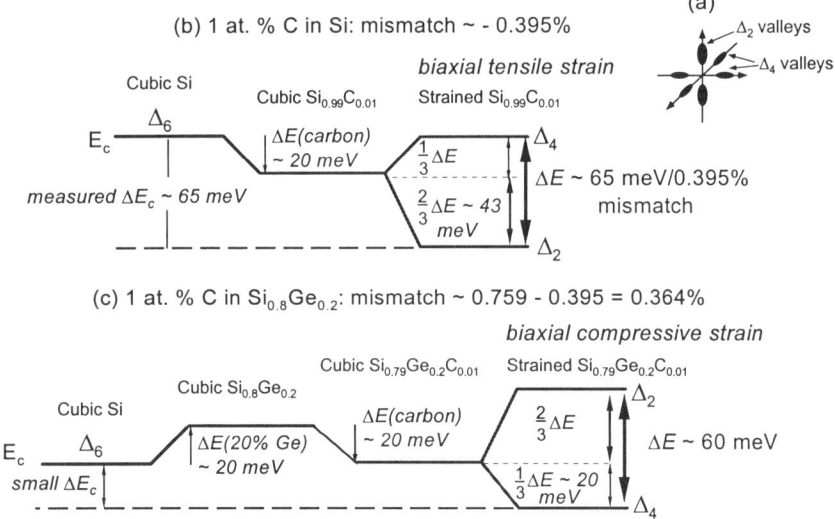

Fig. 8 Schematic illustration of the effects of biaxial strain, C and Ge on the conduction band edge in $Si_{1-y}C_y$ and $Si_{1-x-y}Ge_xC_y$ alloys. The sketch in (a) illustrates the constant energy surface diagram for pseudomorphic $Si_{1-y}C_y$ grown on Si. As discussed in the text, the expected conduction band energy lowering is larger in the case of (b) $Si_{1-y}C_y$/Si compared to (c) $Si_{1-x-y}Ge_xC_y$/Si, due to the combined effects of Ge and compressive strain on the bands in that case.

The addition of C to a $Si_{1-x}Ge_x$ alloy with 20% Ge will result in less downward movement of the conduction band energy for several reasons, as shown schematically in Fig. 8 (c). First, the C is compensating the strain, so that the overall strain splitting ΔE tends to be reduced as C is added to $Si_{1-x}Ge_x$. For 20 % Ge and 1 % C, the mismatch m is $0.759 - 0.395 = 0.364\%$. Using the above formulas, we find the total strain splitting ΔE in this case to be 60 meV. Second, in the compressively strained case, the lowest energy band is now the Δ_4 valley, which moves in response to the deformation at half the rate of the Δ_2 band. Hence, the lowering of the conduction band due to the compressive strain is only $1/3 \cdot \Delta E$, or 20 meV in this case. Combining this with the intrinsic C contribution, we find an estimated band offset of 42 ± 15 meV, assuming that the Ge itself has no effect on the conduction band energy. However, there is evidence that the addition of Ge tends to increase the conduction band energy of Si. The conduction band edge energy of totally relaxed $Si_{1-x}Ge_x$ alloys is higher than that in Si throughout the entire Ge concentration range [36]. From Ref [35], we may estimate the intrinsic effect of Ge to be to raise the conduction band energy by roughly 20 meV for a Ge concentration of 20%. As shown in Fig. 8 (c), this tends to compete with the effect of the C. Hence, for dilute C concentrations, the combined effects of the compressive strain and the Ge itself may reduce the impact of the C on the conduction band in $Si_{1-x-y}Ge_xC_y$ relative to the tensile-strained $Si_{1-y}C_y/Si$ case.

SUMMARY

High quality carbon-containing epitaxial layers have been grown by RTCVD. The fraction of the total C concentration which is substitutional depends upon the details of the growth conditions. For CVD, lower growth temperatures and high silane partial pressures promote substitutional C incorporation, and $Si_{1-y}C_y$ layers with good agreement between the SIMS and XRD-measured C fractions have been grown for C concentrations up to 1.8 atomic %. As C is added to Si to form strained $Si_{1-y}C_y$, the conduction band edge energy is lowered due to the combined effect of the tensile strain splitting, which increases with increasing C in this case, and the intrinsic impact of C on the conduction band. $Si/Si_{1-y}C_y$ and $Si/Si_{1-x-y}Ge_xC_y$ MOS capacitors show well-behaved C-V characteristics for carbon concentrations up to about 1.5 atomic %. Analysis of the C-V characteristics of MOS capacitors indicates that the conduction band energy of $Si_{1-y}C_y$ is lowered at the rate of roughly 65 meV for 1 at. % C. Estimates of the strain contribution show that the intrinsic impact of C on the conduction band energy is roughly -20 ± 15 meV for 1 at. % C. This reduction of the conduction band energy associated with the addition of C to Si is consistent with the predicted theoretical trend. The compressive strain and the intrinsic effect of Ge combine to result in a reduction in the conduction band energy offset for the case of $Si_{1-x-y}Ge_xC_y/Si$ compared to $Si_{1-y}C_y/Si$ heterojunctions. This is consistent with experimental observations for dilute C concentrations. The $Si_{1-y}C_y/Si$ band alignment is complementary to that of strained $Si_{1-x}Ge_x$ on Si and promises new opportunities for silicon-based electronic devices.

REFERENCES

[1] K Eberl, S.S. Iyer, S. Zollner, J.C. Tsang, and F.K. LeGoues, Appl. Phys. Lett. **60**, 3033 (1992).

[2] M. Kim, G. Lippert, and H.J. Osten, J. Appl. Phys. **80**, 5848 (1996).

[3] Z. Atzmon, A.E. Bair, E.J. Jaquez, J.W. Mayer, D. Chandresekhar, D.J. Smith, R.. Hervig, and M.J. Robinson, Appl. Phys. Lett. **65**, 2599 (1994).

[4] J. Mi, P. Warren, P. Letourneau, M. Judelewicz, M. Gailhanou, M. Dutoit, C. Dubois, and J.C. Dupuy, Appl. Phys. Lett. **67**, 259 (1995).

[5] See for example Thin Solid Films, Vol. 294 (1997), and references therein.

[6] A. Demkov and O.F. Sankey, Phys. Rev. B **48**, 2207 (1993).

[7] M.A. Berding, A. Sher, and M. van Schilfgaarde, Phys. Rev. B **56** (7), 3885 (1997).

[8] J.L. Hoyt, C.A. King, D.B. Noble, C.M. Gronet, J.F. Gibbons, M.P. Scott, S.S. Laderman, S.J. Rosner, K. Nauka, J. Turner, and T.I. Kamins, in Thin Solid Films **184**, 93 (1990).

[9] J.L. Hoyt, T.O. Mitchell, K. Rim, D.V. Singh, and J.F. Gibbons, in Chemical Vapor Deposition, Proc. XIV Intl. Conf. and EUROCVD-11, edt. M.A. Allendorf and C. Bernard, (Electrochem. Soc., Pennington, NJ, 1997), pp. 1254-1265.

[10] T.O. Mitchell, J.L. Hoyt, and J.F. Gibbons, Appl. Phys. Lett. **71** (12), 1688 (1997).

[11] J.L. Hoyt, T.O. Mitchell, K. Rim, D.V. Singh, and J.F. Gibbons, to appear in Thin Solid Films, May, 1998.

[12] P.F. Fewster, computer code: High Resolution Simulation Program, (Philips Research Laboratories, Redhill, UK, 1990), see also P.F. Fewster and C.J. Curling, J. Appl. Phys. **62**, 4154 (1987).

[13] Landolt-Bornstein, Numerical Data and Functional Relationships in Science and Technology, (Springer-Verlang, Berlin, 1987), Vol. 22, p. 9-49.

[14] R.W. Olesinski and G.J. Abbaschian, Bull. Alloy Phase Diagrams, **5**, 180 (1984).

[15] P.C. Kelires, Phys. Rev. B **55**, 8785 (1997.)

[16] M. Berti. D. DeSalvador, A.V. Drigo, F. Romanato, J. Stangl, S. Zerlauth, F. Schaffler, G. Bauer, submitted to Appl. Phys. Lett., 1998.

[17] K. Rim, S. Takagi, J.J. Welser, J.L. Hoyt and J.F. Gibbons, in Mat. Res. Soc. Symp. Proc., **342**, E.A. Fitzgerald, J.L. Hoyt, J.C. Bean, and K.Y. Cheng, Editors (Mat. Res. Soc., Pittsburgh, PA, 1994), p. 327.

[18] H.J. Osten, M. Kim, K. Pressel, and P. Zaumseil, J. Appl. Phys. **80** (12), 6711 (1996).

[19] S. Zerlauth, H. Seyringer, C. Penn, and F. Schaffler, Appl. Phys. Lett. **71** (26), 3826 (1997).

[20] T.O. Mitchell, Ph.D. Thesis, Mat. Sci. and Engineering, Stanford Univeristy, in preparation.

[21] L.D. Lanzerotti, A. St. Amour, C.W. Liu, J.C. Sturm, J.K. Watanabe, and N.D. Theodore, IEEE Elec. Dev. Lett. **17** (7), 334 (1996).

[22] L. D. Lanzerotti, J.C. Sturm, E. Stach, R. Hull, T. Buyuklimanli, and C. Magee, in IEEE IEDM Tech. Dig., Dec. 1996, p. 249.

[23] H.J. Osten, G. Lippert, D. Knoll, R. Barth, B. Heinemann, H. Rucker, and P. Schley, in IEEE IEDM Tech. Dig., Dec. 1997, p. 803.

[24] S.K. Ray, S. John, S. Oswal, and S.K. Banergee, in IEEE IEDM Tech. Dig., Dec. 1996, p. 261.

[25] K. Rim, T.O. Mitchell, J.L. Hoyt, G. Fountain, and J.F. Gibbons, in this Volume, (Mat. Res. Soc., Pittsburgh, PA), Spring Meeting, April, 1998.

[26] K. Rim, T.O. Mitchell, D.V. Singh, J.L. Hoyt, J. F. Gibbons, and G. Fountain, Appl. Phys. Lett. **72** (19), 2286, 1998.

[27] S.P. Voinigescu, K. Iniewski, R. Lisak, C.A.T. Salama, J.P. Noel, and D.C. Houghton, Solid-State Electron. **37**, 1491 (1994).

[28] TMA Medici, Version 2.2, Technology Modeling Associates, Sunnyvale, CA 94086.

[29] J.J. Welser, Application of Strained Si/Relaxed SiGe to MOSFETs, Ph.D. Thesis, Electrical Engineering, Stanford Univeristy, Dec. 1994.

[30] D.V. Singh, K. Rim, T.O. Mitchell, J.L. Hoyt, and J.F. Gibbons, in preparation.

[31] W. Faschinger, S. Zerlauth, G. Gauer and L. Palmetshofer, Appl. Phys. Lett. **67**, 933 (1995).

[32] R.L. Williams, G.C. Aers, N.L. Rowell, K. Brunner, W. Winter, and K. Eberl, Appl. Phys. Lett. **72** (11), 1320 (1998).

[33] K. Brunner, K. Eberl, and W. Winter, Phys. Rev. Lett **76**, 303 (1996).

[34] R. People, Phys. Rev. B **32**, 1405 (1985).

[35] C.G. Van de Walle and R.M. Martin, Phys. Rev. B **34** (8), 5621 (1986).

[36] J.F. Morar, P.E. Batson, and J. Tersoff, Phys. Rev. B **47** (7), 4107 (1993).

CARBON INCORPORATION IN Si$_{1-y}$C$_y$ ALLOYS GROWN BY ULTRAHIGH VACUUM CHEMICAL VAPOR DEPOSITION

A. C. MOCUTA, D. W. GREVE
Department of Electrical and Computer Engineering, Carnegie Mellon University, Pittsburgh, PA 15213

ABSTRACT

Thin heteroepitaxial Si$_{1-y}$C$_y$ films have been grown on Si (100) by Ultrahigh Vacuum Chemical Vapor Deposition (UHV/CVD) using silane and methylsilane as silicon and carbon precursors. Carbon incorporation has been studied in the growth temperature range of 550°C to 650°C. The layers have been characterized using high resolution X-ray diffraction and secondary ion mass spectrometry. The total carbon content of the alloys increases linearly with the methylsilane partial pressure and a methylsilane sticking coefficient approximately 2 times higher than that of silane was extracted. Layers with up to 1.34 % substitutional carbon have been obtained at the lowest growth temperature. Fully substitutional carbon can be obtained for levels up to 0.65%. Variations of the growth rate with temperature and carbon content are also discussed.

INTRODUCTION

Incorporation of carbon into silicon and SiGe alloys is of interest as it extends the possibilities for bandgap and strain engineering in the SiGe materials system. Si$_{1-y}$C$_y$ alloys make it possible to have tensile strain and a conduction band offset [1] relative to silicon while the strain compensation properties of carbon allow higher layer thickness and better thermal stability for SiGeC layers.

The main problem in the growth of Si$_{1-y}$C$_y$ alloys is the incorporation of carbon on substitutional lattice sites in amounts of a few percent. Interstitial carbon is known to form electrically active complexes in silicon [2] so alloys with fully substitutional carbon content are of interest. Although the equilibrium solid solubility of carbon in silicon is extremely small (about 10^{-4} at. % at 1420 °C [3]), Si$_{1-y}$C$_y$ alloys with up to 2.1% substitutional carbon have been obtained by nonequilibrium growth methods such as MBE or RTCVD [4-8]. The incorporation of such high carbon fractions is possible because during growth, the surface solubility of carbon plays a more important role rather than the bulk solubility. The solubility of carbon in the few atomic layers at the growth front is increased by orders of magnitude for two reasons. First the presence of a surface partially relieves the strain associated with the atomic size mismatch. Second, there is a stress field near the surface, associated with the atomic reconstruction and as a result the energy of a smaller (carbon) atom is substantially lower on certain sites [9,10]. In this paper we report on the influence of growth temperature on the incorporation and substitutionality of carbon in Si$_{1-y}$C$_y$ alloys grown by UHV/CVD. The effects of carbon concentration on the alloy growth rates are also discussed.

GROWTH AND CHARACTERIZATION

Growths of Si$_{1-y}$C$_y$ alloys have been performed in a UHV/CVD reactor described previously [11], using silane and methylsilane (2% in hydrogen) as source gases for silicon and carbon respectively. CZ silicon (100) substrates were subject to an acid clean prior to loading and to an in-situ thermal clean before growth was initiated, as detailed in [12]. Alloy layers have

Mat. Res. Soc. Symp. Proc. Vol. 533 © 1998 Materials Research Society

been grown at temperatures between 550°C and 650°C, and gas pressure around 1mT. The silane flow was 6 sccm for all layers while the flow of $SiCH_6/H_2$ has been varied from 0 to 6 sccm.

The substitutional carbon concentration has been evaluated using high resolution X-ray diffraction (HRXRD). ω–2θ scans were taken around the (004) reciprocal lattice point using a Philips Materials Research Diffractometer in a four crystal, Ge (220) mode with a 0.8 mm receiving slit. Dynamical simulations of the (004) spectra using the Philips HRS simulation software have been performed to obtain the alloy substitutional carbon content and the layer thickness. The alloy lattice parameter and Poisson ratio have been linearly interpolated between their values for silicon and for diamond. Poisson ratios of 0.278 and 0.104 have been used for Si and C respectively. The (224) rocking curves were used to verify that the layers were pseudomorphic. The accuracy of the substitutional carbon level determination has been estimated at ± 0.01 at. %, by taking into account the uncertainties of the X-ray diffraction measurement and that in determining the peak positions.

Total carbon concentration has been determined by secondary ion mass spectrometry (SIMS) performed by Evans East, using a carbon implanted standard with a peak concentration of 0.4% C. The error in total carbon content is estimated at ± 10%.

RESULTS AND DISCUSSION

Figure 1 shows the HRXRD (004) spectra of samples with up to 1.2% substitutional carbon grown at 600°C. The alloys have been grown using a 6 sccm silane flow and a methylsilane flow as indicated. As the methylsilane flow is increased more carbon is incorporated substitutionally, the alloy lattice constant along the growth direction becomes smaller and the layer peak shifts to higher angles. The crystal quality of the layers is excellent as demonstrated by the presence of well defined thickness fringes and by the excellent fit between the data and the simulated spectra, as shown in Figure 2.

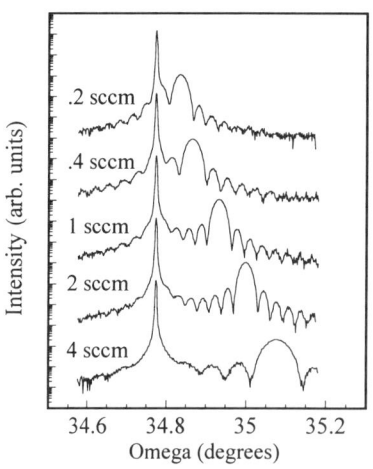

Fig. 1. HRXRD spectra of layers grown at 600 °C using 6 sccm silane flow and methylsilane flow as indicated.

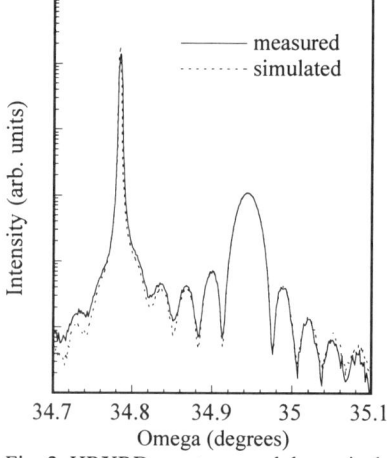

Fig. 2. HRXRD spectrum and dynamical simulation for a sample grown at 600 °C using 6 sccm silane and 2 sccm $SiCH_6$.

Alloys with identical gas flows have been grown at temperatures of 550 °C, 600 °C, 630 °C and 650 °C. Films with up to 3% carbon concentration were 1000-3000 Å thick. They were smooth and defect free when examined by optical or Nomarski microscopy. The data on substitutional and total carbon concentration in these layers obtained by SIMS and HRXRD is summarized in Figure 3. Both total and substitutional carbon fractions increase with the methylsilane flow but while the total carbon level increases linearly with the gas flow at all temperatures, the substitutional carbon level tends to saturate at higher methylsilane flows. A decrease in growth temperature significantly improves carbon substitutionality down to a temperature of 600 °C. For temperatures below 600 °C the level of substitutional carbon for the same gas flows improves only marginally. Layers with up to 0.65% fully substitutional carbon can be obtained at growth temperatures below 600 °C and a maximum of 1.34% substitutional carbon content has been obtained at 550 °C with a 4 sccm methylsilane flow. However, taking into account the high interstitial carbon content of this layer, it is possible that this maximum substitutional carbon level has been underestimated. For methylsilane flows above 4 sccm the layers become of poor crystal quality and exhibit significant surface roughness.

It has been known from MBE experiments [4, 5] that growth at lower temperature and higher growth rate increases the amount of carbon incorporated on substitutional positions, possibly by limiting the formation of Si-interstitial carbon defects at the growth surface [5]. Using MBE at 500 °C and a growth rate of 240 Å/min Iyer et al. [4] have obtained good quality $Si_{1-y}C_y$ alloys with up to 1.5% substitutional carbon while for a growth rate of 12 Å/min and growth temperature of 550 °C films with even 0.1% carbon became islanded and highly twinned. Osten et al. have obtained 1.8 % $Si_{1-y}C_y$ alloys by MBE with 1.6 % substitutional carbon by growing at 400 °C and 48 Å/min growth rate. However, for growth conditions more similar to those in this study - 550 °C growth temperature and 18 Å/min growth rate- the substitutional

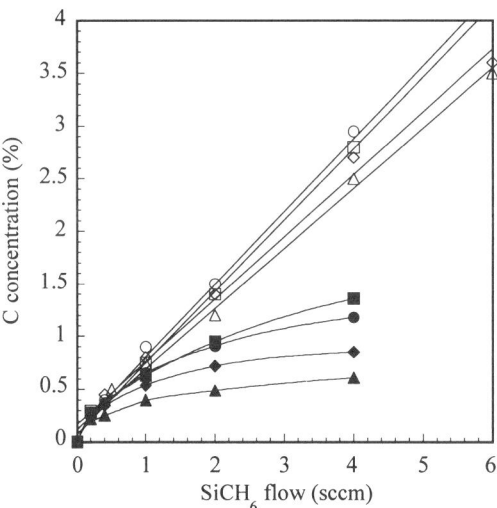

Fig. 3. Substitutional (filled markers) and total (empty markers) carbon concentration in layers grown at 550 °C (squares), 600 °C (circles), 630 °C (diamonds) and 650 °C (triangles).

carbon content was only 0.2% in an alloy with 1.8 % total carbon.

In UHV/CVD lower growth temperatures are available but at the expense of growth rate. The growth rates in the 550-650 °C range are limited by the surface hydrogen coverage. Growth rates of the alloys in this study have been between 3 Å/min at 550°C and 48 Å/min for a 0.3% carbon alloy grown at 650 °C. Using growth conditions similar to those in the MBE studies discussed above, we have been able through CVD growth to obtain good quality alloys and to incorporate significantly more substitutional carbon. We attribute this effect to the hydrogen coverage of the growth surface. The increased surface hydrogen coverage improves both film morphology and the substitutionality of carbon probably by modifying surface kinetics through saturation of free bonds. Our data on carbon incorporation supports this idea: substitutional carbon concentration for the same gas flows increases significantly with lowering of the growth temperature until a substantial hydrogen coverage is reached at 600 °C. Surface hydrogen coverage is around 46% at 600 °C for pure Si growth and it may increase to up to 56% in the presence of carbon. At 550 °C very low growth rates may reduce the effect of further increasing hydrogen coverage on carbon substitutionality.

Incorporation of carbon in silicon by UHV/CVD follows similar stages as in RTCVD [5,7]: for low methylsilane flow all the carbon is incorporated substitutionally, then, for increased flow carbon occupies interstitial positions as well and beyond a limit epitaxy is destroyed. In RTCVD however higher levels of fully substitutional carbon have been obtained (up to 1.8% at 550 °C) using the same reactant gases. It has been demonstrated [8] that a high silane partial pressure during growth increases the substitutional carbon fraction.

In UHV/CVD the alloy content is determined only by the partial pressures of the precursors and their sticking coefficients as gas phase reactions and formation of a boundary layer are not likely to occur at these low pressures [13]. The carbon fraction is given by:

$$y = \frac{s_{SiCH_6} \cdot Z_{SiCH_6}}{s_{SiH_4} \cdot Z_{SiH_4} + 2 \cdot s_{SiCH_6} \cdot Z_{SiCH_6}}$$

where s_{SiCH_6} and s_{SiH_4} are the sticking coefficients and Z_{SiCH_6} and Z_{SiH_4} are the fluxes of molecules at the wafer surface for the respective species. Using the data on total carbon concentration vs. partial pressure ratio a value for the sticking coefficient of methylsilane relative to that of silane can be extracted. The sticking coefficient of methylsilane is found to be approximately two times higher than that of silicon in the temperature range of 550-650 °C.

Growth rate measurements clearly show that incorporation of carbon influences the surface reaction rates. In Fig. 4 $Si_{1-y}C_y$ alloy growth rates ratioed to the pure silicon growth rate at the same growth temperature are presented. The growth rates for $Si_{1-y}C_y$ alloys are smaller than the growth rate of silicon at the same temperature and the decrease is more pronounced for high temperatures. At higher temperatures the growth rate decreases progressively with the increase in carbon content. In this temperature range the growth rate is controlled in UHV/CVD by both the sticking coefficient of silane molecules and the surface hydrogen coverage. The growth rate decrease can be explained qualitatively by the fact that the C-H bond is stronger than the Si-H bond and the hydrogen coverage is increased when carbon is added to a silicon surface. However, the number of carbon atoms present on the surface of these dilute alloys is not expected to be large. So it is likely that the silane sticking coefficient on a $Si_{1-y}C_y$ surface is smaller than that on a pure Si surface, with the ratio $s_{SiCH_6}/s_{SiH_4} \cong 2$.

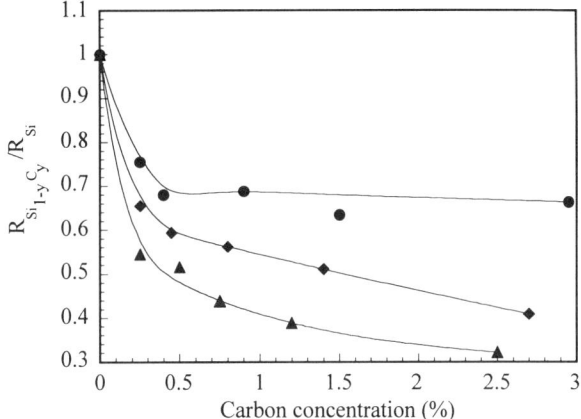

Fig. 4. Growth rates of $Si_{1-y}C_y$ alloys at temperatures of 600 °C (circles), 630 °C(diamonds) and 650 °C (triangles).

SUMMARY

Substitutional carbon incorporation has been studied in $Si_{1-y}C_y$ alloy layers grown by UHV/CVD using silane and methylsilane as source gases. At low growth temperatures (550 °C) carbon is incorporated on substitutional positions up to a concentration of 0.65%. Alloys with concentrations of up to 3% total carbon and 1.34% substitutional carbon have been obtained. At low growth temperatures increased hydrogen coverage plays an essential role in obtaining of smooth films and higher substitutional carbon fraction, by modifying surface kinetics through saturating free bonds. Growth using high silane pressure (~ 1 Torr), as demonstrated in RTCVD may improve carbon substitutionality. From the data on carbon incorporation the sticking coefficient of methylsilane was found to be approximately two times higher than that of silane. The growth rates of $Si_{1-y}C_y$ alloys were lower than the growth rates of silicon at the same temperature and this could be explained by an increase in hydrogen coverage due to the stronger C-H bond and possibly by a decrease in silane sticking probability on a $Si_{1-y}C_y$ surface.

ACKNOWLEDGMENTS

We would like to acknowledge financial support from DARPA through the MAFET program and Northrop Grumman.

REFERENCES

1. K. Brunner, K. Eberl and W. Winter, Phys. Rev. Lett. **76**, 303 (1996)
2. G. Davis and R.C. Newmann in Handbook of Semiconductors, 2nd edition (1994) vol.3
3. R.W. Olesinski and G.J. Abbaschian, Bull. Alloy Phase Diagrams, **5**, 486 (1984)

4. S.S. Iyer, K. Eberl, M.S. Goorski, F.K. LeGoues, J.C. Tsang and F. Cardone, Appl. Phys. Lett. **60**, 356 (1992)

5. H.J. Osten, M. Kim, K. Pressel and P. Zaumseil, J. Appl. Phys. **80**, 6711 (1996)

6. P. Boucaud, C. Francis, A. Larre, F.H. Julien, J.M. Lourtioz, D. Bouchier, S. Bodnar and J.L. Regolini, Appl. Phys. Lett. **66**, 70 (1995)

7. J. W. Strane, H.J. Stein, S.R. Lee, B.L. Doyle, S.T. Picraux and J.W. Mayer, Appl. Phys. Lett. **63**, 2786 (1993)

8. T. O. Mitchell, J.L. Hoyt and J.F. Gibbons, Appl. Phys. Lett. **71**, 1688 (1997)

9. J. Tersoff, Phys. Rev. Lett., **74**, 5080 (1995)

10. P.C. Kelires and J. Tersoff, Phys. Rev. Lett., **63**, 1164 (1989)

11. D.W. Greve and M. Racanelli, J. Vac. Sci. Tech. B **8**, 511 (1990)

12. M. Racanelli, D.W. Greve, M.K. Hatalis and L.J. van Yzendoorn, J. Electrochem. Soc. **138**, 3783 (1992)

13. D.W. Greve, Mater. Sci. Eng., B **18**, 22 (1993)

THE APPLICATION OF NOVEL CHEMICAL PRECURSORS FOR THE PREPARATION OF SI-GE-C HETEROSTRUCTURES AND SUPERLATTICES

DAVID C. NESTING, JOHN KOUVETAKIS, JULIE LORENTZEN AND JOSÉ MENÉNDEZ
Department of Chemistry and Biochemistry, Arizona State University, Tempe, AZ 85287
Department of Physics and Astronomy, Arizona State University, Tempe, AZ 85287

ABSTRACT

We have investigated new synthetic methods to Si-Ge-C materials by UHV-CVD and novel chemical precursors. The reactions of $(SiH_3)_4C$ and $(GeH_3)_4C$ with GeH_4 and SiH_4 produce diamond structured materials with the general formula $(CSi_4)_xGe_y$ and $(CGe_4)_xSi_y$ respectively. These reactions demonstrate the application of novel C-H free precursors that incorporate Si_4C and Ge_4C building blocks to prepare single phase materials containing a significant amount of carbon (5-6 at.%). They also indicate the remarkable degree of compositional control provided by the precursor to incorporate the Si_4C and Ge_4C molecular framework into the solid state. A range of $Si_{1-x-y}Ge_xC_y$ compositions were also prepared pseudomorphically on [1 0 0] Si at 450-500 °C via reactions of $(SiH_3)_4C$ with mixtures of GeH_4 and SiH_4. The $(Si_2Ge)_{31}C_6$ composition appears to display ordering of the Ge and Si atoms in the structure. The incorporation of C in the Si planes seems to provide local strain centers which facilitate the formation of this ordered phase.

INTRODUCTION

Silicon-germanium alloys are well known to form a continuous solid solution[1]. The application of these materials in novel device technology is currently being realized in the form of heterojunction bipolar transistors for use in high frequency devices, infrared detectors and modulation doped field effect transistors (MODFET's)[2,3]. One of the central goals is the incorporation of optoelectronics into existing silicon based microelectronics. The recent development of SiGe based far-infrared cameras which incorporate the SiGe photodetector with a silicon CCD camera on the same substrate demonstrate the advances made in the group IV heterostructures[3]. The quantum efficiency of these devices already exceeds that of the current iridium/silicon detectors and the frequency range is continuously tunable by varying the Ge:Si composition. A remaining challenge is how best to accommodate the ~4% lattice mismatch between silicon and germanium. Although strain is known to decrease the band gap in SiGe/Si heteroepitaxial structures[4], the incorporation of misfit dislocations which are present as threading dislocations extending to the film surface severely degrade device performance.

Two methods are generally used to deal with the effects of strain. The first is the use of a graded GeSi buffer layer, and the second is to incorporate carbon into the film. The graded buffer method effectively dilutes the dislocations so that the buffer layer is silicon rich at the Si interface and approaches the germanium ratio of the device layer at the buffer/SiGe interface[5]. The use of a third element with a lattice parameter smaller than that of Si provides a tunable parameter for lattice matching purposes as well as band gap engineering[4,6-9]. Carbon having a lattice parameter of 3.56 Å is considerably smaller than Si (5.431 Å) or Ge (5.658Å). As a result, the incorporation of carbon corresponding to a Ge:C ratio of approximately 9:1 is required to produce nominally lattice matched SiGeC alloy on Si. The incorporation of carbon into the SiGe film increases the band gap, but this effect is smaller than relieving strain through the use of lower germanium concentrations[4]. The net electronic effect of carbon incorporation is twofold. Although the strain at the interface is reduced by decreasing the lattice parameter, the short Si-C bond induces strain fields in the crystal[5]. The second factor is the net result of incorporating a wide gap material into the structure. In actual device applications, the use of SiGeC alloys could reduce the necessity of thick graded buffer layers in order to achieve high quality heteroepitaxial films.

A second effect observed in these systems is the atomic ordering of the silicon and germanium into layers along <1 1 1> direction. This initial observation of ordering by Ourmazd and Bean[10] consisted of very weak extra Laue spots at $\frac{1}{2}(111)$, $\frac{1}{2}(311)$ and $\frac{1}{2}(331)$ positions. This data indicated ordering of the Si and Ge atoms to double the crystal period in the <111> direction. Of note is that these superlattice structures only seem to form under the kinetically controlled CVD or MBE conditions, and not in bulk alloy crystals obtained from Czochralski growth [11,12].

The presence of long range order has been attributed to strain in lattice-mismatched epitaxial SiGe layers, but ordering in $Si_{0.5}Ge_{0.5}$ was observed in bulk unstrained SiGe films[13-15]. The observed phase consisted of bilayers of Si and Ge along all four <111> directions. This observation is a result of alternating tensile and compressive stress, or microscopic strain which is established in the <100> growth direction as the elements are deposited. The potential to grow ordered heterostructures may have a direct impact on device technologies based upon the MBE growth of strained layer superlattices for the fabrication of light emitters and detectors.

Our research is directed towards the development of new practical routes to prepare silicon-germanium-carbon materials with a wider range of Si:Ge compositions and greater carbon concentrations[6,7,16]. We require an extensive variation in composition and substantial carbon incorporation in order to influence the band structure and to obtain significant reduction of the lattice constant in the SiGeC system. Our methodology emphasizes an integrated approach involving molecular precursors and UHV-CVD to prepare metastable $Si_{1-x-y}Ge_xC_y$ alloys and ordered phases.

EXPERIMENTAL DETAILS

Our UHV-CVD reactor is similar to that developed for epitaxial growth of Si and $Si_{1-x}Ge_x$ at less than 600 °C. The experimental details of the deposition process and the reactor design have been described elsewhere[7]. In this paper we report the characterization of atomic ordering in $(Si_2Ge)_{1-x}C_x$ (x=6 at. %) obtained by chemical vapor deposition of the precursor molecule $(SiH_3)_4C$ with silane and germane at <500 °C.

We investigated the structure by high resolution TEM imaging using a JEOL 4000EX microscope, and selected area electron diffraction (SAED) using a Philips CM200 field emission microscope, and compare the experimental diffraction patterns and images to the proposed structure. Composition was determined by Rutherford backscattering spectroscopy (RBS). The incorporation of substitutional carbon was confirmed using transmission FT-IR and Raman spectroscopy. In addition, SIMS depth profiling was performed to confirm the film homogeneity.

RESULTS AND DISCUSSION

In our approach, we use chemical precursors which incorporate carbon in its tetrahedral form. $C(SiH_3)_4$ and $C(GeH_3)_4$ have molecular arrangements where the carbon occupies a diamond-like tetrahedral site surrounded by four silicon or four germanium atoms respectively [17,18]. Tetrasilylmethane (TSM) was deposited on Si(100) with germane and silane in a hot wall reactor at 450-475 °C, intermixed with a large excess of ultrahigh purity H_2. The resulting films had the diamond cubic structure. However, there were large concentrations of defects such as twins and stacking faults as well as some polycrystalline domains as revealed by cross sectional TEM.

To improve the crystallinity, deposition was followed by several hours *in situ* annealing at 700 °C and slow cooling in a H_2 flow at 10^{-4} Torr. The cross sectional TEM image (Figure 1) revealed a dramatic change in the microstructure. The composition, determined by RBS utilizing both 2 MeV He^{2+} ions and the 4.265 MeV $^{12}C(\alpha,\alpha)^{12}C$ carbon resonance reaction, was determined to be $(Si_2Ge)_{31}C_6$. The TEM image in figure 1 reveals the superlattice stacking in the film. This ordering appears as a tripling of the (111) periodicity in all four directions and is in agreement with a proposed model of two bilayers of silicon doped with carbon followed by a germanium bilayer in the <111> direction. Of note is the absence of twins and the extent of the periodicity extending to the film surface. The model structure yields a new hexagonal unit cell whereby the (111) planes of the diamond cubic structure become (003) planes in the new cell. The proposed structure is shown in figure 2.

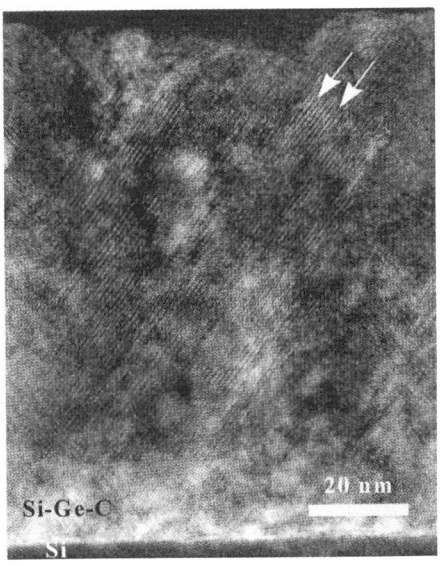

Figure 1. Cross sectional TEM of $(Si_2Ge)_{1-x}C_x$ sample demonstrating that the ordered phase forms throughout the layer. The lattice fringes indicated by the arrows are the superlattice fringes at 9.53 Å.

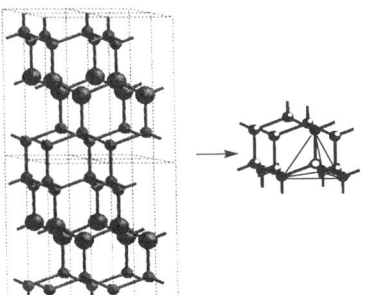

Figure 2. Proposed hexagonal structure showing two layers of Si followed by 1 layer of Ge (Larger Spheres). The inset shows a possible carbon site (Open Circle) as an Si_4C unit.

The incorporation of the Si_4C building block from the TSM molecule drives the formation of a structure with two sheets of Si as shown in the inset of figure 2. The carbon may serve as a strain center while the molecular building block predisposes the structure to incorporate substantial short range order. This structure was evaluated first using a molecular modeling and TEM simulation software package CERIUS. Calculations for a range of thickness of 10 to 30 unit cells and varying microscope defocus were used. The simulated image is shown as an overlay on the experimental image in figure 3. Second, we performed a selected area electron diffraction study (figure 4) in which the recorded diffraction patterns were compared to the predicted model. As shown in figure 4, the agreement between the calculated and observed diffraction is excellent. Additional periodicity in the observed pattern is a result of sampling regions of the film in which the superlattice propagates along more than one of the <111> directions. To confirm that the ordering propagates along all four <111> directions, cross sectional samples were prepared which were cut 90 degrees to each other so that all four sets of (111) planes could be examined in the same sample, in addition to a plan view sample. Optical diffraction from smaller regions shows ordering in one direction only[6].

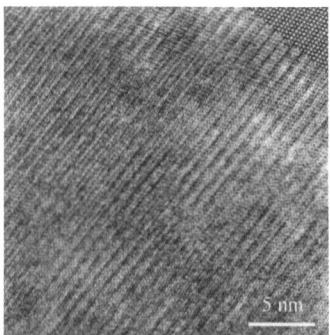

Figure 3. High resolution TEM image with the simulated image inset showing a match to the (111) periodicity for the <110> zone axis.

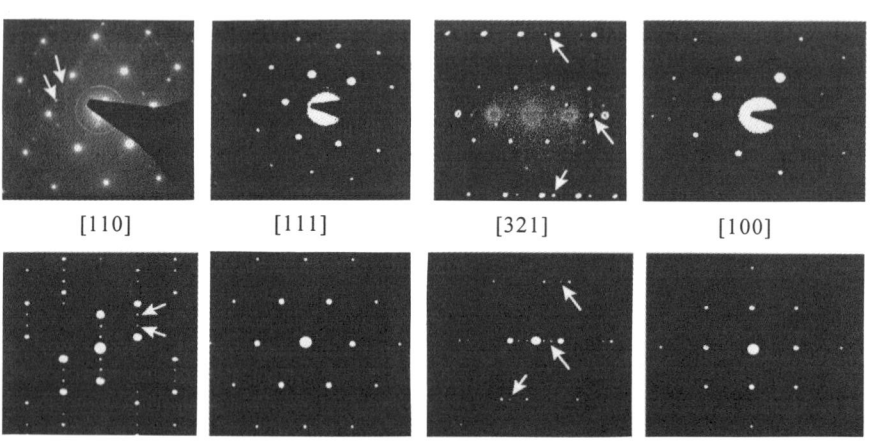

Figure 4. Selected Area Electron Diffraction (top figures) with the corresponding simulations (bottom). Reflections due to double diffraction are not simulated. As expected the periodicity is present on the [110] and [321] zones and absent along [100] and [111].

To examine the composition, Raman spectra were collected to determine that the carbon was incorporated on substitutional sites. The Raman spectrum consists of five features in figure 5. The peaks at 292, 409, and 503 cm^{-1} represent the Ge-Ge, Si-Ge and Si-Si phonons respectively. The 520 cm^{-1} peak is from the Si substrate and the 603 cm^{-1} is the substitutional carbon peak.

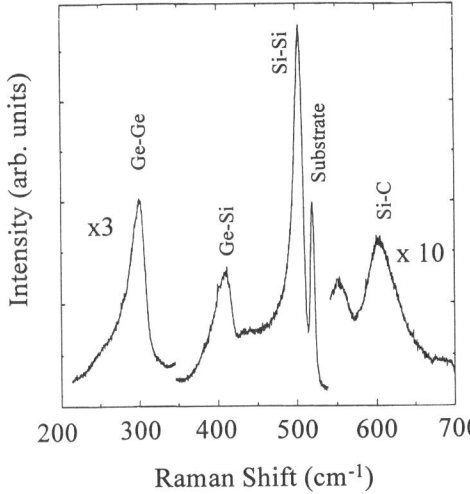

Figure 5. The microfocus Raman spectrum of $(Si_2Ge)_{1-x}C_x$. Spectra were collected using different integration times to be scaled as shown.

The use of the precursor tetrasilylmethane creates a strain center during chemical vapor deposition. The incorporation of this molecular building block imposes short range order in the structure. Previous workers have observed the ordering in 1:1 SiGe systems as doubling along <111>. The observation of tripling seems attributable to the use of a molecular precursor which is robust under the deposition conditions. The linking of the tetrahedral molecules of silicon and carbon with their relatively short bonds would seem to prefer the incorporation of Si atoms rather than Ge to complete the bilayer structure. As shown in figure 2, the incorporation of an Si_4C unit requires that there be adjacent layers of Si as there are no C-Ge bonds.

Single crystal SAED along the different zones indicated tripling along all four <111> directions. This effect is inconsistent with the simple picture of twins along 111, as the lattice fringes exhibit no discontinuities in the ordered regions. We would expect that if there were twinning of the crystal along the 111 plane, diffraction on the 111 zone would show some rotational twinning of the spots and this is not observed.

Finally, the observation of this structure in samples annealed at 700 °C is inconsistent with a defect structure as annealing would be expected to reduce the defect density rather than add to them. An ancillary point is that it would be unusual for the defect structure to grow over such an extended region, from near the interface to the film surface at the above temperatures.

CONCLUSIONS

In summary, we have observed long range ordering in thick unstrained Si-Ge-C layers with the composition $(Si_2Ge)_{1-x}C_x$. The ordering occurs along the <111> direction and corresponds to an alternating double layer of Si-Si (doped with C) and a single layer of Ge. The carbon incorporation is clearly an important factor in the formation of this new phase since similar ordering (2:1) has not been reported in pure Si-Ge alloys. It is known from previous studies that the Si_4C composition and structure from the precursor is retained in the solid state. It is likely that this building block facilitates the formation of the ordered phase observed in the present study. The annealing conditions also appear to play a large role in this ordering. The as deposited films grown at 450 °C display very limited ordering. A systematic study of the effect in both Si_2Ge and Ge_2Si is underway.

ACKNOWLEDGEMENTS

We gratefully acknowledge the support of the National Science Foundation (Grants DMR 9458047 and DMR 9424445). We would also like to thank Professor Ernst Bauer for his helpful discussions.

REFERENCES

1. M. Hansen, Constitution of Binary Alloys, 2nd ed., (McGraw-Hill, New York, 1958), p. 774.
2. G. L. Patton, D. L. Harame, J. M. Strock, B. S. Mayerson and G. S. Scilla, IEEE Electron Device Letters **10**, 534 (1989).
3. B.-Y. Tsaur, C. K. Chen and S. A. Marino, Optical Engineering **33**, 72 (1994).
4. C. L. Chang, A. S. Amour and J. C. Sturm, Appl. Phys. Lett. **70**, 1557 (1997).
5. H. Presting, Mat. Res. Soc. Symp. Proc. **379**, 417 (1995).
6. J. Kouvetakis, D. Nesting and M. O'Keeffe, Chem. Mater. **10**, 1396 (1998).
7. J. Kouvetakis, M. Todd, D. Chandrasekhar and D. J. Smith, Appl. Phys. Lett. **65**, 2960 (1994).
8. G. Abstreiter, Physica Scripta **T49**, 42 (1993).
9. M. A. Berding, A. Sher and M. van Schilfgaarde, Phys. Rev. B. **56**, 3885 (1997).
10. A. Ourmazd and J. C. Bean, Phys. Rev. Lett. **55**, 765 (1985).
11. D. J. Lockwood, K. Rajan, E. W. Fenton, J.-M. Baribeau and M. W. Denhoff, Solid State Comm. **61**, 465 (1987).
12. E. Müller, H.-U. Nissen, M. Ospelt and H. von Känel, Phys. Rev. Lett. **63**, 1819 (1989).
13. J. C. Tsang, V. P. Kesan, J. L. Freeouf, F. K. LeGoues and S. S. Iyer, Phys. Rev. B **46**, 6907 (1992).
14. V. P. Kesan, F. K. LeGoues and S. S. Iyer, Phys. Rev. B **46**, 1576 (1992).
15. F. K. LeGoues, V. P. Kesan and S. S. Iyer, Phys. Rev. Lett. **64**, 40 (1990).
16. M. Todd, P. Matsunaga, J. Kouvetakis, D. Chandrasekhar and D. J. Smith, Appl. Phys. Lett. **67**, 1247 (1995).
17. R. Hager, O. Steigelman, G. Muller, H. Schmidbaur, H. H. Robertson and D. W. Rankin, Angew. Chem. Int. Ed. **29**, 201 (1990).
18. P. T. Matsunaga, J. Kouvetakis and T. L. Groy, Inorg. Chem. **34**, 5103 (1995).

Part VI

Epitaxy of Si Ge
and
Related Materials

LIMITATIONS TO THE USE OF Sb AS A SURFACTANT DURING SiGe MBE

Glenn G. Jernigan, Conrad L. Silvestre, Mohammad Fatemi, Mark E. Twigg, and Phillip E. Thompson, Naval Research Laboratory, Electronics Science and Technology Division, 4555 Overlook Ave. SW, Washington DC 20375

ABSTRACT

The use of Sb as a surfactant in suppressing Ge segregation during SiGe alloy growth was investigated as a function of Sb surface coverage, Ge alloy concentration, and alloy thickness using x-ray photoelectron spectroscopy, x-ray diffraction, and transmission electron microscopy. Unlike previous studies where Sb was found to completely quench Ge segregation into a Si capping layer, we find that Sb can not completely prevent Ge segregation while Si and Ge are being co-deposited. This results in the production of a non-square quantum well with missing Ge at the beginning and extra Ge at the end of the alloy. We also found that Sb does not relieve strain in thin films but does result in compositional or strain variations within thick alloy layers.

INTRODUCTION

Segregation of Ge during the deposition of SiGe materials must be considered during the development of device applications. To take advantage of Ge incorporation in Si processing, control of film composition and film strain are necessary. Surfactant assisted growth has been investigated as a method to inhibit Ge segregation[1-4]. Surfactant assisted growth works by pre-depositing a third material (typically a group V element) which lowers the surface free energy beyond that of a Si or Ge surface[5-7] and which segregates more readily than Ge. The use of elemental Sb and As have been successfully demonstrated in relieving strain in thick Ge films (>1nm) producing flat Ge surfaces and in preventing Ge segregation onto Si capping layers for the growth of stacked structures[1,6,8,9]. These studies, however, have not investigated the effect of the surfactant on the growth of a SiGe alloy. We have reported recently that the segregation of Ge during the start of alloy growth is different from segregation which occurs into a Si capping layer[10].

Prior to our work, all previous studies of Ge segregation have been done on the trailing edge of growth. That is, Ge segregation has been studied after all of the Ge deposition has already occurred and only Si is still being deposited. However, in the growth of a SiGe alloy, there are three regions of interest: a leading edge, a stoichiometric plateau, and a trailing edge. These three regions are shown schematically in Figure 1. The leading edge is the region in the film where the Ge composition is rising. A leading edge exists because Ge segregation prevents all of the deposited Ge from incorporating into the film. The stoichiometric plateau is the region where the composition of the growing film is equal to the composition of the deposition flux. The stoichiometric plateau occurs once the segregated surface Ge reaches steady-state growth. The trailing edge is the region in the film where the Ge composition is decreasing. No additional Ge is deposited during the trailing edge. Ge in the trailing edge comes from the gradual incorporation of Ge segregating in the topmost layers.

In this paper, we report on the use of an Sb surfactant to prevent Ge segregation during the leading edge of alloy growth. The effects of Sb surface coverage, Ge concentration within the alloy, and film thickness were all studied using x-ray photoelectron spectroscopy (XPS). Unlike trailing edge studies, we found that Ge segregation could not be completely suppressed by using an Sb coverage up to and including 1 monolayer. With increasing Ge concentration, the Sb surfactant effect decreased during growth. By growing films of increasing thickness, the abruptness of the leading edge for a SiGe quantum well was determined. X-ray diffraction

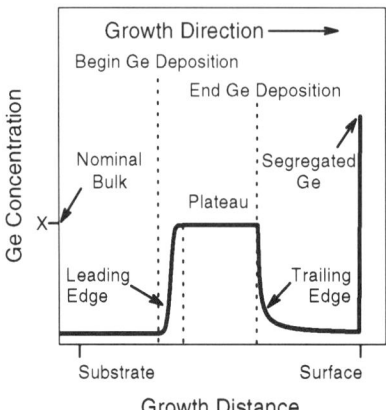

(XRD) and transmission electron microscopy (TEM) were performed to determine if the use of Sb resulted in a change in film strain.

EXPERIMENTAL

Experiments were performed in a VG ESCA lab system that is connected to a VG V80 growth chamber allowing for *in vacuo* transfer of samples. SiGe films were grown on 75 mm Si (100) wafers (p-type 10-20 ohm cm). Each wafer is chemically cleaned using a modified RCA cleaning procedure terminating in a HF dip prior to entrance to the growth system[11]. Elemental Si and Ge sources were electron beam evaporated at rates of 0.06-0.10 and 0.005-0.04 nm/sec, respectively, during the growth of the samples maintaining a total rate of 0.1 nm/sec. Sb was deposited from a Knudsen cell at a temperature of 425 °C (flux 3.0×10^{12} atoms/cm^2/sec). Growth was controlled using a Sentinel III film monitor. Each growth began with a 100 nm Si buffer layer at 650 °C. The temperature was ramped down to 500 °C where 0 to 1 monolayer (6.8×10^{14} atoms/cm^2) of Sb was deposited, concurrent with continued Si deposition. A 10 nm Si spacer was deposited followed by the SiGe layer. The SiGe films were grown in varying thickness of 0-10 nm with nominal compositions of 5, 10, 20, and 40%

Ge. Growth temperature was measured using an optical pyrometer that has been calibrated against the eutectic temperatures of AuSi (363 °C) and AlSi (577 °C). The base pressure of the MBE chamber is 5×10^{-9} Pa and during growth was typically in the low 10^{-7} Pa range.

The ESCA lab consists of a hemispherical analyzer, a twin Al/Mg x-ray anode, and a 75 mm sample stage. XPS was performed on each sample using the Al Kα line ($h\nu = 1486.6$ eV) operated at 510 W (15 kV and 34 mA). Spectra were taken of the Ge 2p$_{3/2}$, Ge 3d, Si 2p, and Sb 3d regions using a 10 eV pass energy. Each region was scanned 10 times at three different spots on the wafer to check for uniformity. The C 1s region, O 1s region, and a wide scan (1400-0 eV) were also collected on each sample to measure any impurities on the surface. Results are presented as peak area ratios to remove any variation in sample-to-detector distance or in electron multiplier sensitivity between runs. Reference spectra were taken on clean Si (100) and Ge (100) wafers to be used to quantitatively determine the Ge concentration in the grown alloys.

X-ray rocking curves were made on a high resolution double crystal diffractometer using a Si(001) first crystal beam conditioner adjusted for the 004 reflection of Cu K$_\alpha$1 radiation. Single crystal (θ-2θ) diffraction patterns were obtained on a powder diffractometer using a Cu x-ray tube. Cross sectional TEM images were taken on room temperature, ion milled samples.

RESULTS AND DISCUSSION

The effect of Sb coverage on steady-state Ge segregation from a 20% Ge, 10 nm alloy is shown in Figure 2. A maximum of 1 monolayer (ML) of Sb was used, because coverages greater than one monolayer have been reported to result in defective crystal growth[12]. We observe that there is a small decrease in the Ge 2p/Si 2p ratio between 0 and 0.3 ML of Sb. At coverages between 0.3 and 0.75 ML, the Ge 2p/Si 2p ratio begins to decrease noticeably. Between 0.75 and 1.0 ML of Sb, we see a plateau is reached in the Ge 2p/Si 2p ratio. No significant improvement in

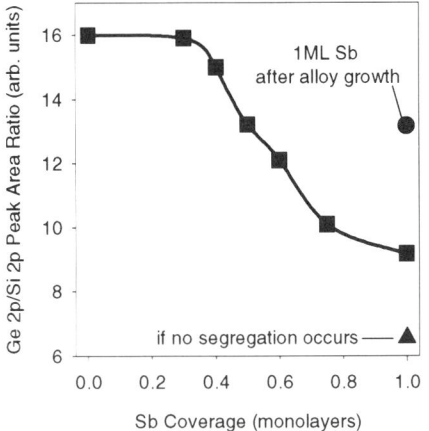

Fig. 2. XPS data of the Sb coverage dependence in preventing Ge segregation in the growth of a 10 nm Si0.8Ge0.2 alloy.

the reduction of Ge segregation is observed for coverages greater than 0.75 ML of Sb. This agrees with the trailing edge study by Fukatsu who reported that Sb has a maximum effect in reducing Ge segregation at a coverage of 0.8 ML[13].

Also shown in Figure 2 are the theoretically predicted Ge 2p/Si 2p ratios for a 20% alloy film without Ge segregation and for 1 ML of Sb deposited after the alloy growth. The measured peak ratio for 1 ML deposited prior to alloy growth is greater than the predicted value if no segregation occurs indicating that some Ge has still segregated. A comparison between the deposition of 1 ML before alloy growth and 1 ML after alloy growth shows that the decrease in the peak ratio is not the result of signal attenuation due to the coverage of Sb. Using our previously reported method of determining the amount of Ge segregation[10], we can calculate that there remains approximately 0.8 ML of excess Ge on the surface of a $Si_{0.8}Ge_{0.2}$ alloy when 1 ML of Sb is used as a surfactant.

Using a fixed 0.75 ML Sb coverage, we studied the surfactant effect as a function of alloy composition. Figure 3 shows the Ge 2p/Si 2p ratio for 10 nm 5, 10, 20, and 40% Ge alloys. Plotted are experimental data with and without

Sb, and the theoretically predicted value for the alloys if no Ge segregation occurs. We can see in all cases that the use of Sb reduces the Ge 2p/Si 2p ratio as compared to a growth without Sb. In none of the alloy films do we see complete suppression of Ge segregation. With increasing Ge concentration, the Sb data line diverges from the theoretical line indicating that Sb's surfactant effect is decreasing.

Ge segregation exists during Sb surfactant assisted growth indicating that the leading edge is not atomically abrupt. This is in contrast with the trailing edge studies where Ge segregation has been shown to be completely suppressed [2,14-16]. However, it is not clear how the segregated Ge accumulated in the leading edge, so a series of alloys with increasing thickness were grown to measure the abruptness of the leading edge. Figure 4 shows how 0.25 and 0.75 ML of Sb affect the build-up of Ge in a 10% alloy. The figure also shows the experimental results for alloys grown without Sb and a theoretically predicted curve for alloy growth without segregation. The curve for films grown with 0.25 ML of Sb is similar to the films without the Sb surfactant. A steady state Ge 2p/Si 2p ratio is reached in both films at approximately 4.5 nm of film thickness. The

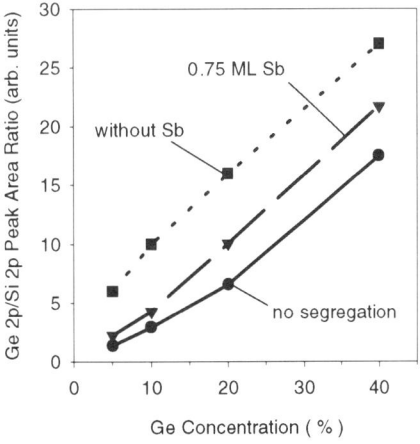

Fig. 3. XPS data showing the effect of 0.75 ML of Sb in suppressing Ge segregation in a 10 nm, 5, 10, 20, and 40% Ge alloy.

steady state value for the 0.25 ML is less than without Sb indicating that some decrease in Ge segregation had occurred. The curve for films grown with 0.75 ML of Sb compares favorably with the theoretical prediction for no segregation. Both curves reach steady state values at approximately 3.0 nm film thickness. The 0.75 ML film shows deviation from the theoretical curve at a thickness of only 1 nm indicating that Ge segregation occurs immediately.

Combining our measurement of Ge segregation in the leading edge with the absence of segregation in the trailing edge, we realize that the compositional profile for a SiGe quantum well is still skewed when an Sb surfactant is used. There is a depletion of Ge at the start of alloy growth and an excess of Ge at the end of alloy growth. Additional studies are needed to quantify the exact compositional profile for SiGe alloys grown with an Sb surfactant.

Why 1 ML of Sb is not able to completely suppress Ge segregation during the growth of a SiGe alloy can be explained by two plausible reasons. The first reason is that the rate of exchange between an Sb and a Ge is not as fast as the rate of exchange between an Sb and a Si. In these experiments Sb is deposited prior to alloy

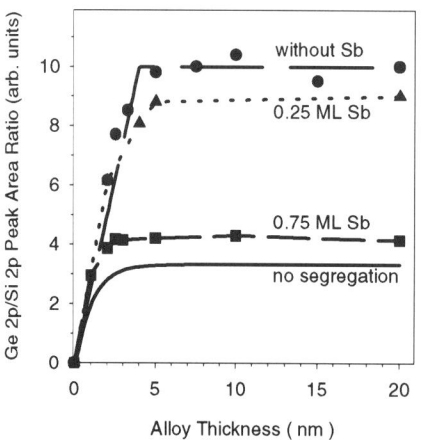

Fig. 4. XPS data showing the build up in segregated Ge for Si0.9Ge0.1 films grown with 0.25 ML, 0.75 ML, and without Sb along with a theoretical curve for a film grown without segregation.

growth, and as Si and Ge are deposited, the Sb must exchange from underneath. The rate of exchange is related to the energetics of creating an Sb terminated surface[17]. The energetics for an exchange between Si and Sb is very favorable as it is well known that Sb strongly segregates in Si[18-20]. The rate of exchange between an Sb and a Ge may not be as energetically favorable. The Sb may preferentially incorporate Si faster than it does Ge. The decrease in Sb's surfactant ability with increasing Ge concentration appears to indicate that the increased flux of Ge atoms does compete with the exchange process. As more Ge is deposited, the time necessary for Sb to exchange with Ge decreases and more Ge is found to segregate.

Another explanation for incomplete suppression of Ge segregation is related to the fact that segregation is not solely a surface phenomena. Morphology of a SiGe film is the result of opposing forces between surface stress and bulk film strain[21]. The use of an Sb surfactant to increase the incorporation of Ge may change the film strain. It has been reported for the growth of pure Ge on Si that an Sb surfactant does result in the early onset of dislocations[22-25]. Additionally, it has been reported that an Sb surfactant also suppresses Si and Ge intermixing which would be a natural way to create a graded buffer within the alloy and to alleviate film strain [26].

To observe if the use of Sb was affecting strain in a thin alloy, we grew a multiple quantum well structure to be studied by x-ray diffraction. The structure consisted of a 2.0 nm $Si_{0.8}Ge_{0.2}$ alloy with a 8.0 nm Si spacer repeated 20 times grown with and without 0.9 ML of Sb. The XRD (powder diffraction) results with and without Sb are shown in Figure 5. The spectra have been normalized to the Si substrate line from the $K_{\alpha 1}$ x-ray at $2\theta = 69.13$. Each peak is a doublet due to the $K_{\alpha 1}$ and $K_{\alpha 2}$ x-rays from the Cu anode. The peak at $2\theta = 68.88$ is due to the average composition within the film, and the satellites at $2\theta = 64.8, 65.8, 66.8,$ and 67.9 are due to the quantum well structure. We see that the peak positions with and without Sb are nearly identical. If there was a change in strain or film composition, we would expect to observe a noticeable shift in θ. We do not see a shifting in the peaks, therefore Sb did not change the composition or strain. However, we do observe that the diffracted intensity from the quantum well structure with Sb is greater. The enhancement of the diffraction is due to an improvement of the interface between the quantum well and the Si spacer. The use of Sb does produce a more abrupt interface.

Changes in strain may be more observable in a thicker alloy layer, so a 100 nm $Si_{0.86}Ge_{0.14}$ with a 100 nm Si cap grown with and without 0.8 ML Sb was studied by XRD (rocking curves) and TEM. This alloy film is below the People-Bean critical thickness transition from a meta-stable film to a dislocated film[27]. The XRD results are shown in Figure 6. We note both spectra have alloy θ values of -1350 (arc sec.) indicating that they contain the same average composition and strain. The sample with Sb has a broader alloy peak and the sample without Sb has satellites. The broadness of the alloy peak with Sb could be due to fluctuations in either composition or strain within the film, and the lack of satellites could be due to a degradation of the alloy-Si interfaces. The TEM image, shown in Figure 7, corroborates the XRD findings. Within the alloy layer, strong contrast is observed that

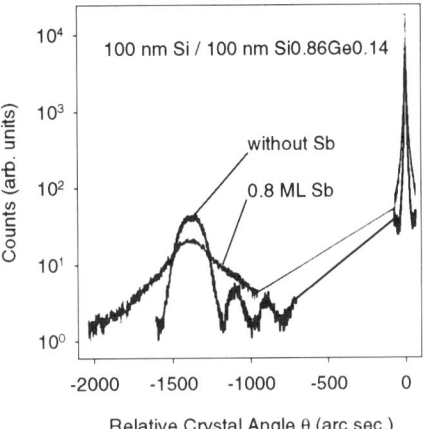

Fig. 6. XRD data for a 100 nm Si0.86Ge0.14 alloy capped with 100 nm of Si with and without 0.8 ML of Sb during alloy growth.

may be due to either Ge concentration or strain differences. The top SiGe-Si interface and the topmost Si surface show film undulations. At the bottom SiGe-Si interface, a few misfit dislocations are observed, but no threading dislocations are observed in either the alloy or the Si cap layer.

CONCLUSIONS

The use of up to 1 monolayer of Sb as a surfactant during SiGe growth was unable to completely suppress Ge segregation. This results in a leading edge and in an excess of surface Ge at the end of the alloy growth which does not segregate into the trailing edge. The Ge

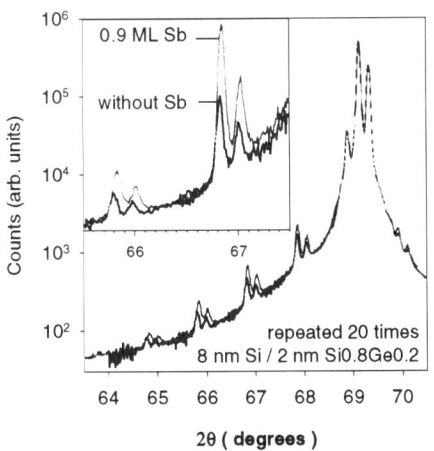

Fig. 5. XRD data for a multiple quantum well structure, 8 nm Si spacer after a 2 nm Si0.8Ge0.2 well repeated 20 times, with and without 0.9ML of Sb.

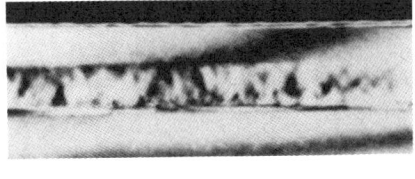

Fig. 7. TEM image of a 100 nm Si0.86Ge0.14 alloy capped with 100 nm of Si using 0.8 ML of Sb during alloy growth.

concentration profile for a quantum well has a region of below nominal Ge concentration at the start of alloy growth and a region of above nominal Ge concentration at the end of alloy growth. The Si-SiGe interfaces, however, did become more abrupt when Sb was used. Sb was not found to relieve strain in the alloy films, as it was reported to do for the growth of pure Ge on Si. In thick films the use of Sb resulted in compositional or strain fluctuations within the alloy.

REFERENCES

1. K. Fujita, S. Fukatsu, H. Yaguchi, T. Igarashi, Y. Shiraki, and R. Ito, Jpn. J. Appl. Phys. **29,** L1981 (1990).

2. M. Copel and R. M. Tromp, Appl. Phys. Lett. **58,** 2648 (1991).

3. X. W. Lin, Z. Liliental-Weber, J. Washburn, E. R. Weber, A. Sasaki, A. Wakahara, and T. Hasegawa, J. Vac. Sci. Technol. B **13,** 1805 (1995).

4. B. D. Yu, T. Ide, and A. Oshiyama, Phys. Rev. B **50,** 14631 (1994).

5. D. J. Eaglesham, F. C. Unterwald, and D. C. Jacobson, Phys. Rev. Lett. **70,** 966 (1993).

6. R. Cao, X. Yang, J. Terry, and P.Pianetta, Appl. Phys. Lett. **61,** 2347 (1992).

7. B. Voigtlander, A. Zinner, T. Weber, and H. P. Bonzel, Phys. Rev. B **51,** 7583 (1995).

8. M. Copel, M. C. Reuter, E. Kaxiras, and R. M. Tromp, Phys. Rev. Let. **63,** 632 (1989).

9. H. J. Osten, J. Klatt, G. Lippert, B. Dietrich, and E. Bugiel, Phys. Rev. Lett. **69,** 480 (1992).

10. G. G. Jernigan, P. E. Thompson, and C. L. Silvestre, Surf. Sci. **380,** 417 (1997).

11. P. E. Thompson, M. E. Twigg, D. J. Godbey, K. D. Hobart, and D. S. Simons, J. Vac. Sci. Technol. B **11,** 1077 (1993).

12. K. D. Hobart, D. J. Godbey, M. E. Twigg, M. Fatemi, P. E. Thompson, and D. S. Simons, Surf. Sci. **334,** 29 (1995).

13. S. Fukatsu, PhD. Thesis, University of Tokyo, 1992.

14. J. Falta, D. Bahr, G. Materlik, B. H. Muller, and M. Horn-von Hoegen, Appl. Phys. Lett. **68,** 1394 (1996).

15. K. Fujita, S. Fukatsu, H. Yaguchi, T. Igarashi, Y. Shiraki, and R. Ito, Mat. Res. Soc. Symp. Proc. **220,** 193 (1991).

16. K. Sakamoto, H. Matsuhata, K. Kyoya, K. Miki, and T. Sakamoto, Jpn. J. Appl. Phys. **33,** 2307 (1994).

17. I. Markov, Phys. Rev. B **50,** 11271 (1994).

18. S. A. Barnett, H. F. Winters, and J. E. Greene, Surf. Sci. **165,** 303 (1986).

19. H. Jorke, Surf. Sci. **193,** 569 (1988).

20. K. D. Hobart, D. J. Godbey, P. E. Thompson, and D. S. Simons, Appl. Phys. Lett. **63,** 1381 (1993).

21. J. E. Guyer and P. W. Voorhees, Phys. Rev. B **54,** 11710 (1996).

22. H. Zhu, et al., J. Cryst. Growth **179,** 115 (1997).

23. M. Horn-von Hoegen, B. H. Müller, and A. Al-Falou, Phys. Rev. B **50,** 11640 (1994).

24. F. K. LeGoues, M. Copel, and R. Tromp, Phys. Rev. Lett. **63,** 1826 (1989).

25. H. J. Osten and J. Klatt, Appl. Phys. Lett. **65,** 630 (1994).

26. M. Katayama, T. Nakayama, M. Aono, and C. F. McConville, Phys. Rev. B **54,** 8600 (1996).

27. R. People and J. C. Bean, Appl. Phys. Lett. **47,** 322 (1985); Appl. Phys. Lett **49** 229 (1986).

INFLUENCE OF GROWTH CONDITIONS ON THE THERMAL QUENCHING OF PHOTOLUMINESCENCE FROM SiGe/Si QUANTUM STRUCTURES

I. A. BUYANOVA, W. M. CHEN, W.-X. NI, G. V. HANSSON AND B. MONEMAR
Dept of Physics and Measurement Technology, Linköping University, S-581 83 Linköping, Sweden

ABSTRACT

In this work we study effects of growth temperature and use of surfactant during growth on thermal quenching of photoluminescence (PL) from SiGe/Si quantum wells (QWs) grown by molecular beam epitaxy (MBE). We show that although all investigated structures demonstrate intense and sharp excitonic emissions from the SiGe QWs at liquid helium temperature, thermal quenching of this PL critically depends on the growth conditions. In particular, the use of low ($\leq$ 550 ^{o}C) growth temperatures or employing Sb as a surfactant during high temperature (620° C) growth considerably degrades the PL thermal quenching behaviour by introducing some competing quenching processes with low activation energies of about 5 meV. The optimum growth conditions judging from the PL thermal behaviour are realised during high temperature growth without surfactant (620° C). Even higher growth temperature is shown to be required during surfactant mediated growth to improve the thermal quenching behaviour. From optically detected magnetic resonance (ODMR) studies, the competing quenching processes are attributed to a thermal activation of non-radiative defects introduced during either low-temperature MBE growth or during surfactant-mediated growth.

INTRODUCTION

The development of modern epitaxial growth techniques during the last two decades has opened a new field of semiconductor physics, that of artificial heterostructures of reduced dimensionality. The possibility of bandgap engineering and built-in strain in heterostructures have offered new possibilities to achieve desired material properties, which were not previously accessible. Strained layer SiGe/Si quantum structures are now attracting much attention due to their potential for new optoelectronic devices based on mature Si technology. The abruptness of the SiGe/Si interface is of extreme importance. Investigations have shown that the real SiGe/Si interface undergoes intermixing of germanium and silicon caused by islanding and segregation of germanium [1].

In order to secure sharp SiGe/Si interfaces and to maintain a pseudomorphic growth in strained SiGe/Si heterostructures epitaxial growth of high quality SiGe/Si structures is usually performed either at rather low (< 550 ^{o}C) growth temperatures [2, 3], or at higher temperatures ($\approx$ 650 ^{o}C) but using different surfactant impurities which suppress severe Ge segregation [4, 5]. Samples grown under the aforementioned conditions usually demonstrate at helium temperatures intense and sharp excitonic emission, considered as a fingerprint of the high quality of the layers. However, the effect of such growth conditions on the thermal behaviour of the photoluminescence (PL) is not known. Such information, however, is of particular importance since a severe thermal quenching of the luminescence below room temperature has so far excluded the SiGe/Si material system from practical applications for light-emitting devices operating at room temperature.

In this paper we study effects of the growth temperature and the use of surfactant on the thermal quenching of photoluminescence from SiGe/Si quantum wells (QWs) grown by molecular beam epitaxy (MBE), aiming at determination of the optimum growth conditions with respect to the PL thermal behaviour. We show that low temperature growth and/or using of surfactant may cause formation of competing non-radiative defects responsible for the premature temperature quenching of PL.

Mat. Res. Soc. Symp. Proc. Vol. 533 © 1998 Materials Research Society

EXPERIMENTAL

The investigated samples include 30 Å wide $Si_{0.8}Ge_{0.2}$ single and multi QWs structures grown at 420, 550 and 620 °C without surfactant or at 620 and 650 °C using Sb as a surfactant. All samples were grown on Si substrates by MBE with a Balzers UMS 630 Si-MBE system. All the structures were capped by a 1000 Å undoped Si layer.

The PL measurements were performed at a temperature range from 2 K up to 150 K in a variable temperature cryostat. PL was excited by the 514.5 nm line of an Ar^+ laser and was detected using a double grating monochromator equipped with a nitrogen-cooled Ge-detector. The optically detected magnetic resonance (ODMR) experiments were done at the X-band (9.23 GHz) using a modified Bruker ESR spectrometer, equipped with a TE_{011} microwave cavity. The PL emissions from the samples, under illumination of the UV multilines of an Ar^+ laser, were monitored by a Ge detector. The ODMR signal was obtained by detecting a synchronous change in the PL with field modulation of the magnetic field. A derivative line shape of the ODMR is observed in this case when the magnetic field is modulated on and off the spin resonance conditions.

RESULTS AND DISCUSSION

Role of growth temperature

Excitonic recombination of SiGe/Si quantum structures usually involves three types of excitonic transitions, i. e. recombination of localised exciton (LE), impurity (X) bound excitons (BE) and free excitons (FE) [6, 7]. The relative contributions of the aforementioned transitions depend on the experimental conditions, such as temperature and excitation power, as well as on the interface roughness of the structure. The LE recombination arises from the recombination of excitons localised at random alloy fluctuations. This PL process, typically with a broad linewidth is characterised by saturation at very low excitation power, less than 0.1 mW/cm^{-2}, when the recombination of the BE becomes dominant emission. Observation of the FE recombination requires a further increase of excitation power and/or measurement temperature.

Typical PL spectra recorded at 2K from the $Si_{1-x}Ge_x$/Si single QW (SQW) structures grown at different temperatures without surfactant are presented in Fig. 1. Independent on the growth temperature, the

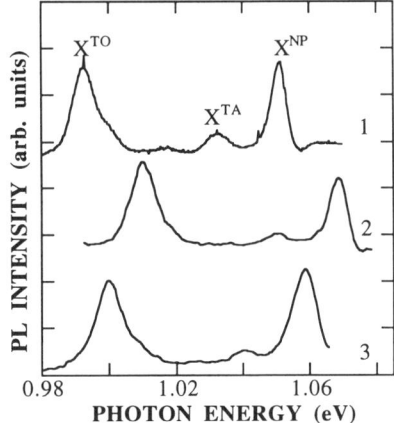

Fig. 1. PL spectra of the SiGe/Si QW samples grown at 420 °C (curve 1), 550 °C (curve 2) and 620 °C (curve 3) without surfactant.

PL spectra are dominated by sharp and intense BE emissions from the SiGe QWs, which correspond to the no-phonon (X^{NP}), transverse optical (X^{TO}) and transverse acoustic (X^{TA}) phonon assisted transitions [6]. The observed slight shift of the PL maximum position between different samples is caused by the variation of the Ge composition within x = 0.20 - 0.22. The observation of intense excitonic PL is usually regarded as an indication of good structural quality of the samples, also supported by high resolution X-Ray diffraction measurements.

The temperature behaviour of the PL, however, changes significantly depending on the growth temperature - see Fig. 2. For the structures grown at 420 °C even a minor temperature

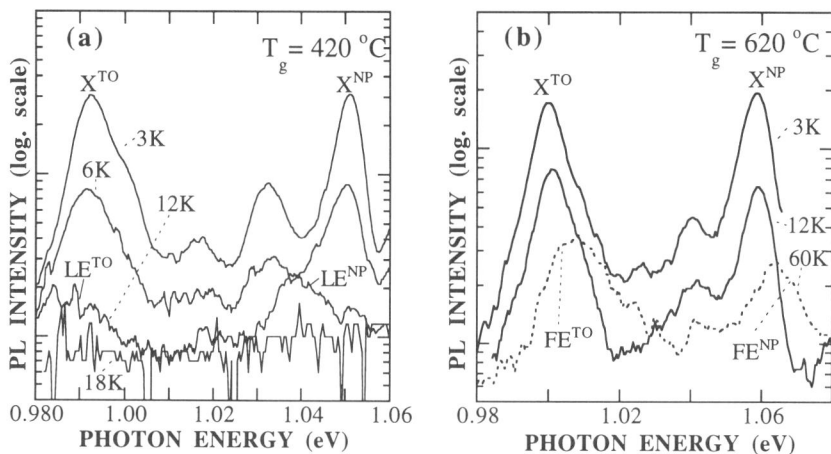

Fig. 2. Temperature dependent PL spectra measured from the SiGe/Si SQW structures grown at 420° C (a) and 620° C (b) without surfactant. Spectra were measured at rather low excitation power of 7.5 mW/cm^{-2} to avoid a possible partial saturation of the competing nonradiative channels.

raise up to 12 K causes a drastical quenching of the BE emissions, accompanied by a shift of the PL peaks towards lower energies by approximately 10 meV - Fig. 2a. The remaining emission at 12 K has a spectral shape typical for the localised exciton (LE) recombination known to be efficient in SiGe/Si structures [7]. A further temperature increase causes a complete quenching of the SiGe related PL. On the other hand, the intensity of the SiGe emissions from the samples grown at 620 °C remains almost unaffected by increasing temperature up to at least 45 K - Fig. 2b. A major PL quenching in these structures occurs at T > 60 K. A temperature increase in this case causes blue shift of the SiGe emissions by ~ 5 meV [6], when the FE recombination becomes the dominant PL process, evident from the observation of the Maxwell-Boltzmann line shape.

Quantitative information about the processes involved in the PL thermal quenching can be obtained from the analysis of an Arrhenius plot of the integrated PL intensity with respect to the reciprocal temperature - Fig. 3. For the structures grown at 420 °C the quenching occurs with a very low activation energy of about 4 meV, responsible for the complete vanishing of the QW-related PL at very low temperatures < 15 K. This low energy activation process is also contributing to the PL quenching of the structures grown at 550 °C. In this case, however, it can be partially saturated by increasing the excitation power, leading to a shift of the "knee" of the Arrhenius curve towards higher temperatures. For the samples grown at 620 °C thermal quenching of the PL is found to be dominated by a process with an activation energy of about 95 meV. The obtained value is very close to the QW depth of 90 meV, estimated from the energy difference between the BE transitions in Si barriers and SiGe QWs. We need to point out, that thermal quenching of the PL from SQW structures grown at low temperatures can be remarkably improved by post growth treatments, such as annealing and hydrogenation [8], making the intrinsic thermal quenching from the QW visible on the high temperature side.

A rapid thermal quenching of the PL from the SiGe/Si QW structures grown by MBE at low temperatures points towards the predominance in these structures of competing non-radiative (NR) recombination, which can easily be thermally activated. Such a behaviour is rather different from the quenching mechanism in the SiGe/Si structures grown by the chemical vapour deposition

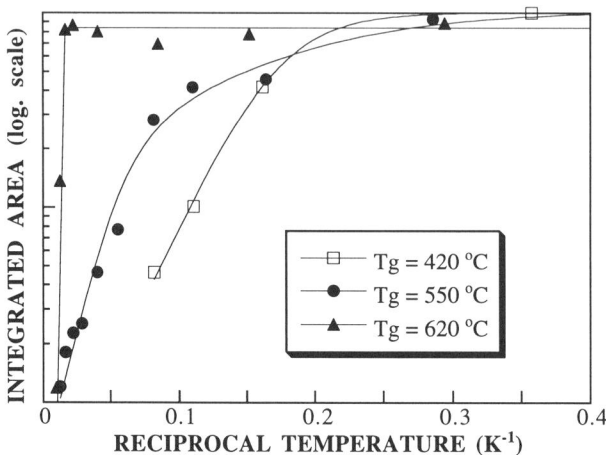

Fig. 3. Arrhenius plots of the integrated PL intensity from SiGe QWs grown without surfactant. Spectra were measured at an excitation power of 150 mW/cm^{-2}. The growth temperature for different structures is specified at the Figure. Symbols represent experimental data. Solid lines represent fits of experimental data using activation energies of 4 ± 2 meV (for $T_g = 420$ °C), 6 ± 2 meV (for $T_g = 550$ °C) and 95 ± 5 meV (for $T_g = 620$ °C), respectively.

(CVD) technique, where the PL efficiency at high temperatures was shown [9] to be controlled by surface recombination rather than by bulk SiGe or Si properties, and can be substantially improved by proper surface passivation. The obtained activation energy of the non-radiative process in low temperature grown SiGe/Si SQW structures of 4 ± 2 meV is very close to the binding energy of excitons to impurities in SiGe QWs, which is ~ 5 meV [6]. Thus the activation process can rather be attributed to the thermal release of excitons from impurities, than to the thermally induced increase of the capture cross-section of the nonradiative defects. This is also supported by the observation that quenching of the BE recombination occurs at lower temperatures as compared to the deeper LE emissions - Fig. 2a.

Using the ODMR method we have earlier shown [10, 11] that one of the dominant NR defects, introduced during the low temperature growth in the Si layers and Si/SiGe structures, is the vacancy-oxygen (V-O) complex. From the ODMR, these defects were shown to be mainly located in the Si barriers and can be efficiently removed either using a high growth temperature or by post growth treatments, such as post-growth annealing at $T \geq 500$ °C or hydrogen passivation [12]. The effect of the growth temperature, as well as post-growth treatments on the NR defects and on the PL thermal behaviour correlates remarkably well, establishing a direct link between these defects and the PL thermal quenching. It is therefore believed that the NR defects in the Si barriers are to a great extent responsible for the rapid thermal quenching of the PL from the SiGe SQWs, via strong competing carrier capture and recombination processes in the barriers.

Role of the surfactant.

In order to suppress Ge segregation which becomes severe with increasing growth temperature, high temperature growth of Si-based heterostructure is often performed using some surfactant species. Various studies of dopant incorporation in Si/Ge MBE growth have revealed that most dopants strongly segregate on to the growing film [1]. In particular, it was shown that Sb is a strong segregant during the MBE growth of both Si on Ge and Ge on Si and consequently is a good candidate for a surfactant. Significant improvements in the abruptness of the Si/SiGe

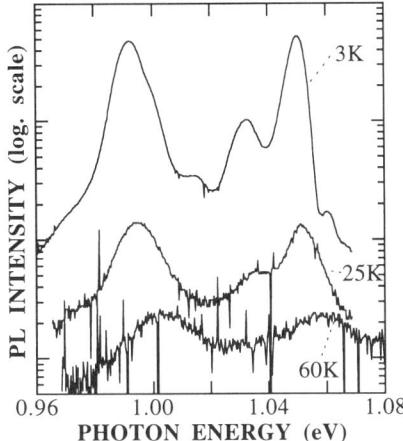

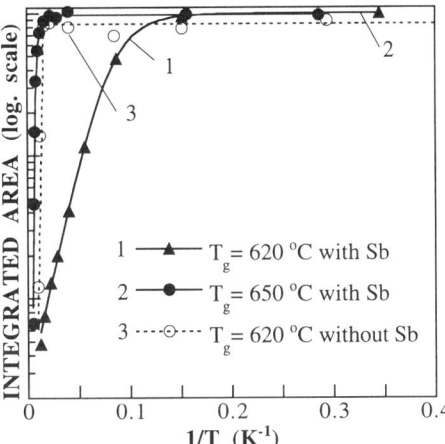

Fig. 4. Temperature dependent PL spectra measured from the SiGe/Si SQW structures fabricated by surfactant assisted growth at 620° C using Sb adlayer. Spectra were measured at an excitation power of 7.5 mW/cm^{-2}.

Fig. 5. Arrhenius plots of the integrated PL intensity from the SiGe/Si SQWs grown under conditions specified in the legend. Symbols represent experimental data. Solid lines are fitting curves obtained using activation energies of 7 ± 2 meV (curve 1) and 95 ± 5 meV (curves 2 and 3), respectively.

interface have indeed been achieved when using an Sb adlayer, as demonstrated by sputter depth profile studies [4] and secondary ion mass spectroscopy [5].

The effect of the surfactant on the PL thermal behaviour can be determined by comparing temperature dependence of the SiGe PL for the structures grown at the same temperature (620 °C in our case) with and without surfactant. Corresponding temperature dependent PL spectra are shown in Fig. 4 and Fig. 2b, respectively. As it is clear from the comparison, use of the surfactant at 620 °C significantly degrades the PL thermal quenching. This is caused by the appearance of a competing process with a low activation energy of about 7 ± 2 meV, which dominates the PL thermal behaviour - Fig. 5. The situation can be significantly improved by a further increase of the growth temperature up to 650 °C, when the intrinsic activation from the SiGe QW becomes the dominant quenching process.

The observation of the quenching process with a low activation energy in the structures fabricated by surfactant assisted growth at 620° C using a Sb adlayer indicates efficient introduction of competing nonradiative centres under such growth. The origin of the corresponding NR defects is different from those introduced during the low temperature growth. Vacancy-oxygen complexes, which are responsible for the premature PL thermal quenching in the low temperature grown structures, should not be efficiently formed at such high growth temperatures. In addition, the NR defects created during surfactant mediated growth are insensitive to hydrogen passivation. The formation of these NR defect can be caused by the partial self-incorporation of Sb atoms into Si barriers during the growth [4, 5]. In this case the concentration of the NR defects should decrease with increasing growth temperature due to a reduced self-incorporation of Sb atoms [5] leading to an improvement in the PL thermal quenching, as indeed observed in our experiments - see Fig. 5.

The formation of the NR defects during surfactant-mediated growth is also confirmed by the ODMR studies, which reveal important contribution of non-radiative recombination channels in corresponding structures. The broader width of the ODMR lines detected from the SiGe/Si samples grown with surfactant, as well as strong overlap by the substrate-related ODMR signal, unfortunately, has prevented us from a definite identification of the NR defects.

CONCLUSIONS

We have shown that growth conditions e.g. growth temperature and the use of surfactant impurities significantly affect the thermal behaviour of the PL emissions from the SiGe/Si quantum structures grown by MBE. In particular the use of a low growth temperature causes a premature PL quenching at $T < 20$ K with a very low activation energy of about 4 meV. The activation process is shown to be attributed to the thermal release of excitons from impurities in SiGe QWs with subsequent capture by competing efficient nonraditive defects, of which one has been identified as the vacancy-oxygen complexes from the ODMR studies. These defects can be efficiently removed either by an increase of growth temperature or by post growth annealing or hydrogen passivation, which lead to a significant improvement in the PL thermal behaviour.

It is also shown that the use of surfactant impurities during the growth can degrade PL thermal quenching, even though the corresponding structures demonstrate a significant improvement in abruptness of the Si/SiGe interface. The formation of the competing NR defects in this case is suggested to be due to the self-incorporation of the surfactant impurities into the Si barriers, and can be suppressed by increasing the growth temperature.

REFERENCES

1. E. Tournie and K. H. Ploog, Thin Sold Films **231**, 43 (1993).

2. W.-X. Ni, A. Henry, M. I. Larsson, K. Joelsson, and G. V. Hansson, Appl. Phys. Lett. **65**, 1772 (1994).

3. M. Wachter, F. Schäffler, H.-J. Herzog, K. Thonke, and R. Sauer, Appl. Phys. Lett. **63**, 376 (1993).

4. N. Usami, S. Fukatsu, and Y. Shiraki, Appl. Phys. Lett. **63**, 388 (1993).

5. H. P. Zeindl, S. Nilsson, J. Klatt, D. Krüger, R. Kurps, J. Crystal Growth, **157**, 31 (1995).

6. J. C. Sturm, H. Manoharan, L. C. Lenchyshyn, M. L. W. Thewalt, N. W. Rowell, J.-P. Noël, and D. C. Houghton, Phys. Rev. Lett. **66**, 1362 (1991).

7. L. C. Lenchyshyn, M. L. W. Thewalt, J. C. Sturm, P. V. Schwartz, E. J. Prinz, , N. W. Rowell, J.-P. Noël, and D. C. Houghton, Appl. Phys. Lett. **60**, 3174 (1992).

8. I. A. Buyanova, W. M. Chen, G. Pozina, W.-X. Ni, G. V. Hansson, and B. Monemar, Appl. Phys. Lett. **71**, 3676 (1997).

9. A. St. Amour, J. C. Sturm, Y. Lacroix, and M. L. W. Thewalt, Appl. Phys. Lett. **65**, 3344 (1994).

10. W. M. Chen, I. A. Buyanova, W.-X. Ni, G. V. Hansson, and B. Monemar, Phys. Rev. Lett. **77**, 4214 (1996).

11. W. M. Chen, I. A. Buyanova, A. Henry, W.-X. Ni, G. V. Hansson, and B. Monemar, Appl. Phys. Lett. **68**, 1256 (1996).

12. W. M. Chen, I. A. Buyanova, W.-X. Ni, G. V. Hansson, and B. Monemar, Appl. Phys. Lett. **70**, 369 (1997).

LOW-ENERGY PLASMA ENHANCED CHEMICAL VAPOR DEPOSITION

CARSTEN ROSENBLAD[1], THOMAS GRAF[1], ALEX DOMMANN[2], HANS VON KÄNEL[1]
[1]Laboratorium für Festkörperphysik, ETH Zürich, CH-8093 Zürich, Switzerland
[2]Neu-Technikum Buchs, CH-9470 Buchs, Switzerland

ABSTRACT

We discuss a new method for plasma enhanced chemical vapor deposition, applied to the epitaxial growth of Si and of Si-Ge heterostructures. Growth rates up to 5 nm/s become possible at substrate temperatures below 600° C, by utilizing very intense but low energy plasmas to crack the reactive gases, SiH_4 and GeH_4, and to speed up the surface kinetics. The method is applied to the synthesis of step-graded Si-Ge buffer layers, exhibiting the well known cross-hatched surface morphology.

INTRODUCTION

Deposition techniques which can be used at comparatively low substrate temperatures are required in order to suppress, e.g., islanding of strained-layer semiconductor heterostructures and dopant segregation. The methods most commonly used in Si-Ge heteroepitaxy are ultrahigh-vacuum chemical vapor deposition (UHV-CVD) [1] and molecular beam epitaxy (MBE) [2]. While material of excellent quality can be produced by both, neither is well suited for production. Thus MBE is hampered by the frequent need to replenish the source materials for the electron beam evaporators, while in UHV-CVD the growth rates become exceedingly low at substrate temperatures below 600° C. Plasma enhancement has been recognized long ago to be a possibility to overcome this shortcoming [3, 4]. Ion-induced damage is, however, often a serious problem, unless the energy of ions impinging on the growing film is carefully controlled [5]. In this paper we describe a new process for plasma-enhanced CVD, based on a low-voltage DC arc discharge. The process is called low-energy plasma enhanced chemical vapor deposition (LEPECVD), in order to emphasize the exceptionally low ion energies created in this kind of plasma. While ion energies are low, the plasma densities at the position of the substrate can be exceedingly large, leading to very efficient cracking of the reactive gases and to greatly enhanced surface kinetics. In this way, exceptionally high growth rates of several nm/s become possible at substrate temperatures below 600° C. We have applied LEPECVD to the synthesis of homoepitaxial Si films and of step-graded Si-Ge buffer layers used as virtual substrates for the modulation doped field effect transistor (MODFET).

EXPERIMENTAL

The plasma source used for LEPECVD has been used previously for hydrogen plasma cleaning of Si substrates [6]. The essence of it is a low voltage arc discharge sustained by a hot Ta filament. The plasma source is connected to the UHV deposition chamber through an orifice $\sim$ 1 cm in diameter. Typical voltages between the filament and the grounded chamber walls are in the range of 20-30 V, leading to arc currents of up to 70 A. This, together with an externally applied substrate bias [7], limits the energy of ions impinging on the substrate to typically 10 eV or even less. In order to gain better control over the plasma distribution, the deposition chamber is equipped with an auxiliary grounded anode. Furthermore, the plasma is compressed by the magnetic field generated by coils surrounding the chamber. The discharge gas, Ar, is fed directly into the plasma source, while the reactive gases, H_2, SiH_4, GeH_4, and PH_3 diluted with Ar, are introduced into the deposition chamber through a gas dispersal ring placed below the radiatively heated substrate. The substrates used for this study were lightly boron doped Si(001) wafers. They were subjected to an HF dip before introducing them into the UHV system. After degassing in the deposition chamber for 1 h at 350°C a hydrogen plasma clean was performed in two steps, in which the substrate was kept at 350 and 600°C, respectively. Epitaxial growth usually commenced by depositing a 100 nm thick Si buffer layer at a substrate temperature of 600°C at a silane flow of 1 sccm, resulting in a growth rate of $\sim$ 0.3 nm/s at an arc current of 40 A and a substrate bias of - 8.5 V with respect to ground. Such buffer layer surfaces exhibit very sharp 2 × 1 RHEED patterns, indicating epitaxial growth in the step-flow regime, as was confirmed by scanning tunneling microscopy (STM). Homoepitaxial Si films were grown by LEPECVD at substrate temperatures between 400 and 600° C, and at growth rates up to 5 nm/s. Doping was achieved by mixing PH_3 diluted in Ar into the reactive gas phase. One application, in which the high growth rates achievable by LEPECVD are particularly attractive, are step-graded Si-Ge buffer layers. We present first results achieved with buffers graded up to a final Ge concentration of 30 % at growth rates exceeding 1 nm/s. The crystal quality was examined by high-resolution X-ray diffraction, defect etching, transmission electron microscopy and scanning force microscopy, as well as optical interference contrast microscopy. High resolution X-ray reciprocal space mapping in the vicinity of the Si(004) and Si($\bar{2}\bar{2}4$) Bragg peaks was carried out to determine the depth strain profile. To this end, high-resolution X-ray rocking curves were measured at a step omega-scan of 0.002°, using a 4-circle high-resolution diffractometer equipped with a Ge (220) monochromator in the $(+ - - +)$ arrangement (Philips MPD 1880), and a channel cut analyzer in the CuK_{α_1} diffracted beam. Both monochromator and analyzer used Ge(220) setting.

RESULTS

Even though very little is known to date about the plasma chemistry going on in the intense plasma of the arc discharge used for LEPECVD, two main factors could be shown to affect the epitaxial growth of Si [8]. The first one is bulk damage induced by impinging ions with energies above $\sim$ 15 eV, and the second one is hydrogen coverage and its effect on adatom mobility. Both effects could be shown to lead to the formation of stacking faults after a certain thickness of defect-free growth. There appears to be a close correlation between the defect formation and the surface roughness accumulating during growth. Ion damage could be entirely eliminated by raising the substrate potential sufficiently above the floating potential, i.e. from ~ -16 V to ~ -9 V, such that the energy of ions striking the substrate is below $\sim$ 15 eV [7]. On the other hand, the bombardment of the surface with low-energy ions is very much desired, as it leads to a substantial increase of the surface kinetics. Most importantly, the hydrogen coverage of the surface becomes lower at high plasma densities, as a result of which epitaxial growth rates can be greatly enhanced. An example for the case of Si homoepitaxy carried out at 550° C is shown in Fig. 1. Here, the Si growth rate is displayed as a function of the silane flow, for an arc current of 60 A. All films were grown to thicknesses of several μm, and all of them were epitaxial, as evidenced by the sharp Kikuchi pattern observed by RHEED. To our knowledge, these rates for epitaxial growth are the highest ever achieved at such low growth temperatures. Hence LEPECVD might

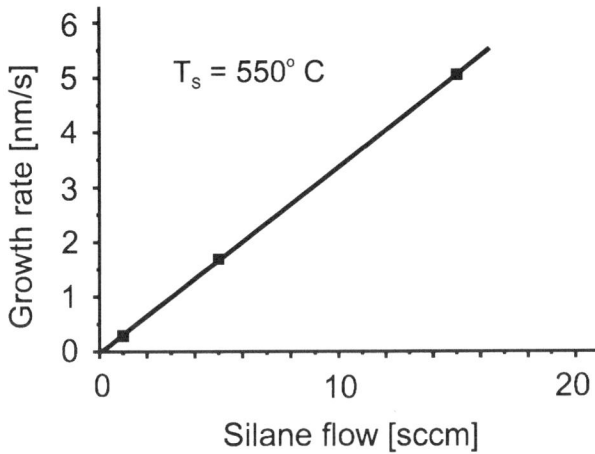

Figure 1: Growth rate of Si films deposited by LEPECVD as a function of the silane flow at a substrate temperature of 550° C. The plasma conditions used (arc current 60 A, substrate bias -8.5 V) resulted in epitaxial growth up to the highest rate of 5 nm/s.

become the method of choice wherever fast deposition offers a major advantage, provided of course that the material can be shown to be of sufficient electronic quality. A first indication stems from defect etching, revealing etch pit densities below 10^6 cm^{-2} even for films grown at 5 nm/s. Secondly, some p-i-n junctions were fabricated by growing first undoped Si films, several μm in thickness, on p$^+$ substrates. A doped cap layer was subsequently grown by mixing dilute phosphine into the reactive gas phase.

One of the most interesting applications of LEPECVD to Si-Ge heteroepitaxy is the synthesis of relaxed buffer layers for MODFETs, since here the use of rather thick layers is highly desirable [2]. Such buffer layers were grown at rates exceeding 1 nm/s and at a substrate temperature of 500° C. Fig. 2 shows an example of a high resolution X-ray reciprocal space map (RSM), represented by intensity contour plots, in the vicinity of the symmetrical (004) and asymmetrical ($\bar{2}\bar{2}4$) diffraction peaks. The RSM was obtained on a Si$_{1-x}$Ge$_x$ buffer step-graded at 5 % per μm up to a final Ge concentration of nominally $x = 0.3$, given by the ratio of germane and silane flows. A 3 μm thick film with constant composition was grown on top of the graded part. The average growth rate was approximately 1.7 nm/s. The spread in perpendicular lattice constant due to the grading can readily be seen from the (004) map in Fig. 2. A small crystallographic tilt of the graded epilayer with respect to the substrate can also be observed in the right part of Fig. 2. It amounts to ~ 0.2 degrees. From the ($\bar{2}\bar{2}4$) map the degree of lattice relaxation can be deduced. The parallel q-vector of a pseudomorphic layer would be expected to lie on the vertical line through the reflection of the Si substrate (left hand side of Fig. 2). The inclined line on the other hand marks the expected peak positions for a completely relaxed film. As can be seen, the graded part of the buffer is relaxed to a large extent, whereas the cap layer of constant composition is still partly strained. In order to evaluate the residual strain quantitatively, the ($\bar{2}\bar{2}4$) map

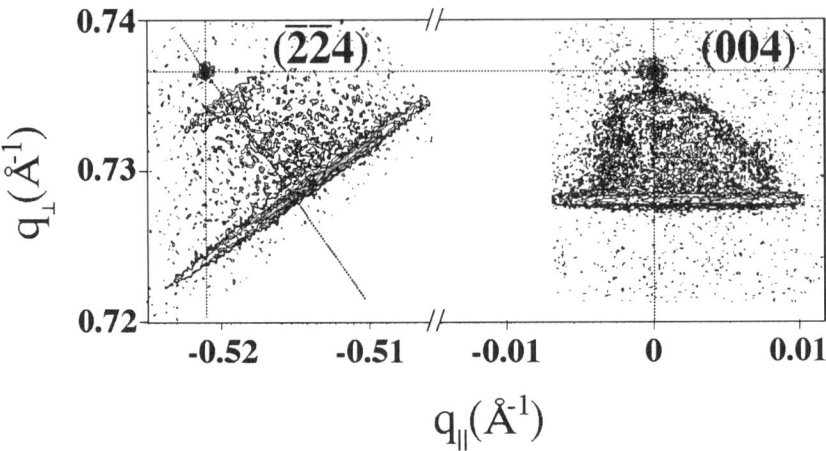

Figure 2: X-ray reciprocal space map of a 9.2 μm thick step-graded Si$_{1-x}$Ge$_x$ buffer layer grown by LEPECVD at a rate of 1.7 nm/s. The sample was graded to a final Ge concentration of $x = 0.286$ at a grading rate of 5% μm^{-1}. The thickness of the Si$_{0.714}$Ge$_{0.286}$ cap is 3 μm.

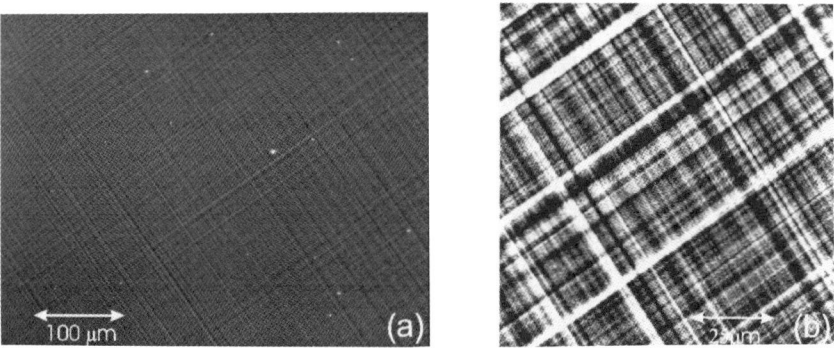

Figure 3: (a) Normarski intereference contrast micrograph of a 9.2 μm thick Si$_{1-x}$Ge$_x$ buffer layer step-graded up to $x = 28.6$ % at a grading rate of 5%/μm and capped with 3 μm of Si$_{0.714}$Ge$_{0.286}$. The growth rate was 1.7 nm/s and the substrate temperature 500° C. (b) AFM image of the same layer. The rms roughness evaluated across the entire image was 4 nm. Growth parameters were: Substrate bias - 8.5 V, arc current 60 A, Ar flow 50 sccm, total reactive gas flow (SiH$_4$ + GeH$_4$) 5 sccm.

was first corrected for the tilt deduced from the (004) map. Using the procedure outlined in Ref. [9], the actual Ge concentration in the cap layer was determined to be $x = 0.286$ and the residual parallel strain $\epsilon_\parallel = 0.2$ %. This strain is hence only 4 times larger than the one of comparable MBE material, grown at much lower rates and much higher temperatures [9]. The buffer layer exhibits the usual cross-hatch morphology due to dislocation glide along $\{111\}$ planes [2, 10]. Fig. 3(a) shows a Nomarski intereference contrast micrograph of such a surface, and Fig. 3(b) a corresponding smaller scale AFM image. The defects found most frequently in step-graded buffer layers grown by LEPECVD are clusters of stacking faults, leading to heaviliy distorted regions up to tens of microns in size. The formation of these defects appears to be correlated with the relaxation mechanism, as the defects are often aligned parallel to the cross-hatch lines. On surfaces exhibiting a cross-hatch the density of these clusters of defects is of the order of 10^5 cm^{-2}. They are, however, far less frequent on smooth surfaces, such as homoepitaxial Si or SiGe alloys grown at constant composition, i.e., without grading. It is hence likeley that the surface roughness plays an important role in the formation of these defects, as previously observed in the case of ion damage or limited surface mobility [8]. A detailed interpretation will, however, have to await a more detailed study of the influence of the grading rate, growth rate and substrate temperature on the formation of these defects.

CONCLUSIONS

A new technique, called low-energy plasma enhanced chemical vapor deposition (LEP-ECVD) has been investigated. At substrate temperatures below 600° C the method has been shown to yield exceptionally high growth rates, of several nm/s, in both Si homoepitaxy and SiGe heteroepitaxy. Step-graded $Si_{1-x}Ge_x$ buffer layers could be shown to relax in the usual way even when grown at rates above 1 nm/s by LEPECVD. Provided that the electrical properties of the material turns out to be satisfactory, LEPECVD might become the method of choice whereever high growth rates are required at low substrate temperatures.

ACKNOWLEDGMENT

The authors wish to thank J. Schulze from the Universität der Bundeswehr, München, for the defect etching on relaxed buffer layers, and to T. Meyer and M. Kummer for their help with the analysis of the RSM maps. Financial support by the Swiss Priority Program in Micro & Nano System Technology (MINAST) is gratefully acknowledged.

References

[1] B.S. Meyerson, F.K. LeGoues, T.N. Guyen, and D.L. Harame, Appl. Phys. Lett. **50**, 113 (1987)

[2] For a recent review, see F. Schäffler, Semicond. Sci. Technol. **12**, 1515 (1997)

[3] T.J. Donahue and R. Reif, J. Appl. Phys. **57**, 2757 (1985)

[4] I. Nagai, T. Tahakagi, A. Ishitani, H. Kuroda, and M. Yoshikawa, J. Appl. Phys. **64**, 5183 (1988)

[5] H.-S. Tae. S.-H. Hwang, S.-J. Park, E. Yoon, and K.-W. Whang, Appl. Phys. Lett. **64**, 1021 (1994)

[6] J. Ramm, E. Beck, A. Züger, A. Dommann, and R.E. Pixley, Thin Solid Films **222**, 126 (1992)

[7] The plasma potential has been shown to be close to ground, such that a substrate bias referenced to ground reflects directly the energy of ions impinging on the substate. See N. Korner, E. Beck, A. Dommann, N. Onda, and J. Ramm, Surface and Coatings Technology **76-77**, 731 (1995)

[8] C. Rosenblad, H.R. Deller, A. Dommann, T. Meyer, P. Schroeter, and H. von Känel, unpublished

[9] J.H. Li, V. Holy, and G. Bauer, J. Appl. Phys. **82**, 2881 (1997)

[10] R.M. Feenstra, M.A. Lutz, F. Stern, K. Ismail, R.P. Mooney, F.K. LeGoues, C. Stanis, J.O. Chu, and B.S. Meyerson, J. Vac. Sci. Technol. B **13**, 1608 (1995)

LOW TEMPERATURE EPITAXY OF Si/Si$_{1-x}$Ge$_x$/Si MULTILAYERS BY LOW PRESSURE RTCVD FOR VERY THIN SOI APPLICATIONS

D.W. McNEILL, D.L. GAY[†], X. LI[‡], B.M. ARMSTRONG and H.S. GAMBLE
Northern Ireland Semiconductor Research Centre, Dept. Electrical and Electronic Engineering, The Queen's University of Belfast, Belfast, BT9 5AH, N. Ireland, U.K.
† Now with BCO Technologies Ltd., Glen Road, Belfast.
‡ Now with Applied Materials, Newbridge, Midlothian, Scotland.

ABSTRACT

The growth by rapid thermal chemical vapour deposition of Si/Si$_{1-x}$Ge$_x$/Si multilayer structures, suitable for thin bond and etch-back silicon-on-insulator fabrication has been investigated. Surface topography was studied by scanning probe microscopy, and layer contamination by secondary ion mass spectrometry. Smooth layers are only achieved at high growth temperatures (>700°C), and when surface oxide contamination is reduced by a combination of ex-situ HF vapour treatment and in-situ high temperature H$_2$ bake. A surface peak-to-peak roughness of 15nm for a Si/Si$_{1-x}$Ge$_x$/Si multilayer structure has been achieved by reducing the growth time at 700°C or less. Further improvement is possible, especially if carbon contamination can be reduced.

INTRODUCTION

Silicon-on-insulator (SOI) substrates present an attractive alternative to bulk silicon substrates for low-power, high-frequency applications. Increased packing densities compensate for the initially higher substrate costs.

The SIMOX (separation by implanted oxygen) fabrication technique, developed by Izumi et al in 1978 [1], produces thin SOI layers with very low total thickness variation (ttv), due to the uniformity of the implantation process. Bond and etch-back SOI (BESOI) fabrication techniques, reviewed by Maszara [2], utilise selective etch-stop layers, during final thinning of the active wafer, to ensure a low ttv can be achieved. The use of a Si$_{1-x}$Ge$_x$ etch-stop layer for BESOI preparation was first reported by Godbey et al [3], using molecular beam epitaxy (MBE). The present authors have already reported BESOI fabrication using an epitaxial Si$_{1-x}$Ge$_x$ etch-stop layer grown by rapid thermal chemical vapour deposition (RTCVD) [4].

BESOI FABRICATION

In its simplest form, the BESOI fabrication process begins with the epitaxial growth of a thin Si$_{1-x}$Ge$_x$ etch-stop layer on a clean silicon wafer, immediately followed by the growth of an overlying silicon layer, which will eventually constitute the active SOI layer. This wafer is then bonded to an oxidised handle wafer, and thinned using grinding and polishing techniques, resulting in the cross-section of Figure 1.

Mat. Res. Soc. Symp. Proc. Vol. 533 © 1998 Materials Research Society

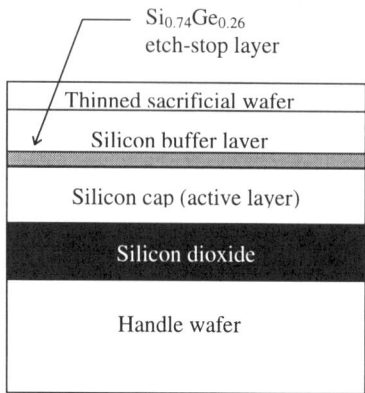

Figure 1: Structure of bonded substrate after thinning

The silicon may now be etched in EPW etch[†] at 100°C, followed by selective $Si_{1-x}Ge_x$ removal in HHA etch[‡] at room temperature. For a germanium fraction of x=0.26, the selectivity of the first etch is 90:1, whilst that of the second etch is in excess of 200:1 [4]. The excellent combined selectivity of the double etch process ensures that the surface topography left after the polishing process is not transferred to the final SOI layer.

Therefore, the final ttv of the active SOI layer is totally dependent on the epitaxial growth process, which needs to demonstrate thickness uniformity across the substrate and minimal surface roughness.

EXPERIMENTAL RESULTS: BESOI STRUCTURES

Epitaxial layers were grown by RTCVD in a load-locked system [5], with SiH_4 and GeH_4 as the source gases. The quartz-walled growth chamber accommodates a single process wafer, which is heated by tungsten-halogen lamps.

In the originally reported BESOI fabrication process [4], substrates were cleaned ex-situ using an HF vapour treatment, designed to create an oxide-free, hydrogen-passivated surface. No in-situ clean was performed prior to growth of the multi-layer structure depicted in Figure 2. Silicon layers were grown at 700°C; the $Si_{1-x}Ge_x$ layer at 600°C. The low growth temperatures were dictated by the need to avoid relaxation of the pseudomorphic $Si_{1-x}Ge_x$ alloy, in which the germanium fraction was x=0.26.

Despite the fact that the grown layers had a specular appearance, an early indication of surface roughness was that initial attempts to bond to oxidised handle wafers were unsuccessful. The epitaxial layers had first to be coated with a low temperature CVD oxide followed by a CVD polysilicon layer. After polishing, bonding was successfully accomplished, and BESOI structures fabricated.

[†] Composition of EPW etch: 200ml ethylenediamine, 27g pyrocatechol, 40ml water
[‡] Composition of HHA etch: 1 part 48% hydrofluoric acid, 2 parts 30% hydrogen peroxide, 3 parts 98% acetic acid

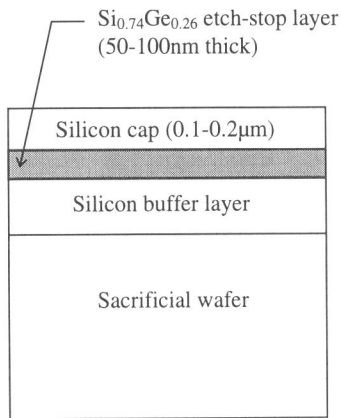

Si$_{0.74}$Ge$_{0.26}$ etch-stop layer
(50-100nm thick)

Silicon cap (0.1-0.2μm)

Silicon buffer layer

Sacrificial wafer

Figure 2: Epitaxial multi-layer etch-stop structure

The final thin SOI layers were punctuated by a high density of voids, again indicating surface roughness, but samples prepared for transmission electron diffraction (TED) demonstrated the single-crystalline structure of the layers. The TED pattern of Figure 3, although indicating some crystal defects, shows a well-defined epitaxial structure.

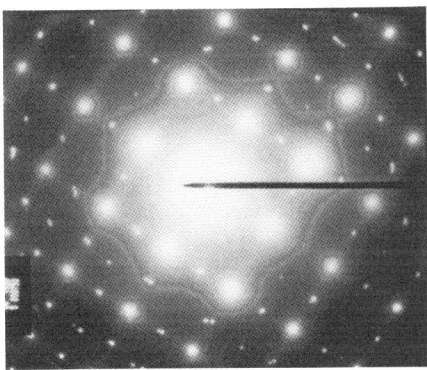

Figure 3: TED pattern of BESOI layer, showing single-crystalline structure

EXPERIMENTAL RESULTS: EPITAXIAL GROWTH

To investigate the origin of the suspected surface roughness, a series of epitaxial structures was grown by RTCVD. These included multilayer structures incorporating Si$_{1-x}$Ge$_x$ etch-stop layers, as well as single-layer structures for comparison. Surface topography was studied by scanning probe microscopy (SPM), and layer contamination by secondary ion mass spectrometry (SIMS).

All wafers were pre-cleaned using a 30 sec. 10:1 HF dip, followed by a 10 min. H_2O_2:H_2SO_4 clean, and finishing with a 10 min. megasonic clean in SC1 solution. In the majority of cases, there followed an HF vapour treatment immediately before loading, to create a hydrogen-passivated surface. This entails holding the wafer upside-down over a solution of 2:1 HF:H_2O for 1 minute.

After a 10 minute pump-down and transfer into the growth chamber, an optional high temperature H_2 bake was carried out, followed by layer growth. The growth conditions for a range of samples are detailed in Table 1, along with surface roughness values obtained by SPM.

Sample	Wafer orientation	Layer description	Growth temperature (°C)	In-situ H_2 bake Temp (°C)	Time (min.)	Surface Roughness (nm)
A	(100)	Si/SiGe/Si	700/600/700	None		≈150
B	(111)	Si/SiGe/Si	700/600/700	None		≈50
C	(100)	Si	700	None		≈40
D	(100)	Si	990	None		0.8
E	(100)	Si	990	990	1	0.5
F	(100)	Si	870	870	15	0.9
G	(100)	Si	700	870	15	10
H	(100)	Si/SiGe/Si	990-700/600/700	990	1	15

Table 1:   **Process conditions and surface peak-to-peak roughness values for RTCVD epitaxial samples**

The multi-layer structure of sample A was the same as that used for BESOI fabrication. An SPM image of the surface (Figure 4) reveals island growth, orientated to the (100) substrate. Wang et al found similar results [6], which they attributed to polycrystalline growth. However, the TED evidence would not support this. Aligned island growth was also evident on sample B, which differed only in the (111) substrate orientation. Even when the $Si_{1-x}Ge_x$ layer and silicon cap layer were omitted (sample C), rough substrate-aligned layers were observed, as shown in Figure 5. SIMS analysis revealed that these samples had high residual concentrations of both oxygen and carbon at the original substrate surface. It seems likely, therefore, that the roughness is caused by selective epitaxial growth on a partially oxide-covered surface.

Despite the surface contamination, reasonably smooth silicon layers could be grown at higher temperatures. Sample D, grown at 990°C, exhibited a peak-to-peak roughness of only 0.8nm over a 1μm square. Nevertheless, it was felt necessary to introduce an in-situ H_2 bake to the process, in order to obtain an oxide-free surface, and restore the prospect of smooth epitaxial growth at lower temperatures.

Earlier work had shown the detrimental effect of relying solely on an in-situ H_2 bake for oxide removal [4]. The long time required at high temperature (typically 5 min. at 950°C) was sufficient to cause localised substrate etching, and subsequently grown epitaxial layers were non-specular, with a high density of square defects. (The presence of surface carbon contamination or particulates may determine the density and size of the defects.) It was, therefore, decided to combine the ex-situ HF vapour treatment with a reduced in-situ H_2 bake.

Sample E, grown at 990°C after a 1 min. H_2 bake at the same temperature, showed an improved roughness figure (compared to sample D) of 0.5nm. SIMS confirmed that the in-situ bake had reduced the residual oxygen contamination by an order of magnitude. When the bake temperature was reduced to 870°C, a longer time of 15 min. was necessary to achieve comparable oxide removal. Sample F, grown at 870°C after a 15 min. H_2 bake at the same temperature, exhibited a roughness figure still below 1nm. However, when the same bake was followed by silicon growth at 700°C (sample G), the roughness figure rose to 10nm, even though SIMS proved that oxide removal had been unaffected. This result clearly shows the importance of surface mobility in determining the roughness of epitaxial silicon growth, even on an oxide-free surface. The critical temperature for smooth layer growth is likely to be determined by surface carbon contamination, which remains high and could seriously reduce surface mobility.

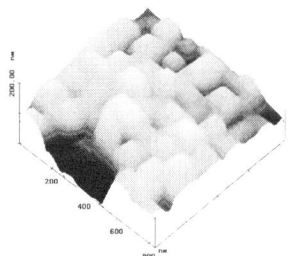

Figure 4: SPM surface morphology for sample A

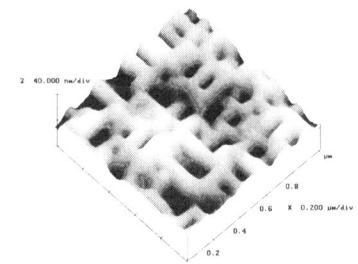

Figure 5: SPM surface morphology for sample C

The implication for $Si_{1-x}Ge_x$ etch-stop layer growth in the present RTCVD system is that any growth performed at 700°C or less will inevitably cause surface roughening. However, if the time spent at or below 700°C is minimised, the surface roughness may be reduced to acceptable levels.

Finally, a new $Si_{1-x}Ge_x$ etch-stop structure was grown (sample H). This sample had the same $Si_{1-x}Ge_x$/Si surface layer growth conditions as sample A, but the pre-clean and silicon buffer growth were both modified, in the light of the results presented here. The ex-situ HF vapour treatment was now followed by a H_2 bake at 990°C for 1 min., to reduce surface oxide contamination. Instead of a 5 min. silicon buffer growth at 700°C, which can be expected to severely roughen the surface, silicon growth was now commenced at 990°C, and a reduced time of 1 min. was spent at 700°C, immediately prior to $Si_{1-x}Ge_x$ growth at 600°C and silicon cap growth at 700°C. The total growth time spent at 700°C or less was now 2 min. compared to 6 min. for sample A. SPM showed the surface roughness to be 15nm across a 1µm square, compared to approximately ten times that value for sample A. A SIMS depth profile of sample H is shown in Figure 6, revealing low oxygen contamination levels, but a high residual carbon contamination level at the substrate surface.

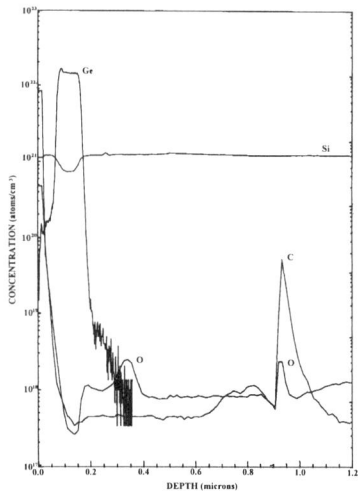

Figure 6: SIMS depth profile of final $Si/Si_{1-x}Ge_x/Si$ multilayer structure (Sample H)

CONCLUSIONS

The surface roughness of RTCVD epitaxial silicon layers has been shown to deteriorate if surface oxide removal is incomplete prior to growth. An ex-situ HF vapour treatment followed by a short high temperature H_2 bake gives adequate oxide removal without detrimental substrate etching.

Growth at temperatures of 700°C or below has been shown to cause surface roughening even after oxide removal. Carbon contamination levels may be responsible for reducing surface mobility, and should be reduced if smooth growth at lower temperatures is sought.

A tenfold reduction in the surface roughness of $Si/Si_{1-x}Ge_x/Si$ etch-stop structures has been achieved by reducing the growth time spent at 700°C or less. Further improvement could result from totally eliminating the silicon buffer growth step at 700°C, although the $Si_{1-x}Ge_x$ growth and silicon cap growth must still be carried out at low temperature to avoid relaxation. BESOI fabrication, using a $Si_{1-x}Ge_x$ etch-stop layer grown by RTCVD, remains a viable means of producing thin SOI layers with low ttv.

REFERENCES

[1] K. Izumi, M. Doken and H. Ariyoshi, Electronics Lett., **14(18)**, pp.593-594, (1978).

[2] W.P. Maszara, J. Electrochem. Soc., **138(1)**, pp.341-347, (1991).

[3] D. Godbey, H. Hughes, F. Kub, M. Twigg, L. Palkuti, P. Leonov and J. Wang, Appl. Phys. Lett., **56(4)**, pp.373-375, (1990).

[4] X. Li, D.L. Gay, D.W. McNeill, B.M. Armstrong and H.S. Gamble, presented at 1997 Joint International Meeting of Electrochemical Society in Paris, shortly to be published in Proceedings.

[5] D.W. McNeill, B.M. Armstrong and H.S. Gamble, Mat. Res. Soc. Symp. Proc., **326**, pp.187-192, (1994).

[6] C.L. Wang, S. Unnikrishnan, B.Y. Kim, D.L. Kwong and A.F. Tasch, J. Electrochem. Soc., **143(7)**, pp.2387-2391, (1996).

LOADING EFFECTS DURING LOW-TEMPERATURE SEG OF Si AND SiGe

W.B. DE BOER, D. TERPSTRA AND R. DEKKER
Philips Research Laboratories, Prof. Holstlaan 4, 5656 AA Eindhoven, The Netherlands

ABSTRACT

Selective Epitaxial Growth (SEG) of Si and SiGe suffers from pattern sensitivity. The growth rate and layer composition change with the pattern and the window size. In relation to growth at atmospheric pressure, the sensitivity to the window size is suppressed at reduced pressure, whereas some growth rate effects of a more global nature become more visible. The Si growth rate decreases when the area of exposed silicon on the wafer and the susceptor decreases. SiGe shows the opposite behavior: its growth rate increases with decreasing silicon area. Another difference between Si and SiGe is the range over which the loading effects are active. The influence of a large silicon area on the Si growth rate can be felt inches away, whereas the SiGe growth rate is affected over a much shorter distance. In common epi reactors the wafer rests on a susceptor, which extends beyond the wafer, exposing a large Si-coated surface area around the circumference of the wafer. Consequently, the Si growth rate varies unacceptably across the wafer. A sacrificial polysilicon layer has been successfully applied to improve the growth rate uniformity across the wafer.

INTRODUCTION

SEG of Si and SiGe at temperatures ranging from 600 to 800°C is a preferred way of depositing the base of a heterojunction bipolar transistor (HBT) self-aligned, enabling the growth of abrupt and complicated Ge and dope profiles. Full selectivity can be obtained on oxide as well as on nitride. In spite of the fact that impressive results have been achieved using this technology [1-4], its implementation in production is not as straightforward as the early results seemed to indicate. The growth rate and layer composition of doped Si and SiGe epi layers varies strongly when the patterns on the wafer are changed. This so-called loading effect is well-known when the growth takes place at atmospheric pressure. The growth rate in the windows depends on the size of the window and is usually higher in small windows and lower in large windows. The layer deposited in the windows is also uneven: thicker at the edges and thinner at the centre of the window. These effects are so severe that the use of atmospheric pressure SEG to grow e.g. the base of an HBT is precluded.

Growth at reduced pressure overcomes these local loading effects as the increased diffusion lengths of the reactants in the gas phase have very beneficial averaging effects [5]. In this paper we report on the loading effects remaining when SEG is effected at a reduced pressure and low temperature.

EXPERIMENTAL

All the growth experiments were performed in an Epsilon One reactor. This is a commercially available epi reactor for large-scale production, characterized by loadlocked wafer entry and single-wafer processing. The wafer is heated by two banks of tungsten-halogen lamps and is supported inside a quartz reaction chamber on a SiC-coated susceptor. A laminar flow of process gas flows parallel to the surface across the wafer. In order to obtain a good layer uniformity, the wafer is rotated at 20-30 rpm. The reactor is capable of running conventional epi processes at temperatures of 1150-1200 °C as well as low-temperature processes at 600-700 °C. The runs can be made at atmospheric or reduced pressure.

In this study, the SEG runs were carried out at reduced pressure (20 Torr) in an H_2 ambient at 700°C. The main H_2 flow was 10 slm (standard liter/minute). The Si source was SiH_2Cl_2, the Ge source GeH_4, and PH_3 and B_2H_6 were used as doping gases. In order to maintain the selectivity towards oxide and nitride some HCl was added to the gas flow during the growth of Si or SiGe. Prior to deposition the silicon surfaces were subjected to a bake in a hydrogen ambient at 900 °C in order to remove the native oxide and to be able to grow device-quality epi layers.

Unless otherwise indicated, the Si and SiGe layers were grown selectively on (100) oriented Si wafers with a cross-section of 150 mm in windows in a nitride- or oxide-covered surface. The

Mat. Res. Soc. Symp. Proc. Vol. 533 © 1998 Materials Research Society

stack to be deposited was an epitaxial SiGe layer, containing 10-15 at % Ge, capped by a Si layer. The layer thicknesses and composition were measured by means of SIMS (secondary ion mass spectroscopy) .

RESULTS AND DISCUSSION

The growth rate of SiGe and Si on patterned wafers is different from the growth rate obtained on bare substrates. This is no surprise as it is known that the exposed free silicon area has an influence on the growth rate [6]. This difference is illustrated in fig.1, which shows the SiGe and Si growth rates under identical conditions on four wafers with different free silicon areas. The exposed silicon area was 100% on a bare substrate, 50% on a wafer with a checkerboard pattern of 10x10mm², 15% on a patterned wafer with silicon exposed in the scribe lines and approximately 1% on a patterned wafer with oxide in the scribe lines. The layer thickness was measured at the centre of the wafer. In the area below 20% coverage the relative variations in free silicon area are large and the growth rate variations are also large. Going from exposed to oxide-covered scribe lines, the Si growth rate drops by a factor of almost three and the SiGe growth rate increases by a factor of 1.5. This indicates that any pattern change will be accompanied by a change in growth rate and that several iterations are needed before the layer thicknesses are back to

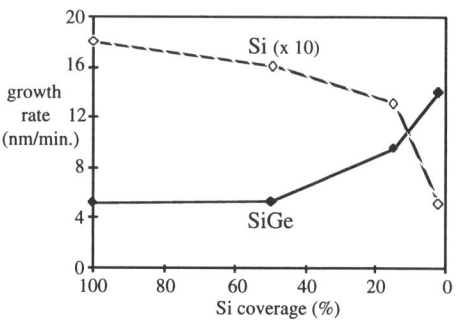

Fig.1 - The Si and SiGe growth rate depend strongly on the exposed Si area on the wafer. Note that the Si growth rate has been multiplied by a factor of 10 to better demonstrate the different behavior of Si and SiGe.

specification. In practise this is no problem as long as the layer uniformity across the wafer is sufficient.

The fact that the Si and SiGe growth rates behave contradictory is somewhat surprising. Similar behavior has been reported for the growth rate of SiGe by Ito et al.[7]. The decrease in growth rate with increasing silicon coverage, which one would expect, can be reversed to an opposite trend by increasing the HCl flow, according to their study. The selective Si growth shown in fig.1 is apparently in the latter regime, while the selective SiGe growth is still in the 'normal' regime. Note that the Si growth rate is about an order of magnitude lower than the SiGe growth rate. This is due to the growth rate enhancement of GeH₄.

In the experiment described above the exposed and covered Si surfaces were distributed evenly across the wafer on a macroscopic scale. However, when the susceptor is taken into account this is no longer the case. The susceptor which holds the wafer has a cross-section of 175 mm. Silicon is

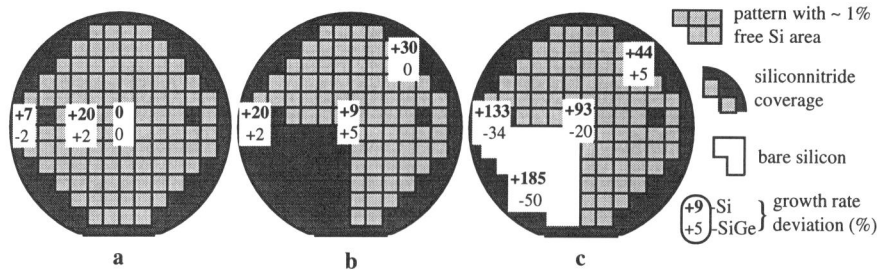

Fig.2 - Si and SiGe growth rate variations across the wafer for different pattern configurations. The centre of the fully patterned wafer (fig.a) serves as the reference point. The growth rate variations are given in percentages with respect to the reference point.

deposited on the susceptor and the large susceptor area just outside the wafer acts like a Si surface during SEG. The difference in exposed Si area between a patterned wafer with 1% exposed Si area and the susceptor area just outside the wafer which is 100% Si, is so great that centre-to-edge growth rate variations are to be expected. This was checked in a growth experiment using partly patterned wafers. The same double layer, SiGe with a Si cap, was deposited under selective conditions on all the wafers. The SiGe and Si growth rates were measured at several spots on the wafer and compared with the growth rate at the centre of a fully patterned wafer. Fig. 2 shows the results of three wafers used in the test. The first one was fully patterned (fig. 2a), the second 75% patterned, the remaining quarter being covered with nitride (fig. 2b), and the third one also 75% patterned while the remaining quarter was uncovered (fig. 2c).

Fig. 2a suggests that the growth rate variations across the wafer are within reasonable limits. Closer examination shows that this could be due to the fact that the measurement at the wafer edge was taken close to the alignment marker, which was surrounded by a large nitride area. Possible loading effects may also be blurred by growth rate deviations and measurement errors. Fig. 2b resembles 2a: the growth rate variations are relatively small. The Si growth rate seems higher at the wafer edge. In fig. 2c the loading effects are more pronounced. At the silicon/pattern boundary the Si growth rate is significantly higher than at the nitride/pattern boundary in fig. 2b, and the SiGe growth rate is lower. The influence of exposed Si versus nitride coverage is clear. Comparison of the different growth rates shows that the Si growth rate deviates more than the SiGe growth rate, at least under the SEG conditions used in this experiment. As expected, the outside susceptor ring causes a high Si growth rate at the wafer edge. An effect that seems to be strengthened by the additional free Si area (fig. 2c). Another conclusion that can be drawn from this experiment is that the range over which these loading effects are active is very large, well over 5 cm.

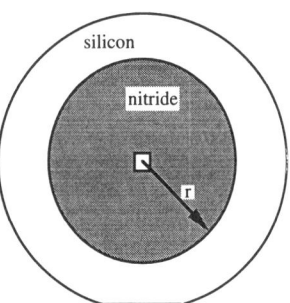

Fig.3 - Test pattern used to investigate global loading effects.

To determine more accurately the active range of the loading effects, a special test geometry was used. Rather than growing on patterned wafers of which a quadrant was left unpatterned, a circular symmetry was chosen. With a rotating wafer this leads to centre-to-edge variations only, which simplifies the interpretation. An inner circle of the wafer was covered with siliconnitride, except for the very centre of the wafer, where an area of 0.5 x 0.5 mm^2 was left uncovered, as sketched in fig. 3. The area between the edge of the wafer and the nitride-covered ring also was left uncovered, exposing bare Si. The effects of a varying Si coverage on the growth rate could now be studied by growing selectively on wafers with a nitrided area and measuring how the growth rate in the centre square of the wafer changes with a varying radius of the nitrided ring.

SiGe/Si layers were grown under selective conditions on a set of wafers with nitride radii ranging from 3 to 75 mm. It should be borne in mind that the susceptor area just outside the wafer was

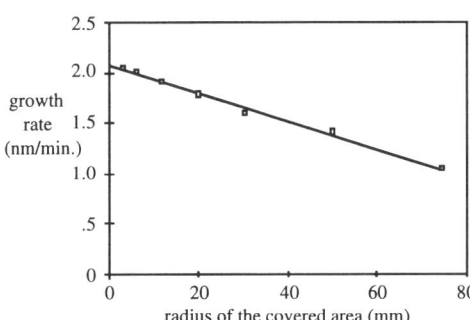

Fig.4 - The Si growth rate under selective growth conditions, measured at the centre of the wafer, decreases with increasing radius of the nitride-covered area.

a Si-covered area. Fig. 4 shows how the Si growth rate in the centre decreases with a factor of two, going from a bare silicon wafer to a fully nitride-covered surface. This confirms the earlier results obtained using patterned wafers. Our first impressions concerning the long range of the loading effects are also confirmed. Even the nitride coverage at the very outer edge of the wafer has an influence on the growth rate at the centre: the centre of the wafer still senses what is happening at the wafer edge and probably beyond. The loading effects are active over a distance of at least the radius of the wafer, which is 75 mm.

Fig. 5 shows the results of the same experiment using SiGe. Now the growth rate increases with increasing nitride coverage in contrast to the Si growth rate. Another interesting difference is the range over which this loading effect is active. The curve has a tendency to saturate at 30 - 40 mm nitride radius. The growth rate at the centre is hardly influenced by what happens outside this radius on the wafer or on the susceptor. Although the loading effects are active over a considerable distance, their range is smaller for SiGe than for Si. This means that especially the Si growth rate will vary across the wafer due to the fact that the susceptor area acts as a large surface of bare Si, whereas within the wafer usually only a few percent of the area is exposed. The shorter range of the effect for SiGe allows an area at the centre of the wafer to grow at a constant rate. This is in agreement with the fact that in figs. 1 and 2 the

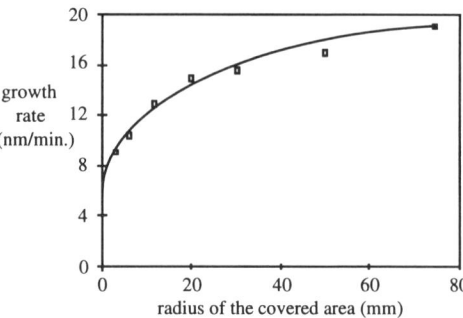

Fig.5 - The SiGe growth rate under selective growth conditions, measured at the centre of the wafer, increases with increasing radius of the nitride-covered area and ultimately saturates.

growth rate variations caused by the changing pattern are more severe for Si than for SiGe. Loading effects disturb the layer thickness uniformity across the wafer, but the dopant and Ge incorporation into the Si layer also suffer from centre-to-edge variations in this selective growth regime. This leads to variations in the transistor parameters which are largely uncontrolled and which cannot be accepted in a production process.

The problem was tackled by expanding the Si area on the wafer, to make the wafer look like the susceptor with its 100% Si coverage. This was done by covering the top of a patterned wafer with a polysilicon layer during SEG of Si and SiGe. In the windows the growth will be epitaxial on monocrystalline (100) Si and polysilicon will grow on the poly layer. In the case of the transistor structures the growth still has to be selective to avoid growth on the siliconnitride side walls of the seeding windows. After the deposition the polysilicon layer is etched away. This sacrificial poly layer can be applied without adding too much complexity to the process and without adverse effects on the integration aspects.

The selective growth in the presence of a sacrificial poly layer was characterized in the same way as before, using the circular test pattern. The circle of varying radius now was covered with polysilicon instead of siliconnitride. Fig. 6 shows how the growth rate in the small square at the centre of the wafer varied with the increasing polysilicon radius. The differences with respect to figs. 4 and 5 are remarkable. The Si and SiGe growth rates both decreased in the same way when the poly-covered area was increased. Both growth rates now saturated when the covered area

reached a 20 mm radius and the saturated growth rate was lower than the growth rate on a bare (100) substrate. In comparison with the former experiment with nitride, the poly coverage reduced the loading effects considerably. Especially the variation of the Si growth rate is back to manageable proportions. The conditions in a practical situation are such that growth will occur in the saturated region if a sacrificial poly layer is used. Even at the edge of the wafer the growth rate will no longer be position-dependent. In a practical situation the application of a sacrificial polysilicon layer will improve the centre-to-edge uniformity across the wafer sufficiently to make low-temperature SEG of Si and SiGe possible in production, but it is not the ultimate solution one might expect it to be. In spite of the fact that the (poly)Si coverage is 100% on the wa-

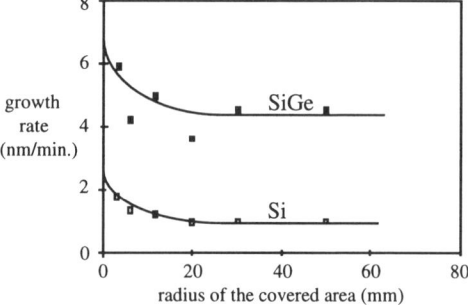

Fig.6 - Growth rate decrease of Si and SiGe at the centre of a (100) oriented substrate with increasing radius of the polysilicon-covered area around the centre. The growth took place under selective conditions.

fer as well as on the susceptor, the experimental results shown in fig. 6 still reveal a persisting loading effect. The poly/mono silicon ratio has an influence: growing on monocrystalline (100) silicon is different from growing on polysilicon. The cause is well-known, albeit that common Si epitaxy is usually effected at high temperatures in the masstransport-limited growth regime where the growth rates on different crystal orientations are the same. In this case low process temperatures were used during SEG of Si and SiGe and the growth rate was determined by surface kinetics. A (100) and a (111) wafer grown at high temperatures (> 1000°C) under identical conditions will grow at the same rate, but in the surface-kinetics-controlled regime this is no longer the case. To demonstrate this, a (100) and a (111) wafer were subjected to the same SEG process as used in all the former experiments to deposit a SiGe layer capped by a B-doped Si layer. The layer thicknesses of the SiGe layer and the top Si layer can be determined from the Ge concentration vs. depth plot in fig. 7 . The growth rate on a (111) sur-

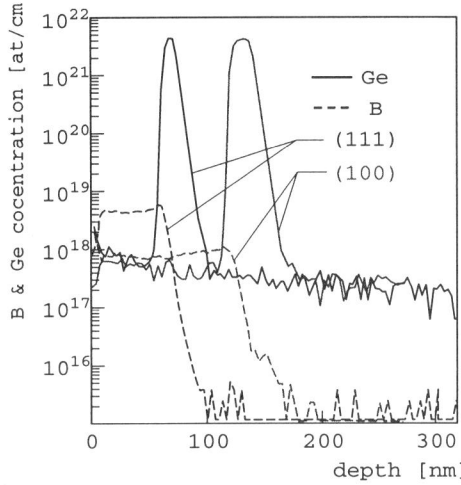

Fig.7 - A SIMS plot of the Ge and B concentrations vs. the depth of the Si and SiGe layers grown at 700°C under identical conditions on a (100) and a (111) surface. Differences in the Si and SiGe growth rates and in the degree of B incorporation are observable.

face was in case of both SiGe and Si reduced by a factor of approximately two relative to the equivalent growth rates on (100) silicon. There is also a factor of five difference in the degree of B incorporation into Si, the B concentration being higher on the (111) surface grown more slowly.

On a polysilicon surface various crystal orientations are exposed and the resulting overall growth rate is considerably lower than the growth rate on (100) Si under the processing conditions used in these experiments, due to the kinetically controlled growth regime. This explains why a loading effect still occurs in a window exposing (100) Si when the remainder of the surface is covered with poly-silicon. The resulting growth rate on a small exposed (100) area is very close to the poly growth rate, indicating that the growth rate is governed by some averaged chemistry on and above the wafer. In comparison with a situation in which the growth is totally inhibited, as was the case on a nitride surface, the application of a polysilicon layer is a substantial improvement. The difference in growth rate between windows at the centre and at the edge of a nitride-covered wafer disappears when the nitride is replaced with or covered by polysilicon. The solution is not perfect due to the difference in growth rates between polysilicon and (100) silicon. If (111) material is chosen for the substrate the loading effects will be reduced further, because the growth rate on poly-silicon is very close to the growth rate on (111) silicon.

The results of the experiment illustrated in fig. 2 can be understood better knowing these effects. The loading effects in fig. 2c are more severe at the exposed silicon/pattern boundary than at the edge of the wafer. Originally it was thought that this was due to the fact that the wafer edge was separated from the poly-coated susceptor by a thin region of nitride around the circumference of the wafer. With the insight gained in the last experiment it is concluded that the real cause is that the growth rate on the polysilicon-coated susceptor is lower than the growth rate on the quarter of the wafer with exposed (mono) silicon, resulting in a different averaged chemistry. In comparison with the growth rate on nitride, which is zero, the low growth rate of polysilicon on the susceptor just outside the wafer causes less adverse effects than the higher growth rate on the exposed quarter of the wafer, which is (100) material.

The mechanism that causes the SiGe growth rate to increase and the Si growth rate to decrease with increasing nitride coverage of the surface, as observed in fig. 1, is difficult to unravel. The decomposition of SiH_2Cl_2 generates HCl and a decreasing amount of HCl, resulting from an increasing nitride coverage, could qualitatively explain the increasing SiGe growth rate in fig. 4.

However, the same logic would also apply to fig. 3 showing the opposite phenomenon, a decreasing Si growth rate. It is clear that the addition and local generation of HCl plays an important role. The inverted behavior of the SiGe growth rate with increasing poly coverage relative to nitride, as shown in fig.4, could also be caused by the generation of HCl on the increasing poly surface. This assumption would be in accordance with the results of the experiments of Ito et al.[7]. On the other hand, the growth rate even drops below the growth rate on a silicon (100) surface, the extrapolated growth rate at zero poly coverage in fig. 4, in spite of the higher HCl generation on a (100) surface. This indicates that more factors have to be considered than HCl generation alone. Gas dynamics and surface kinetics are important and intermediate reaction compounds probably also play a role.

The growth experiments described in this investigation were performed with the addition of HCl to the gasstream in order to get full selectivity on patterned surfaces. It is to be expected that also when no HCl is added, e.g. during the low-temperature growth of blanket Si and SiGe layers on (100) Si wafers, centre-to-edge thickness and composition variations will occur.

SUMMARY AND CONCLUSION

It has been demonstrated that during the selective growth of Si and SiGe epi layers long-ranged loading effects can seriously disturb the on-wafer layer uniformity in terms of thickness and composition. Local variations in the growth rate across the surface of the wafer due to a changing (100) Si vs. nitride or poly ratio give rise to averaging effects in the growth-determining chemistry. A detailed description of the reaction mechanism explaining the growth phenomena is lacking, but the application of a sacrificial polysilicon layer has proven to be a powerful instrument in suppressing loading effects during SEG to a level at which the growth rate and layer composition across the wafer are sufficiently uniform.

ACKNOWLEDGMENTS

This work was partially supported by the European Commission (Esprit project 23229-BETA). The authors would like to thank J.G.M. van Berkum and W.M.van de Wijgert for extensive SIMS support and M.A. van den Berg for assistance in processing the wafers.

REFERENCES

1. A. Pruijmboom, D.Terpstra, C.E.Timmering, W.B.de Boer, M.J.J.Theunissen, J.W. Slotboom, R.J.E.Hueting and J.J.E.M.Hageraats, IEEE IEDM95 Technical Digest, 747 (1995).

2. T.F. Meister, H. Schäfer, M. Franosch, W. Molzer, K. Aufinger, U. Schler, C. Walz, M. Stolz, S. Boguth and J. Böck, IEEE IEDM95 Technical Digest, 739 (1995).

3. D.Terpstra, W.B. de Boer and J.W. Slotboom, Solid State El. **41**, 1493 (1997).

4. K. Washio, E. Ohue, K. Oda, M. Tanabe, H. Shimamoto and T. Onai, IEEE IEDM97 Technical Digest, 745 (1997).

5. T.I.Kamins, J. Appl. Phys. **74**, 5799 (1993).

6. C.I.Drowly and M.L.Hammond, Solid State Technol. **33**,135 (1990).

7. S.Ito, T.Nakamura and S.Nishikawa, J. Appl. Phys. **78**, 2716 (1995).

A SELECTIVE/NON–SELECTIVE EPITAXY PROCESS FOR A NOVEL SIGE HBT ARCHITECTURE

J.SCHIZ, J.M. BONAR, P. ASHBURN
Department of Electronics and Computer Science, University of Southampton, Southampton, SO17 1BJ, UK.

Abstract

A selective/non-selective epitaxy process for a novel self-aligned SiGe HBT structure is described. It is shown that damage in the silicon substrate induced by reactive ion etching can be effectively removed and that this allows the growth of high quality selective and non-selective epitaxy. Control of the base thickness and doping can be maintained during selective and non-selective epitaxy, thereby demonstrating the feasability of producing the self-aligned HBT structure.

Introduction

The device performance of silicon-germanium (SiGe) heterojunction bipolar transistors (HBT) has been improved to such an extent that they are now suitable for applications in the microwave and millimetre-wave regime. Cut-off frequencies as high as 130GHz [1] and maximum oscillation frequencies as high as 160GHz [2] have been achieved. In order to achieve high bipolar circuit speeds, it is essential to minimise the C/B capacitance and base resistance. The former requires self-aligned fabrication techniques and the latter requires high base doping. Low temperature epitaxy is an attractive technique for the fabrication of narrow SiGe bases with abrupt doping transitions. It also allows for the base to be more heavily doped than the emitter and avoids the problem of the channeling tail which is often obtained with ion implantation [3]. In this paper it is shown that low temperature selective silicon epitaxy and non-selective SiGe epitaxy can be performed in one process step which allows the realisation of a self-aligned SiGe HBT with minimised parasitic capacitances and resistances.

Device structure and technology

In the past, mesa structures, blanket epitaxy, differential epitaxy, and selective epitaxy have all been used to produce SiGe HBTs. However, many of these approaches require additional process steps for LOCOS or trench device isolation. Furthermore, growth interfaces are often in or close to the depletion regions of the device which might lead to leakage currents. A cross-sectional view of the proposed device architecture, which is intended to overcome some of the constraints mentioned above, is shown in figure 1. Selective epitaxy is used to grow the n-type collector in a window previously etched into an oxide layer. Hence the oxide isolation is an intrinsic part of the device structure. The growth conditions are then changed from selective to non-selective epitaxy for the formation of the p-type SiGe base and the n-type

321

Mat. Res. Soc. Symp. Proc. Vol. 533 [c]**1998 Materials Research Society**

silicon low-doped emitter. As will be shown later, all these layers can be grown in the same process step. This has the advantage of avoiding impurities at interfaces in the depletion regions which might occur if the deposition of layers was interrupted. During the non-selective growth, material is deposited on the oxide isolation as polysilicon. This polysilicon layer is implanted with boron to form the p^+ extrinsic base contact. The extrinsic base is self-aligned to the polysilicon emitter contact, and hence the C/B capacitance is only limited by the alignment tolerance between the emitter window and the n^+ polysilicon. Oxide isolation is an intrinsic part of the device structure and hence no separate shallow trench isolation or LOCOS isolation is required. All the epitaxial layers are in-situ doped which has the advantage that no implantation into single crystal material is needed. Hence, damage due to ion implantation and thus transient enhanced diffusion can be avoided.

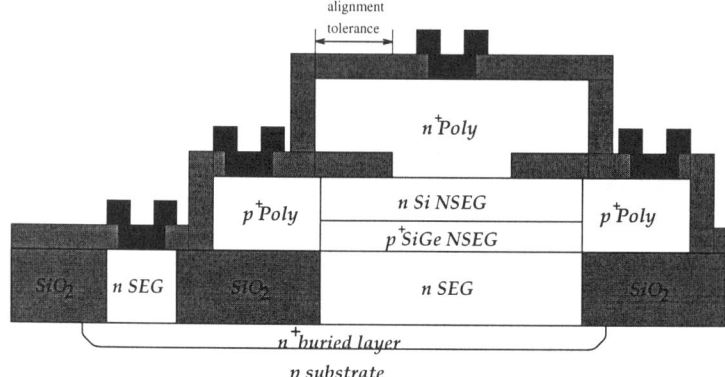

Figure 1: *Schematic of novel SiGe HBT*

The processing of such a device is very challenging since defect-free epitaxy requires an undamaged silicon surface. However, anisotropic dry etching, which is needed to produce vertical sidewalls, is known to damage the underlying silicon substrate, therefore an etch damage removal treatment (EDR) is necessary. The switch from selective to non-selective growth is achieved by increasing the growth pressure using a throttle valve. While the pressure is increasing the growth rate will vary, thus care is needed to ensure that the layers grown during the time the pressure is changing are the correct thickness. It is crucial not to exceed the critical thickness of the SiGe layer but also to have a sufficiently thick p-type layer to incorporate a sufficient amount of boron for a high base doping. Furthermore, a continuous polysilicon layer is required over the oxide edge to form the base contact.

Experimental Procedure

In this experiment, (100) CZ p-type silicon wafers were used. The wafers were oxidised at 1100°C in order to obtain a field oxide with a thickness of 400nm. An anisotropic dry etch using CHF_3 and Ar was used to etch seed holes into the oxide to provide windows for the epitaxial silicon growth. The dry-etched wafers were etched for 15 or 30% longer than needed to remove 400nm of oxide to produce an over-etch of 15 or 30%. This is necessary to account for variations in etch rate across the wafer. Some wafers received an etch damage removal treatment (EDR), consisting of reoxidation at 950°C in dry O_2 for 20 minutes, which produced about 20nm of oxide. The oxide was then removed with a short etch in BHF. Before the growth, the wafers received an RCA clean and a short etch in dilute HF, leaving the wafer oxide-terminated. The oxide was desorbed in the chamber of the LPCVD reactor using a heat treatment at 900°C in H_2. After the desorbtion of the oxide, the temperature was dropped and the growths were performed at 800°C. The process gases were H_2, SiH_4, PH_3, GeH_4 and B_2H_6. Hence, no chlorinated gases were used to increase selectivity. This avoids potential problems with undercutting of the oxide mask which are often encountered when chlorinated gases are used. The quality of the epitaxial layers was evaluated using SEM and TEM.

Results and Discussion

Figure 2 shows a SEM micrograph of epitaxy grown on a wafer which received a 15% over-etch and no EDR. The centre of the device area shows very rough epitaxy which looks very similar to the polysilicon layer on the oxide around the perimeter of the feature. After Secco defect etching [4] a high density of crystallographic defects was seen around the perimeter of the defective device area [5]. The material in the feature centre appears polysilicon-like, however, diffraction measurements made in cross-section TEM demonstrate that the region is heavily faulted silicon epitaxy [6]. Cross-section TEM micrographs of the defective region have shown [5] that the epitaxy surface is very rough and that twin defects are emanating from the near-interface region. Such an epitaxial layer is clearly not suitable for device fabrication.

Figure 3 shows a SEM micrograph of epitaxy grown on a wafer which had a 30% over-etch and an EDR treatment. This wafer does not show heavily faulted epitaxy in the centre of the feature. In this case the epitaxy appears completely flat and defect-free. After Secco defect etching [4], a low density of defects was seen [5]. The density of defects was comparable to that obtained for epitaxy on a wet-etched test wafer, i.e. a wafer with undamaged silicon surface. This indicates that a dry etch with 30% over-etch followed by an etch damage removal treatment is suitable for device fabrication.

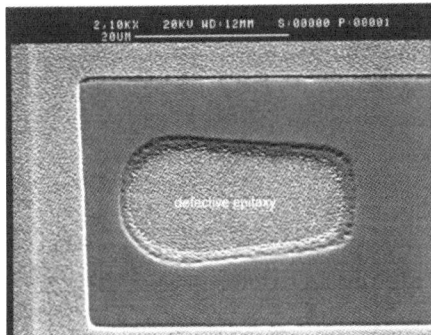 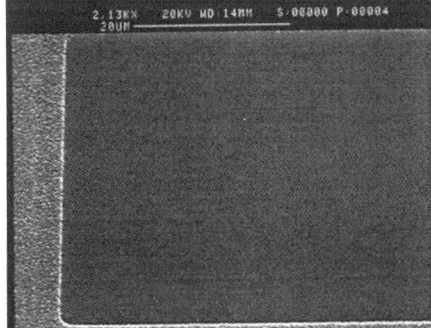

Figure 2: *SEM micrograph of epitaxy on a dry-etched surface without EDR treatment*   Figure 3: *SEM micrograph of epitaxy on a dry-etched surface with EDR treatment*

An SEM micrograph of a wafer after the selective/non-selective epitaxy is shown in figure 4. The left hand side of the micrograph shows the Si/SiGe/Si epitaxy stack. The interface between the epitaxy and the substrate cannot be seen. On the right hand side, the substrate, oxide layer and polysilicon layer resulting from the non-selective epitaxy can be seen. A continuous polysilicon layer exists over the oxide edge which implies that the polysilicon layer grown during the non-selective epitaxy can be used as an extrinsic base contact. From figure 4 it is clear that the trench was indeed filled with selective epitaxy. Had the growth been non-selective from the beginning, a step would be observed between the polysilicon on top of the oxide and the single crystal epitaxy. This indicates that the polysilicon growth on the oxide only started when the growth conditions were changed to non-selective growth by increasing the growth pressure.

The cross-section TEM micrograph in figure 5 shows the Si/SiGe/Si epitaxy stack. The dark horizontal band is the strained SiGe layer. The vertical line on the left hand side is not a defect but a crack in the TEM sample. This TEM micrograph confirms that the epitaxy grown is defect-free. Furthermore, it can be seen that the SiGe layer is planar and that there are no signs of self-organised strain-relief. The layer thicknesses are very close to specification which suggests that the layer thickness can be well controlled despite the fast growth rate in the LPCVD equipment used and despite the change in growth rate during the switch from selective to non-selective growth.

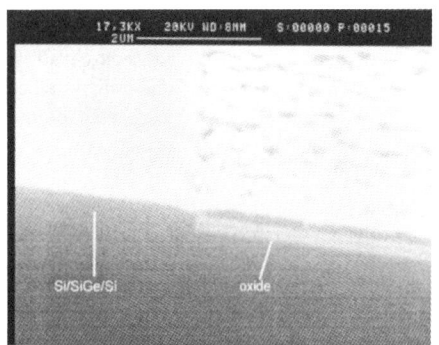

Figure 4: *SEM micrograph of epitaxy in the etched seed hole and on the oxide isolation*

Figure 5: *TEM micrograph of a defect-free Si/SiGe/Si epitaxy stack*

As mentioned above, one potential difficulty with the proposed device architecture is the accurate control of the thickness of the spacers on either side of the base and the base itself while the pressure is changing due to the switch from selective to non-selective growth. Figure 6 shows the B and Ge depth profiles obtained by SIMS analysis. The profiles were obtained from a region which received both the selective and non-selective growth. In measuring the data, a region was considered to be SiGe when the Ge concentration was above $1 \times 10^{20} \mathrm{cm}^{-3}$, which is 0.2% Ge. The Ge concentration in the base layer is 12.8% and fairly uniform. The emitter doping is intended to be $1 \times 10^{18} \mathrm{cm}^{-3}$, thus the intrinsic E/B spacer is the region of SiGe for which the B doping is below $1 \times 10^{18} \mathrm{cm}^{-3}$. The collector doping is intended to be $2 \times 10^{17} \mathrm{cm}^{-3}$, therefore the intrinsic C/B spacer is the region of SiGe for which the B doping is below $2 \times 10^{17} \mathrm{cm}^{-3}$. The C/B spacer is measured to be about 10nm which exactly meets the specification and the E/B spacer is measured to be about 8nm which is slightly thinner than specified. The width of the doped base is about 50nm which is slightly thicker than specified. However, it is very encouraging that the specifications were nearly met and by adjusting the growth times even better matching of the specifications is expected. The growth time of the doped base was obviously sufficient to incorporate enough B to ensure a base doping above $1 \times 10^{19} \mathrm{cm}^{-3}$. The base doping in figure 6 is about $2 - 3 \times 10^{19} \mathrm{cm}^{-3}$ which is even above the specification of $1 - 2 \times 10^{19} \mathrm{cm}^{-3}$.

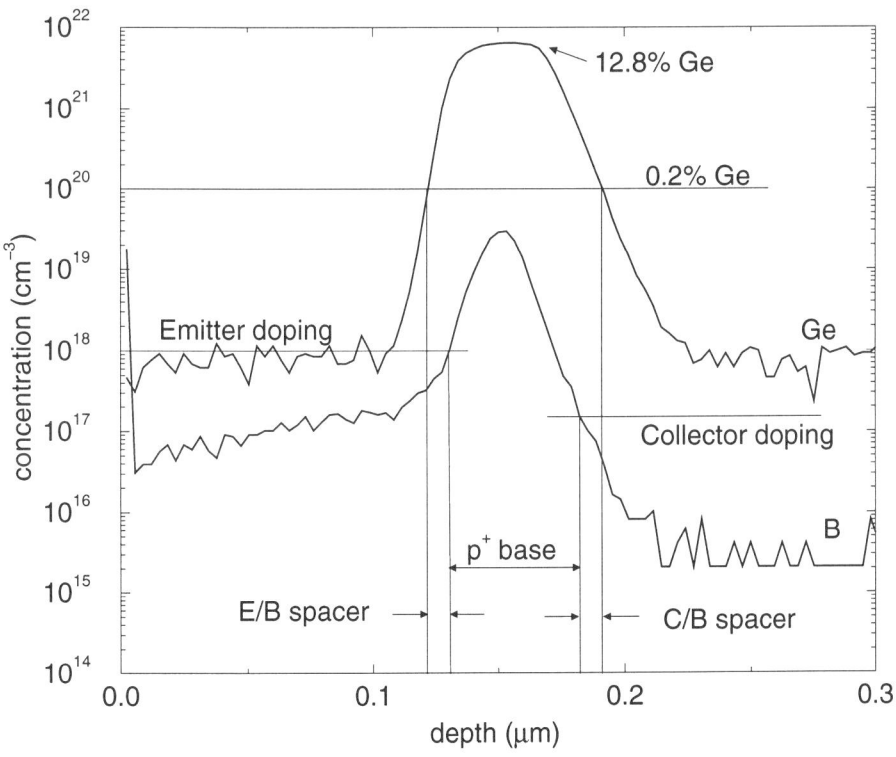

Figure 6: *B and Ge SIMS profiles of a completed device structure*

Conclusion

A novel self-aligned SiGe HBT architecture has been demonstrated. It has been shown that a dry etch with 30% overetch followed by an EDR treatment can be used to provide a damage-free surface which allows the growth of virtually defect-free epitaxy. Furthermore it has been shown that the growth conditions from selective to non-selective growth can be changed in one process step, and that it is possible to control layer thicknesss even while the pressure is changing due to the switch from selective to non-selective growth.

References

[1] K. Oda, E. Ohue, M. Tanabe, H. Shimamoto, T. Onai, and K. Washio, "130 GHz f_T SiGe HBT technology," in *IEEE International Electron Device Meeting*, pp. 791–794, 1997.

[2] A. Schüppen, U. Erben, A. Gruhle, H. Kibbel, H. Schumacher, and U. König, "Enhanced SiGe heterojunction bipolar transistors with 160GHz f_{max}," *IEEE International Electron Device Meeting*, pp. 743–746, 1995.

[3] T. Sugii, T. Yamazaki, T. Fukano, and T. Ito, "Epitaxially grown base transistor for high-speed operation," *IEEE Electron Device Letters*, vol. 8, no. 11, pp. 528–530, 1987.

[4] D. G. Schimmel, "A comparison of chemical etches for revealing <100> silicon crystal defects," *Journal of the Electrochemical Society*, vol. 123, no. 5, pp. 734–741, 1976.

[5] J. Schiz, J. Bonar, and P. Ashburn, "A novel self-aligned SiGe HBT structure using selective and non-selective epitaxy," in *1997 Workshop on High Performance Electron Devices for Microwave and Optoelectronic Applications*, pp. 255–260, 1997.

[6] J. Bonar, J. Schiz, and P. Ashburn, "Improved epitaxial quality following etch damage removal on plasma etched silicon surfaces," in *Microsc. Semicond. Mater. Conf.*, vol. 157 of *Inst. Phys. Conf. Series*, pp. 407–410, 1997.

PHOTOLUMINESCENCE OF SiSnC ALLOYS GROWN ON (100) Si SUBSTRATES

N. Wright [*], A.T. Khan [*], P.R. Berger [*], F.J. Guarin [**], S.S. Iyer [**]
[*]Department of Electrical and Computer Engineering, University of Delaware, Newark, DE 19716
[**]IBM Microelectronics Division, 1580 Route 52, Hopewell Junction, NY 12533

ABSTRACT

Photoluminescence (PL) of SiSnC alloys grown on Si (100) substrate by molecular beam epitaxy (MBE) has been investigated. The following epitaxial layers were investigated. The samples were similar in structure, and consisted of a 200 Å Si buffer layer grown on a (100) Si substrate followed by the respective alloy layer, a 4500 Å $Si_{0.955}Sn_{0.03}C_{0.015}$, a 1500 Å $Si_{0.96}Sn_{0.04}$, or a 1500 Å $Si_{0.985}C_{0.015}$. The layer composition was measured by Rutherford back scattering spectrometry (RBS) and confirmed by x-ray diffraction analysis (XRD). X-ray diffraction measurements of the layers confirmed the Sn and C were substitutional and the layers were pseudomorphic and coherently strained. The $Si_{0.955}Sn_{0.03}C_{0.015}$ alloy layer was found to be strain compensated. PL spectra of all the layers revealed band edge luminescence as well as a very sharp peak at 0.767 eV superimposed on a very broad peak that exhibited excitonic behavior. The addition of C to the alloys resulted in a reduction of the bandgap, contrary to what is predicted by Vegard's Law. However, the addition of Sn results in a reduction in the bandgap, which was attributed to the bulk alloy effect and residual strain. The luminescence feature at 0.767 eV was found to be much more intense in alloys that contain carbon. We have observed quite similar deep-level well resolved PL spectra for SiGe, SiC, SiSn, SiGeC, SiSnC, and SiGeSnC alloy layers grown by MBE on Si (100) substrates. Previous studies on PL of silicon have reported a peak at 0.767 eV to be associated with oxygen (P line) and the broad peak to carbon/oxygen complexes.

INTRODUCTION

The growth of alloys of group IV elements has been under investigation recently as potential material systems for Si-based heterostructure device applications [1]. In the past decade extensive effort has been directed toward the SiGe material system, which has led to the recent commercial availability of a few SiGe devices and circuits. There has been some interest recently in the growth of Sn-based and C-based group IV semiconductor alloys. The carbon based alloys of SiC and SiGeC have received the most attention. There have also been a few investigations directed toward the Sn-based alloys. High quality pseudomorphic crystalline layers of Sn and C based alloys have been grown by a few groups. The Sn-based alloys have been grown by MBE [2], and C-based alloys have been grown by chemical vapor deposition (CVD) [3] and MBE [4]. These studies indicate that only alloys containing a few percent of C or Sn can be grown pseudomorphically on a Si substrate. This is due mainly to the following factors. The alloys suffer from metastability problems, large lattice and bond length mismatches, and low Sn and C solid solubilities.

Alloy layers that exceed a few percent of C were found to be plagued by defects, non-planar growth, or SiC precipitates. This can be directly attributed to the extremely low solubility of C in Si ($\sim 3 \times 10^{18}$ cm^{-3}) [5], and the large difference in the Si and C bond lengths. Si has a bond length of 2.35 Å, diamond a bond length of 1.55 Å, and zinc-blende β-SiC phase has a bond length of 1.89 Å [6]. The bond length mismatch of about 40 % causes large local strain and according to the Si-C phase diagram, stoichiometric SiC is the only stable compound. Previous work on the Sn-based alloys indicated that, it is very difficult to stabilize Sn in the Si alloy lattice. This is mainly due to the large difference in the Si and Sn lattice parameters. Grey tin has a lattice parameter of 6.489 Å, and

Si has a lattice parameter of 5.431 Å, resulting in a 19.48 % lattice mismatch. Grey tin (α-Sn) crystallizes in the diamond structure, and at 13.2 °C it transforms into the tetragonal structure of metallic white tin (β-Sn). Sn has a very low solid solubility in crystalline silicon (~5 x 10^{19} cm^{-3}) [5]. Therefore the substitutional incorporation of C and/or Sn above the solid solubility limit requires far from thermodynamic equilibrium growth conditions, such as MBE. The incorporation of isoelectronic Sn or C into silicon will potentially offer new possibilities, for bandgap engineering, silicon compatible wide or narrow bandgap material, and large band offsets. This could lead to changes in the optical and electrical properties of the material. According to recent theoretical calculations [1], $Si_{1-x-y}Sn_xC_y$ alloys would ideally span the energy range of 0.08 eV to 5.48 eV. These alloys offer the possibility of tuning the bandgap energy over an enormous range by varying the compositions of their respective constituents. Additionally, the bandgap was predicted [7] to be direct for certain compositions of SiSn and SnC. They also have the potential for exact lattice matching to Si. Strained layers of SiSnC alloy semiconductors would find numerous applications in electronic and optoelectronic heterostructures. These alloys also offer the potential for high electron and hole mobilities. To the best of our knowledge the optical luminescence properties of pseudomorphic $Si_{1-x-y}Sn_xC_y$ alloys grown on Si substrates have not been reported in any previous studies.

EXPERIMENT

The details of the solid source MBE system used for the growth of the SiSnC layers are described elsewhere [2]. A thin 200 Å Si buffer layer was grown initially on a (100) Si substrate followed by the respective alloy layer. The $Si_{0.985}C_{0.015}$ and the $Si_{0.96}Sn_{0.04}$ layer was grown to a thickness of 1500 Å, and the $Si_{0.955}Sn_{0.03}C_{0.015}$ layer was grown to a thickness 4500 Å. The growth conditions were optimized to maximize the incorporation of C and/or Sn on Si lattice sites. The layer compositions were measured by Rutherford backscattering spectrometry (RBS) and x-ray diffraction analysis (XRD), which confirmed the Sn and C were substitutional and the layers were pseudomorphically strained. XRD of the SiSnC alloy layer showed it to be strain compensated and lattice matched to Si, the details are described elsewhere [2].

Photoluminescence spectra was recorded in standard lock-in configuration, using a dispersive 1-m high-resolution Jarell-Ash (Czerny-Turner) monochromator and detected by a liquid-nitrogen cooled Ge p-i-n photodetector (North Coast EO-817L). The samples were mounted on a cold finger in a temperature variable helium-flow cryostat. The excitation was provided by a multi-line cw Argon laser (488-514 nm) focused to a sample area of approximately 2 mm^2, with pump intensities between 0.3 and 5 W/cm^2. Data collection, and lock-in amplification were controlled by a desktop computer.

RESULTS

Low temperature PL spectra at 6 K for the $Si_{0.955}Sn_{0.03}C_{0.015}$ sample is presented in Fig. 1. The luminescence consists of a very intense deep-level luminescence and a few edge peaks superimposed on a broad peak. The most intense peak at 1.040 eV is ascribed to a no-phonon (NP) transition, the peak located at 1.022 eV is red-shifted by 18 meV and can be identified with the momentum-conserving (MC) transverse acoustic (TA) phonon replica in silicon, the peak located at 0.982 eV is red-shifted by 58 meV and corresponds to its transverse optic (TO) Si-Si phonon replica. Additional peaks were observed at 1.006 eV, 0.966 eV, and 0.933 eV. These peaks were red-shifted from the ascribed NP line by 34 meV, 70 meV, and 107 meV respectively. The PL spectrum for the $Si_{0.96}Sn_{0.04}$ and $Si_{0.985}C_{0.015}$ samples is presented in Fig. 2. The luminescence from the $Si_{0.96}Sn_{0.04}$

sample consists of a low intensity deep-level luminescence and two edge peaks. The most intense peak at 1.055 eV is ascribed to a NP transition whereas the second peak at 0.994 eV is red-shifted by 61 meV and corresponds to its transverse optic (TO) Si-Si phonon replica. This is to our knowledge the first observation of luminescence from a SiSn alloy, and the details are reported elsewhere [8]. The luminescence from the $Si_{0.985}C_{0.015}$ sample consists of an intense deep-level luminescence and two edge peaks. The most intense peak at 1.060 eV is ascribed to a no-phonon (NP) transition whereas the second peak at 1.002 eV is red-shifted by about 58 meV and corresponds to its transverse optic (TO) Si-Si phonon replica, for more details refer to the following paper [9]. The luminescence peak at 1.1089 eV was observed in all the samples and is attributed to a Si TO phonon transition within the Si substrate. Also, a deep-level broad band luminescence is observed in all the samples around 0.778 eV. The deep-level feature was much more intense in the alloys that contained carbon, and consisted of a very sharp peak located at 0.767 eV superimposed on a broad peak. Previous studies on Si have reported a sharp peak located at 0.767 eV to be associated with oxygen (P line) [10]. Similar deep-level broad band luminescence has been observed in $Si_{1-x}Ge_x$, $Si_{1-x-y}Ge_xC_y$, and $Si_{1-x}C_x$ layers grown by MBE, and were attributed to localized excitons in a strain field created by Ge platelets [11], the emission of deep pseudo-acceptors [12], or to carbon-oxygen complexes [3], respectively.

Soref[7], using a linear interpolation between the constituent elements, predicted the addition of Sn to Si would result in a red shift of the bandgap, and the addition of carbon a blue shift in the bandgap. The bandgap of all the alloys investigated were found to be red shifted. The $Si_{0.955}Sn_{0.03}C_{0.015}$ alloy NP transition is shifted 113 meV below that observed in pure Si [13], and is additionally

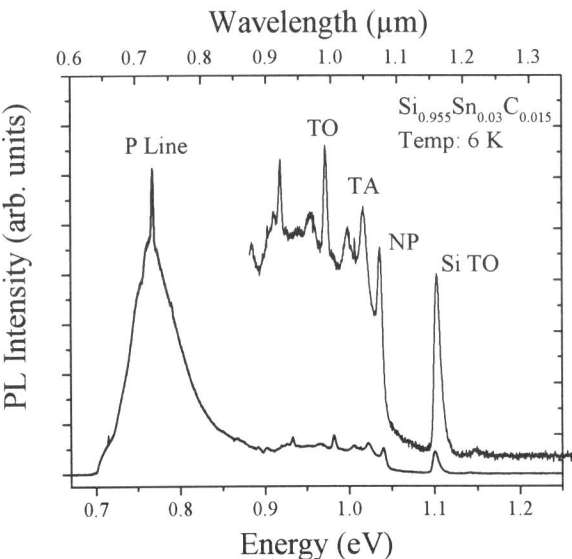

Fig. 1. PL spectra at 6 K of a strain compensated $Si_{0.955}Sn_{0.03}C_{0.015}$ alloy layer grown by MBE on (100) Si substrate, showing band-edge and deep-level luminescence. Inset shows an expanded view of the spectra between the energy range 0.9 - 1.25 eV.

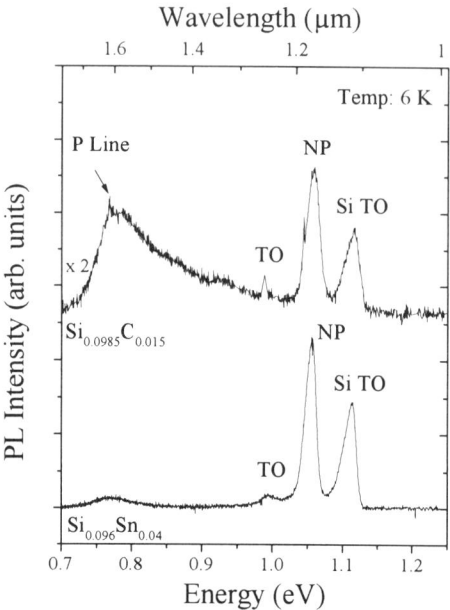

Fig. 2. PL spectrum of a $Si_{0.96}Sn_{0.04}$ and a $Si_{0.985}C_{0.015}$ alloy layer showing bandedge and deep-level luminescence.

shifted about 146 meV below an expected 1.186 eV luminescence position using Vegard's Law. XRD results [2] revealed a strain compensated and lattice matched $Si_{0.955}Sn_{0.03}C_{0.015}$ alloy layer. We attribute this red shift to the lattice distortion within the alloy due to highly localized strain around the carbon and Sn atoms. The NP transition of the $Si_{0.96}Sn_{0.04}$ alloy is shifted 98 meV below that observed in pure Si [13]. We attribute this red shift to the reduction in band gap of the binary SiSn alloy. However, the observed NP transition is also shifted 55 meV below an expected 1.110 eV luminescence position using Vegard's Law for a SiSn alloy composition with 4% Sn. This additional red shift we attribute to reduction in the bandgap caused by residual strain in the pseudomorphic layer and highly localized strain around the Sn atoms. The NP transition of the $Si_{0.985}C_{0.015}$ alloy is shifted 93 meV below that observed in pure Si [13], and is additionally shifted about 159 meV below an expected 1.2182 eV luminescence position using Vegard's Law for a $Si_{1-x}C_x$ alloy composition with 1.5 % C. We attribute this red shift to both the lattice distortion within the alloy due to highly localized strain around the carbon atoms, and residual strain between the pseudomorphic alloy layer and the substrate. The lowering of the bandgap due to the pseudomorphic strain energy in the layer is not enough to explain the large red shift alone. The bandgap reduction in the carbon based alloys seems to follow the trend suggested by recent theoretical investigations [14] on the bandgap of $Si_{1-x}C_x$ alloys. They indicate that the addition of small carbon concentrations reduces the $Si_{1-x}C_x$ bandgap below Si, it turns semi-metallic around 10% C, and then increases above Si beyond 10% carbon.

The laser power dependence of the designated NP lines and the deep-level feature of the alloys were investigated. The NP peaks of the $Si_{0.96}Sn_{0.04}$ and the $Si_{0.985}C_{0.015}$ alloys reveal a linear increase in intensity with increasing laser power, while the $Si_{0.955}Sn_{0.03}C_{0.015}$ alloy reveal a near linear dependence as is shown in Fig. 3. PL laser power dependence of this nature is characteristic of free exciton (FE) recombination. The deep-level feature, the line designated as the P line and the broad peak, exhibit a sub-linear almost square-root dependence, which indicates that a recombination center may be involved in the recombination process and have previously been attributed to carbon-oxygen complexes [3].

Temperature dependence of the designated NP peaks in Figs. 1 and 2 were also investigated. It was observed that the luminescence of the $Si_{0.96}Sn_{0.04}$ alloy layer persisted past 50 K, and the $Si_{0.955}Sn_{0.03}C_{0.015}$ alloy layer luminescence persisted up to 100 K, but was degraded in intensity. The luminescence of the $Si_{0.985}C_{0.015}$ alloy layer however persisted past 200 K but was degraded in intensity. A further temperature increase resulted in the labeled NP and Si TO peaks merging into a single broad peak which persisted up to room temperature. As the temperature was increased, the designated NP peaks of the alloys all showed a decrease in PL intensity and a broadening of the linewidth in the direction of higher energy. This type of linewidth behavior is characteristic of the Maxwell-Boltzman distribution. PL temperature dependence of this nature is characteristic of free exciton (FE) recombination. Therefore, we attribute the labeled NP peaks as due to no phonon free exciton recombination. The full width at half maximum of the NP peak at 6 K of the $Si_{0.96}Sn_{0.04}$ and the $Si_{0.985}C_{0.015}$ layer were 18 meV and 19 meV respectively, which is rather broad compared to the intrinsic FE thermal line width, but this could be attributed to statistical fluctuations in the atomic distributions of the alloy as discussed by Robbins et al [15] and the lattice distortion of the alloy due to the large bond length difference between Si, Sn, and C as discussed by Demkov et al [14].

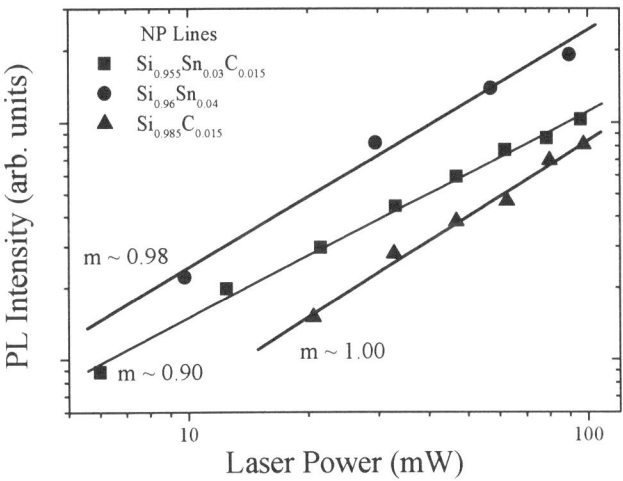

Fig. 3. PL laser power dependence of the assigned NP lines in the $Si_{0.96}Sn_{0.04}$, $Si_{0.985}C_{0.015}$, and $Si_{0.955}Sn_{0.03}C_{0.015}$ alloy layers.

CONCLUSIONS

In conclusion, we have reported low temperature photoluminescence of a strain compensated $Si_{0.955}Sn_{0.03}C_{0.015}$ alloy layer, a compressively strained pseudomorphic $Si_{0.96}Sn_{0.04}$ alloy layer and a tensially strained pseudomorphic $Si_{0.985}C_{0.015}$ alloy layer grown on Si (100) substrates by solid source MBE. Two general features have been observed, a deep-level broad band luminescence and band-edge luminescence consisting of a NP and TO replica in the $Si_{0.96}Sn_{0.04}$ and $Si_{0.985}C_{0.015}$ alloy layers, and a NP, TA and TO replicas in the $Si_{0.955}Sn_{0.03}C_{0.015}$ alloy layer. The deep-level luminescence of the alloys containing carbon was very intense and consisted of a very sharp peak located at 0.767 eV which was attributed to oxygen (P line) superimposed on a very broad peak. Temperature dependent PL analysis of the band-edge feature of the alloy layers indicated an exponential temperature dependence, which is characteristic of free excitonic recombination. The band-edge feature of the $Si_{0.96}Sn_{0.04}$ and $Si_{0.985}C_{0.015}$ layers exhibited a linear power dependence, and the $Si_{0.955}Sn_{0.03}C_{0.015}$ layer exhibited a near linear power dependence, which is also characteristic of free exciton recombination. The deep level feature in all the layers showed a sub-linear almost square-root dependence, which indicates that a recombination center may be involved in the recombination process. The energy gap of all the alloy layers was red-shifted with respect to Si and was attributed to lattice distortion within the alloy and residual strain between the alloy layer and the substrate where applicable.

REFERENCES

1. R.A. Soref, Proc. IEEE **81**, 1687 (1993)

2. F.J. Guarin, S.S. Iyer, B.A. Ek, and A.R. Powell, Appl. Phys Lett. **68**, 3608 (1996)

3. P. Boucaud, C. Francis, A. Larré, F.H. Julien, J.-M. Lourtioz, D. Bouchier, S. Bodnar, and J.L. Regolini, Appl. Phys. Lett. **66**, 70 (1995)

4. S.S. Iyer, K. Eberl, M.S. Goorsky, F.K. LeGoues, J.C. Tsang, and F. Cardone, Appl. Phys. Lett. **60**, 356 (1992)

5. A.G. Milnes, Deep Impurities in Semiconductors, Wiley, New York, (1973)

6. H.J. Osten, E. Bugiel, and P. Zamuseil, J. Cryst. Growth **142**, 322 (1994)

7. R.A. Soref, J. Appl. Phys **70**, 2470 (1991)

8. A.T. Khan, P.R. Berger, F.J. Guarin, and S.S. Iyer, Appl. Phys. Lett. **68**, 3105 (1996)

9. A.T. Khan, P.R. Berger, F.J. Guarin, and S.S. Iyer, Thin Solid Films **294**, 122 (1997)

10. J. Weber, and R. Sauer, Mat. Res. Soc. Symp. Proc., **14**, 165 (1983)

11. J.-P. Noël, N.L. Rowell, D.C. Houghton, and D.D. Perovic, Appl. Phys. Lett. **61**, 690 (1992)

12. J. Denzel, K. Thonke, J. Spitzer, R. Sauer, H. Kibbel, H.-J. Herzog, and E. Kasper, Thin Solid Films **222**, 89 (1992)

13. P.J. Dean, J.R. Haynes, and W.F. Flood, Phys. Rev. **161**, 711 (1967).

14. A.A. Demkov, and O.F. Sankey, Phys. Rev. B **48**, 2207 (1993)

15. D.J. Robbins, L.T. Canham, S.J. Barnett, A.D. Pitt, and P. Calcott, J. Appl. Phys. **71**, 1407 (1992)

A NOVEL LAYER-BY-LAYER HETERO-EPITAXY OF GERMANIUM ON SILICON (100) SURFACE

S. Sugahara, M. Matsuyama, K. Hosaka, K. Ikeda, Y. Uchida* and M. Matsumura

Department of Physical Electronics, Tokyo Institute of Technology, 2-12-1 O-okayama, Meguro-ku, Tokyo 152-8550, Japan
*Department of Electronics and Information Science, Teikyo University of Science and Technology, Uenohara-machi, Yamanashi 409-0193, Japan

ABSTRACT

Layer-by-layer hetero-epitaxy of Ge has been successfully demonstrated on the Si(100) surface by combining the initial 1ML-Ge film growth on the Si surface and the successive Ge atomic-layer-epitxy (ALE), for the first time. The former was achieved using the substrate temperature modulation with alternate exposures of $GeCl_4$ and atomic H, and the later was established by cyclic exposures of $(CH_3)_2GeH_2$ and atomic H under isothermal conditions. XPS measurements confirmed a discrete and uniform increase in the grown Ge film thickness with one monolayer/cycle step up to the critical Ge thickness, and no C contamination at the Ge/Si interface. Critical exposure for the saturated Ge adsorption was different from that for the homo-ALE on the bulk Ge surface.

INTRODUCTION

Electronic structures of atomic layer superlattices (ALSs) have an interesting dependence on the periodicity of their slabs, since the electronic structure is modulated by the zone-folding effect very strongly when the thickness of each slabs is reduced to a few monolayers[1-4]. Thus, Si/Ge ALSs are expected very attractive for Si-based ultra-large-scale integrated circuits, since they are compatible with well-refined silicon technology. To produce desired electronic structures in the Si/Ge ALS, the critical adjustment is required for the thickness of each Si and Ge slabs, and also for the abruptness of their interface, on an atomic scale.

Atomic layer epitaxy (ALE)[5-10] is considered as the most promising technique to create the ALS by following reasons: The ALE has inherently a precise controllability for film thickness on an atomic scale determined only by the number of execution cycles, since the film growth takes place in a self-limiting layer-by-layer manner due to perfectly controlled surface reactions, i.e., the saturated adsorption of precursors and the complete desorption of ligands from adsorbates.

For fabrication of the Si/Ge ALS based on ALE, both the Ge hetero-ALE on the Si surface and the Si hetero-ALE on the Ge surface could be established. This paper reports the former, since the later has been successfully demonstrated in another paper[11].

The Ge ALE on the Ge(100) substrate, called "Ge homo-ALE", has been established by alternate exposures of atomic H and dimethylgermane ($(CH_3)_2GeH_2$; DMG) with a wide temperature window between 420°C and 530°C[9]. However, this Ge homo-ALE cannot be applied directly to the Si(100) surface due to undesirable contamination of the surface by C atoms in DMG[12]. Thus, the Ge hetero-ALE must be divided into two steps, i.e., the 1ML growth of the Ge film on the Si substrate surface using C-free source gases and the successive Ge homo-ALE on the pre-grown Ge top film above the Si underlayer.

We have already succeeded in the C-contamination-free 1ML growth of the Ge film on the Si surface based on a repeated precursor-adsorption and ligand-desorption (RPALD) method[13]. The method consists of the substrate temperature modulation with alternate exposures of

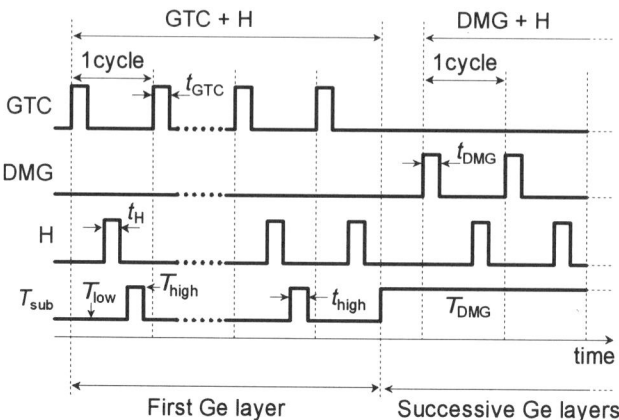

Fig. 1. Time sequence for the Ge hetero-ALE method.

germanium tetrachloride (GeCl$_4$; GTC) and atomic H for more than 15 cycles. The temperature window was as wide as 60°C between 470°C and 530°C, that overlapped with the window of the Ge homo-ALE using DMG[9].

In this paper, the Ge hetero-ALE on the Si(100) surface has been demonstrated by combining the previously established techniques. It was found that the Si underlayer has a strong influence on the monolayer adsorption of Ge precursor on the pre-deposited Ge top film.

EXPERIMENTAL

Time sequence for the Ge hetero-ALE is shown in Fig. 1, which is divided into a 1ML-growth step of the Ge film on the Si surface[13] and successive Ge homo-ALE steps on the pre-grown 1ML Ge-film surface[9].

There are three important phases for the 1ML-growth of the Ge film on the Si surface: a source gas exposure phase at low temperatures for the self-limiting adsorption of GTC, an atomic H exposure phase for the abstraction of surface-terminating Cl atoms, and a surface H desorption phase, where the substrate temperature is elevated, for the complete thermal desorption of surface H that has adsorbed in the pregoing H exposure phase. These phases are repeated for about 15 cycles for 1ML growth of the Ge film on the Si surface, since Ge precursors can chemisorb at only one third of Si dangling bonds on the surface. After the formation of the 1ML Ge film on the Si(100) surface, the Ge homo-ALE is done by cyclic exposures of DMG and H under isothermal conditions, since the surface has been fully covered by Ge and there is no possibility of C contamination.

Experiments were carried out using an ultra-high vacuum system described in detail elsewhere[9,13]. Sample structure used in experiments and their cleaning procedure were also described there. A standard set of experimental parameters is shown in Table I. These parameters were used throughout experiments unless otherwise mentioned.

The Ge coverage was evaluated from the intensity ratio I_G/I_S between Si$_{2p}$ and Ge$_{3d}$ core level spectra obtained by ex-situ X-ray photoelectron spectroscopy (XPS). The ratio was calibrated by the previously obtained one for a 1/2ML Ge-film adsorbed Si surface formed by DMG exposure at room temperature[12]. The inelastic mean free path of photoelectrons was assumed as escape depth[14]. We must note, however, that there was an under-estimation in the

Table I. Experimental condition.

(a) 1ML-Ge growth on Si(100)

GTC pressure	P_{GTC}=30 mTorr
GTC exposure time	t_{GTC}=3 s
Atomic H exposure time	t_H=20 s
T_{sub} at the GTC and H exposure steps	T_{low}=300°C
T_{sub} at the H desorption step	T_{high}=500°C
H desorption time	t_{high}=30 s
Number of execution cycles	n=15

(b) Ge homo-ALE on 1ML-Ge/Si(100)

DMG pressure	P_{DMG}=30 mTorr
DMG exposure time	t_{DMG}=5 s
Atomic H exposure time	t_H=30 s
Substrate temperature	T_{DMG}=500°C

previously calibrated value and that we used, in this paper, the more realistic one which is 1.1 times larger than the previously reported one[13].

RESULTS AND DISCUSSION

Solid circles and open squares in Fig. 2 shows the intensity ratio I_G/I_S as a function of DMG exposure time t_{DMG} for the 2nd and 3rd Ge layer growth, respectively, where the Ge homo-ALE with only one or two execution cycles was done on the pre-grown 1ML-Ge/Si(100) surface. Dashed lines are the theoretically predicted ratios from the 1ML-Ge/Si(100), 2ML-Ge/Si(100) and 3ML-Ge/Si(100) surfaces, respectively. A data shown at t_{DMG}=0 (open circle) is the ratio for the 1ML-grown Ge film on the Si(100) surface by the RPALD method. For the 2nd Ge layer growth, the Ge coverage gradually increased with elongating t_{DMG} and saturated at about 2ML, i.e. twice of the saturated value for the 1st step, for t_{DMG} of more than 4s. This indicates that the 2nd Ge layer grew in a self-limiting layer-by-layer manner as desired. For the 3rd Ge layer growth, pre-grown 2ML-Ge/Si(100) surface was exposed to DMG. Namely, t_{DMG} for the 2nd Ge growth cycle was fixed at 5s to form the 2ML-Ge layer on the Si surface and t_{DMG} for the 3rd Ge growth cycle was changed. The Ge coverage also increased with elongating t_{DMG} and saturated at 3ML which was equal to triple of the saturated value for the 1st step. The saturation of Ge coverage was achieved within shorter t_{DMG} (2s) compared with that for 2nd Ge growth cycle.

Our previous experiments[8,9] showed that the saturated adsorption of DMG was achieved within the DMG exposure time of 0.3 s on the Ge substrate surface under the same temperature and DMG pressure conditions as those in the present work. Measured results presented here, however, showed no saturated behavior while t_{DMG} was less than 4s for the 2nd layer, although the Si surface had been already covered by the 1ML-Ge top layer. This may be caused by the slow desorption rate of surface H from the 1ML-Ge/Si(100) surface compared with that from the bulk Ge surface as follows: DMG adsorbs dissocietively on the Ge surface and H bonded with precursor moves to the Ge surface, i.e., both H and Ge(CH₃)ₓ species are deposited on the surface. Since for the 1ML-adsorption of Ge(CH₃)ₓ, all sites having been occupied initially by H should bond with Ge(CH₃)ₓ by successive thermal desorption of H and following adsorption of Ge(CH₃)ₓ, the adsorption kinetics of Ge(CH₃)ₓ is governed by the thermal desorption kinetics of surface H. Boishin and Surnev have demonstrated that the Si underlayer causes an upward TPD temperature shift for hydrogen desorbed from the top Ge layer compared with that from the pure Ge surface[15], and explained that the shift is induced by electronic effects generated from the Ge-Si backbonds. Thus slow desorption of hydrogen from the 1ML-Ge/Si surface compared

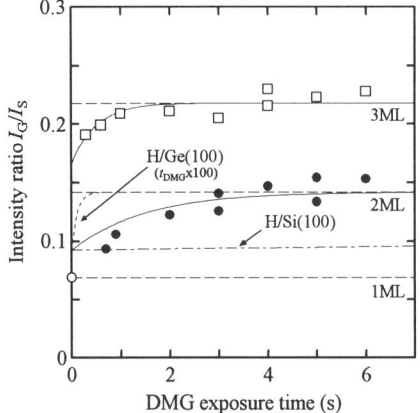

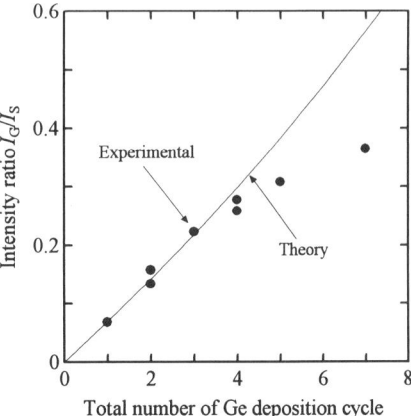

Fig. 2. Intensity ratio I_G/I_S as a function of DMG exposure time for the 2nd and 3rd Ge layers.

Fig. 3. Intensity ratio I_G/I_S as a function of total number of Ge deposition cycle.

with that from the bulk Ge surface is the origin of the long DMG exposure for the saturated 2ML-Ge adsorption. The effect of the Si underlayer becomes minor for the thick Ge toplayer. Therefore, the saturation coverage of 3rd Ge layer can be achieved within a shorter DMG exposure duration.

The sharp curve in the figure are the intensity ratio for the 2nd and 3rd Ge layer calculated by the following rate equation for hydrogen coverage θ_H, in the unit of ML, as:

$$\frac{d\theta_H}{dt} = -k_H\theta_H + \frac{2}{3}k_H\theta_H,$$

where k_H expresses the rate constant for hydrogen desorption from the 1ML-Ge/Si(100) or 2ML-Ge/Si(100) surface. The second term in the right hand side is the successive adsorption of H from the H-desorbed site. Note that two third of desorbed sites are re-occupied by H from the newly arrived DMG. Ge(CH$_3$)$_x$ coverage is expressed by (1-θ_H). Good coincidence has been obtained when k_H=2.1s^{-1} and k_H=6.0s^{-1} for the 2nd and 3rd Ge layer growth, respectively, as shown in the figure. Dotted-dashed curve and dotted curve show the calculated intensity ratio when k_H has the same value as that for the bulk Si(100) and Ge(100) surfaces[12,16], respectively. The k_H value for the 2nd Ge layer growth is as small as 7×10^{-4} times that for the bulk Ge(100) surface and as large as 70 times that for the bulk Si(100) surface.

Figure 3 shows the dependence of the ratio I_G/I_S on total number of the Ge deposition cycle n, where the 1st Ge layer was prepared by the RPALD method and the Ge layers for more than the 2nd layer was grown by the homo-ALE. The sharp curve is the theoretically predicted ratio from the ideal nML-Ge/Si(100) structure. The experimental results coincided with theoretical results up to n=3 and hence the Ge coverage increased in proportion to the number of growth cycles, with the growth rate of 1ML/cycle, for early ALE stages. Thus, we speculated that the ideal layer-by-layer growth is established as mentioned previously.

For more than n=4, the rate of increase in the grown Ge coverage was less than 1ML/cycle. Thus the critical Ge layer thickness by the hetero-ALE is considered to be 3ML. A surface morphology change due to the island growth seems to affect the growth rate.

Uniformity of the grown layer was evaluated by the angle-resolved XPS. Figure 4 shows the intensity ratio as a function of the take-off angle of photoelectrons for n=2. The calculated ratio

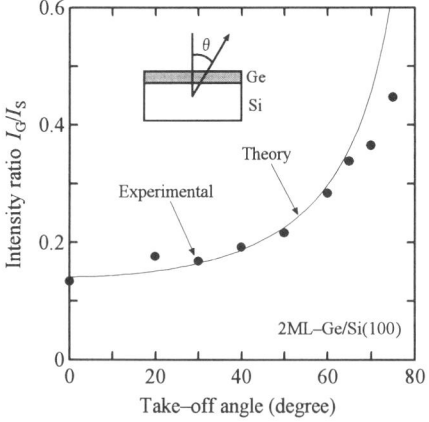

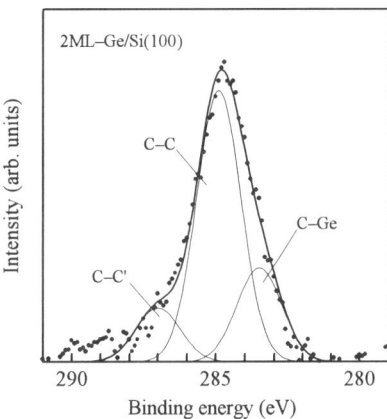

Fig. 4. Intensity ratio I_G/I_S as a function of take-off angle of photoelectrons.

Fig. 5. C_{1s} spectrum after 2nd Ge layer growth.

is shown by a sharp curve in the figure for the 2ML-Ge/Si(100) structure. The experimental data agreed well with the estimated curve even at the large take-off angle. Thus, it was confirmed that Ge grew uniformly.

Since a C-containing source, DMG, was used in the homo-ALE step, there was a room of undesirable Si-C formation. Figure 5 shows the C_{1s} spectrum after the 2nd Ge layer growth by using DMG, where atomic H was not exposed after the DMG exposure. The spectrum had a non-Gaussian form, indicating that it consists of a few chemically shifted components. Decomposed spectra were also shown in the figure, where we used the deconvolution procedure described elsewhere[12]. The strong component by C-C bonds and a weak chemically shifted (+1.4eV) component denoted by C-C' in the figure had been confirmed, by the preceding experiments, to come from the surface contaminants caused by air exposure. A chemically shifted (-1.4eV) component induced by C-Ge bonds was clearly decomposed as shown in the figure, which is originated from the Ge-CH₃ bond in the adsorbed DMG. No chemical shift (of -2.0eV) due to the Si-C formation, however, was observed. This means that by the help of the C-free sources, GTC, and H for the 1ML Ge-film growth on the Si surface, undesirable C contamination was low enough although the C-containing source, DMG, was used as the successive homo-ALE step.

CONCLUSION

We have successfully demonstrated the novel hetero-epitaxy of Ge on the Si(100) surface based on atomic layer epitaxy. The hetero-ALE method was divided into two steps by taking into account different chemical conditions of the surface, i.e., the 1ML growth of the Ge film on the Si (100) surface and the successive Ge homo-ALE on the pre-grown Ge top-layer above the Si substrate. The 1ML growth of the Ge film was achieved by a combination of the substrate temperature modulation and alternate exposures of GTC and atomic H. And the homo-ALE was done by cyclic exposures of DMG and atomic H. It was confirmed by the XPS measurements that there is a step-like increase in the grown Ge film thickness with the 1ML/cycle rate up to the critical Ge thickness of 3ML, but with less than the 1ML/cycle rate for more than the critical

337

thickness. The ALE window was, however, different from that of the home-ALE. The origin of this difference was attributed to the difference in thermal desorption rate of surface terminating H from the 1ML-Ge/Si surface and that from the bulk Ge surface.

REFERENCES

1. S. Satpathy, R. M. Martin and C. G. Van de Walle, Phys. Rev. **B38**, 13237 (1988).
2. U. Schmid, N. E. Christensen, M. Alouani and M. Cardona, Phys. Rev. **B43**, 14597 (1991).
3. M. Ikeda, T. Terakura and T. Oguchi, Phys. Rev. **B48**, 1571 (1993).
4. S. Sugahara, O. Sugiura and M. Matsumura, Jpn. J. Appl. Phys. **32**, 384 (1993)
5. S. Imai, T. Iizuka, O. Sugiura and M. Matsumura, Thin Solid Films **225**, 168 (1993).
6. S. Sugahara, E. Hasunuma, S. Imai and M. Matsumura, Appl. Surf. Sci. **107**, 161 (1996).
7. E. Hasunuma, S. Sugahara, S. Hoshino, S. Imai and M. Matsumura, J. Vac. Sci. Technol. **A16**, 679 (1998).
8. S. Sugahara, T. Kitamura, S. Imai and M. Matsumura, Appl. Surf. Sci. **82/83**, 380 (1994).
9. S. Sugahara, M. Kadoshima, T. Kitamura, S. Imai and M. Matsumura, Appl. Surf. Sci. **90**, 349 (1995).
10. S. Sugahara and M. Matsumura, Appl. Surf. Sci. **112**, 176 (1997).
11. K. Ikeda, S. Sugahara, Y. Uchida, T. Nagai and M. Matsumura, Jpn. J. Appl. Phys. **37**, 1311 (1998).
12. S. Sugahara, T. Kitamura, S. Imai, Y. Uchida and M. Matsumura, Appl. Surf. Sci. **107**, 137 (1996).
13. S. Sugahara, Y. Uchida, T. Kitamura, T. Nagai, M. Matsuyama, T. Hattori and M. Matsumura, Jpn. J. Appl. Phys. **36**, 1609 (1997).
14. S. Tanuma, C.J. Powell and D.R. Penn, Surf. Interface Anal. **21**, 165 (1994).
15. G. Boishin and L. Surnev, Surf. Sci. **345**, 64 (1996).
16. M.P. D'Evelyn, Y.L. Yang and S.M. Cohen, J. Chem. Phys. **101**, 2463 (1994).

COMPARATIVE GROWTH KINETICS OF SIGE IN A COMMERCIAL REDUCED PRESSURE CHEMICAL VAPOUR DEPOSITION EPI REACTOR AND ANOMALIES DURING GROWTH OF THIN SI LAYERS ON SIGE

MATTY CAYMAX*, ROGER LOO*, BERT BRIJS*, WILFRIED VANDERVORST*, DAVID J. HOWARD*+, KENJI KIMURA**, KAORU NAKAJIMA**
*IMEC, Kapeldreef 75, B-3001 Leuven, Belgium, caymax@imec.be
+ present address : Rockwell Semiconductor Systems, 4311 Jamboree Road, Newport Beach, Ca 92660-3095.
** Dept. of Engineering Physics and Mechanics, Kyoto University, Kyoto 606-8501, Japan

ABSTRACT

A short discussion about growth kinetics of Si and $Si_{1-x}Ge_x$ epitaxial layers in a reduced pressure CVD reactor using both dichlorosilane and silane is presented. Through careful observations of the growth of very thin Si layers on SiGe, an anomaly in the Si growth ratewas detected such that the thinner the Si layer, the higher the Si growth rate on SiGe. Due to the difficult nature of very thin film characterization, several analysis techniques were used. A possible explanation based on TEM observations is put forward.

INTRODUCTION

The past 20 years has seen the development of very advanced $Si/Si_{1-x}Ge_x$ (Si/SiGe further on) multi-layer systems consisting of layers widely ranging in Ge content and film thickness, depending upon application. An interesting new application is the growth of ultra-thin sacrificial Si capping layers of only a few nm on top of SiGe [1]. For this application, a very accurate determination of Si growth rates and layer thicknesses is a prime requirement. In the first part of this paper it is interesting to we compare some aspects of the growth kinetics. In a second part, we discuss ultra-thin layers layer growth with nm control, and some difficulties encountered due to anomalous growth rates in the very first atomic layers.

EXPERIMENTAL

The AP/RP-CVD system used in this work is an Epsilon-One epi-reactor, built by ASM America. It is essentially a horizontal, single wafer, load locked reactor, with a lamp heated graphite susceptor in a rectangular quartz tube, generally operated at atmospheric pressure, but equipped with a dry pump for reduced pressure operation (RP). In this work, a growth sequence consists of i) loading regular 150 mm substrates from the wafer manufacturer's box *via* the loadlock into the reactor; ii) baking the wafer at 1050 °C in H_2 for 30 sec. to remove the native oxide; (no buffer layer is grown) ; iii) lowering the temperature and depositing the layer(s) at 40 Torr (for blanket layers, operation at 1 bar or 40 Torr is rather similar). It is worthwhile to mention that, standard thickness and Ge content uniformities across 150 mm wafers are in the order of 3 % (1 sigma) or better. In this work, we used everywhere a H_2 carrier gas flow of 20 slm at a pressure of 40 Torr. As Si-source gas, we use SiH_2Cl_2 (DCS) or SiH_4 as pure gases, whereas GeH_4 is supplied as a mixture of 1 % GeH_4 in H_2.

Characterization of thicker layers was done by means of conventional Rutherford Backscattering Spectrometry (RBS) and Secondary Ion Mass Spectrometry (SIMS) for Ge content determination and thickness measurements, and step profilometry for thickness and thickness uniformity measurements. Very thin layers were measured by means of High Resolution RBS (HRBS) [2]. HRBS differs from conventional RBS mainly in its detection : HRBS makes use of a high-resolution energy spectrometer consisting of a 90° magnet (300 mm radius, 1200 mm dispersion) in combination with a position sensitive ($\delta x = 0.13$ mm) detector. The geometry of the measurement set-up is quite important . Statistical fluctuations in the energy loss process due to energy straggle can widen the FWHM of the backscattered ion energy distribution appreciably, but, because the widening of the FWHM from backscattered ion distribution is an inverse function of the exit angle, a set-up with a grazing exit angle solves this problem. Transmission Electron Microscopy (TEM) was used for general characterization of multilayer systems. X-ray Photoelectron Spectroscopy (XPS) and Spectroscopic Ellipsometry was used for some special investigations of oxidized thin layers.

RESULTS

Kinetics

The overall deposition rate of Si in an CVD epitaxial reactor, as is well known, is controlled by limited gas phase mass transfer at high temperatures (due to diffusion limitations), and by the limited rate of thermally activated homogeneous or heterogeneous silicon surface chemical reactions at lower temperatures. The transition point between these two regimes is determined by the kind of Si source gas and by the deposition conditions. Typical values are 950 °C for DCS and about 890 °C for SiH_4 at a pressure of 1 bar and a Si source gas concentration of 0.1 % [3]. It is important to realize, however, that it is also possible to find a mass-transfer controlled regime at low temperature, if the supply of reactive species towards the surface is insufficient for the rate at which they are consumed by the surface reactions. This can be caused *e.g.* by a limited input gas flow.

An important factor governing the kinetically controlled regime, then, is the so-called Arrhenius activation energy. Fig. 1 shows a determination of the activation energy for both DCS and silane at different gas flows.

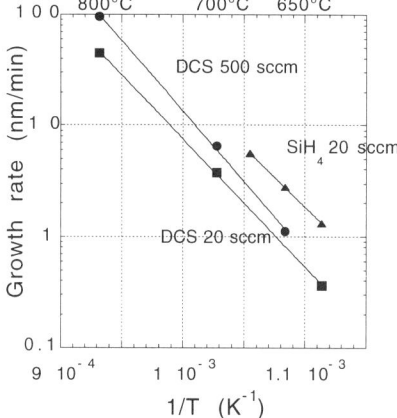

Fig. 1 Temperature dependence of growth rate for DCS and SiH₄ at different flows

For DCS, the activation energy is 52.5 and 57.8 kcal/mole at flows of 20 and 500 sccm, respectively. It is clear that the lower value is biased by the influence of the mass transfer control typically having an activation energy of about 5 kcal/mole. The activation energy for SiH_4 determined in this way is 48.9 kcal/mole for a SiH_4 flow of 20 sccm. Probably this value is underestimated, as is analogous to the 20 sccm case for DCS in view of the even higher deposition rates. Still, the activation energy for SiH_4 based epi growth is lower than for DCS, which can be explained by the difference in bond energies of H resp. and Cl with respect to Si (77.5 vs. 109 kcal/mole [4-5]). The desorption of the reaction by-products H and/or Cl from the surface is the rate-determining process in the kinetically controlled regime.

The addition of GeH_4 to the growth mixture in order to deposit $Si_{1-x}Ge_x$ alloy layers, changes the deposition kinetics considerably, and, as is well known, increases the growth rate considerably [5]. The reason for this lies in the less strong bonds of H and Cl to Ge (69.1 resp. and 82 kcal/mole) [4-5], which lowers the activation energy accordingly. This trend is shown in Fig. 2 which plots growth rate vs. inverse temperature for gas mixtures with 20 sccm DCS and various GeH_4 flows. It is interesting to note that for the highest GeH_4 flow, the activation energy decreases at higher temperatures, indicating again an increasing control over growth rate by insufficient mass transfer. An analogous experiment for SiH_4/GeH_4 kinetics results in Fig. 3. The resulting activation energies are summarized in Table I. The same shift in the rate controlling step as in the DCS/GeH$_4$ case can be observed, i.e. transition in H$_2$ desorption H desorption from Si-H to Ge-H sites.

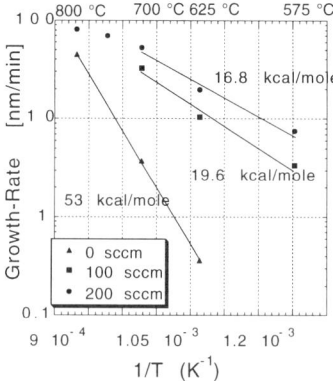

Fig. 2 Arrhenius plot of growth rates of SiGe with DCS = 20 sccm, p = 40 torr and various GeH$_4$ flows as indicated

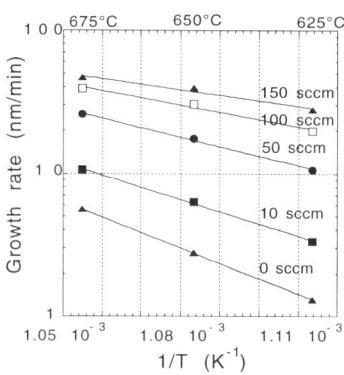

Fig. 3 Arrhenius plot of growth rates of SiGe with SiH$_4$ = 20 sccm, p = 40 torr and various GeH$_4$ flows

Table I Activation energies (Kcal/mole) for SiGe growth in SiH$_4$/GeH$_4$ mixtures

GeH$_4$ flow	Activation Energy
0	48.9
10	39.1
50	29.9
100	22.9
150	17.5

Growth of ultra-thin SiGe layers

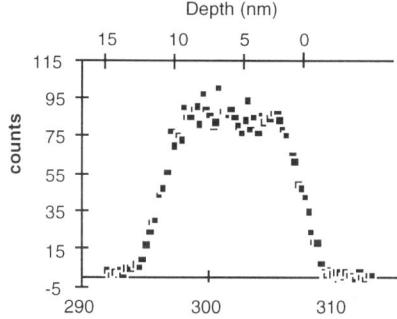

Fig. 4 Ge signal in HRBS measurement of Si$_{0.7}$Ge$_{0.3}$ layer

The growth rate data outlined above were used to formulate recipes for ultra-thin layers of Si and SiGe. In a first attempt to deposit an ultra-thin layer of SiGe on Si, a sample was prepared at 575 °C with a deposition time of 1 min, 20 sccm of DCS and 200 sccm of GeH$_4$.
Fig. 4 shows the HRBS spectrum of the Ge signal. The depth scale in the figure is estimated with a stopping power of pure Si given by Ziegler. Measurement conditions were 0.3 MeV He+ ions, scatter angle 100 °, exit angle 30 °, total beam dose 3.5 µC. From the Ge spectrum, a Ge content of 33 % and a thickness of 10-11 nm can be extracted. This Ge content is in excellent agreement with a conventional RBS measurement on a second, 48 nm thick sample, grown in identical conditions, giving 31 %.

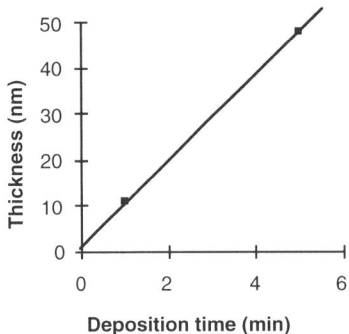

Fig. 5 Layer thickness vs. time (575 °C, 20 sccm DCS, 200 sccm GeH₄)

Fig. 5 Layer thickness vs. time (575 °C, 20 sccm DCS, 200 sccm GeH$_4$)

In Fig. 5, we plot the layer thickness of these two samples vs. deposition time. By extrapolation to zero time, it shows that there is no incubation time visible on this time scale (and within experimental error). It should be mentioned that the growth rate as derived in this experiment deviates slightly from the previously discussed data. This is mainly due to the time lapse between the two experiments, in which the deposition equipment underwent some modifications such that the overall temperature profile was slightly changed

Fig. 6 x-TEM view of sample # 532/5 the dark bands are SiGe layers, the top light band is glue

Growth of ultra-thin Si layers

In order to accurately measure very thin Si layers by means of a step profilometer, a number of multiple SiGe/Si layer stacks were deposited in two separate experiments, see Table II. The stacks consist of five (or ten) consecutive SiGe/Si bi-layer combinations with different Si deposition times, resulting in ten (or twenty) sublayers bi-layers making a thickness determination by a step profilometer more accurate. The Si layer thickness was also measured by TEM (Fig. 6, showing sample 532/5), and is in good agreement with the step profilometer. The nominal growth rate for Si was determined

Table II Si layer thickness measurement by Profilometer on multi-layer samples

Sample	# of Si layers	# SiGe layers	dep. time / Si layer	SiGe [1] thickness	full stack thickness	Si [2] thickness	TEM Si thickness	Si dep. rate [3]
532/3	5	5	8'23"	4.9	45.3	4.2	4.8	0.50
532/4	5	5	16'45"	4.9	64.3	8.0	7.9	0.48
532/5	5	5	25'08"	4.9	80.9	11.3	10.7	0.45
606/2	9	10	25'08"	4.1	166.8	14.1		0.56
606/3	9	10	4'12"	4.1	64.8	2.7		0.65

Remarks :
1. as calculated based on growth rate measurement on single SiGe layer sample grown immediately before these samples ; all thicknesses in nm
2. Si thickness = measured stack thickness - calc. SiGe thickness per layer
3. as calculated from profilometer measurements

by profilometry over a single Si layer sample with a deposition time of 120 min, resulting in a growth rate of 0.36 nm/min. The data of 606/2-3 are somewhat suspect due to a likely problem with the temperature profile in the reactor (also clear from the less good thickness uniformity over the wafer). They are neglected for further analysis, but included here for completeness.

Another kind of sample, containing one SiGe/Si bilayer, was grown with four different SiGe layer thicknesses to be measured by means of HRBS. The results are summarized in Table III. Whereas there is a very good agreement between the target SiGe layer thickness and the measured value, the measured Si layer deposited on SiGe clearly is again thicker than the nominal Si thickness.

Table III HRBS measurements on Si/SiGe bilayer samples

Sample	Nominal thickness SiGe (nm)	Nominal thickness Si (nm)	Measured thickness SiGe (nm) (HRBS)	Measured thickn. Si (nm) (HRBS)
556/5	0.54 (4 ML)	1.08 (8 ML)	< 0.85	
556/6	1.08 (8 ML)	1.08	1.2	
556/7	1.62 (12 ML)	1.08	1.5	
556/8	4.86 (36 ML)	1.08	4.4	2.3

An additional experiment for information about the thickness of a Si cap layer involved thermal oxidation of a thin Si/SiGe bi-layer and measurement of the resulting oxidized bilayer layer combination my means of XPS and Spectroscopic Ellipsometry (SE). As is well known [7], SiGe oxidizes faster than Si. By carefully choosing the Si cap layer thickness and the oxidation conditions (which means, the amount of Si which will be consumed by the oxidation process), it is possible to get some idea about the Si thickness as the oxidation behaviour will change immediately when the oxidation front passes beyond the Si/SiGe interface. Table IV lists the nominal thicknesses of the original Si capping layer and the resulting oxide layer, as well as the measurement results of XPS and SE. From SE, both the SiO_2 layer thickness and the remaining Si under the SiO_2 (if any) are measured. For samples 626/1 and 563/9 the signal for GeO_2 was detected, indicating that the oxidation front crossed the Si/SiGe interface. A comparison of the nominal Si thickness and the measured SiO_2 thickness indicates that the real Si thickness must have superseeded the nominal value, confirming

Table IV Nominal, measured and calculated thicknesses of Si and SiO_2 layers

Sample	nominal Si thickness	nom. SiO_2 thickness	SiO_2 by XPS	SiO_2 by SE	Si under SiO_2 by SE	Si consumed (XPS)	Si consumed (SE)	calc'ted Si thickness
563/1	1	2	1.6			0.73		
563/2	2	2	1.6			0.73		
563/3	3	2	0.5			0.73		
563/5	1	4	4.6			2.28		≥ 2.28
563/6	2	4	3.9			1.77		
563/7	3	4	3.9	4.5	3.6	1.77	2.05	5.65
563/9	1	6	--(GeO_2)	31				
563/10	2	6	5.7	6.2	1.3	2.59	2.82	4.12
563/11	3	6	5.7	6.7	2.3	2.59	3.05	5.35
626/1	0.5	4	--(GeO_2)	19	< 0			
626/2	1.1	4	4.2	5.7	< 0	1.91	2.59	1.91-2.59
626/3	1.6	4	4.2	5.3	< 0	1.91	2.41	2.41
626/4	2.2	4	4.0			1.82		
626/5	2.7	4	4.1			1.86		

again the previously stated anomaly in growth rate of Si on SiGe layers. The last column lists the "real" Si cap layer thickness as can be calculated from the XPS and SE data ; where appropriate, we summed up the thickness of the Si layer consumed by the oxidation and the amount of Si left under the oxide.

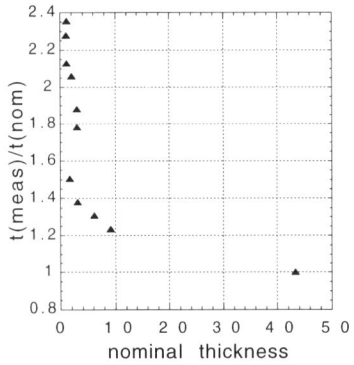

Fig.7 Measured thickness for ultra-thin Si layers vs. nominal thickness

In Fig. 7, all relevant data of real Si cap layer thicknesses are included in a ratio of measured to nominal Si thickness plotted as a function of nominal Si based on thick Si films. In this figure, we don't give the mere growth rate as such, but rather the ratio of measured over nominal thickness. This shows that indeed the Si growth rate increases considerably for layers approaching monolayer thicknesses. For thicker layers, of course, this ratio goes to one as this influence disappears in the experimental error.

A possible explanation for this observation might be found in the x-TEM picture in Fig.6. Although the Si/SiGe layer transition is pretty sharp, the next SiGe/Si layer transition looks somewhat blurred, as if the Ge atoms diffused very slightly into the next Si layer. If this is the case, then it might be that these Ge atoms increased the growth rate during the very first atomic layers of each new Si layer.

CONCLUSION

The growth of ultra-thin SiGe layers on Si appears to behave according to the overall kinetics known from thicker films. Si growth on SiGe however shows an increased growth rate in the first nm of the film, which can be more than 2 times higher than for thicker layers. This can possibly be explained based on TEM observations by a slight intermixing of Ge into the first monolayers of Si, so increasing the growth rate.

ACKNOWLEDGEMENTS

The authors are indebted to the European Commission (CEC) for financial support in the framework of the ESPRIT IV LTR *VAHMOS 2000* Project nr. 22495. One of us (R.L.) is indebted to the CEC for granting him a fellowship in the TMR Network *AApples* nr. FMRXCT96-0029

REFERENCES
1. R. Loo, M. Caymax, P. Verheyen, T. Conard, H. Bender, W. Vandervorst, N. Collaert, K. De Meyer, I. Szendrö and P. Alkemade ; submitted to the 28th ESSDERC Conference, 1998.
2. K. Kimura and M. Mannami, Nucl. Instrum. and Methods B, 113, 270 (1996)
3. J. Bloem and L.J. Giling, in : *Current Topics in Materials Science*, Vol. I, Ed. E. Kaldis, North-Holland Publishing Co. (1978).
4. J.Phys.Chem.Ref. Data 11, Suppl. 2 (1982).
5. Handbook of Chemistry and Physics, 65th Ed. R.C. Weast Editor, CRC Press (1984)
6. W.B de Boer and D.J. Meyer, Appl. Phys. Lett. 58 (12), 1286 (1991).
7. M. Nicolet and Wen-Shu Liu, Microelectronic Engineering 28 (1995) 185-191

THERMAL STABILITY OF SiGe/Si
QUANTUM WELL STRUCTURES GROWN BY APCVD

Q.X. Zhao[a], O. Nur[a], U. Södervall[a], C.J. Patel[a], M. Willander[a], P.O. Holtz[b] and W.B. de Boer[c]
a) Physical Electronics and Photonics, Department of Microelectronics and Nanoscience, Chalmers University of Technology and Göteborg University, S-412 96 Göteborg, Sweden.(Zhao@fy.chalmers.se)
b) Department of Physics, Linköping University of Technology, S-583 81 Linköping, Sweden
c) Philips Research Laboratories, Prof. Holstlaan 4, 5656 AA Eindhoven, The Netherlands

ABSTRACT

Single and double $Si_{1-x}Ge_x$/Si quantum well (QW) structures, which were grown by atmospheric pressure chemical vapor deposition (APCVD), are characterized by photoluminescence and secondary ion mass spectrometry. Systematic post-growth annealing treatments were carried out at temperatures between 600 °C and 1100 °C in pure N_2 ambient. The interdiffusion between the Si layer and the $Si_{1-x}Ge_x$ well layers occurs at the annealing temperature around 900°C. The diffusion coefficient is deduced at different temperatures from SIMS measurements for single QW structures. The activation energy is about 3.9 eV in the temperature range between 950 °C and 1100 °C. The double QW structures show a similar value, but the accurate value is more difficult to obtain because it is more complicated to analyze the SIMS profile of the double QW structures. The intensity of the exciton recombination related to carriers confined in the double QW structures decreases with increasing annealing temperatures and becomes strongly suppressed at 750 °C. When the annealing temperature is increased further, the intensity of the QW emission recovers. The results indicate that nonradiative centers were generated at annealing temperature of about 750 °C.

INTRODUCTION

Investigation of SiGe/Si structures is motivated by both the fundamental interest and the potential application in Si based devices, such as using SiGe/Si heterostructures to enhance the mobility of holes [1-3] since they may have potential to be used in a CMOS device. A quantum well (QW) structure is typically used as a channel with confined hole charge carriers. Thus it is very important to know the structure stability and the quality of the material at different post-growth heat treatments. In the past several years, there have been many reports concerning the stability of SiGe/Si QW structures grown by molecular beam exiptaxy (MBE) and ultrahigh vacuum chemical vapor deposition (UHVCVD) [4-10] under thermal annealing treatments. The results show that the photoluminescence from SiGe/Si QWs exhibits a blue shift as a function of annealing temperatures due to the Ge diffusion [5, 7, 8]. It has also been reported recently that the quality of the well which is deeper from surface in the double SiGe/Si QW structure at annealing temperatures lower than the growth temperature is improved presumably due to non-equilibrium growth condition [5]. There are significantly less studies on the SiGe/Si structures grown by atmospheric pressure chemical vapor deposition (APCVD).

In this report, we present an experimental work on systematic annealing treatment of single and double SiGe/Si quantum well structures grown by APCVD. We investigate the optical properties of the various QW structures and the profile of Ge atoms versus annealing temperatures. The results show that the excitons are quenched at the annealing temperature of 750 °C in our samples, and they recover to a certain extent at higher annealing temperatures in those structures. The SIMS profile of the Ge concentration indicates that the structures grown by APCVD show a good thermal stability up to 900 °C annealing in a pure N_2 furnace for 30 minutes. The diffusion coefficient was deduced from analyzing SIMS profiles.

SAMPLES AND EXPERIMENTAL

The samples used in this investigation are grown by APCVD in an ASM Epsilon One. The growth temperature is 660°C for the SiGe layer and 700°C for the Si capping layer. The 10 nm Si capping layer is n-type with approximately $10^{16}/cm^3$ phosphorous donors. The SiGe/Si QW structures were grown on an intentionally undoped 2 μm Si buffer layer. Fig.1 shows a schematic drawing of the structures and the information of the single and double QW structures (labeled No.1 and No.2) used in this investigation. The post annealing were carried out in a well calibrated three-zone furnace. The samples were annealed in pure N_2 for 30 minutes to six hours at temperatures of 600, 750, 900, 950, 1000, 1050 and 1100 °C, respectively. To achieve the same annealing conditions, the two different samples were heat treated together at each desired temperature.

PL measurements were carried out in a low temperature cryostat at 2.0 K. A 514.5 nm line from an Ar-laser was used as the excitation source. The emissions from the samples were focused into a double monochromator and detected by a liquid nitrogen cooled Ge-detector. Atomic Ge and Si concentrations were measured by secondary-ion mass spectrometry (SIMS).

RESULTS AND DISCUSSIONS

Sample No.1

Sample No.2

| 10 nm Si caping layer |
| 10 nm SiGe (x=0.3) |
| 5 nm Si layer |
| 10 nm SiGe (x=0.15) |
| Si Buffer layer and substrate |

| 10 nm Si caping layer |
| 10 nm SiGe (x=0.3) |
| Si Buffer layer and substrate |

Fig.1 Schematic drawing of the samples No.1 and No.2 used in this study.

The profiles of Ge and Si atoms across the structures were measured by SIMS before and after annealing. Fig.2 shows a typical Ge-calibrated profile spectrum (for Si the scale only serves as an indication) of samples No.1 and No.2 annealed 30 minutes at different temperatures. The Ge calibration is carried out by comparing the SIMS intensity with a well characterized $Si_{0.86}Ge_{0.14}$ epilayer. First we can see from the figure that both the absolute value and the ratio of the Ge concentration between the two QWs in the double QW structures are consistent with the aimed values with an accuracy of about 10% of the value, which is the SIMS resolution. At the annealing temperature of 1000°C the double SiGe QWs become a single QW, and in the single QW structures the Ge distribution broadens with a factor of three in comparison with the as-grown samples. This indicates that the Ge atoms strongly diffuse into the Si barrier regions from the SiGe well layers. The significant interdiffusion of Ge and Si atoms occurs

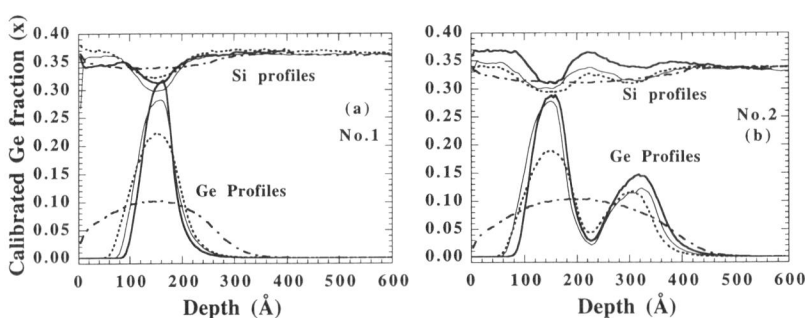

Fig.2 The Si and Ge SIMS profiles of the as-grown and the thermal annealed samples of (a) No.1 and (b) No.2. The scale is calibrated for the Ge fraction x of $Si_{1-x}Ge_x$. Thick solid lines: as grown samples, thin solid lines: 750°C annealing, dotted lines:900°C annealing, dot-dashed lines:1000°C annealing

already at the annealing temperature of 900°C as can been seen from the figure. The diffusion coefficient can be deduced by fitting the SIMS profiles. Since the diffusion coefficient is independent on the annealing time, we have annealed the samples between 30 minutes and 6 hours at each temperature to get more accurate value of the diffusion coefficient. SIMS profile was made using low energy O_2^+ ions with approximately 2 keV implant energy and analyzing the positive secondary ions of ^{30}Si and ^{74}Ge. From a simple model assuming an error function distribution in the low concentration region ($<1x10^{20}$ atoms /cm³), the Ge diffusion coefficients (denoted D in Fig.3) in Si are calculated. The results are shown in Fig.3. The values are similar as reported by Hettich *et al* [11]. The deduced activation energy is about 3.9 eV in temperature range between 950°C and 1100°C. The double QW structures show a similar value, but the value is less accurate in this study, since it requires a more complicated model to fit the SIMS profiles of the double QW structures. It was not possible to obtain the diffusion coefficients for temperatures below 950°C since it is difficult to obtain the accurate thermal broadening of the Ge profile in such narrow QW structures.

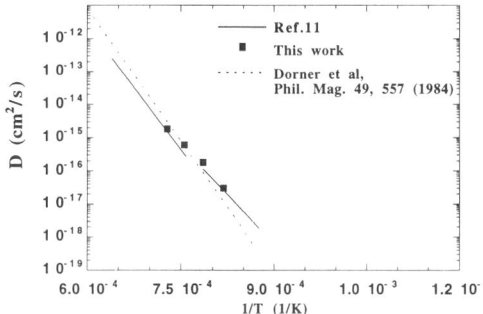

Fig.3 Ge diffusion coefficients in SiGe/Si QW structures.

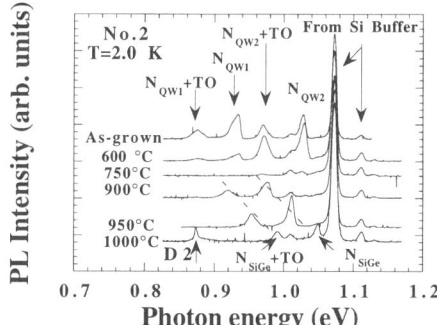

Fig.4 The PL spectra from the as-grown and the thermal annealed samples of No.2 at the excitation density of 300 W/mm².

Fig.4 shows the PL spectra of the sample No.2 before and after 30 minute annealing at different temperatures. The dominant exciton transition is from the Si buffer layer. The temperature and excitation intensity dependencies of the PL spectra, it seems that the excitons from Si layer are related to impurity-bound excitons. The chemical origin of the impurity is unclear at the moment. Since we are mainly interested to study the properties of SiGe/Si QW structures, we will instead focus on the transitions associated with the QW. Several excitonic transitions related to the free exciton (no-phonon line) in the QW and its phonon replica (TO), labeled N_{QW} and $N_{QW}+TO$, respectively, appear in the PL spectra. The evolution of the SiGe/Si QW structures under different annealing treatments can also be seen in this figure. There are several effects which can be noticed in the PL spectra. First the intensity of the QW excitons are dramatically reduced at the annealing temperature of 750°C. However, the intensity of the excitonic recombination starts to increase at the annealing temperature of 900°C. The transitions N_{QW1} and $N_{QW1}+TO$ show a strong blue shift as indicated by dashed lines in Fig.4. At 1000°C the interdiffusion between the Si and SiGe well layer is considerable, and we label the excitonic transition as N_{SiGe} in the PL. Secondly, the dislocation recombination line labeled D2 can been clearly seen at the annealing temperature of 1000°C. The transition related to the SiGe layer can not be observed anymore when the annealing temperature exceeds 1050°C. The single QW sample No.1 shows a similar behavior.

We should emphasize that the Ge and Si profiles across the structures do not show any dramatic difference for the 750°C annealed samples in comparison with the as-grown samples. This leads us to conclude that the quenching of the QW exciton at the annealing temperature of 750°C is not correlated with the smearing out of the SiGe/Si QW structure, but is correlated with the formation of nonradiative centers in the QW region. The dependence of the exciton transition on the excitation intensity also supports this conclusion. When a high excitation density was used, the non-radiative centers were saturated to a certain degree and the QW excitons become visible again. The PL spectra obtained for the samples annealed at the different temperatures, imply that such non-radiative centers can also be significantly suppressed at the annealing temperature of 900°C or higher. It has been reported that two different types of deep traps are formed for partially relaxed and fully relaxed SiGe/Si heteroepitaxial films, respectively [12]. Though there is no doubt about the creation of defects during the relaxation of the SiGe/Si heterostructures at the thermal treatments in this study, we are unable to identify the chemical nature of the nonradiative defect centers created in the structures. A more extensive study has to be done in order to draw any conclusion on the origin of the nonradiative defects (such as the chemical species and the geometric structures of the defects) appearing in our structures or to provide any correlation between such nonradiative defects and the deep traps reported in SiGe/Si heteroepitaxial films [13].

In summary, we have presented a study of the optical properties of different SiGe/Si QW structures grown by APCVD by using optical characterization and SIMS profile measurements. The diffusion of Ge atoms into the Si layer starts at the annealing temperature of 900°C. The diffusion coefficient is deduced at different temperatures from SIMS measurements. We can conclude that the SiGe/Si QW structures grown by APCVD can not be sustained during the conventional thermal treatment in a furnace at a temperature above 1000°C for 30 minutes. The rapid thermal annealing technique for high temperature treatments has to be used. Our optical results also indicate that a large numbers of the nonradiative defect centers created at annealing temperature of 750°C start to be suppressed at an annealing temperature of 900°C.

REFERENCES

1   P.M. Garone, V. Venkataraman, and J.C. Sturm, IEEE Electron Device Lett. **12**, 230 (1991).
2   S.S. Iyer, P.M. Solomon, V. O. Kesan, A.A. Bright, J.L. Freeouf, T.N. Nguyen, and A.C. Warren, IEEE, Electron Device Lett. **12**, 246 (1991).
3   S. Verdonckt-Vandebroek, E.F. Crabbe, B.S. Meyerson, D.L. Harame, P.J. Restle, J.M.C. Stork, and J.B. Johnson, IEEE Trans. Electron Device **41**, 90 81994).
4   P. Boucaud, L. Wu, C. Guedi, F.H. Julien, I. Sajnes, Y. Campidelli, and L. Garchery, J. Appl. Phys. **80**, 1414 (1996).
5   H.M. Latuske, U. Mantz, K. Yhonke, R. Sauer, F. Schaffler, and H.J. Herzog, <u>22nd International conference on the Physics of semiconductors</u>, Vol.2, (World Scientific, Singapore, 1995)p.1233
6   H.H. Radamson, W.X. Ni, and G.V. Hansson, Appl. Surf. Science Vol. 102, 82 (1996).
7   H. Lafontaine, D.C. Houghton, N.L. Rowell, and G.C. Aers, Appl. Phys. Lett. **69**, 1444 (1996).
8   H.P. Zeindl, S. Nilsson, and E. Bugiel, Appl. Surf. Science Vol. 92, 552 (1996).
9   N. Usami, Y. Shiraki, and S. Fukatsu, Appl. Phys. Lett. **68**, 2340 (1996).
10   H. Sunamura, S. Fukatsu, N. Usami, and Y. Shiraki, Appl. Phys. Lett. **63**, 1651 (1993).
11   G. Hettich and H. Mehrer, <u>Materials Science Series: Diffusion in Crystalline Solids</u>, ed. by Graeme E. Murch and Arthur S. Nowick, (Academic Press,1984), p81.
12   D.C. Houghton, G.C. Aers, S.R. Eric Yang, E. Wang, and N.L. Rowell, Phys. Rev. Lett. **75**, 866 (1995).
13   Y. Mera, K. Maeda, and Y. Shiraki, Materials Science forum Vol.196-201, 365 (1995).

LOW-TEMPERATURE EPITAXIAL GROWTH OF
IN-SITU HEAVILY B-DOPED $Si_{1-x}Ge_x$ FILMS USING ULTRACLEAN LPCVD

A. MORIYA*+, M. SAKURABA*, T. MATSUURA*, J. MUROTA*
I. KAWASHIMA** and N. YABUMOTO**
* Laboratory for Electronic Intelligent Systems, RIEC, Tohoku University,
  Sendai 980-8577, Japan, murota@riec.tohoku.ac.jp
** Atsugi Material Analysis Center, NTT Advanced Technology Corp., Atsugi 243-0124, Japan

ABSTRACT

In-situ heavy doping of B into $Si_{1-x}Ge_x$ epitaxial films on the Si(100) substrate have been investigated at 550°C in a SiH_4(6.0Pa)-GeH_4(0.1-6.0Pa)-B_2H_6(1.25×10^{-5}-3.75×10^{-2}Pa)-H_2(17-24Pa) gas mixture by using an ultraclean hot-wall low-pressure CVD system. The deposition rate increased with increasing GeH_4 partial pressure, and it decreased with increasing B_2H_6 partial pressure only at the higher GeH_4 partial pressure. As the B_2H_6 partial pressure increased, the Ge fraction scarcely changed although the lattice constant of the film decreased. These characteristics can be explained by the suppression of both the SiH_4 and GeH_4 adsorption/ reactions in a similar degree due to B_2H_6 adsorption on the Si-Ge and/or Ge-Ge bond sites. The B concentration in the film increased proportionally up to $10^{22}cm^{-3}$ with increasing B_2H_6 partial pressure.

INTRODUCTION

Heteroepitaxy of $Si_{1-x}Ge_x$ alloys on Si by LPCVD has attracted much interest for the fabrication of novel heterostructure devices in silicon based technology[1,2]. The extensive and exact control of the doping concentration and the Ge fraction x in the $Si_{1-x}Ge_x$ layer is a key to realizing devices of high performance such as $S^3EMOSFET$ where selective CVD $Si_{1-x}Ge_x$ is used as a self-aligned source/drain electrode material[3,4]. However, in most of the previous papers, the in-situ doping properties and the deposition mechanism for $Si_{1-x}Ge_x$ layers were limited to those of low concentration dopants with x around 0.25 or less at a deposition temperature above 600°C [5-7].

In order to apply the $Si_{1-x}Ge_x$ films to the CMOS structure, high concentration doping control of both n-type and p-type dopants is necessary. In our previous paper, we already reported in-situ high concentration P doping into the $Si_{1-x}Ge_x$(0.2<x<0.8) epitaxial films on Si(100) by using a SiH_4, GeH_4, PH_3 and H_2 gas system at 550°C by the ultraclean LPCVD[8].

+ On leave from Kokusai Electric Co., Ltd., Sendai Research Lab., Sendai 981-3206, Japan.

Mat. Res. Soc. Symp. Proc. Vol. 533 © 1998 Materials Research Society

In this paper, we describe in-situ heavily B-doped $Si_{1-x}Ge_x(0.15<x<0.85)$ epitaxial films grown on Si(100) at 550°C using ultraclean LPCVD with SiH_4, GeH_4, B_2H_6 and H_2 as process gasses, and discuss the relationships among the deposition rate, the Ge fraction and the B concentration in the films.

EXPERIMENTAL

The epitaxial growth of the B-doped $Si_{1-x}Ge_x$ films on the Si(100) substrates was carried out at 550°C in a SiH_4-GeH_4-B_2H_6-H_2 gas mixture by using an ultraclean hot wall LPCVD system, schematically shown in Figure 1[9]. This system was made ultrahigh vacuum compatible with gate valves and a turbo molecular pump system. The turbo molecular pump was connected to the reactor with several sizes of vacuum exhaust tubes, and the reactor could be directly evacuated by the turbo molecular pump from an atmospheric pressure. To minimize air contamination into the reactor during wafer load and unload, a N_2 purged transfer chamber was combined with the reactor inlet. Wafers were transported into the reactor under an ultraclean N_2 atmosphere through the transfer chamber. After closing the gate valve, the reactor tube was purged with high purity H_2 gas. In order to prevent any contamination from the exhaust line, the purge gas was supplied continuously during vacuum pumping before deposition.

The in-situ doping was performed by introducing B_2H_6 gases. The total deposition pressure was about 30 Pa, and the partial pressures of SiH_4, GeH_4 and B_2H_6 were in the range of 6.0 Pa, 0.1-6.0 Pa and 1.25×10^{-5}-3.75×10^{-2} Pa, respectively. The substrates used were n-type Si wafers of 3-5 ohm-cm with mirror polished (100) surfaces. Before loading the wafers into the transfer chamber, they were cleaned in several cycles in a 4:1 solution of H_2SO_4 and H_2O_2, high purity DI water, and 1-2% HF with the final rinse in DI water.

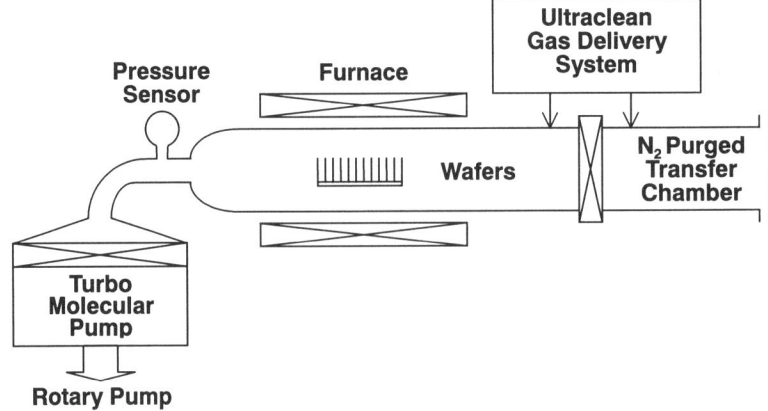

Figure 1 : Schematic diagram of the ultraclean hot-wall LPCVD system.

The film thickness was measured by using a step profiler. The Ge fraction in the film was estimated from the lattice constant measured by x-ray diffraction(XRD) and also was evaluated from secondary ion mass spectroscopy(SIMS). The B concentration in the $Si_{1-x}Ge_x$ film was determined by SIMS. The surface structure was evaluated by reflection high-energy electron diffraction(RHEED).

RESULTS AND DISCUSSION

Figure 2 shows the dependence of the deposition rate of the $Si_{1-x}Ge_x$ films on the B_2H_6 partial pressure. It is remarkable that the deposition rate decrease with increasing B_2H_6 partial pressure only at the higher GeH_4 partial pressure, but it is nearly independent of the B_2H_6 partial pressure at the lower GeH_4 partial pressure. The deposition rate increases with increasing GeH_4 partial pressure.

Figure 3 shows the dependence of the Ge fraction in the $Si_{1-x}Ge_x$ films on the B_2H_6 partial pressure. The Ge fraction x is almost constant independently of the B_2H_6 partial pressure, although we have observed by XRD that the lattice constant of the film decreased at the higher B_2H_6 partial pressure, for example by about 0.4% at the B_2H_6 partial pressure of 3.75×10^{-2}Pa and the GeH_4 partial pressure of 2.0Pa.

Figure 4 shows the dependence of the B concentration in the $Si_{1-x}Ge_x$ films on the B_2H_6 partial pressure. The B concentration in the film increases almost proportionally up to 10^{22}cm^{-3} with increasing B_2H_6 partial pressure, and it decreases with increasing GeH_4 partial pressure.   It is

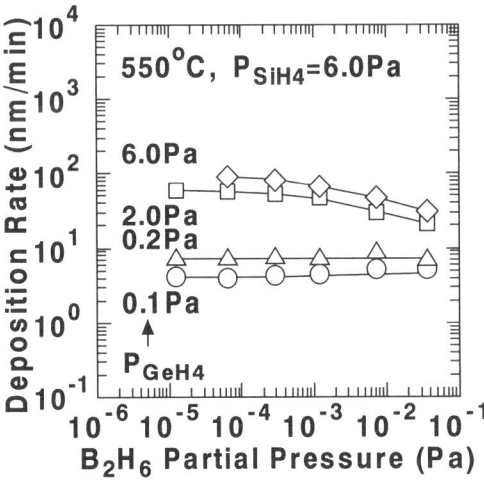

Figure 2 : Dependence of the deposition rate of $Si_{1-x}Ge_x$ films on the B_2H_6 partial pressure.

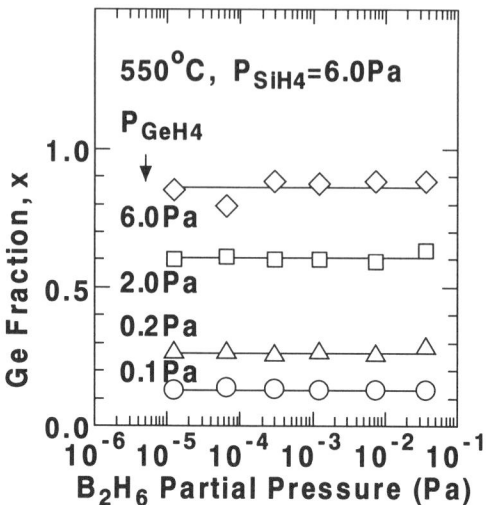

Figure 3 : Dependence of Ge fraction in the $Si_{1-x}Ge_x$ films on the B_2H_6 partial pressure.

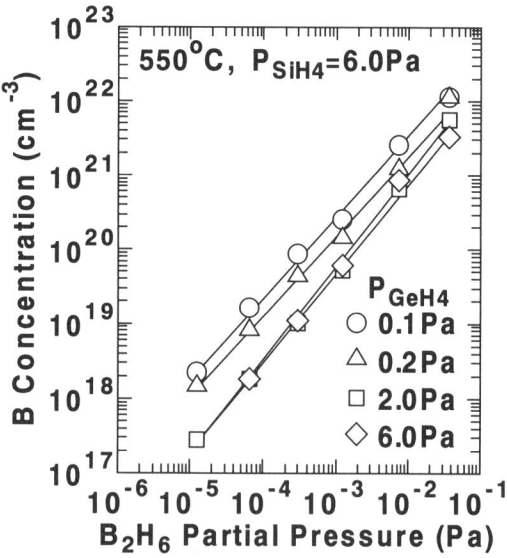

Figure 4 : Dependence of the B concentration in the $Si_{1-x}Ge_x$ films on the B_2H_6 partial pressure.

noteworthy that these all films were single crystals although their quality at high B concentrations and high Ge fractions was degraded by generation of misfit dislocation due to the larger mismatch between Si and $Si_{1-x}Ge_x$ as well as by island growth.

Figure 5 shows the incorporation rate of B atoms into the $Si_{1-x}Ge_x$ films, which is obtained directly by the product of the deposition rate and the B concentration. The incorporation rate of B atoms is linear with the B_2H_6 partial pressure, and increases with increasing GeH_4/SiH_4 partial pressure ratio, in other words, Ge fraction. This means that the incorporation rate is higher at the Ge-related site than at the Si-related site. This also suggests that incorporation of B into the $Si_{1-x}Ge_x$ films is not limited by the mass-transport of B-hydride in the gas phase, but is controlled by the surface reactions. Since B concentration was nearly equal to the carrier concentration at least below $2\times10^{20}cm^{-3}$[10], it is clear that B is monatomic in the films. Assuming that B_2H_6 molecules are decomposed completely into BH_3 in the gas phase[11], it is proposed that monatomic B is incorporated into the film according to Henry's law, because the the B concentration is linear to the B_2H_6 partial pressure.

The decrease of the deposition rate at the higher B_2H_6 partial pressure only at the higher GeH_4 partial pressure is considered to be caused by suppression of both the SiH_4 and GeH_4 adsorption/ reactions due to the higher surface coverage of B-hydride on the Si-Ge and/or Ge-Ge bond sites than on the Si-Si bond site. And their suppression effects are considered to be similar for SiH_4

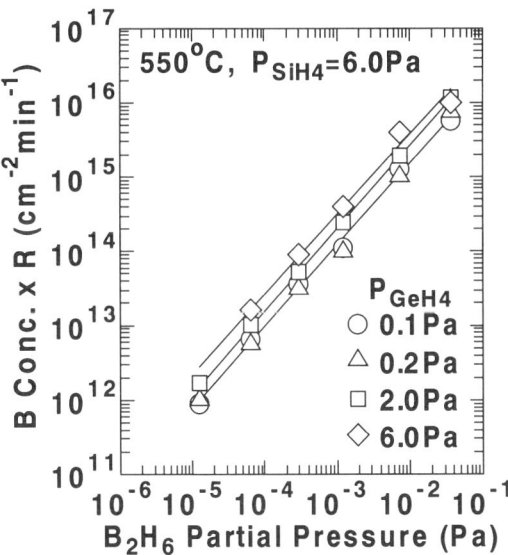

Figure 5 : Dependence of the incorporation rate of B atoms into the $Si_{1-x}Ge_x$ films on the B_2H_6 partial pressure.

and GeH_4 adsorption/reactions, because the Ge fraction is almost constant independently of the B_2H_6 partial pressure.

CONCLUSIONS

In-situ heavy doping of B into $Si_{1-x}Ge_x$ epitaxial films on the Si(100) substrate by ultraclean LPCVD at 550°C was investigated. With increasing B_2H_6 partial pressure, the deposition rate decreased only at the higher GeH_4 partial pressure and the B concentration in the film increased proportionally up to $10^{22}cm^{-3}$. The Ge fraction scarcely changed with B_2H_6 addition. These doping characteristics can be explained by suppression of both the SiH_4 and GeH_4 adsorption/ reactions in a similar degree due to B_2H_6 adsorption on the Si-Ge and/or Ge-Ge bond sites.

ACKNOWLEDGMENTS

This study was carried out in the Superclean Room of the Laboratory for Electronic Intelligent Systems, under the Cooperative Research Project Program of the Research Institute of Electrical Communication, Tohoku University. The CVD reactor was provided by Kokusai Electric Co., Ltd. This study was partially supported by a Grant-in-Aid for Scientific Research from the Ministry of Education, Science, Sports and Culture of Japan, the Research for the Future Program(No. JSPS-RFTF97P00202, Atomic-Scale Surface and Interface Dynamics) from the Japan Society for Promotion of Science.

REFERENCES

[1] G.L.Patton, J.H.Comfort, B.S.Meyerson, E.F.Crabbe, G.J.Scilla, E.D.Fresart, J.M. C.Stork, J.Y.C.Sun, D.L.Harame and J.N.Burghartz, IEEE Electron Device Lett. **EDL-11** 171 (1990).
[2] J.C.Sturm, E.J.Prinz and C.W.Magee, IEEE Electron Device Lett. **EDL-13** 56 (1992).
[3] K.Goto, J.Murota, F.Honma, T.Matsuura and Y.Sawada, in ULSI Science and Technology/1995, Electrochemical Society, **PV95-5**, p.512(1995).
[4] J.Murota, M.Ishii, K.Goto, M.Sakuraba, T.Matsuura, Y.Kudo and M.Koyanagi; in 27th ESSDERC'97, ed. H. Grünbacher, p.376 (1997).
[5] C.Tsai, S.M.Jang, J.Tsai and R.Reif, J.Appl.Phys. **69** 8158 (1991).
[6] H.Kühne, Appl.Phys.Lett. **62** 1967 (1993).
[7] S.M.Jang, K.Liao and R.Reif, Appl.Phys.Lett. **62** 1675 (1993).
[8] C.J.Lee, M.Sakuraba, T.Matsuura and J.Murota, in CVD-X IV/1997, Electrochemical Society, **PV97-25**, p.1356(1997).
[9] J.Murota, N.Nakamura, M.Kato, N.Mikoshiba and T.Ohmi, Appl.Phys.Lett. **54** 1007 (1989).
[10] J.Murota, F.Honma, T.Yoshida, K.Goto, T.Maeda, K.Aizawa and Y,Sawada, J.Phys.IV France **3** C3-427(1993).
[11] S.H.Bauer, A.Shepp and R.EmcCoy, J. Am. Chem. Soc., **78**, 5775(1956).

GROWTH AND CHARACTERIZATION OF EPITAXIALLY STABILIZED PSEUDOMORPHIC α-Sn/Si HETEROSTRUCTURES

Kyu Sung Min* and Harry A. Atwater
Thomas J. Watson Laboratory of Applied Physics, California Institute of Technology, Pasadena, California 91125.
*ksmin@daedalus.caltech.edu

ABSTRACT

Growth and structural characterization of ultrathin, coherently strained α-Sn/Si quantum well heterostructures have been performed. Severe Sn segregation to the surface during growth, which prevents growth of these structures at ordinary Si epitaxy temperatures, has been minimized by substrate temperature and growth rate modulations during molecular beam epitaxy. Single Sn/Si quantum wells grown with Sn coverage up to 1.4 ML have been verified to be pseudomorphic by transmission electron microscopy and X-ray rocking curve analysis. Similarly, pseudomorphic superlattices with up to 10 periods of 1 ML Sn/7.7 nm Si have been verified to be free of extended defects.

INTRODUCTION

Development of optoelectronic devices and functions monolithically integrated with Si-based ULSI circuits remains a challenging frontier in microelectronics. The biggest challenge comes from the fact that while Si and Ge are the only known group IV elements known to form a solid solution, both are indirect semiconductors. One possible route for realization of a direct gap group IV alloy system involves alloying Si or Ge with Sn to form epitaxially stabilized diamond cubic Sn_xGe_{1-x}/Ge and Sn_xSi_{1-x}/Si heterostructures. Diamond cubic α-Sn, which is thermodynamically stable below $13.2^{\circ}C$, is a zero band gap semiconductor with degenerate conduction and valence bands at the Γ point ($\vec{k}$ =<0,0,0>) [1]. When Si or Ge is alloyed with Sn, the conduction band at L point ($\vec{k}$ =<1,1,1>) decreases in energy with increasing Sn concentration but at a slower rate than the conduction band at Γ point [2,3]. As a result, both Sn_xGe_{1-x} and Sn_xSi_{1-x} are predicted to have direct and tunable optical energy gaps for compositional range exceeding some critical Sn concentration [2,3], thus opening up a possibility for growth of Si-based semiconductor with a direct-gap tunable from 0 eV to few tenth of an eV for potential application in fabrication of long-wavelength detectors and light emitters. Transition to a direct band gap at lower Sn concentration than theoretically predicted values has recently been demonstrated for the Sn_xGe_{1-x} system [4]. Recently, growth of dilute (x<0.06) pseudomorphic $Si_{1-x}Sn_x$ alloy films have also been reported [5,6]. In this report, we investigate growth of coherently strained, epitaxially stabilized ultrathin α-Sn/Si quantum well heterostructures. Growth of such heterostuctures is motivated by the following predictions. First, by growing very thin α-Sn layers in Si, one may potentially take advantage of the quantum carrier confinement. And by growing multiple periods of α-Sn/Si with appropriate thickness, one may also potentially take advantage of superlattice effects to further tune the band gap of the resulting heterostructure over a significant portion of the infrared spectrum. Band gap tunability of short-period α-Sn/Ge superlattices has already been demonstrated [7]. One might also take advantage of a thin channel of α-Sn layer with high carrier mobilities for potential application in modulation doped heterostructures [8].

Mat. Res. Soc. Symp. Proc. Vol. 533 © 1998 Materials Research Society

Despite potentially promising electronic and optical properties, growth of α-Sn/Si heterostructures has been difficult to date due to large lattice mismatch between α-Sn and Si (19.5%), very low solid solubility of Sn in crystalline Si (~$5x10^{19}$ cm^{-3}) [9], and severe Sn segregation to the surface during growth at ordinary Si epitaxy temperatures (T>~ 400°C). In order to incorporate large amounts of Sn while maintaining a high epitaxial quality, we employ temperature and growth rate modulations in molecular beam epitaxy, a technique used previously for growing Sb delta-doped layers in Si [10] and Ge/α-Sn heterostructures on Ge [11].

EXPERIMENT

All samples were grown in a custom-built molecular beam epitaxy chamber with a base pressure of $2x10^{-10}$ torr and equipped to perform reflection high energy electron diffraction (RHEED) for *in situ* surface analysis. The substrates were (100) Si initially coated with 100 nm thermal SiO$_2$ films. The substrates were prepared by first chemically cleaning the oxide coated substrate in 5:1:1 H$_2$O:H$_2$O$_2$:NH$_4$OH and subsequently completely removing the oxide in 10% HF-H$_2$O to obtain a dihydride-terminated Si (100) surface. After transferring into the ultrahigh vacuum chamber, the substrates were baked *in situ* at 200°C for 2 hours to desorb hydrocarbon contaminants. Just prior to growth of the Si buffer layer, the substrates were heated to 550°C to obtain a clean (2x1) surface reconstructed surface. An electron beam evaporation source and an effusion cell was used for Si and Sn deposition, respectively. The growth rates were controlled using a crystal thickness monitor for Si deposition and by controlling the temperature of the effusion cell for Sn deposition. The temperature of the Sn effusion cell was varied between 675-755°C to control the deposition rates between 0.003-0.02 ML/sec (1 ML = $6.79x10^{14}$/cm^3). For all samples, the Si buffer layers were grown at 550°C at 0.05 nm/sec to obtain a smooth Si surface prior to deposition of Sn layers. Sn deposition immediately followed at 550°C at a deposition rate of 0.02 ML/sec ($1.4x10^{13}$/cm^2/sec). After Sn deposition, the growth was interrupted to cool the substrate to 170-220°C, which took approximately 25-35 min. Then, in order to cap the Sn layer with minimal Sn segregation to the surface, the initial Si overlayer growth was commenced at 160°C at growth rates of 0.01-0.03 nm/sec. The initial low growth rate is important in order to maximize the thickness of the Si overlayer that could be deposited at such a low temperature before it starts undergoing a crystal-to-amorphous transition. It has been shown that the critical thickness h_{epi} at which an epitaxial film undergoes a crystal-to-amorphous transition is related to the growth temperature by an Arrhenius relationship, and that the activation energy is dependent on the growth rate [12]. The activation energy is smaller for lower growth rates and therefore h_{epi} is larger for lower growth rates. Therefore, in order to ensure that the Si overlayer grown at low temperatures is thick enough to suppress Sn segregation to the surface by means of diffusive exchange, it is important to grow the low-temperature Si overlayer at a low growth rate. After deposition of 3-6 nm of Si overlayer at 160°C, the substrate temperature is gradually increased such that the temperature is returned to 550°C before the overlayer thickness reaches the desired period thickness (8-27 nm). After the substrate temperature reaches 550°C, the growth rate is also increased back to 0.05 nm/sec. This process ensures that each Sn layer is deposited on a smooth Si surface. After growing the Si overlayer with a desired thickness, another Sn layer is deposited and the whole process is repeated for desired number of times.

The samples were analyzed by cross sectional transmission electron microscopy (TEM), Rutherford backscattering spectrometry (RBS) with a 2 MeV He^{++} beam, and high-resolution X-ray rocking curve analysis (XRD) using Cu K$_{\alpha 1}$ radiation.

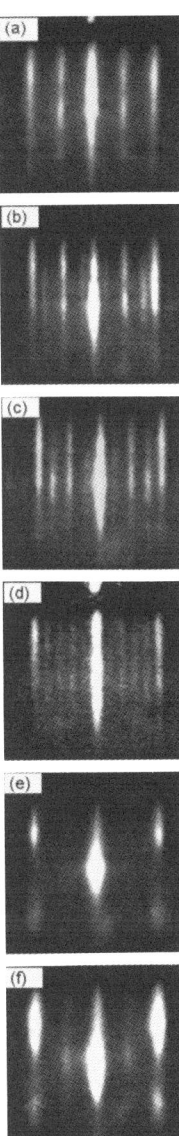

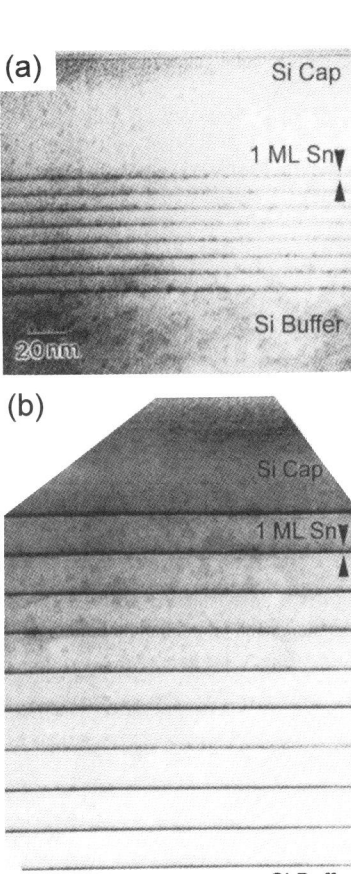

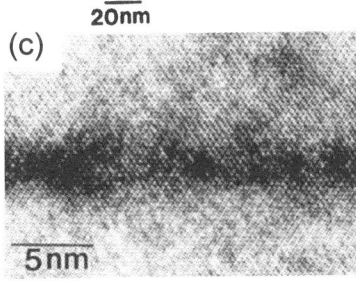

FIG. 1 *In situ* RHEED patterns at various stages during growth: (a) a smooth (2x1) reconstructed Si (100) surface at 550°C right before the deposition of Sn; at various Sn coverage on Smooth Si (100): (b) (6x2) at converges of 0.375-0.5 ML, (c) c(4x8) at 0.5-1.0 ML, and (d) (5x1) at 1.0-1.5 ML; after deposition of (e) 1.5 nm of initial Si overlayer (160°C) and (f) 7.5 nm of Si overlayer (350°C).

FIG. 2. Bright field transmission electron micrographs taken under (400) 2-beam excitation of (a) an 8-period superlattice of 1 ML Sn/7.7 nm Si and (b) a 10-period superlattice of 1 ML Sn/20.5 nm Si. A high-resolution image of one of the Sn layers of the sample in (b) is shown in (c).

357

RESULTS AND DISCUSSION

It has been demonstrated from a previous study on samples consisting of single Sn/Si quantum well structures that initial Sn coverage in excess of about 1.4 ML segregates to the surface during the Si overlayer growth even at temperatures as low as 160°C [13]. It has also been demonstrated that the Si overlayer relaxes when the deposited Sn layer thickness exceeds about 1.4 ML. This suggests that 1.4 ML represent a kinetic critical thickness t_C for commensurate growth of the individual Sn layers. In strained-layer superlattice growth, however, there are two critical thickness [14]: the critical thickness t_C for the individual strained layers and the critical thickness t_{SL} for the overall superlattice. Even if the individual strained layers are below t_C, the superlattice could still relax if the strain energy integrated over the entire film exceeds the t_{SL}. This suggests that one should be able to grow commensurate Sn/Si quantum well superlattice structures up to t_{SL} for individual Sn layer thickness less than t_C=1.4 ML. The superlattice films investigated consisted of 7-10 periods of 1.0 ML Sn sandwiched between 8-27 nm Si spacer layers.

Surface morphology and reconstruction was analyzed using *in situ* reflection high energy electron diffraction for the growth of Sn and Si overlayers. A typical cycle of surface evolution during growth of the Sn layer on smooth Si followed by a Si overlayer growth is shown in Fig. 1. Figure 1(a) is the RHEED pattern of a smooth Si (100) surface at 550°C right before the deposition of Sn, showing a (2x1) reconstruction characteristic of smooth and clean Si (100). Figures 1(b)-1(d) show RHEED patterns for different amounts of Sn coverage: (6x2) for 0.375-0.5 ML, c(4x8) for 0.5-1.0 ML, and (5x1) for 1.0-1.5 ML, respectively. The observed Sn-induced reconstructions are in agreement with previous studies [15]. After Sn deposition, growth is interrupted and the sample is cooled down to 160°C. Si overlayer growth commences at 160°C, with a rough surface as indicated by the spotty RHEED pattern shown in Fig. 1(e). After deposition of about 3-6 nm of Si at 160°C, the temperature is increased again. Beyond about 350°C, the overlayer surface begins to smoothen and (2x1) Si (100) surface reconstruction returns, as shown in Fig. 1(f). As the temperature is increased higher the surface morphology returns back to smooth Si (100) as shown in Fig 1(a).

Crystal quality of selected samples was analyzed by cross-sectional transmission electron microscopy. Figures 2(a) and 2(b) show bright field images taken under (400) 2-beam excitation of an 8-period superlattice of 1 ML Sn/7.7 nm Si and a 10-period superlattice of 1 ML Sn/20.5 nm Si, respectively. No growth-related extended defects are visible within the imaged electron-transparent area (dislocation density $\sim$<$1x10^6$cm^{-2}). Figure 2(c) is a high-resolution image of one of the Sn layers of the sample in Fig. 2(b) taken in the [110] projection. The Sn layers appear thicker than a monolayer, most likely due to two reasons. First, the embedded Sn layer is highly strained, giving rise to high strain contrast in addition to mass-thickness contrast in the vicinity of the Sn layer. Second, as demonstrated previously [11] for the case of short period α-Sn/Ge superlattices, the Sn layer may not be atomically abrupt; the Sn atoms may be distributed over few atomic layer in an exponentially decaying profile. From the X-ray rocking curve analysis, however, it is evident that the distribution cannot be spread over more than a few monolayers [13].

Figure 3 shows a representative high-resolution X-ray rocking curve of the superlattice structures analyzed. The dotted curve shows a high-resolution scan around Si (004) of the 10-period 1 ML Sn/20.5 nm Si superlattice shown in Figs. 2(b) and 2(c). Well-defined fundamental (SL_0) and higher order (SL_{-1}, $SL_{\pm 2}$) superlattice peaks as well as interference fringes are clearly identified. From the position of the peaks, an accurate measurement of the average periodicity can be obtained. The periodicity of the superlattice can be accurately determined from the angular spacing $\Delta\theta$ between adjacent superlattice peaks through the relationship

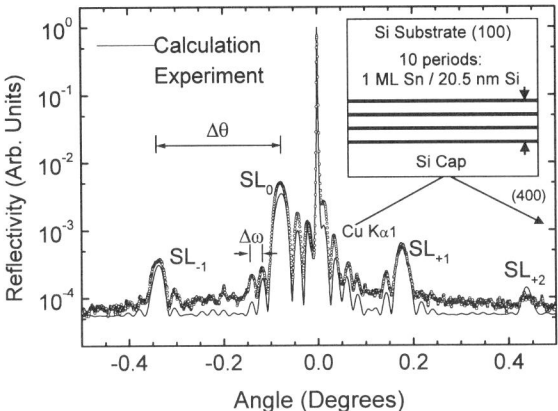

FIG. 3. A representative symmetric (004) reflection high-resolution X-ray rocking curve of the superlattice structures. The dotted curve is a scan around Si (004) of the 10-period 1 ML Sn/20.5 nm Si superlattice shown in Figs. 2(b) and 2(c). The dotted curve and the solid curve represent the experimental scan and the dynamical simulation, respectively.

$$\Delta\theta = \frac{\lambda |\gamma_h|}{t \sin 2\theta_B} \qquad (1)$$

where t is the thickness of one period, λ is the wavelength of the radiation, θ_B is the Bragg angle, and γ_h is the direction cosine of the diffracted beam. The superlattice periodicity can be determined to within ±0.1 nm using Eq. (1). For the film in Fig. 3, the superlattice periodicity was determined to be 20.7 nm. In addition, the number of Pendellosung oscillations of angular spacing $\Delta\omega$ within the superlattice periodicity $\Delta\theta$ gives the number of periods. From the position of the "SL_0" peak, an equivalent average concentration can be determined. For the film in Fig. 3, the equivalent strain is that of 20.7 nm film of 0.6% average Sn concentration. From a similar analysis of the superlattice film with the shortest period shown in Fig. 2(a), the highest equivalent strain achieved was determined to be that of 61.6 nm 1.8 % average Sn concentration (0.4 % strain).

The angular distance between superlattice peaks and Pendellosung fringes is independent of the physical properties of the sandwiched layer. However, the angular position and the intensity of the interference fringes near the (004) Si reflection depend strongly on the factors that cause dephasing of X-rays scattered between the superlattice layers and the Si substrate. Because of the large atomic size difference between Sn and Si, analysis of the angular position and the intensity of the interference fringes offer a very sensitive technique for quantitatively characterizing low coverage Sn incorporation, enabling thickness resolution within a fraction of a monolayer. The amount of incorporated Sn can be measured from the best theoretical fit to the experimental curve. The solid curve in Fig. 3 is the best dynamical simulation curve of a model structure of 10 periods of 1 ML Sn/20.5 nm Si, capped with 59 nm Si. The simulation was performed under the assumption of atomically abrupt interfaces and tetragonally distorted α-Sn unit cell with the bulk lattice constant of 0.64890 nm (in-plane lattice constant of 0.54310 nm and perpendicular lattice constant of 0.73875 nm) and bulk elastic constants. The peak positions as well as the shape and intensities of the satellite peaks are in very good agreement with the experimental curve. The fact that both the peak positions as well as the relative intensities match well suggests that the interfaces are sharp, with most Sn atoms localized within about one unit cell thickness.

CONCLUSIONS

Growth of pseudomorphic single quantum well structures of up to 1.4 ML Sn/Si and α-Sn/Si superlattice structures with periods as short as 7.9 nm (Sn_1Si_{57} with average Sn concentration of ~2%) has been demonstrated to be possible by employing temperature and growth rate modulations in molecular beam epitaxy. The films analyzed have been verified to be free of extended defects by cross sectional transmission electron microscopy and the interfaces have been verified to be sharp within the resolution of X-ray rocking curve analysis.

ACKNOWLEDGEMENTS

This work was supported by the National Science Foundation (DMR 95-03210) and Intel Graduate Fellowship (K.S.M).

REFERENCES

1. S. Groves and W. Paul, Phys. Rev. Lett. **11**, 194 (1963).

2 R. A. Soref and C. H. Perry, J. Appl. Phys. **69**, 539 (1991).

3. H. A. Atwater, G. He, and K. Saipetch, Mat. Res. Soc. Symp. Proc. **355**, 123 (1995).

4. G. He and H. A. Atwater, Phys. Rev. Lett. **79**, 1937 (1997).

5. S. Yu. Shiryaev, J. Lundsgaard Hansen, P. Kringhoj, and A. Nylandsted Larsen, Appl. Phys. Lett. **67**, 2287 (1995);

6. Al-Sameen T. Khan, Paul R. Berger, Fernando J. Guarin, Subramanian S. Iyer, Appl. Phys. Lett. **68**, 3105 (1996).

7. J.Olajos, P. Vogl, W. Wegscheider, and G. Abstreiter, Phys. Rev. Lett. **67**, 1991.

8. C.H.L. Goodman, IEEE Proc. **129**, 189 (1982).

9. F. A. Trumbore, C. R. Isenberg, and E. M. Porbansky, J. Phys. Chem. Solids **9**, 60 (1959).

10. H. P. Zeindl, T. Wegehaupt, I. Eisele, H. Oppolzer, H. Reisinger, G. Tempel, and F. Koch, Appl. Phys. Lett. **50**, 1164 (1987).

11. W. Wegscheider, J. Olajos, U. Menczigar, W. Dondl, and G. Abstreiter, J. Cryst. Growth **123**, 75 (1992).

12. D.J. Eaglesham, H.-J. Gossmann, and M. Cerullo, Phys. Rev. Lett. **65**, 1227 (1990).

13. K. S. Min and Harry A. Atwater, to be published in Appl. Phys. Lett. **72**, 1998.

14. R. Hull, J.C. Bean, F. Cerdeira, A.T. Flory, and J.M. Gibson, Appl. Phys. Lett. **48**, 56 (1986).

15. K. Ueda, K. Kinoshita, and M. Mannami, Surf. Sci. **145**, 261 (1984); D.H. Rich, T, Miller, A. Samsavar, H.F. Lin, and T.-C. Chiang, Phys. Rev. **B 37**, 10221 (1988); A. A. Baski and C. F. Quate, Phys. Rev. **B 44**, 11167 (1991).

EVIDENCE FOR SUBSTITUTIONAL C, ORDERING EFFECTS AND INTERDIFFUSION IN EPITAXIAL GE-C AND GE-RICH GE-SI-C ALLOYS

BI-KE YANG, W. H. WEBER[†] AND M. KRISHNAMURTHY
Department of Metallurgical and Materials Engineering, Michigan Technological University
Houghton, MI 49931
†Physics Department, Ford Research Laboratories, Dearborn, MI 48121-2053

ABSTRACT

We report on the epitaxial growth of Ge-C and Ge-Si-C alloys (C<10%) grown on Si(100) and Ge(100) substrates using low temperature (~200°C) molecular beam epitaxy. Thin films (50-70 nm) were characterized *in-situ* by RHEED and *ex-situ* by transmission electron microscopy, x-ray diffraction, and Raman spectroscopy. The films were annealed at 750°C and 850°C in an Ar atmosphere to study interdiffusion effects.

Raman spectroscopy of Ge-C on Ge indicates the existence of a Ge-C local mode at 530cm^{-1} and is direct evidence for the presence of substitutional C in Ge. The GeSiC alloys grown on Ge do not show the Ge-C local mode, consistent with preferential Si-C bonding. There is evidence for strain enhanced solubility of C based on a comparison of the substitutional C content in Ge-C films on Si (~1 at %) and on Ge substrates (~0.1 at %). Silicon interdiffusion in annealed Ge-C samples is strongly suppressed by the presence of C. A simple diffusion model is used to illustrate Si indiffusion in Ge.

INTRODUCTION

SiGe alloys have been extensively studied in order to exploit the promise of bandgap engineering on silicon [1]. In the past few years, there has been a particularly strong interest in SiGe alloys containing carbon [2]. The main reason is that strain-compensation by C addition is expected to circumvent the critical thickness limitations inherent in SiGe technology. Furthermore, C in the SiGe lattice makes it possible to use both alloy concentration and strain as a variable for bandgap engineering. However, several obstacles must be overcome. Firstly, C is relatively insoluble in both Ge and Si, unlike the completely miscible SiGe system. Secondly, the formation of various forms of SiC is thermodynamically favored at higher growth temperatures [3]. Despite these problems, low concentrations of C in Ge-C [4-6] and Si-Ge-C [7-9] have been successfully incorporated using a variety of techniques including chemical vapor deposition (CVD)[7, 9] and molecular beam epitaxy (MBE) [4-6, 8, 10-15].

We have recently reported on the microstructural evolution and optical properties of $Ge_{1-x}C_x$ alloys (x <0.1) grown at ~200°C [11, 12], and the effect of Si additions to $Ge_{1-x}C_x$ films [13, 14]. Here, we focus our discussions on our observations of the Ge-C local bonding mode using Raman spectroscopy, and Si interdiffusion behavior in Ge-C films annealed at high temperatures. Characterization techniques include *in situ* reflection high-energy electron diffraction (RHEED), *ex situ* transmission electron microscopy (TEM), x-ray diffraction (XRD), and Raman spectroscopy. Detailed experimental procedures are described elsewhere [12-14].

Mat. Res. Soc. Symp. Proc. Vol. 533 © 1998 Materials Research Society

RESULTS

1. Ge-C and Ge-Si-C Grown on Si(100)

We have studied the low temperature (~200°C) epitaxial growth of Ge-C [11, 12] and Ge-Si-C alloys [13, 14] on Si(100). Detailed results have recently been published and we only summarize our findings here. For C concentration less than 2-3at.%, the films grow as 2D layers, while films with higher C content form 3D islands after initial layer growth. The typical microstructural features consist of short interfacial misfit dislocations, threading dislocations and planar defects (e.g. twins and stacking faults) as observed with plan-view and cross-section TEM [11-14]. In particular, it was noted that while the threading dislocation density decreased, the planar defect density increased with increasing C content in the Ge-C and GeSiC films. Furthermore, the planar defect density was higher for the GeSiC films than for Ge-C films with nominally the same C content [13, 14].

Most samples were annealed at 750°C and 850°C for about 1 hour in an Ar ambient to study interdiffusion effects. Figure 1 shows the X-ray diffraction (XRD) data showing normalized changes in d_{400} spacing of the Ge-C alloys relative to Si. Data from both the as-grown and annealed samples are shown. Two points are noted. First, the lattice parameter of the as grown alloy decreases with increasing nominal C content, to a value below that of the bulk relaxed Ge. This is a clear indication of substitutional C. It is estimated (based on Vegard's law) that about 1 at.% C may have been substitutionally incorporated in Ge-C films.

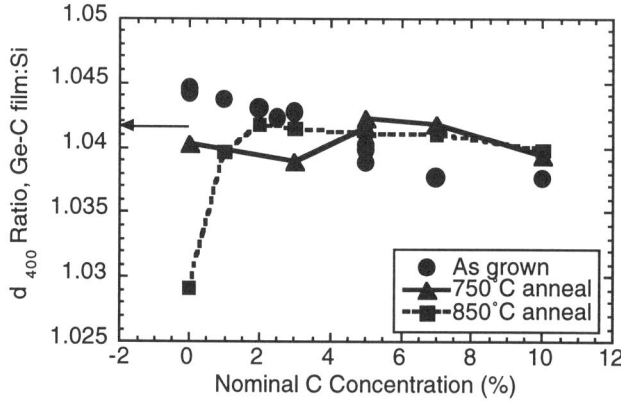

Figure 1. The ratio of the interplanar spacing d_{400} for the Ge-C films relative to the Si-substrates obtained from XRD, and plotted against the nominal C concentration. Notice the decrease of lattice parameter with increasing C content for the as-grown films. The arrow indicates the value for the relaxed bulk Ge. After anneal, the pure Ge shows a decrease in the lattice spacing (below bulk Ge value) while the lattice spacing of the Ge-C films almost remains at the bulk relaxed Ge value.

Second, the annealed Ge samples show a steep drop in the lattice parameter below that of bulk Ge. This effect is due to Si interdiffusion into the films, as also confirmed by Raman spectroscopy [12]. The interesting observation, however, is that the annealed Ge-C films show no significant reduction in the lattice parameter indicating that the Si interdiffusion is mostly suppressed. This effect is generally consistent with Raman spectroscopy measurements [14]. The behavior of C as a Si interdiffusion barrier is discussed later. The XRD data of GeSiC films (not shown here) indicate that a maximum of ~2-3 at.% C is substitutional [14].

2. Ge-C and Ge-Si-C Alloys Grown on Ge(100)

To evaluate the influence of strain on the solubility of C in Ge, Ge-C alloys were also grown on clean Ge(100) substrates. After *in situ* oxide desorption of the Ge substrate and the growth of a thin buffer layer, a sharp (2x1) reconstruction pattern was typically observed in RHEED. The co-deposition of Ge and C leads to initial several monolayers of layer growth followed by formation of islands. An interesting feature was the development of sharp {311} facets on the islands as indicated by RHEED and TEM studies [12, 13]. From XRD, no extra Ge-C peak (apart from the Ge substrate peak) was detected. TEM plan view images showed no sign of moiré fringes; Instead, strain contrast from quasi-periodic modulations and some precipitate contrast were observed [13].

3. Raman Studies: Ge-C Local Mode and the Si-C Preferential bonding

Raman spectroscopy was performed on all Ge-C and Ge-Si-C films grown on Si and Ge substrates. As previously reported [12], the Raman spectra from Ge-C films grown on Si showed no clear evidence of substitutional C. It is important to note, however, that the expected peak position (based on reduced masses and expected lattice parameter) lies very near the strong Si substrate peak . However, in the Ge-C films grown on Ge, the Ge-C local mode was observed as a sharp line at 530 cm^{-1}[15]. Figure 2 shows the Raman spectra from Ge-C samples with different C content, compared with a Ge wafer.

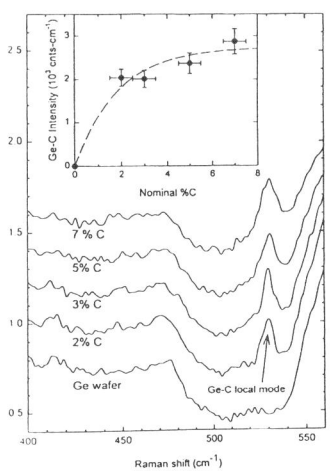

Figure 2. Raman spectra of films with different C concentrations and compared to a Ge wafer. The inset shows the integrated intensity of the Ge-C local mode vs. nominal %C.

The spectra were recorded with a scattering geometry for which the first-order c-Ge line and the Ge-C local mode are allowed and the second-order scattering is largely suppressed. In the plot, the broad feature peaking near 575 cm^{-1} is the second-order band from the Ge optical branch. All the features in the spectra, excepting the narrow line at 530 cm^{-1} also appear in the spectrum from a pure Ge(100) wafer. This line is absent from the Ge-wafer spectrum, its strength increases with C content and it follows the same selection rules as the c-Ge peak. In addition, its frequency is consistent with that expected from the known position of Si-C local mode at 605 cm^{-1}. The linewidth increases about 25% at the highest C concentration, but the integrated line strength, shown in the inset, increase by only about 50%. This result suggests that the substitutional C concentration saturates at a value well below the nominal C content expected from the growth process.

In ternary Ge-Si-C alloy films with 10 and 20 at.% Si and nominally 3% C, the usual Ge-Ge and Ge-Si lines were found, but *no evidence for the Ge-C local mode* was found. A likely explanation for this effect is that the strong preference for C to form Si-C bonds leads to every C being paired with a Si atom. While this effect has been predicted [16], we believe our results may be the first experimental verification of this ordering phenomenon. More detailed experiments are currently in progress to further evaluate these effects.

DISCUSSION

Based on the results presented, several interesting issues can be addressed. From Figure 1, it can be noted that the d_{400} spacing of the as-grown Ge-C alloys has a tendency to decrease with increasing C content. The decrease in the lattice parameter below that of relaxed pure Ge (shown by the arrow) suggests that the C decreases the lattice parameter of the films by substitutional incorporation (~1 at %C). However, the substitutional C content in Ge-C films grown on Ge(100) [15] was estimated to be less than 0.1at.% using integrated intensity ratios [15]. This suggests that the presence of epitaxial strain may have helped enhance the solubility of C in Ge.

From the d ratio plot in Fig. 1, a decrease of the lattice parameter of the post-annealed Ge thin film (when compared with the as-grown films) was noted. This is attributed to inter-diffusion of Si from the substrate to the thin films during annealing, as confirmed by Raman spectroscopy [14]. The amount of Si that has interdiffused into the Ge thin films was estimated from XRD (according to Vegard's Law) to be ~3 at.% and 30 at.%, for the 750°C and 850°C anneal respectively. In the presence of C, however, there is a dramatic drop in Si content indicating that C in Ge acts as an effective diffusion barrier for Si.

The modeling of the Si interdiffusion into pure Ge is discussed below. In the literature, there are to our knowledge, only two reports on impurity-bulk diffusion of Si in Ge [17, 18], both giving similar values of the diffusion coefficient at 750°C and 850°C. The diffusion equation was solved numerically [19] using an activation energy of 2.88eV [17] and the resulting Si concentration profile is shown in Figure 3. The estimated average Si concentration (the gravity center of the two curves for the 50 nm thick thin film range) is about 20% and 60% for 750°C and 850°C respectively, which are much larger than the values determined from the XRD data. However, by assuming an activation energy of 3.2 eV and using the calculated diffusion profile shown in Fig. 4, the average Si concentration is estimated to be about 3% and 45% for 750°C and 850°C respectively.

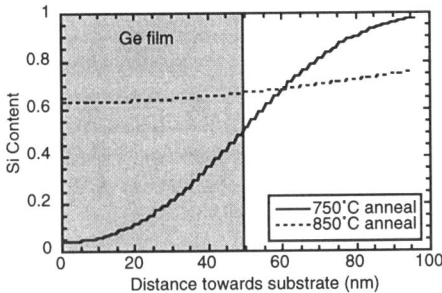

Figure 3. Simulated Si concentration profile after anneal, activation energy =2.88 eV.

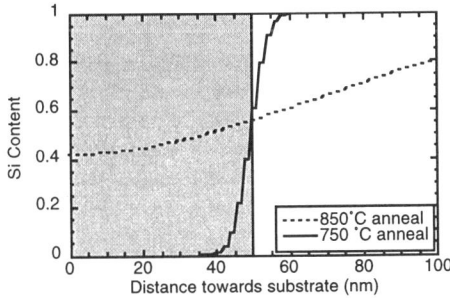

Figure 4. Simulated Si concentration profile after anneal, activation energy =3.2 eV.

These values are in much better agreement with the XRD data. The activation energy of 3.2 eV used in our model is higher than 2.88 eV quoted in Ref. [17] and lower than 3.47 eV measured in Ref. [18]. We argue that the low value in Ref. [17] is possibly related to the fact that the implantation technique used may have induced a high density of point defects. In Ref. [18], a Ge wafer with a dislocation density less than 3000 cm^{-2} was used and may reflect a different diffusion mechanism. In our case, the Ge thin film has an estimated dislocation density of $\sim 10^{10}$ cm^{-2} and the diffusion may be dominated by the dislocations.

The important question then is: why does the presence of C suppress Si interdiffusion significantly? Possible reasons are: (i) C occupies the short circuit paths (e.g. dislocations, interstitial sites) for Si interdiffusion; (ii) the reduced threading dislocation density in Ge-C films suppresses Si diffusion (it then implies that the higher planar defect density in Ge-C films is less important for diffusion), and (iii) the strain-enhanced Si interdiffusion is less pronounced since carbon relaxes the residual strain.

In summary, we present a variety of results from the studies of epitaxial Ge-C and Ge-Si-C alloys on Si and Ge substrates. Raman spectroscopy of Ge-C films on Ge show the first evidence for a Ge-C local mode at 530 cm^{-1}. The local mode was not observed in the SiGeC films indicating a preferred Si-C bonding tendency. On annealing the Ge-C/Si samples at high temperature, Si interdiffusion takes place. An important observation was that the presence of C dramatically suppresses Si indiffusion. Various factors such as reduced threading dislocation density and the possible role of C within defects as a diffusion blocker are discussed. A simple diffusion model is used to demonstrate Si indiffusion into the Ge film.

ACKNOWLEDGMENTS

Financial support for research at MTU was provided by ONR (N00014-96-1-0793).

REFERENCES

[1]   J. C. Bean, Proc. IEEE 80, 571 (1992).
[2]   R. A. Soref, Proc. IEEE 81, 1687 (1993).
[3]   A. R. Powell, F. K. LeGoues and S. S. Iyer, Appl. Phys. Lett. 64, 324 (1994)
[4]   M. Todd, J. Kouvetakis, and D. J. Smith, Appl. Phys. Lett. 68, 2407 (1996).
[5]   J. L. Regoloni, S. Bodnar, J.C. Oberlin, F. Ferrieu, M.Gauneau, B. Lambert and P. Boucaud, J. Vac. Sci. Technol. A 12, 1015 (1994)
[6]   J. Kolodzey, P. A. O'Neil, S. Zhang, B. A. Orner, K. Roe, K. M. Unruh, C. P. Swann, M. M. Waite and S. Ismat Shah, Appl. Phys. Lett. 67, 865 (1995).
[7]   Z. Atzmon, A. E. Bair, E. J. Jaquez, J. W. Mayer, etc., Appl. Phys. Lett. 65, 2559 (1994).
[8]   A. R. Powell, K. Eberl, B. A. Ek and S. S. Iyer, J. Cryst. Growth 127, 425 (1993)
[9]   H. J. Osten, E. Bugiel and P. Zaumseil, J. Cryst. Growth 142, 322 (1994).
[10] M. Krishnamurthy, J. S. Drucker and A. Challa, J. Appl. Phys. 78, 7070 (1995)
[11] M. Krishnamurthy, Bi-Ke Yang and W. H. Weber, Appl. Phys. Lett. 69, 2572 (1996).
[12] B.-K. Yang, M. Krishnamurthy, and W. H. Weber, J. Appl. Phys. 82, 3287(1997).
[13] Bi-Ke Yang, M. Krishnamurthy, and W. H. Weber, Structure and Evolution of Surfaces, Mat. Res. Soc. Symp. Proc., Vol.440, 371(1997).
[14] Bi-Ke Yang, M. Krishnamurthy, and W. H. Weber, to be published.
[15] W. H. Weber, Bi-Ke Yang, and M. Krishnamurthy, to be published.
[16] P. C. Kelires, Phys. Rev. Lett. 75, 1114(1995).
[17] J. Raisanen, J. Hirvonen, and A. Antilla, Solid State Electron., 24, 333(1981).
[18] U. Sodervall and M. Friesel, Defect and Diffusion Forum, 143-147, 1053(1997).
[19] Bi-Ke Yang, Ph.D. Thesis, Michigan Technological University, 1998.

THE FORMATION OF ABRUPT N+ DOPING PROFILES USING ATOMIC HYDROGEN AND Sb DURING Si MBE

P. E. THOMPSON*, C. SILVESTRE**, M. TWIGG*, G. JERNIGAN*, AND D. S. SIMONS***

*Code 6812, Naval Research Laboratory, Washington, DC 20375-5347
(202)404-8541, Fax: (202)404-7194, email: thompson@estd.nrl.navy.mil
** ASEE Post-Doctoral Fellow, Naval Research Laboratory, Washington, DC 20375-5347
*** NIST, Gaithersburg, MD 20899 USA

ABSTRACT

Previously, atomic hydrogen has been shown to be effective in reducing the segregation of Sb on Si(100) during solid source molecular beam epitaxy growth. In this work we have investigated the electrical activation of the Sb. Using Hall measurements, spreading resistance profilometry, and secondary ion mass spectrometry, we have demonstrated that the co-deposition of atomic hydrogen during Sb doping of Si at 500 °C produced well-defined doping spikes. Comparing the sheet carrier concentration obtained by Hall measurements to the Sb atomic concentration obtained by SIMS, the overall activation of the Sb was greater than 50%.

INTRODUCTION

While Sb is a standard n-type dopant for Si molecular beam epitaxy (MBE), it has also been used as a surfactant during the growth of Ge on Si[1]. This dual nature of Sb (dopant and surfactant) makes it difficult to fabricate spatially-confined n-type layers required for device applications, such as high electron mobility transistors (HEMTs), PIN diodes, or Esaki diodes. Previously this group has investigated the surface segregation of Sb in Si(100) films as a function of growth temperature in the interval of 325 to 550 °C[2, 3]. In that work, it was shown that a monolayer of Sb deposited at a growth temperature of 500 °C would segregate along the growth front and slowly be incorporated into the Si layer. After growing a 400 nm Si cap, the resulting doping concentration was $2 \times 10^{18}/cm^3$.

In order to control the segregation of Sb, we investigated the use of atomic hydrogen (AH) during Sb doping of Si [4]. It had been demonstrated that AH was effective in reducing the segregation of Ge during the growth of $Si_{1-x}Ge_x$ quantum wells on Si [5-10]. In our investigation we established a growth-temperature/AH-flux regime where AH is effective in reducing Sb segregation, Fig. 1. We have demonstrated that the use of AH (10^{-3} Pa) during Si MBE growth at 500 °C is effective in reducing the surface segregation of Sb by several orders of magnitude. However, we also reported some deleterious effects when using AH. Cross-sectional transmission electron microscopy (XTEM) revealed that extended defects formed if AH and Sb were used in layers having thicknesses greater than 20 nm [4]. Additionally, capacitance-voltage (C-V) profiling was used to measure a maximum carrier concentration of only $3 \times 10^{17}/cm^3$. We postulated that this low activation may be due, in part, to the narrow geometry of the doping spike. In this paper we investigate the activation of the incorporated Sb using Hall measurements, spreading resistance profilometry (SRP), and secondary ion mass spectrometry (SIMS).

EXPERIMENTAL PROCEDURES

Epitaxial growth was achieved with a specially designed MBE growth system[11] using elemental Si in an e-beam source and elemental Sb in a Knudsen cell. AH was obtained with a commercial source which employed a tungsten filament at 2150 °C to crack the H_2 gas controllably leaked into the chamber. The H_2 cracking efficiency was estimated to be 7%[12]. The structures were grown on 75 mm, 10 - 20 ohm-cm, B-doped Si(100) wafers. Prior to growth the wafers were prepared using a cleaning technique previously described[13]. Base pressure of the MBE

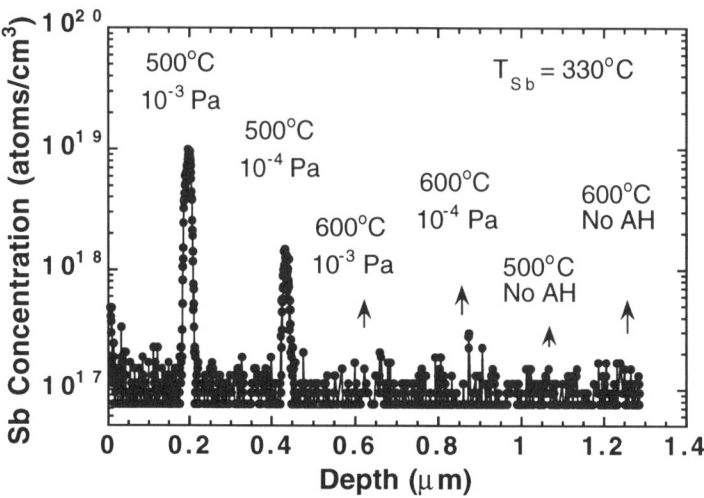

Fig. 1. SIMS Sb atomic concentration profile of a Si epitaxial structure used to establish the conditions during MBE growth that maximize Sb incorporation. Three AH doses (no AH, 10^{-4} Pa, and 10^{-3} Pa) and two substrate temperatures (500 °C and 600 °C) were used. AH was only used during selective Sb doping spikes, and not during the Si spacer layers.

growth system was 5×10^{-9} Pa and typical pressure during growth without atomic hydrogen was 5×10^{-7} Pa. Typical pressures during growth with atomic hydrogen, measured with an ionization gauge, were 10^{-3} Pa. After the H_2 valve was closed, the system pressure reduced to 10^{-5} Pa in 100 s and 1.2×10^{-6} Pa in 1440 s. For all of the samples a total growth rate of 0.1 nm/s was maintained. The structure was a 100 nm Si buffer layer grown at 650 °C, 24 nm of Si grown while the substrate temperature cooled to 500 °C, 20 nm of Sb-doped Si grown coincident with the AH flux, and a 200 nm Si cap. Two different Sb dopant flux were used. The Sb fluxes at Knudsen cell temperatures of 400 and 350 °C are 7×10^{11} and 6.5×10^{10} (Sb/cm^2)/s, respectively, determined by SIMS, based on Sb implant calibrations. After the samples were removed from the growth system, portions of the samples were annealed in a nitrogen atmosphere at temperatures from 600 to 800 °C, for time intervals of 10 min. We have used Hall measurements, SRP, and SIMS to characterize the activation of the Sb. The measurement techniques are complementary to one another. Since we have grown n-type layers on a p- substrate, Hall measurements yield the total n-type sheet carrier concentration, as well as an average value for mobility. SRP measures the resistivity as a function of depth. Assuming bulk carrier mobilities, a carrier concentration profile is obtained. SIMS measures the atomic concentration of the Sb. The Hall measurements were performed at room temperature using a magnetic field of 0.2 T. The SIMS was performed with a high performance magnetic sector secondary ion mass spectrometer. The primary ion beam species was O_2^+, with an impact energy of 3 keV and at an incidence angle of 52° from the normal. The uncertainty in the Hall sheet carrier concentration, the SRP carrier concentration, and the SIMS atomic concentration is estimated to be ±25%.

RESULTS AND DISCUSSION

The sample grown with the Sb cell at 400 °C will be discussed in detail, followed by a brief summary for the sample with the Sb cell at 350 °C. The SRP profiles for samples grown with the Sb cell at 400 °C, with subsequent furnace anneals, are presented in Fig. 2. A summary of the SPR results along with Hall measurements of equivalent samples is presented in Table I. Several observations can be made from the SRP profiles in Fig. 2. The n^+ doping spike grown with Sb and AH is well defined. Thermal anneal for 10 min. at temperatures from 600 to 800 °C did not dramatically increase the activation of the Sb. The anneal at 800 °C for 10 minutes appeared to cause diffusion of the Sb. As noted above, the carrier concentration is obtained from SRP profiles by assuming that the mobility is known. Since we had measured the sheet carrier concentration independently by Hall measurements, we can scale the SRP carrier profiles so that the integrated carrier concentration is equal to the Hall sheet carrier concentration. The SRP maximum carrier concentrations scaled by the Hall measurements are presented in Table I. It is observed in Table I. that the Hall sheet carrier concentration increased with anneal temperature, but that the peak carrier concentration, even when scaled by the Hall measurements, did not increase with anneal temperature. The SIMS profiles of companion samples are shown in Fig. 3. SIMS parameters are summarized in Table II. R_{sheet} is the ratio of the Hall carrier sheet concentration to the SIMS Sb sheet concentration. R_{max} is the ratio of the peak carrier concentration obtained by SRP to the SIMS peak Sb concentration. The SRP carrier concentration was scaled with respect to the sheet carrier concentration measured by Hall. The SIMS atomic concentration profiles show that there is negligible diffusion of the Sb even after the 800 °C anneal, although the Sb concentration is the cap layer is elevated compared with the as-grown sample. Therefore, the in-diffusion shown in the SRP carrier profile after the 800 °C anneal may be an artifact of the measurement. The SIMS peak

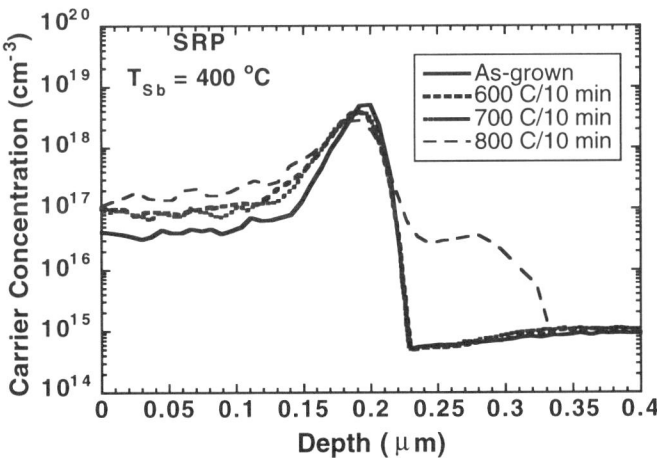

Fig. 2. SRP profiles for Si wafer having a 20 nm Sb doping spike grown using AH. The Sb flux was 7×10^{11} Sb/cm^2/s. A 200 nm undoped Si cap was grown on top of the Sb-doped layer. Post-growth anneals were investigated.

Table I. Summary of Hall measurements and SRP results for the Si wafer having a 20 nm Sb doping spike grown using AH. T_{Sb} was 400 °C. Furnace anneals were performed in an N_2 atmosphere.

	Hall Sheet Carrier Concentration (cm^{-2})	Hall Mobility (cm^2/Vs)	SRP Sheet Carrier Concentration (cm^{-2})	SRP Maximum Carrier Concentration (cm^{-3})	SRP Maximum Carrier Conc. Scaled by Hall (cm^{-3})
As-Grown	1.76×10^{13}	156	1.46×10^{13}	5.06×10^{18}	6.12×10^{18}
600 °C 10 min	1.81×10^{13}	208	1.57×10^{13}	4.01×10^{18}	4.61×10^{18}
700 °C 10 min	2.15×10^{13}	206	1.50×10^{13}	3.87×10^{18}	5.73×10^{18}
800 °C 10 min	2.22×10^{13}	210	1.49×10^{13}	2.78×10^{18}	4.14×10^{18}

profiles are substantially more abrupt than the SRP carrier concentration profiles. The full-width-half-maximum values of the SIMS profiles are less than half those of the SRP profiles. This appears to be due to the difference in the measuring techniques, rather than a true difference in the distributions. The step size for the SIMS had an average value of 2.4 nm, while the average step for the SRP was 8 nm. The Sb activation percentage, given by R_{sheet}, has an average value greater than 50%. The average Sb activation value using the maximum of the distributions, R_{max}, is 23%, but this probably is low due to the averaging of the carrier distribution as measured by the SRP technique.

The sample grown with the lower Sb flux (6.5×10^{10} $(Sb/cm^2)/s$) had similar characteristics. Considering the sample annealed at 800 °C for 10 min, the SIMS maximum was $4 \times 10^{18}/cm^3$ and the sheet Sb concentration was $7.08 \times 10^{12}/cm^2$. The SRP carrier maximum was $7.8 \times 10^{17}/cm^3$ and the integrated sheet carrier concentration was $3.8 \times 10^{12}/cm^2$. The Hall sheet carrier concentration was $3.6 \times 10^{12}/cm^2$, with an average mobility of 330 cm^2/Vs. The Sb activation percentage based on R_{sheet} is 51%, while the activation based on the maximum values was 20%.

It must be noted that the Sb activation percentages cited above are for the Sb "incorporated" into the Si. While the Sb segregation is retarded by the AH, some Sb has still migrated to the surface during growth. During the 200 s that the Sb shutter is open, 1.4×10^{14} and 1.3×10^{13} Sb/cm^2 is deposited with the Sb cell temperatures of 400 and 350 °C. Comparing the SIMS sheet Sb concentrations to these values, incorporation rates of 27% and 54% are established.

SUMMARY

We have demonstrated that the use of AH permits the formation of abrupt n-type doping layers in Si. Previously we had shown that the use of AH during Si MBE growth at 500 °C is effective in reducing the surface segregation of Sb by several orders of magnitude. In this paper we have determined that much of the incorporated Sb is electrically active. Since we restricted the layers to a thickness of 20 nm or less, due to the deleterious effect of AH on thicker layers, it was very difficult to determine the carrier profiles using SRP. Based on the Hall and the SIMS data, we have established that 50% of the incorporated Sb is electrically active. The activation percentage at the carrier peak may be less, or may appear to be less due to perceived profile spreading.

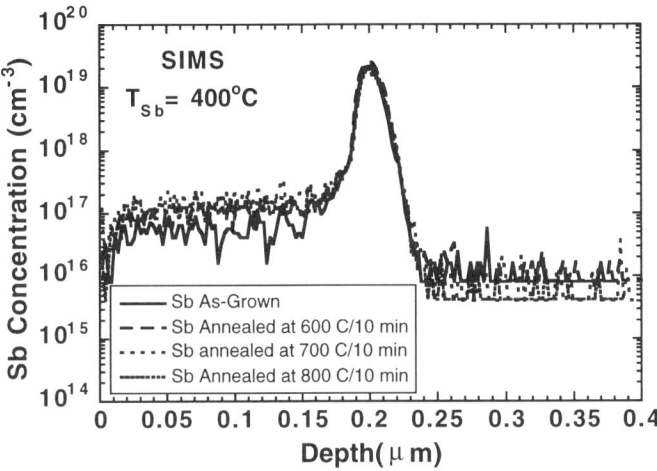

Fig. 3. SIMS profiles for Si wafer having a 20 nm Sb doping spike grown using AH. The samples used for SIMS were adjacent to the samples used for the SRP (Fig. 2) and were annealed at the same time.

Table II. SIMS results for Si wafer having a 20 nm Sb doping spike grown using AH. R_{sheet} is the ratio of the Hall carrier sheet concentration to the SIMS Sb sheet concentration. R_{max} is the ratio of the peak carrier concentration, obtained by SRP, scaled with respect to the sheet carrier concentration measured by Hall, to the SIMS peak Sb concentration.

	SIMS Sheet Sb Concentration (cm^{-2})	R_{sheet}	SIMS Peak Sb Concentration (cm^{-3})	R_{max}	FWHM SIMS (nm)	FWHM SRP (nm)
As-grown	3.55×10^{13}	0.5	2.13×10^{19}	0.29	16.0	24.3
600 °C 10 min	4.63×10^{13}	0.39	2.64×10^{19}	0.17	16.1	30.6
700 °C 10 min	3.85×10^{13}	0.56	2.18×10^{19}	0.26	15.4	31.3
800 °C 10 min	3.34×10^{13}	0.66	1.95×10^{19}	0.21	16.0	37.5

ACKNOWLEDGMENTS

This work was sponsored by the Office of Naval Research and the Air Force Office of Scientific Research. We acknowledge the technical expertise of Mr. Larry Ardis in Hall analysis.

REFERENCES

1. M. Horn-von Hoegen, F. K. LeGoues, M. Copel, M. C. Reuter, and R. M. Trump, *Phys. Rev. Lett.*, 67 (1991) 1130.

2. K. D. Hobart, D. J. Godbey, P. E. Thompson, and D. S. Simons, *Appl. Phys. Lett.*, 63 (1993) 1381.

3. K. D. Hobart, D. J. Godbey, M. E. Twigg, M. Fatemi, P. E. Thompson. and D. S. Simons, *Surf. Sci.,* 334 (1995) 29.

4. P. E. Thompson, C. Silvestre, M. Twigg, G. Jernigan, and D. Simons, accepted for publication in *Thin Sol. Films.*

5. A. Sakai and T. Tasumi, *Appl. Phys. Lett.*, 64 (1994) 52.

6. K. Nakagawa, A. Nishida, Y. Kimura, and T. Shimada, *Jpn. J. Appl. Phys.*, 33 (1994) L1331.

7. G. Ohta, S. Fukatsu, Y. Ebuchi, T. Hattori, N. Usami, and Y. Shiraki, *Appl. Phys. Lett.*, 65 (1994) 2975.

8. K. Nakagawa, A. Nishida, Y. Kimura, and T. Shimada, *J. Cryst. Growth*, 150, (1995) 939.

9. M. Okada, T. Shimizu, H. Ikeda, S. Zaima, and Y. Yasuda, *Appl. Surf. Sci.*, 113/114 (1997) 349.

10. S.-J. Kahng, J. Y. Park, K. H. Booh, J. Lee, Y. Khang, and Y. Kuk, *J. Vac. Sci. Technol.*, A15 (1997) 927.

11. E. D. Richmond, J. G. Pellegrino, M. E. Twigg, S. Qadri, and M. T. Duffy, *Thin Solid Films*, 192 (1990) 287.

12. A. Sutoh, Y. Okada, S. Ohta, and M. Kawabe, *Jpn. J. Appl. Phys.*, 34 (1995) L1379.

13. P. E. Thompson, M. E. Twigg, D. J. Godbey, K. D. Hobart, and D. S. Simons, *J. Vac. Sci. Technol.*, B11 (1993) 1077.

AUTHOR INDEX

SUBJECT INDEX

STM, 183, 209
strain, 203, 281
    compensation, 251
    enhanced solubility, 361
    relaxation, 49
    symmetrization, 171
strained, 43, 145
    α-Sn/Si quantum-well
        heterostructures, 355
    Si, 31
substitutional, 43
surface, 307
    instability, 215
surfactant, 295, 367

TEM, 77
thermal
    quenching, 295
    reaction, 49
    stability, 345
thin-film transistors (TFTs), 145
3D to 2D transition, 209
three-dimensional (3D)
    islanding, 177
    islands, 209

Ti silicide islands, 197
transient-enhanced diffusion, 105
transmission electron microscopy, 289
triple-axis x-ray diffraction, 55

UHV/CVD, 275, 281
ultrathin layers, 339

valence-band offset, 245
velocity overshoot, 31
vicinal
    Si(001) surfaces, 165
    substrates, 77

waveguide, 139
waveguiding, 171

x-ray
    diffraction, 77, 257, 289
    photoelectron spectroscopy, 289

zirconium, 49